Benchmark Papers in Geology Series

Editor: Rhodes W. Fairbridge, Columbia University

A selection from the published volumes in this series

SEAFLOOR SPREADING CENTERS: Hydrothermal Systems / *Peter A. Rona and Robert P. Lowell*
MEGACYCLES: Long-Term Episodicity in Earth and Planetary History / *G. E. Williams*
RIFT VALLEYS: Afro-Arabian / *A. M. Quennell*
OROGENY / *John G. Dennis*
GEOSYNCLINES: Concept and Place Within Plate Tectonics / *F. L. Schwab*
OPHIOLITIC AND RELATED MELANGES / *G. J. H. McCall*
EARTH TIDES / *J. C. Harrison*
CONTINENTAL DRIFT / *James H. Shea*
PLATE TECTONICS / *James H. Shea*

A complete listing of volumes published in this series begins on page 397.

MANTLE FLOW AND PLATE THEORY

Edited by

ZVI GARFUNKEL
Hebrew University of Jerusalem
Israel

A Hutchinson Ross Benchmark® Book

 VAN NOSTRAND REINHOLD COMPANY
New York

Manufactured in the United States of America.

Published by Van Nostrand Reinhold Company Inc.
135 West 50th Street
New York, New York 10020

Van Nostrand Reinhold Company Limited
Molly Millars Lane
Wokingham, Berkshire RG11 2PY, England

Van Nostrand Reinhold
480 Latrobe Street
Melbourne, Victoria 3000, Australia

Macmillan of Canada
Division of Gage Publishing Limited
164 Commander Boulevard
Agincourt, Ontario MIS 3C7, Canada

15   14   13   12   11   10   9   8   7   6   5   4   3   2   1

Library of Congress Cataloging in Publication Data
Main entry under title:
Mantle flow and plate theory.
   (Benchmark papers in geology; v. 84)
   "Hutchinson Ross Benchmark book."
   Includes indexes.
   1. Earth—Mantle—Addresses, essays, lectures.   2. Plate tectonics—Addresses, essays,
lectures.   I. Garfunkel, Zvi. II. Series.
QE509.M27   1985   551.1'3   84-25620
ISBN 0-442-22734-5

CONTENTS

Contents

SERIES EDITOR'S FOREWORD

The philosophy behind the Benchmark Papers in Geology is one of collection, sifting, and rediffusion. Scientific literature today is so vast, so dispersed, and, in the case of old papers, so inaccessible for readers not in the immediate neighborhood of major libraries that much valuable information has been ignored by default. It has become just so difficult, or so time consuming, to search out the key papers in any basic area of research that one can hardly blame a busy person for skimping on some of his or her "homework."

This series of volumes has been devised, therefore, as a practical solution to this critical problem. The geologist, perhaps even more than any other scientist, often suffers from twin difficulties—isolation from central library resources and immensely diffused sources of material. New colleges and industrial libraries simply cannot afford to purchase complete runs of all the world's earth science literature. Specialists simply cannot locate reprints or copies of all their principal reference materials. So it is that we are now making a concerted effort to gather into single volumes the critical materials needed to reconstruct the background of any and every major topic of our discipline.

We are interpreting "geology" in its broadest sense: the fundamental science of the planet Earth, its materials, its history, and its dynamics. Because of training in "earthy" materials, we also take in astrogeology, the corresponding aspect of the planetary sciences. Besides the classical core disciplines such as mineralogy, petrology, structure, geomorphology, paleontology, and stratigraphy, we embrace the newer fields of geophysics and geochemistry, applied also to oceanography, geochronology, and paleoecology. We recognize the work of the mining geologists, the petroleum geologists, the hydrologists, and the engineering and environmental geologists. Each specialist needs a working library. We are endeavoring to make the task of compiling such a library a little easier.

Each volume in the series contains an introduction prepared by a specialist (the volume editor)—a "state of the art" opening or a summary of the object and content of the volume. The articles, usually some twenty to fifty reproduced either in their entirety or in significant extracts, are selected in an attempt to cover the field, from the key papers of the last century to fairly recent work. Where the original works are in foreign languages, we have endeavored to locate or commission translations. Geologists, because of their global subject, are often acutely aware of the oneness of our world. The selections cannot therefore be restricted to any one country, and whenever possible an attempt is made to scan the world literature.

To each article, or group of kindred articles, some sort of "highlight commentary" is usually supplied by the volume editor. This commentary should serve to bring that article into historical perspective and to emphasize its particular role in the growth of the field. References, or citations, wherever possible, will be reproduced in their entirety—for by this means the observant reader can assess the background material available to that particular author, or, if desired, he or she too can double check the earlier sources.

A "benchmark," in surveyor's terminology, is an established point on the ground that is recorded on our maps. It is usually anything that is a vantage point, from a modest hill to a mountain peak. From the historical viewpoint, these benchmarks are the bricks of our scientific edifice.

RHODES W. FAIRBRIDGE

PREFACE

The study of the solid earth gained the status of an independent scientific discipline only in the last two centuries. During this period research shifted from mainly observation and description to interpretation and synthesis. Nevertheless, despite many attempts, a really unifying concept emerged only recently with the formulation of the plate tectonic model (which is an outgrowth of earlier continental drift theories). It is still too early to assess to what extent the current model will be modified, but there can be little doubt that plate tectonics marks a turning point in the development of earth sciences. Like all really important scientific concepts, plate tectonics not only allowed integration and interpretation of existing knowledge, but it also raised new questions and motivated research in new directions. Study of the earth's internal dynamics, and of mantle flow in particular, is one such field in which new vistas were opened and research became very vigorous. As this is one of the most fundamental parts of earth-sciences, it was difficult to resist the offer to edit a Benchmark volume summarizing developments in this field. When selecting the papers, very technical works were avoided; we attempted to select works covering a wide range of topics and important directions and methods of research. The present collection presents the situation in the late seventies.

Like all undertakings of this kind, some factors proved to be more severe than initially foreseen. The limited scope of the volume was one such factor. Another difficulty arose because though the topic has not yet matured, the number of relevant works is quite large and progress is fast. Therefore it is sometimes difficult to identify in each subject the work that will have the greatest impact. The works chosen are believed to mark significant steps in important directions of research. The papers included in this volume were all originally published in English, reflecting the fact that many of the ideas presented here developed in their modern form in English-speaking countries. Researchers from other countries also tend to publish in English.

I am happy to acknowledge the advice given by Rhodes Fairbridge, series editor of the Benchmark Papers in Geology series. I am also very grateful to the staff of Van Nostrand Reinhold Company for their assistance.

ZVI GARFUNKEL

CONTENTS BY AUTHOR

MANTLE FLOW AND PLATE THEORY

INTRODUCTION

Proponents of early continental drift theories invoked the possibility of convection currents in the earth's mantle as a likely mechanism that could drive the crustal blocks. As plate tectonics became widely accepted, superseding the early drift theories, the notion of large-scale mobility of the earth's interior became well established. These ideas motivated many of the recent spectacular advances in the earth sciences.

The plate tectonic regime, with continuous recycling of oceanic lithosphere, requires a large-scale return flow from subduction zones to constructive plate margins. In addition, current knowledge about the earth indicates that its interior is stirred by thermally driven convection. Given such a dynamic earth, it is expected that the mass movements in the mantle will influence all processes of deep-seated origin, such as the motions of the plates and their evolution, tectonism, and magmatism. However, while the behavior of the surficial plates is fairly well known, the deep mass motions are still incompletely understood; the geometry of the flow is unknown, and many problems remain concerning its dynamics, pattern, and relations with the history of the earth's surficial layers. The main difficulty is that direct observation is not possible. Though mantle flow, present and past, is one of the most fundamental processes in the earth, it can be studied only indirectly.

Observable results of mantle flow, such as the motions of the plates, their evolution, igneous activity, gravity anomalies, and so forth, reveal some properties of the flow. Investigation of the kinematics and dynamics of plate motion is especially important, because overturn of oceanic lithosphere is a part of the flow in the earth's interior. However, the behavior of all plates, including oceanic ones, is known quantitatively only for the last 100–150 m.y. Older phenomena resulting from mantle flow can be deciphered from the geologic record, but this is available only from continents, which constitute a small portion of the earth's surface. Therefore, the geologically young periods and the present state of the earth have provided most of the information. Flow in the mantle can also be studied theoretically by modeling the mantle as a fluid. Unfortunately, the models studies so far are much simpler than the natural conditions, and the relevant material properties are known only approximately. Despite these limitations, the results obtained from such fluid-dynamic studies, supplemented by experimental work, provide important insights into fundamental physical properties of convection in the earth and its relation with the rigid lithospheric plates.

With the fast growth of space research, comparative planetology has an increasing impact on the interpretation of terrestrial phenomena. The earth's behavior can be evaluated in a broad context of the basic processes of planetary evolution. Combined, all these lines of research contribute to the understanding of the dynamics of the earth's interior, and of how deep-seated processes influence the planet's external parts.

The present collection of papers attempts to present major problems and directions of study of mantle flow and of its consequences, as well as some important conclusions. The topic is intimately interwoven with plate tectonics, but only articles directly related to problems of deep flow are included. The choice was limited to papers published up to 1975, but later developments until 1979 are also briefly mentioned. However, as research on mantle flow is rapidly expanding and evolving, a comprehensive summary cannot be attempted yet. The choice of papers was guided by the belief that exposition of the correct problems and of the most promising directions of study will go a long way toward a solution. Emphasis is given to papers that show the relevance of mantle flow to various geologic phenomena. Detailed fluid-dynamic studies, though of fundamental importance, are avoided in order not to make this collection too technical.

The development of the main ideas is presented first, followed by some theoretical studies that reveal important physical properties of convection. The next section includes papers that discuss the nature and effects of localized flow in ascending plumes in the mantle. Next are papers discussing the rheology of the mantle. The following group of papers deals with stresses originating in the lithosphere and the mechanics of plate motions, and with their influence on deeper motions. Problems related to the pattern of mantle flow and its possible expressions on the surface constitute the last section.

Part I

DEVELOPMENT OF IDEAS

Editor's Comments
on Papers 1 Through 4

Modern thinking about the internal dynamics of the earth is tightly linked to plate tectonic theory. However, many fundamental ideas were proposed some time before this theory evolved. Often it is difficult to trace the origins of ideas (e.g., Meyerhoff, 1968), but it seems fair (or is common practice) to credit those who phrased their thoughts most clearly and in a context that made these ideas useful and stimulating for further research. The intellectual environment in which ideas develop and their history tend, regrettably, to be largely forgotten. Convection in the earth was proposed in the later half of the nineteenth century (e.g., Osmond Fisher, 1881), and since then it was repeatedly evoked by tectonicians who advocated large-scale motions of crustal blocks. Thus, Otto Ampferer (1906), who first developed the subduction theory ("Verschluckungstheorie"), proposed that convection dragged Africa and pushed it against Europe to crush between them the strongly folded and compressed Alpine chain. This concept was adopted by many Alpine geologists (e.g., Kraus, 1958), but apparently it did not lend strong support to the theory of continental drift. Strangely, Wegener, (1929) the father of modern continental drift theory, did not recognize convection as a likely driving force required by his theory, relying instead on inadequate energy sources related to planetary rotation, though convection was suggested by Schwinner (1920), Bull (1921) and Holmes (1928). Later, Du Toit (1937), who strongly supported continental drift, also did not consider convection as a suitable driving mechanism.

Schwinner (1920) and especially Arthur Holmes (1928) should be credited as the fathers of modern thinking about thermal convection as a fundamental process in the earth and as a possible driving mechanism of continental drift. Holmes (1928) argued that such currents could arise in the earth's interior as a result of the decay of radionucleids, because the heat so generated could not be effectively conducted away, and therefore thermal gradients should grow until convection began. The long-term viscosity at depth, as determined from the postglacial rebound of Fennoscandia (Haskell, 1935), was shown to allow convection in the mantle as a result of lateral or vertical temperature gradients (Pekeris, 1935; Hales, 1936). At the same time, Vening–Meinesz (1932) interpreted the large negative gravity anomalies that he had discovered over trenches along island arcs to indicate that the trenches were maintained dynamically. He proposed that they were developed over descending convection currents whose drag balanced the mass deficiency.

These developments in the thirties allowed Griggs to propose his "model of mountain building" (Paper 1), a remarkable precursor of many modern ideas. Griggs combined the knowledge gained experimentally about the behavior of rocks at high temperatures and pressures with geophysical data to propose a model of thermally driven convection in the mantle, which provided the forces of orogenesis. In view of the episodic nature of tectonic activity and its changing patterns, he envisaged intermittent flow rather than steady-state convection.

Holmes developed his own ideas, and in 1944 presented a model of convection currents that drive continents (Paper 2). His figures are very close to modern concepts of plate tectonics and sea-floor spreading: He proposed that continents would be broken up and pushed apart by ascending limbs of convection currents, over which new oceans with central ridges would develop. The descending currents would produce compressional mountain ranges; the adjacent basaltic oceanic crustal material would thus become down-buckled and compressed and thereby changed to heavy eclogite that would sink into the mantle.

Thus, fundamental ideas about the possibility of a mobile mantle were developed during the years when Wegener's ideas on continental drift were rejected by most earth scientists. Only acquisition of new data after the Second World War revived mobilistic views of the earth: Paleomagnetic studies confirmed continental drift (Blackett et al., 1960; Runcorn, 1962), and marine geophysical studies revealed the global extent of the mid-ocean rift-ridge system, its dilatational origin, and the equality of continental and oceanic heat flows, all of which suggested a mobile mantle (Hess, 1954; Heezen, 1960; also see review of Bullard, 1975). The new observations, and the availability of a plausible driving mechanism, that is, convection currents, set the stage for recognition by Hess (1962) and Dietz (1961) of sea-floor spreading and overturn of the oceanic crust and lithosphere. They also stressed the point that many continental margins were passive features that are carried along with the adjacent oceans. These ideas are incorporated into Wilson's "hypothesis of the earth's behaviour" (Paper 3). This remarkable paper foresaw many important developments ahead of their general acceptance. Wilson exposed the logic relating the distribution of mid-ocean ridges, aseismic ridges, and trenches to the post-Paleozoic drift of continents. He indicated the existence of older junctures (sutures) where preexisting continental fragments once collided, and he recognized repeated opening and closing of oceans. Wilson also depicted overturn of oceanic lithosphere, but did not refer to a cellular return flow. Rather, he considered irregular combinations of mantle flow and sought their expression in relative plate motions. He also stressed his previous interpretation of linear chains of islands as recording the motion of lithosphere over deep sources of magmas.

The following years saw the formulation of the plate tectonic theory. Vine and Matthews (1963) explained the origin of the striped oceanic magnetic anomalies first discovered by Mason (1958). Wilson (1965) recognized the transform nature of major fracture zones, especially those offsetting magnetic anomalies in the oceans (Vacquier, Raff, and Warren, 1961). Bullard, Everett, and Smith (1965) and McKenzie and Parker (1967) applied Euler's theorem to describe the present motion of plates. An integrated formulation of the new concepts of global tectonics was given by Morgan (1968), Heirtzler et al. (1968), LePichon (1968), Isacks, Oliver, and Sykes (1968) and McKenzie (1969). With increasing recognition of the mobility of the earth's outer layers, the probability of important flow, most likely thermal convection, in its interior was emphasized (Runcorn, 1962, 1965; Knopoff, 1964; Tozer, 1965). However, it was not clear how such flow in the mantle was related to phenomena apparent on the earth's surface. In fact, the plate tectonic theory was successful to a great extent because it presented a precise kinematic model related to specific observations, independent of any particular driving mechanism. It became recognized that the complex surficial tectonic patterns depend primarily on the strength of the lithosphere and are not necessarily direct expressions of mass movements in the deeper mantle; certainly a pattern of simple periodic convection cells, such as were considered in theoretical treatments, could not be recognized. Therefore, the relation between the surficial phenomena and the flow at depth cannot be as simple as envisaged by Griggs, Holmes, and Hess. The nature of mantle flow and its relations to the motions and development of the plates must now be discussed in an entirely new context. This remains a cardinal problem of earth science.

Knopoff (Paper 4), in a paper written just before the plate tectonic model became generally accepted, summarized the main features of the new model, and also the major problems related to flow underneath the lithospheric plates, such as: Does the flow occur in the entire mantle or only in the upper 600–700 km? What is the role of phase changes? What is the role of lateral temperature gradients versus heating from within or from below, and what is the time scale for excitation of any flow pattern? How are the mid-oceanic ridges and trenches, whose relative positions change, related to the deep flow? Are they tied to ascending and descending limbs of cells? Does the episodic nature of tectonism imply episodic convection? Today it can be said that, though much progress has been achieved, there are as yet no complete answers to any of these problems.

REFERENCES

Ampferer, O., 1906. Uber das Bewegungsbild von Faltengebirge: *Geol. Bundesanst., Wien, Jahrb.* **56:**539–622.

Blackett, P. M. S., J. A. Clegg, and P. H. S. Stubbs, 1960, An Analysis of Rock Magnetic Data, *Roy. Soc. London Proc.* ser. A, **256:**291–322.

Bull, A. J., 1921, A Hypothesis of Mountain Building, *Geol. Mag.* **58:**364–367.

Bullard, E. C., 1975, The Emergence of Plate Tectonics: A Personal View, *Ann. Rev. Earth and Planetary Sci.* **5:**1–30.

Bullard, E. C., Everett, J. E., and A. G. Smith, 1965, The Fit of the Continents around the Atlantic *Royal Soc. (London) Philos. Trans.* ser. A, **258:**41–51.

Dietz, R. S., 1961, Continent and Ocean Basin Evolution by Spreading of the Sea Floor, *Nature* **190:**854–857.

Du Toit, A., 1937, *Our Wandering Continents*, Oliver and Boyd, Edinburgh, 366p.

Hales, A. L., 1936, Convection Currents in the Earth, *Royal Astron. Soc. Monthly Notices*, Geophys. Suppl. **3:**372–379.

Fisher, O., 1881, *Physics of the Earth's Crust*, 2nd ed., Macmillan, London.

Haskell, N. A., 1935, The Motion of a Viscous Fluid under a Surface Load, *Physics* **6:**265–269.

Heezen, B. C., 1960, The Rift in the Ocean Floor, *Sci. American* **203**(4):99–110.

Heirtzler, J. R., G. O. Dickson, E. M. Herron, W. C. Pitman, and X. LePichon, 1968, Marine Magnetic Anomalies, Geomagnetic Field Reversals and Motions of the Ocean Floor and Continents, *Jour. Geophys. Research* **73:**2119–2136.

Hess, H. H., 1954, Geological Hypothesis and the Earth's Crust under the Oceans *Royal Soc. (London), Proc.*, ser. A, **222:**341–348.

Hess, H. H., 1962, History of the Ocean Basins, in *Petrologic Studies: A Volume in Honor of A. F. Buddington,* A. E. J. Engle, H. L. James, and B. F. Leonard, ed., Geological Society of America, Washington, D.C., p. 599–620.

Holmes, A., 1928, Radioactivity and Earth Movements, *Geol. Soc. Glasgow Trans.,* **18:**559–606.

Isacks, B. L., Oliver, J., and L. R. Sykes, 1968, Seismology and the New Global Tectonics, *Jour. Geophys. Research* **73:**5855–5900.

Knopoff, L., 1964, The Convection Current Hypothesis, *Rev. Geophysics* **2:**89–122.

Kraus, E., 1958, Funfzig Jahre Unterstromungs-Theorie, *Geologie* **7:**261–283.

LePichon, X., 1968, Sea-floor Spreading and Continental Drift, *Jour. Geophys. Research* **73:**3661–3697.

Mason, R. G., 1958, A magnetic survey of the west coast of the United States between latitudes 32° and 36°N, longitudes 121° and 128°W, *Royal Astron. Soc. Geophys. Jour.* **1:**320–329.

McKenzie, D. P., 1969, Speculations on the Consequences and Causes of Plate Motions, *Royal Astron. Soc. Geophys. Jour.* **18:**1–32.

McKenzie, D. P. and R. L. Parker, 1967, The North Pacific: an Example of Tectonics on a Sphere, *Nature* **216:**1276–1280.

Meyerhoff, A. A., 1968, Arthur Holmes: Originator of Spreading Ocean Floor Hypothesis, *Jour. Geophys. Research* **73:**6563–6565.

Morgan, W. J., 1968, Rises, Trenches, Great Faults, and Crustal Blocks, *Jour. Geophys. Research* **73:**1959–1982.

Pekeris, C. L., 1935, Thermal Convection in the Interior of the Earth, *Royal Astron. Soc. Monthly Notices,* Geophys. Suppl., **3:**346–367.

Runcorn, S. K., 1962, Paleomagnetic evidence for continental drift and its geophysical cause, in *Continental Drift,* S. K. Runcorn ed. Academic Press, New York, pp. 1–40.

Runcorn, S. K., 1965, Changes in the convection pattern in the earth's mantle and continental drift evidence for a cold origin of the earth: *Royal Soc. London Philos. Trans.* ser. A **258:**228–251.

Schwinner, R., 1920, Vulkanismus und Gebirgsbildung, *Zeitschr. Vulkanologie,* **5.**

Tozer, D. Z., 1965, Heat Transfer and Convection Currents, *Royal Soc. London. Philos. Trans.* ser. A, **258:**252–271.

Vening-Meinesz, F. A., 1932, *Gravity Expeditions at Sea, 1923–1930,* Waltman, Delft.

Vaquier, V., Raff, A. D., and Warren, R. E., 1961, Horizontal Displacements in the Floor of the Northwestern Pacific Ocean, *Geol. Soc. America. Bull.* **72:**1251–1258.

Vine, F. J. and D. H. Matthews, 1963, Magnetic Anomalies over Oceanic Ridges, *Nature* **199:**947–949.

Wegener, A., 1929, The Origin of Continents and Ocean Basins, trans. J. Biram, Dover, New York, 246p.

Wilson, J. T., 1965, A New Class of Faults and Their Bearing on Continental Drift, *Nature* **207:**343–347.

A THEORY OF MOUNTAIN-BUILDING.

DAVID GRIGGS.

The fact that a natural phenomenon admits of one mechanical explanation is proof that there is an infinity of such explanations.

(Paraphrase from Poincaré.)

TABLE OF CONTENTS.

Introduction.
Fundamental Orogenic Phenomena.
 The Diastrophic Cycle.
 The Crustal Downfold, or Tectogene.
Forces Available for Mountain Building.
 Primary and Secondary Diastrophic Forces.
 The Tidal Force.
 The Polflucht Force.
 The Coriolis Force.
 Compression due to Thermal Contraction of the Earth.
 The Force due to Viscous Drag of Subcrustal Convection Currents.
Depth of the Convecting Shell.
The Convection-Current Cycle.
 Solid Flow of Rocks.
 Causes of Cyclic Convection.
 Phases of the Cycle.
A Dynamic Model of the Earth's Outer Shells.
 Dimensional Analysis of Kuenen's Model.
 The Dynamic Convection Model.
Correlation of the Convection and Orogenic Cycles.
Application of the Cyclic Convection Theory to Earth Structures.
 Alternative Interpretations of Major Structures.
Conclusion.
Acknowledgments.

ABSTRACT. A theory of mountain-building by cyclic convection, thermal and not chemical in origin, is synthesized from: (1) the suggestions of Holmes, (2) the mathematical analyses of Pekeris, Vening Meinesz, and Hales, (3) the writer's experiments on solid flow of rocks, (4) thermal experiments and calculations, and (5) a dynamically similar model to demonstrate the action of cyclic convection currents. The way in which this theory predicts the intermittence of mountain-building is discussed, and its ability to explain the diastrophic cycle.

Previous theories of orogenesis are briefly reviewed and some of their points of inadequacy are discussed.

                    David Griggs.

Introduction.

IN the following, some arguments are presented which tend to make more plausible the suggestion that convection currents may have operated in the earth's substratum to cause the development of our mountain systems. It is clearly recognized that any such attempt must be highly speculative because of the enormous and indeterminate complexities of the situation. The reasoning is based on a few credible assumptions as to the composition and behavior of the earth's shells. The validity of the argument, therefore, depends entirely on geological and geophysical verification of these assumptions.

Any adequate hypothesis of mountain-building must satisfy the three following conditions:

1. It must provide a tangential component of force sufficiently great to fold and thicken the continental crust.
2. It must provide locally sufficient contraction to account for the shortening of the crust in the regions of mountain-building inferred from geological evidence.
3. It must explain the intermittent nature of orogenic processes and the succession of three phases in the mountain-building cycle:
    a. Geosynclinal subsidence and sedimentation.
    b. Compression of the crust and folding of the geosynclinal filling.
    c. Later elevation of the compressed and folded mass.

The literature of geology contains abundant criticisms of all the prevalent theories of orogenesis, based mainly on their inability to satisfy one or more of these conditions. It is not the purpose of this paper to evaluate critically the theories or the objections to them, but to suggest a mechanism which seems physically competent to meet these three conditions.

The basic physical principle on which the present hypothesis depends is not new to geology. Thermal convection currents in the earth's substratum were suggested a century ago by Hopkins (1839).* Since that time, several types of convection have been invoked to explain various features of the earth's history. The theory that our mountain systems are the result of convection currents has been the subject of much discussion by European geologists, but has received relatively little attention in this country.

* Names followed by dates in parentheses refer to references listed at end of this paper.

Holmes (1932) has been led by his study of the thermal history of the earth to conclude that cooling of the earth has been inadequate to produce the contraction necessary for orogenesis. He further concludes that thermal convection is to be expected in the earth's substratum, and that the convection currents may be adequate to cause mountain-building. Vening Meinesz (1934) suggests that subcrustal convection currents compress and fold the crust, developing the "Tectogenes" which are thought to underlie the negative-anomaly bands observed in the East and West Indies. He develops simplified equations of convection and calculates the potential stress which they may exert on the crust. Pekeris (1936) develops more rigorously the convection equations, and shows that on the most probable assumptions of temperature, viscosity, and strength of the substratum, convection is to be expected if, in addition, the substratum is homogeneous enough to permit convection. He also calculates the tangential stress developed by such currents, using an entirely different method from Vening Meinesz's. Hales (1936) develops still another mathematical analysis of convection, which indicates the temperature gradient necessary for the maintenance of the currents. This analysis shows that the temperature gradient most commonly assumed by seismologists to explain the observed increase in seismic velocities will lead to the development of convection currents in a homogeneous substratum. Bull (1929) performed an interesting experiment which suggests that the action of convection currents is one of the few mechanisms which will produce nappe structure.

Many other papers have been published by European geologists dealing with various aspects of the convection-current theory of mountain-building, but the papers above referred to are among the most complete, and afford a representative discussion of the hypotheses of the convection mechanism. Most of the discussion of convection has involved the tacit assumption that a molten condition is necessary for convection. The mathematical analyses have been based on the assumption that steady-state convection has been established. The present paper presents some new ideas concerning the way in which subcrustal convection may cause orogenesis, and in particular, suggests the existence of a convection-current cycle, correlated with the mountain-building cycle.

Fundamental Orogenic Phenomena.

THE DIASTROPHIC CYCLE.

It has long been recognized that in spite of the numerous irregularities of the earth movements culminating in the formation of mountain chains, this activity follows a cycle which has been intermittently repeated during the history of the earth. Bucher (1933) expresses this as Law 20 in his study of·the "Deformation of the Earth's Crust": "The typical orogenic cycle begins with a geosynclinal depression and ends with a major uplift. The interval between these limiting events comprises two phases. The first phase is one essentially of quiet sinking, only occasionally interrupted by uplifts; the second phase consists of crustal foldings separated by diminishing epochs of renewed geosynclinal sinking."

For the purposes of the present discussion, it is convenient to divide this generalized cycle of diastrophism into three phases:

1. The geosynclinal phase, in which a strip of the earth's crust is gradually depressed, receiving sediments from the adjacent, slightly elevated land masses, on one or both sides of the trough.
2. The period of crustal folding, in which the contents of the geosyncline are compressed and folded, without major elevation of the area above sealevel. The forces during this period of folding act mainly in the horizontal direction, but with a sufficiently great downward component to prevent the thickened mass from attaining isostatic equilibrium.
3. The elevation of this folded and thickened mass until it reaches isostatic adjustment. The forces acting during this period are dominantly upward; compressive folding is secondary.

There are minor deviations from the average trend in all of these phases in the mountain ranges of the world, but it seems to be the consensus of geological opinion that the history of all the major mountain systems has followed this broad outline of events.

THE CRUSTAL DOWNFOLD, OR TECTOGENE.

One of the greatest contributions to the understanding of tectonics during the twentieth century has been Vening Meinesz'

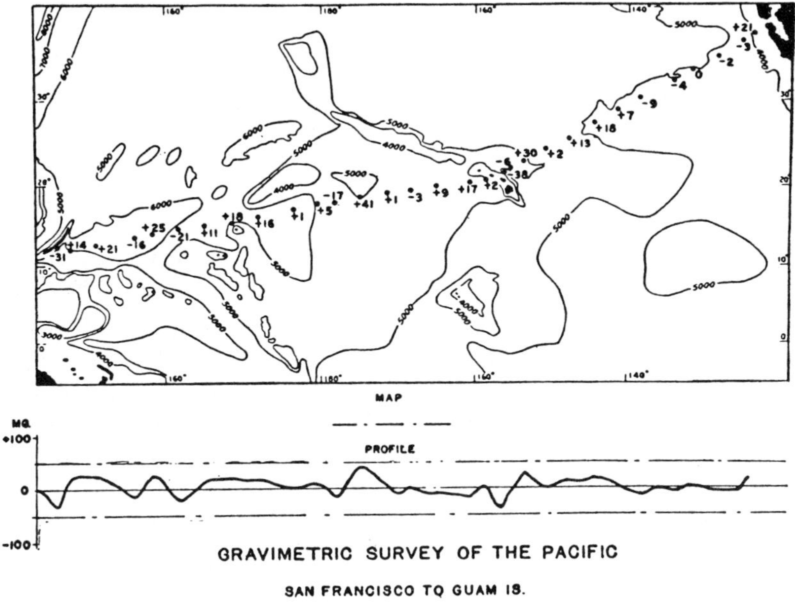

Fig. 1. Series of Measurements of Gravity Anomalies Across the Pacific Ocean after Isostatic Reduction. Figure reproduced through the kindness of R. A. Daly.

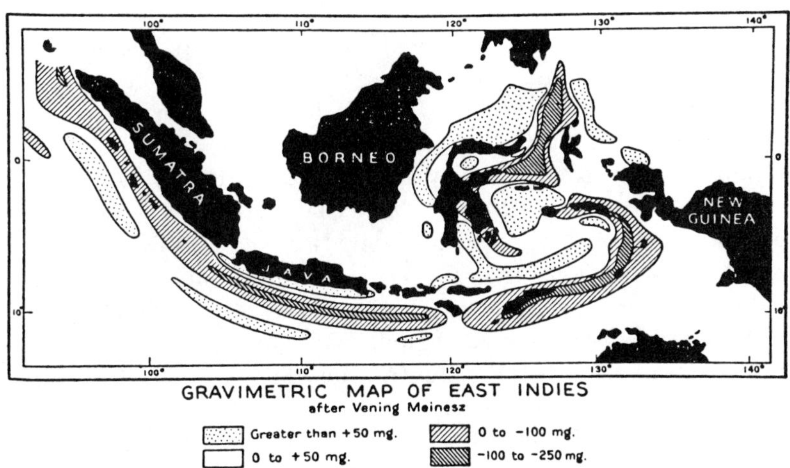

Fig. 2. Gravity Anomalies in the East Indies, Showing Belt of Gravity Deficiency Peripheral to the East Indian Archipelago, Flanked by Bands of Positive Anomalies.

discovery of the great bands of gravity deficiency in the East and West Indies. His careful and painstaking measurements of gravity over the broad expanse of the Pacific Ocean basin disclosed no deviation from normal gravity of more than 50 milligals, as shown in Fig. 1. Peripheral to the East Indian archipelago, however, a persistent gravity deficiency of much greater magnitude was discovered.

Fig. 2 shows the distribution of the gravity anomalies observed by Vening Meinesz (1934) from the Malay Peninsula around the north coast of Australia to the Philippines. Fig. 3 shows the results of a similar set of measurements made in the West Indies by the coöperative enterprise of Vening Meinesz, Hess, Brown, Ewing, and Hoskinson. In each of these island arcs, there is a narrow strip of strong negative anomalies just outside the island festoon. Also in each case, this strip is flanked by irregular patches of less strong positive anomalies. Fig. 4 shows a profile across latitude 15° N. in the West Indies, which illustrates strikingly the intensity of the gravity deficiency and its relation to the topography of the ocean floor.

These negative anomaly bands can be explained only on the assumption of a mass deficiency in the earth's crust at this point. Further, the narrowness of the bands makes it imperative to assume that this mass deficiency is located in the outer 100 km. of the earth's crust. The only explanation that seems geologically possible is that the lighter crust of the earth has been downfolded into the heavier substratum. Fig. 5 (prepared by Hess) shows diagrammatically the way in which such a downfold may account for the observed gravity deficiency.

Both in the West and the East Indies, the distribution of these negative-anomaly strips coincides with the distribution of the most intense late Tertiary folding and thrusting (Vening Meinesz, 1934; Hess, 1938). Each island arc is a branch of the great Tertiary mountain system. It seems fair to conclude that the observations indicate crustal downfolds which were formed as a terminal event of the Laramide-Alpine revolution. It would appear that these regions have just passed through the second phase of the mountain-building cycle. On this assumption it is to be expected that they will, in the geologically near future, be uplifted to form minor mountain masses.

It is suggested that the crustal downfold postulated in these two areas is a universal development of the second phase of the diastrophic cycle. Kuenen (1936) has termed the downfold a

"Tectogene."   If conditions during the first phase of the orogenic cycle are such that geosynclinal sedimentation takes place, then this sedimentary mass will be carried on the back

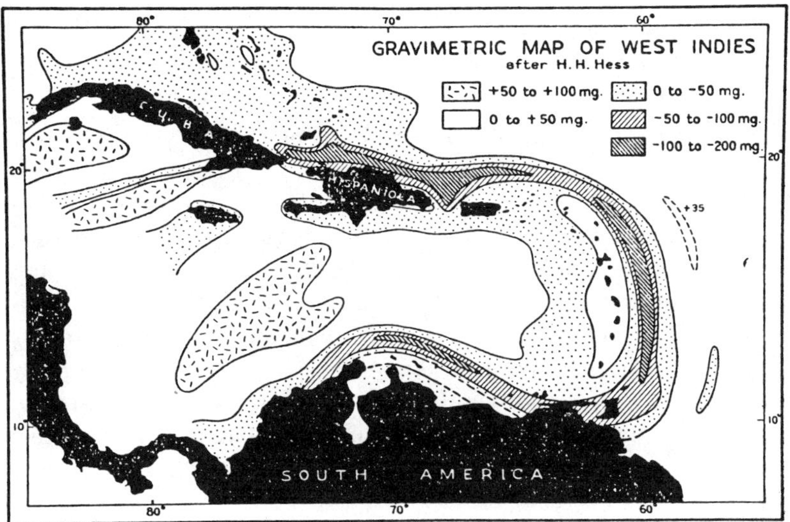

Fig. 3. Gravity Anomalies in the West Indies, Showing Similarity to the East Indies.

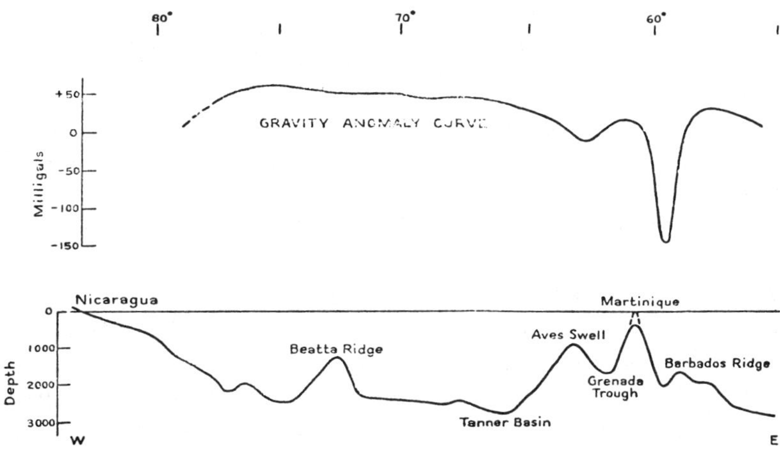

Fig. 4. Gravity and Topography Profiles Along Latitude 15° N, in the West Indies.

of the folding crust and will itself be much more intricately folded.   Such a relation has been suggested as the primary movement of mountain-building, as illustrated in Fig. 6 from Hess).   If this relationship be accepted, one must next search for a force adequate to compress the crust and cause this great downfold and the concomitant folding of the sedimentary filling of the geosyncline.

Forces Available for Mountain-Building.

PRIMARY AND SECONDARY DIASTROPHIC FORCES.

The forces most easily observed or inferred from geological observations are the so-called secondary forces which tend toward the attainment of equilibrium. Thus, the forces of erosion would in a short time reduce the surface of the earth to a flat featureless plain, if there were not primary forces of earth deformation which continually distort the earth's face. Similarly, any deviation from isostatic adjustment develops forces which tend to restore equilibrium. All of these secondary forces depend on primary deformative forces for their existence, and hence cannot be considered the ultimate cause of mountains.

Five types of primary forces have been suggested in the various theories of orogenesis:

1. The tidal force.
2. The Polflucht force.
8. The Coriolis force.
4. Compression due to thermal contraction of the earth.
5. Viscous drag of convection currents in the substratum.

Before proceeding to a discussion of the magnitude of these forces, let us consider the probable strength of the earth's crust and estimate the size of the compressive stress which is necessary to fold it as a unit. The normal compressive strength of granitic rocks tested in the laboratory is 2,000 to 3,000 kg./cm.^{2} It has been shown that high confining pressure such as exists in the earth's crust will greatly increase this strength, if all other conditions are held constant (Adams, 1917; Karman, 1911; Griggs, 1936). Other factors acting simultaneously—high temperature, solutions, and long periods of stress application—undoubtedly change the strength of rocks, but these effects have not been adequately investigated. It is yet too early to deduce from laboratory experiments the probable average strength of the earth's crust.

Studies of the local and regional departures from isostasy serve as the most reliable basis for estimating the strength of the continental crust and substratum. These observations admit of various interpretations, but in general they seem to indicate that the continental crust has a strength somewhere between 100 and 2,000 kg./cm.^{2} For the purposes of the present discussion, it seems that the assumption of an average strength of the continental crust of 1,000 kg./cm.^{2} will not lead to serious error, and will probably be as accurate, relatively, as many of the other values of physical properties which will be used in calculations.

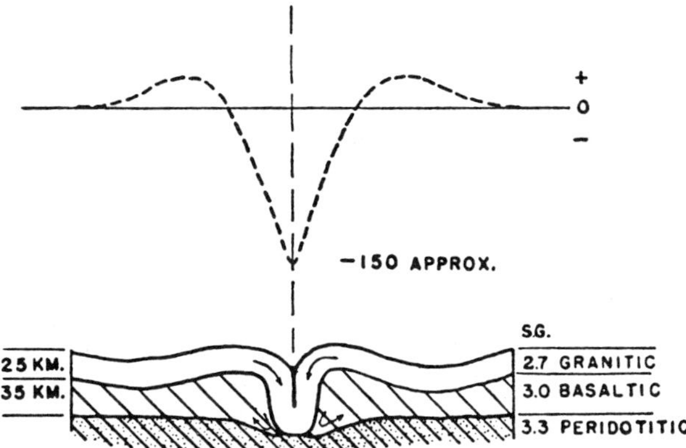

Fig. 5. Crustal buckle, specific gravity distribution and resultant anomaly curve. From H. H. Hess.

In contrast to this strength of the crust, the observed flow of the substratum under the load of the glacial ice-cap indicates that it is either lacking in strength altogether, or possesses a strength of less than 10 kg./cm.^{2} (Daly, 1938, p. 408).

THE TIDAL FORCE.

Some of the advocates of continental drift hypotheses have invoked the westward tidal pull of the moon and sun on the earth's surface to account for westward drift of the continents and consequent mountain formation (Wegener, Joly, etc.). The most telling criticism of this theory is a calculation of the force which can be developed in this manner. This calculation has been made by several people, notably Jeffreys (1929, p. 304), who states, "Tidal currents at their strongest give a bottom drag of the order of 40 dynes/cm.^{2} (4 x 10^{-5} kg./cm.^{2}); but this is abnormal, and is reversed in direction in every tide where it does occur. The mean secular tidal friction producing the slowing down of the earth's rotation corresponds to a westward stress of the order of only 10^{-4} dynes/cm.^{2} (10^{-11} kg./cm.^{2}) over the earth's surface."

This westward stress acts tangentially over the whole bottom of the continental mass. Hence the compressive stress in the crust is greater than this: roughly in proportion to the ratio between the area of the continent and its cross-sectional area. This factor would be at a maximum of the order of 100:1, which would make the compressive stress in the continents 10^{-9} kg./cm.^{2}

The fact that this stress of tidal drift is only one trillionth (10^{-12}) of the stress necessary to compress and fold the crust seems to preclude its serious consideration as a cause of mountain-building, or even as a modifying influence.

THE POLFLUCHT FORCE.

Staub (1928) has been the chief exponent of the hypothesis of mountain formation as a result of equatorward drift of the continents in response to the "Polflucht" force. Because the continents are lighter than the substratum and float in it, they are acted on differentially by the centrifugal force of the earth's rotation. This centrifugal force has a component parallel to the earth's surface which tends to make the center of gravity of the continents move toward the equator. It is easy to calculate this component. Using Lambert's (1921) equations, we find for a continent 100 km. broad and 6 km. deep, a compressive stress of 5×10^{-4} kg./cm.^{2} For a continent 15 km. thick, Jeffreys (1929, p. 301) calculates the stress to be 4×10^{-3} kg./cm.^{2}

Thus we see that the maximum Polflucht force is of the order of one ten-thousandth of that required for mountain-building. Hence it would seem that this force may also be excluded from serious consideration as a primary cause of mountain-building.

THE CORIOLIS FORCE.

Because the earth's axis of rotation is inclined to the line of gravitational attraction of the sun, it wobbles, like a gyroscopic top that gets a little out of the vertical. The forces that cause this wobbling, or precession, act differentially on parts of different density, and produce a component of force which tends to make the continents move. Jeffreys calculates this force and concludes (1929, p. 304), "The stress in precession that makes the whole of the earth precess at the same rate, instead of different shells precessing at different rates, is at most of order 60 dynes/cm.^{2} (6 x 10^{-5} kg./cm.^{2}), and this must be mainly alternating in direction." So we see that this Coriolis force is of the same order of magnitude as the tidal force, and much too small to cause mountain-building deformation of the earth's crust.

COMPRESSION DUE TO THERMAL CONTRACTION OF THE EARTH.

Cooling of the earth will produce a compressive stress in the crust limited only by the strength of the shell resisting the shrinkage so developed. Although there is no question that a force so derived might be sufficient to deform the crust, its adequacy must be examined on other grounds. Two considerations are of prime importance in a discussion of the theory that our mountain systems originated by a wrinkling of the earth's crust resulting from thermal contraction:

1. Is the earth cooling rapidly enough to produce the amount of contraction observed in the Tertiary mountain belt?
2. Can the compressive stress developed by uniform shrinking of the earth's interior be transmitted through the crust for thousands of miles to produce the localized deformation observed in the Tertiary mountain belt?

1. Our present knowledge of temperature gradients and the distribution of radioactive, heat-producing materials renders the answer to the first question indeterminate. Holmes (1932, p. 171) has shown that, if the radioactive elements were distributed throughout the earth in the proportion in which they occur in the average rocks of the crust, and if radioactivity proceeded at all depths in the earth, this in itself would provide a supply of heat fifty times as great as the heat lost from the earth's surface by radiation. The discovery of the amount of radioactivity in rocks makes it impossible to estimate how much the earth is cooling at the present time, and even renders possible the conclusion that the earth may not be cooling at all, but may have reached an equilibrium in which all the heat lost is supplied by radioactivity.

It is of interest to calculate the amount of cooling necessary to account for the observed contraction of the Tertiary mountain belt, assuming that all the contraction of the earth is taken up in the shortening observed in this belt. For this calculation, it is necessary to know the coefficient of thermal expansion for the materials of the earth under the pressures and temperatures which exist at depth. This we do not know, but if we assume that they have the same value as rocks at the surface, we find that an average cooling of the order of 1,500° for the whole earth is necessary. It seems probable that the pressure in the earth would greatly lower the coefficient of expansion, so that this figure would be too low. Such an amount of cooling seems excessive for the last three per cent of the earth's history.

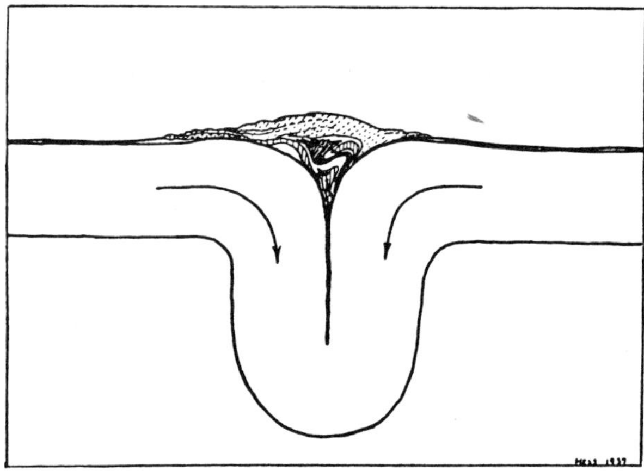

Fig. 6. General section of the Alps superimposed on the tectogene. Both features drawn to the same scale with no vertical exaggeration. From Hess.

2. The thermal contraction theory necessitates that the compressive stress in the crust be transmitted over a large fraction of the earth's circumference to cause the folding localized in the Tertiary mountain belt. If the substratum be assumed to have low strength compared with the crust, it will tend to contract uniformly in a radial direction, and stress differences will be equalized by flow. There will be little tendency to develop tangential motion in the substratum. It follows that the localization of deformation in the crust involves a tangential movement of the crust over the substratum, varying from zero in the center of the shields to a maximum at the mountain abutments. This over-riding motion of the crust is hindered by the viscous drag of the substratum, and it would

seem that the condition for transmission of compressive stress through the crust is that this force of viscous drag be small in comparison with the compressive stress.

In order to calculate the viscous drag of the substratum, we must know the amount of crustal motion (shortening of the crust in the mountain regions), the time required for the movement, the viscosity of the substratum, and the distribution of the cooling which is responsible for the motion. The last of these depends on solution of the thermal conditions in the cooling earth, which we have seen is inadequately known. The other three variables are known to a first degree of approximation. From simplifying assumptions of all these conditions, the writer has made a rough calculation that the compressive stress in the crust developed by the viscous drag in the substratum would be of the order of 750 kg./cm.^{2} In the light of our poor knowledge of the thermal conditions, however, this must be considered as at best an enlightened guess. If it be true, then it would seem that the drag of the substratum would provide sufficient resistance to the motion of the over-riding crust to prevent transmission of stress over large areas, and would cause more widespread distribution of deformation instead of localization into two mountain belts over the whole face of the earth.

A rough approximation to the conditions to be expected from thermal contraction can be attained in a dynamic model of the earth's outer shells. We shall see later, in the discussion of the dynamic model, that it provides some measure of confirmation of the conclusion reached above.

Many other criticisms of the contraction theory are to be found in the literature. In the opinion of the writer, none of the criticisms rests on sufficient evidence to disprove the hypothesis, but in toto they present significant difficulties which stand in the way of its acceptance.

THE FORCE DUE TO VISCOUS DRAG OF SUBCRUSTAL CONVECTION CURRENTS.

Convection currents are familiar to anyone who has watched a bowl of soup being heated, a pot of coffee, or a maple syrup condenser. When heat is applied to the bottom of any body of liquid in excess of a certain amount, the hotter lower part becomes less dense than the upper part and instability is set up.

Convection currents arise at the first disturbance of this equilibrium. In a broad expanse of liquid these currents have the
tendency to form definite cells of more or less polygonal outline. Fig. 7 shows a vertical section through some experimental
convection cells. The breadth of the cell tends to be about
three times the depth, for plane surfaces of heating and cooling.

Any liquid heated from below has a tendency to lose heat
both by conduction and by convective transfer of material.
Conditions favoring convection are: (1) low viscosity, (2) low
conductivity of the material (increase in size has the same
effect as lowering the conductivity), and (3) a large temperature gradient. The effectiveness of convective transfer of
material depends on the rate of flow. The high viscosity in the
earth hinders rapid flow, but the great size favors it. The
velocity of convective flow decreases inversely in proportion to

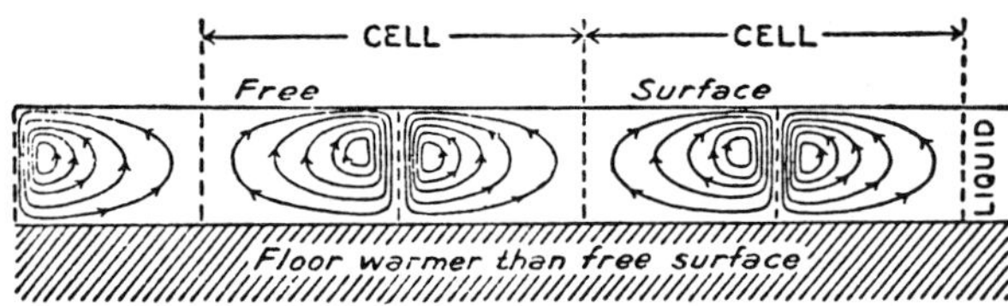

Fig. 7. Section through Experimentally Developed Convection Cells.
After H. Bénard.

the viscosity of the medium, but increases in proportion to the
square of the size of the convecting cell. Another factor favoring convection in the earth is the fact that the amount of heat
transferred by conduction in unit time decreases in proportion
to the square of the thickness of the mass through which it
travels. A third factor may be of cardinal importance in pro-
moting convection—the substratum may not behave as a viscous liquid, but as a "pseudoviscous" solid. This will be discussed in detail later.

Setting up the equations for convection of the earth's
substratum as a viscous liquid, Pekeris (1936) has calculated
the drag which the current may exert on the continental crust.
His equations are simplified in that the term for adiabatic temperature change is not included, and the effect of variation of
the earth's gravitational field with depth is neglected. The
corrections due to these effects, however, would probably not
change the order of magnitude of the result.

For convection cells of the approximate size of the continents and extending down to the 2,900 km. discontinuity at the earth's core, Pekeris has calculated the velocity of the currents and the tangential stress developed on the continents. Fig. 8 shows the stream lines and the velocities of flow in the case of hypothetical currents rising under polar continents and sinking under equatorial oceans. These velocities were calculated on

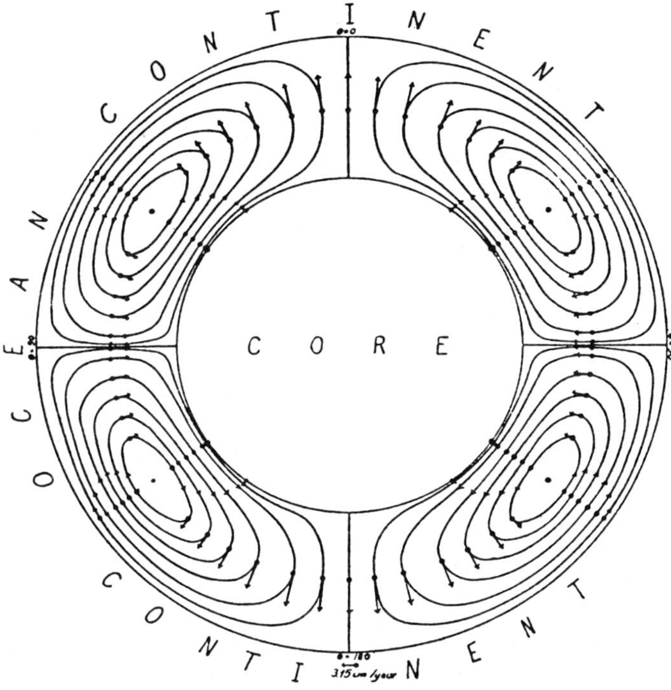

Fig. 8. Convection Current Streamlines and Velocities of Flow (Velocity proportional to the length of the arrows). Under a Crust made up of two Polar Continents and an Equatorial Ocean. After C. L. Pekeris.

the basis of Jeffreys' assumption that the viscosity of the substratum decreases from 10^{22} units at the surface to 10^9 at the core. The calculation of stress is independent of the viscosity assumed. Pekeris finds tangential stress (drag) on the continents of the order of 50 kg./cm.^{2} Because this acts on the whole area above the convection cell, and is concentrated into compression at the junction of the descending currents from two adjacent cells, the compressive stress is many times greater than this tangential stress. If the crust behaved as a perfectly rigid shell, but were free to move away from the center of the cell, the ratio between the compressive and tangential stresses would be the same as the ratio between the surface area of the convection cell and the area of cross-section of the crust peripheral to the cell. For Pekeris' case (Fig. 8) this ratio would be 200:1, and the compressive stress 10,000 kg./cm.^{2} Because the crust is not perfectly rigid, and because some force must be expended in thinning it at the center of the cell, the actual force would be considerably less than this. It is estimated that the actual compressive stress would be of the order of 3,000 kg./cm.^{2}

Vening Meinesz (1934, pp. 54-63) used a different mathematical approach to the same problem in connection with the interpretation of the gravity anomalies in the East Indies. Assuming convection cells extending only down to 1,200 km., Vening Meinesz calculates an effective compressive stress in the crust of the order of 5,000 kg./cm.^{2}

Pekeris also calculates the lateral temperature perturbation necessary to initiate convection on the assumption that the substratum has a threshold strength below which flow does not occur. He concludes that a lateral temperature perturbation of a few tens of degrees will start convection even if the substratum has a threshold strength of as much as 50 kg./cm.^{2}

Hales (1936) calculates the amount of heat transfer necessary to maintain convection in the substratum, assuming it to have a viscosity of 10^{22} and a coefficient of thermal expansion similar to that at the surface. It appears probable from his calculations that an initial temperature gradient only .1°/km. in excess of the adiabatic gradient will be sufficient to develop convection currents.

In discussing the source of heat for this convection, Hales mentions Holmes' suggestion of deep-seated radioactivity with the comment, "It seems doubtful whether the assumption of a deep-seated layer of radioactivity is necessary. If the conductivity in the core is greatly in excess of that in the shell, then even with the lower temperature gradient the heat brought to the lower surface of the shell by conduction would be more than could be carried away by conduction through the shell. This supply of heat would therefore maintain the convection currents in the shell. The conductivity of the core is probably sufficiently large for this."

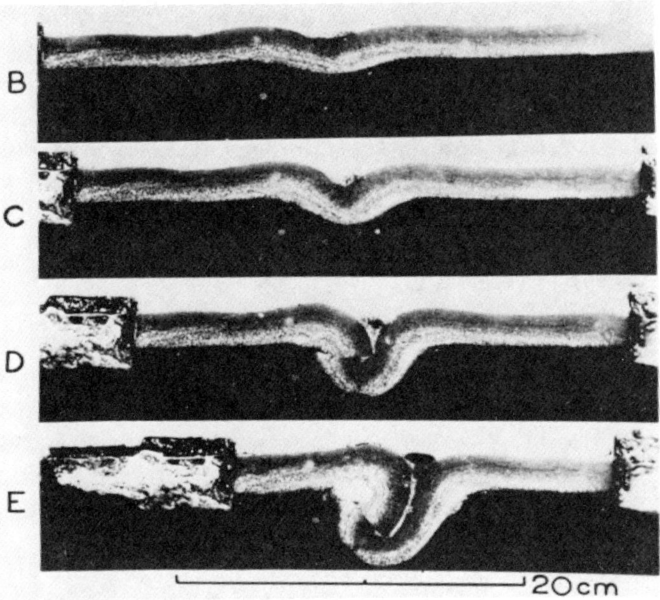

Successive Stages in the Development of a Tectogene During one of Kuenen's Experiments.

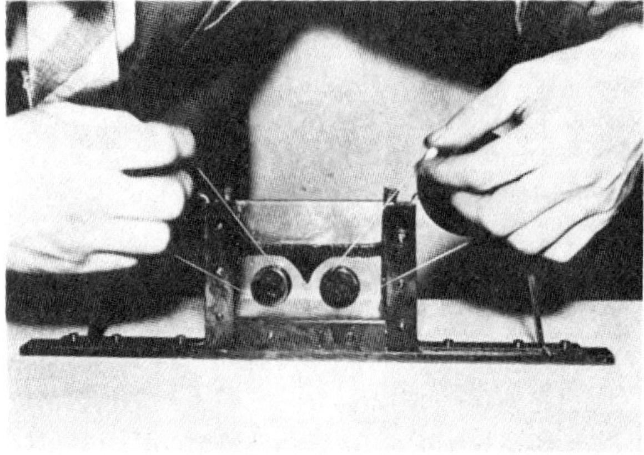

Fig. 1.

Fig. 2.

Fig. 1. Small Dynamic Model to Simulate the Action of Subcrustal Convection Currents and the Response of the Plastic Crust. Photograph Shows Revolving Drums Simulating Convection Currents and the Consequent Development of a Crustal Downfold.

Fig. 2. Large Dynamic Model after Development of Crustal Downfold and Two Underthrusts in the Crust.

Depth of the Convecting Shell.

Among the most fundamental problems of a convection current hypothesis of orogeny is that of the thickness of the convecting layer. Since the convection here in question is due only to the thermal differences in density, any great density discontinuity would act to prevent these density currents from crossing that boundary.

Seismology gives us our only clue as to discontinuities in the earth's shells. Only two discontinuities have stood the test of time in seismological research—the Mohorovičić discontinuity at the bottom of the continental crust, and the first-order discontinuity at the boundary of the central core of the earth (2,900 km. deep). (The discontinuities within the crust are not considered here because it is thought that they act only as secondary modifying influences on the reaction of the crust to the subcrustal convection. It is entirely possible that this is erroneous, and that the crustal discontinuities play a fundamental rôle.)

Various discontinuities have been suggested between the crust and the core: at 475, 1,000, 1,200, 1,700, 2,000 and 2,450 km. The most recent work favors the retention of only the 475 and 1,000 km., and it is interesting to see that one of the masters in this field, Macelwane (1936, p. 231), says, "No one would be more prepared than those who have done the respective pieces of research to admit the tentative and provisional character of the picture of the earth's interior which has been presented . . . Other interpretations may conceivably be developed which will satisfy equally well the data we now have; and tomorrow new facts may be discovered which will sweep away much of our present interpretation."

It must be pointed out that certain types of discontinuities may be no bar to density currents. Thus, if a discontinuity is due to a change in compressibility with pressure, or to polymorphism, it may have little or no effect on a density current crossing it. In the case of polymorphic transitions, the temperature of the material crossing the boundary will be changed slightly by the energy of transition, but this will be the only effect on the current. Jeffreys (1937) suggests that the 475 km. discontinuity is due to a polymorphic transition. The 1,000 km. (Repetti) discontinuity is not a sharp change in properties, but a gradual change in the rate of increase of velocity with depth, and thus might be connected with a change in the compressibility with depth.

The depth of the convecting shell is of primary importance in the distribution of the convection cells over the surface of the earth. Thus, if the cell extends to the core, it is reasonable to expect that the surface extent would be comparable in size to the whole Pacific basin, but if it extended only to the 475 km. discontinuity, the cells would be much smaller. The tangential stresses developed decrease rapidly with the size of the convection cell. Vening Meinesz and Pekeris, however, have calculated that convection in a 1,200 km. shell will be adequate to develop crustal folds. The discussion that follows is based on the arbitrary assumption that the whole substratum down to the core is subject to convection. Other depths of convection will change the magnitude of the calculated effects, but will probably not affect the principles deduced.

The Convection-Current Cycle.

Holmes (1932) suggested a hypothesis of thermal cycles in the earth's history based on a chemical fluxing action of the molten substratum. This hypothesis presents thermal difficulties because melting is involved, and necessitates additional hypotheses as to the composition of the rocks of the substratum. Because of the lack of data as to the composition of the substratum, it seems impossible to work out quantitatively any details of this suggested chemically-cyclic convection and so test the theory.

In contrast to this suggested periodicity, the mathematical treatments of Vening Meinesz, Pekeris, and Hales have all been based on the assumption that the equilibrium state of steady transfer of material is attained, as in ordinary laboratory experiments. The convection equations have involved the further assumption that the flow of the substratum follows the laws of viscous flow.

Consideration of the effects of size, the rate of heat conduction, and the nature of rock flow observed in the laboratory have led the writer to the conclusion that convection currents in the substratum will not attain a steady state, but will be periodic in nature. This hypothesis of cyclic convection, thermal in origin, involves no assumption of melting, and the composition of the convecting shell is assumed to be uniform.

SOLID FLOW OF ROCKS.

It is suggested that the substratum does not behave as a viscous liquid, but as a "pseudo-viscous" solid. In a recent paper the writer experimentally demonstrated pseudo-viscous flow in crystalline rocks subject to small stresses for long periods of time in the laboratory (Griggs, 1939). Two types of experiment have now been performed in which it is possible to measure the velocity of this pseudo-viscous flow as a function of shear stress. Figs. 9 and 10 show the results of these two

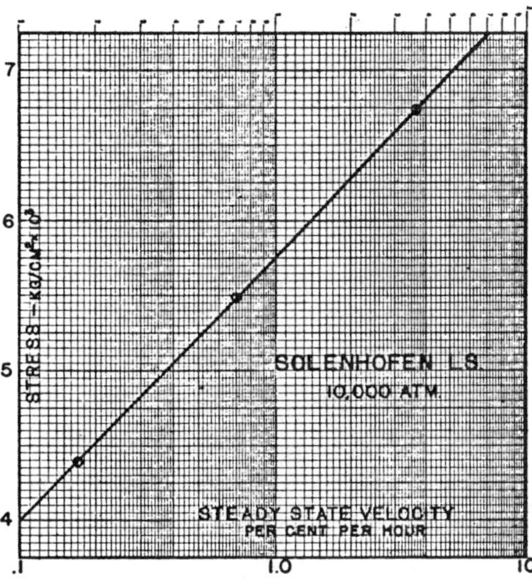

Fig. 9. Velocity of Pseudo-Viscous Flow of Solenhofen Limestone as a Function of Compressive Stress from Creep Tests under 10,000 Atmospheres Confining Pressure.

measurements. Fig. 9 presents observations of the pseudo-viscous flow in Solenhofen limestone under a confining pressure of 10,000 atmospheres (equivalent to that at 22 miles deep in the earth's crust). The data are taken from the creep curves published in an earlier paper (Griggs, 1936, p. 562). The shear stresses in this experiment were high and the deformation rapid. Fig. 10 shows a different type of deformation—pseudo-viscous flow of alabaster under small stresses and in the presence of its own saturated solution.* It has been suggested that this flow is due to solution and recrystallization (Griggs, 1939, p. 249). Both of these widely different types of experi-

* The latter experiments were performed with the aid of a grant from the Pennrose Fund of the G. S. A.

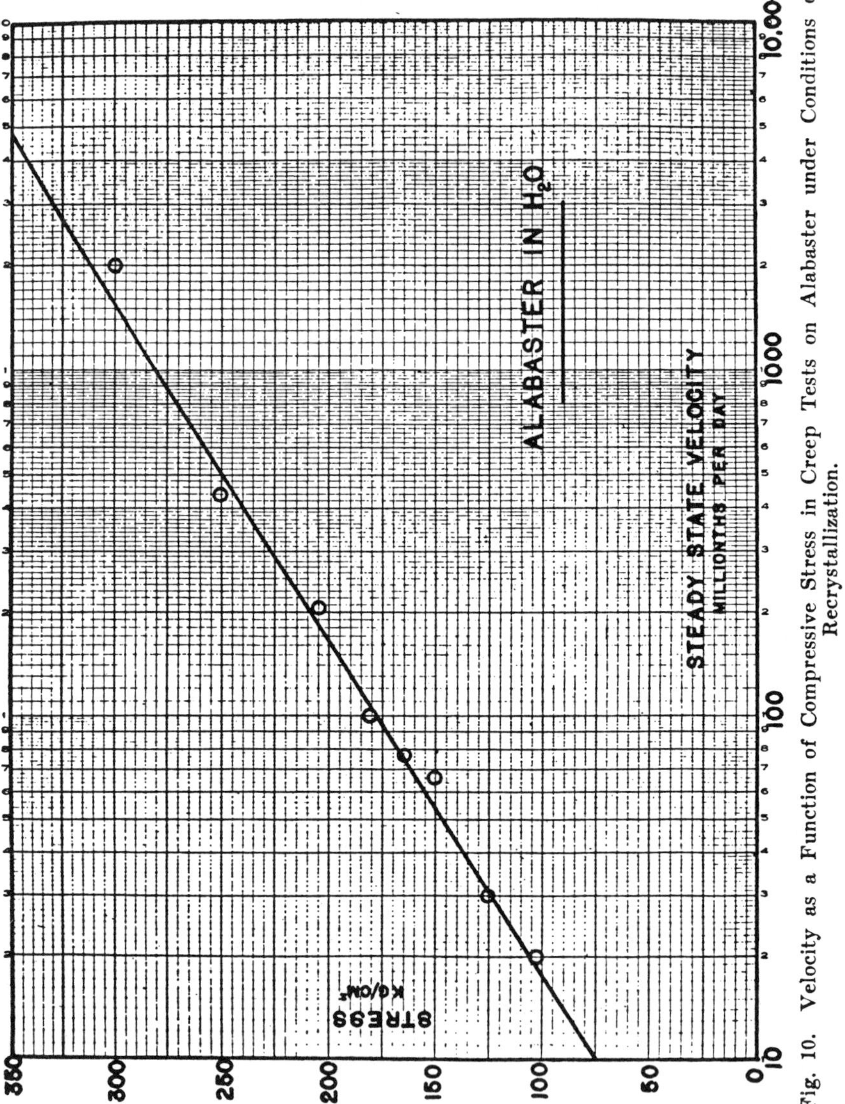

Fig. 10. Velocity as a Function of Compressive Stress in Creep Tests on Alabaster under Conditions of Recrystallization.

ment show that the velocity of pseudo-viscous flow (v) is related to the stress (σ) as follows:

$$\ln v = k\sigma, \text{ or } v = e^{k\sigma} \tag{1}$$

whereas in viscous flow, the velocity is directly proportional to the stress. True viscosity is invariant with stress, but pseudo-viscosity (η) varies with stress as follows:

$$\eta = \frac{k'\sigma}{v} = \frac{k'\sigma}{e^{k\sigma}} \tag{2}$$

Present experiments indicate that there may be a threshold stress below which no continuing flow occurs. This would be the "fundamental strength" of the rock under the conditions of pressure, temperature, and solutions which obtained in the experiment (Griggs, 1936, p. 564).

The importance of this type of flow in the present discussion is that as the stress is increased, the velocity of flow increases exponentially. The way in which this affects convection will be suggested in the next section.

CAUSES OF CYCLIC CONVECTION.

The condition for the *steady* transfer of material by convection is that the temperature of the transferred mass be materially changed by conduction during the time involved in the transfer. This maintains a temperature gradient which serves as a continuous driving force for the convection currents. Steady convection currents are most commonly observed in the laboratory.

Two factors in the earth, however, tend to violate this condition and to make the currents flow periodically:

1. The great size of the cells which makes possible the development of large forces from small temperature gradients, and permits the transfer of material faster than it can change temperature by conduction.
2. Pseudo-viscous rather than viscous flow of the substratum. This involves an extremely high viscosity for low-stress differences, which makes it possible for a temperature gradient to be set up before convection begins. It further involves a low viscosity at high stress-differences which facilitates rapid transfer of the material.

The driving force for convection is gravitational, due to the difference in density between the rising and sinking columns of

a convection cell. It is at once obvious that if the flow of material proceeds more rapidly than cooling by conduction, the hot
material of the rising column will be carried over into the sinking column before it has time to cool by conduction, and similarly the cool material of the sinking column will be carried
under and into the rising column before it can be heated by
conduction. This bodily transfer of hot and cold masses
decreases the driving force of convection, and if the transfer is
complete enough, it follows that the convection current must
stop.

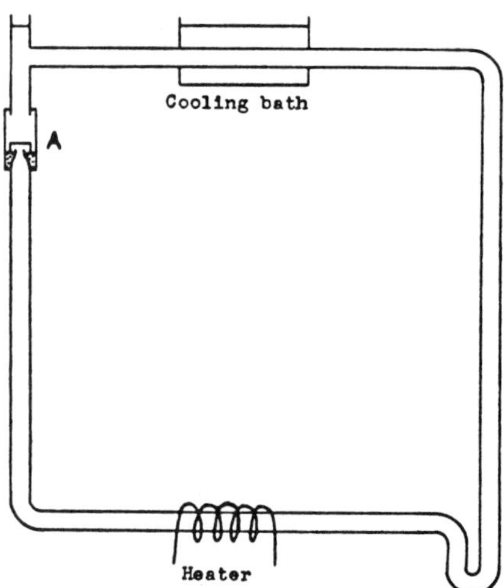

Fig. 11. Experimental Cell to Demonstrate Periodic Convection in a System Subject to Constant Heating and Cooling.

Physicists instinctively distrust the idea of developing a
periodic convective overturn from a steady supply of heat. In
order to demonstrate that it is possible to produce periodic
flow from constant heating and cooling sources, the writer has
set up an experimental convection cell, illustrated in Fig. 11,
in which the conditions are favorable to the production of
periodic convection.* In this cell the heat source at the bottom
and the cooling source at the top are shown, but the boundary
conditions are artificial in that the rising and sinking columns
are localized in the two side tubes of the cell. The unique feature
of this cell is the valve at A. When this valve is removed and a
constant temperature-difference applied to the top and bottom

* This experiment was suggested by Professor L. J. Henderson.

of the cell, normal convection may be observed. When the currents first start they accelerate, as the hot material from the region of the heat source is transferred up into the rising column, and simultaneously the material from the region of cooling moves into the sinking column, thus increasing the difference in density between these two columns. As flow proceeds, however, it slows down, and with a few oscillations the velocity approaches a constant rate.

The valve at A—a light inverted cup floating on mercury— opens when the driving force of convection exceeds a certain amount, and closes when the driving force is less than this amount. The introduction of this valve into the system simulates the behavior of a material which has a threshold strength below which flow does not occur, and thus more nearly reproduces the conditions in the earth's crust than convection of a viscous fluid.

When constant temperature-difference of the right magnitude is applied to this new cell, flow does not begin until the left-hand column of the cell has been heated and the right-hand column cooled sufficiently to develop a driving force large enough to open the valve. The valve then opens suddenly; the flow throughout the cell accelerates rapidly for a short period, and then decelerates just as rapidly and stops as the valve closes. After this sudden transfer of material no flow occurs until the temperature difference of the columns is re-established. This period of quiescence is just ten times as long as the period of flow, in the experiment here described. The cycle is repeated continuously with a constant period.

This serves as experimental verification of the suggestion that when flow is rapid enough to transfer hot and cold masses so quickly that they cannot change temperature by conduction, the driving force of convection will vary periodically.

It is instructive to make some rough calculations for the purpose of comparing the rate of convective transfer of material in the substratum with the rate of temperature change in the moving mass by conduction. Both Pekeris and Vening Meinesz have calculated the velocity of convective flow, assuming steady state convection. Pekeris calculates a maximum velocity of the order of 5 cm./yr. for a convection cell 2,900 km. deep; Vening Meinesz calculates 1 cm./yr. for a cell 1,200 km. deep. Two things would make the peak velocity in convection more rapid than that of the steady-state convection assumed in making these calculations: (1) the initial period of acceleration when convective overturn is begun—laboratory experiments develop a peak velocity three times the steady-state velocity; (2) pseudo-viscous flow, which would make the velocity under the maximum stress differences considerably greater than that calculated from the assumption of viscous flow. From these considerations, it would seem likely that the peak velocity would be of the order of 50 cm./yr., and the average velocity of the fastest moving mass during the convective overturn would be of the order of 10 cm./yr.

On the basis of this estimated velocity, the convective transfer of material from the earth's core to the surface would take 30 million years. For comparison with this velocity of transfer, let us calculate the probable percentage temperature change of the transferred mass during this time. Since only the outer shell of the convection cell is moved rapidly, it would seem that we might estimate the order of magnitude of the heat loss by assuming that a layer 200 km. thick were displaced from the bottom to the top instantaneously and held there for 30 million years. Lovering (1935) developed graphically the heat equations applying to the problem of an extensive dike of uniform thickness, which is essentially our problem. If we assume that this 200 km. shell corresponds to the injection of a dike at the base of the continents, then we can use his derivations to calculate the percentage temperature change of this mass. Assuming a diffusivity of .012, a 200 km. dike would retain 75% of its average excess temperature at the end of 40 million years.

If, during convective overturn, the transferred masses lose less than half of their temperature difference, then it follows that the temperature gradient is reversed and the convective driving force drops to zero. For example, if we assume a temperature difference (in excess of the adiabatic temperature gradient) of 500° C. between the core and the surface, and suppose convective transfer of the top and bottom layers within a period of 40 million years, then on the basis of the above rough calculation, the hot mass which rose to the top will still be 375° hotter than its surroundings and the cool mass which sank to the bottom will be 375° cooler than its surroundings. That is to say, the top mass will be 250° warmer than the bottom mass, and we may expect a static equilibrium to be attained and persist until the unstable temperature gradient (hotter at the bottom) is re-established by conduction.

In these calculations the effect of radioactive heating has been omitted. Holmes (1932) has shown that if radioactive elements were distributed throughout the earth in the proportions in which they exist in the crust, the heat so developed would be fifty times as great as that lost from the earth by radiation. Our assumption of convection in the substratum depends on a fairly uniform distribution of radioactivity in the substratum, and so we may conclude that a probable maximum of such activity would be one per cent of that in the rocks of the crust. Taking Holmes' figures for the average of rocks, this gives an annual radioactive heat output of the order of 3×10^{-8} cal./gm. Assuming a specific heat for the rocks of the substratum of .3, this means that in 100 million years the temperature rise due to radioactivity would be only 10° C., which indicates that its omission in the above calculation involves no serious error.

We may now make a rough estimate of the time required to re-establish a temperature gradient which will permit convection to recur. Using the same thermal equations and assumptions as before, we find that the transferred masses will lose 70 per cent of their temperature difference by conduction in about 700 million years. Radioactivity during this period will increase the temperature of the bottom mass by about 70°. From this it seems that an estimate of 500 million years for the period of static equilibrium would be of the right order of magnitude.

These calculations and experiments lead the writer to believe that convection in the substratum must be periodic. Hypothetical phases of the convection current cycle are illustrated in the next section.

PHASES OF THE CYCLE.

Fig. 12 represents successive stages in convection in one cell extending from the surface to the core of the earth. Before convection begins, heat loss from the core sets up a temperature gradient roughly as shown in Fig. 12-1. In a uniform shell this constitutes an unstable equilibrium in which the hot material tends to rise and the cool surface material to sink. When lateral temperature variations become great enough, currents will start. Pekeris has calculated the magnitude of such variations needed to initiate convection and concludes that even in the case of a substratum having a strength of 50 kg./cm.², "a

35

temperature contrast of a few tens of degrees is sufficient to overcome the above-mentioned strength of the asthenosphere and to start horizontal flow" (Pekeris, 1936, p. 348).

Once started, the temperature instability greatly accelerates the currents. As the hot material rises and the cold material sinks, the central column becomes progressively less dense and the outer column heavier, thus increasing the driving force of the currents (Fig. 12-2). Because of the nature of solid flow

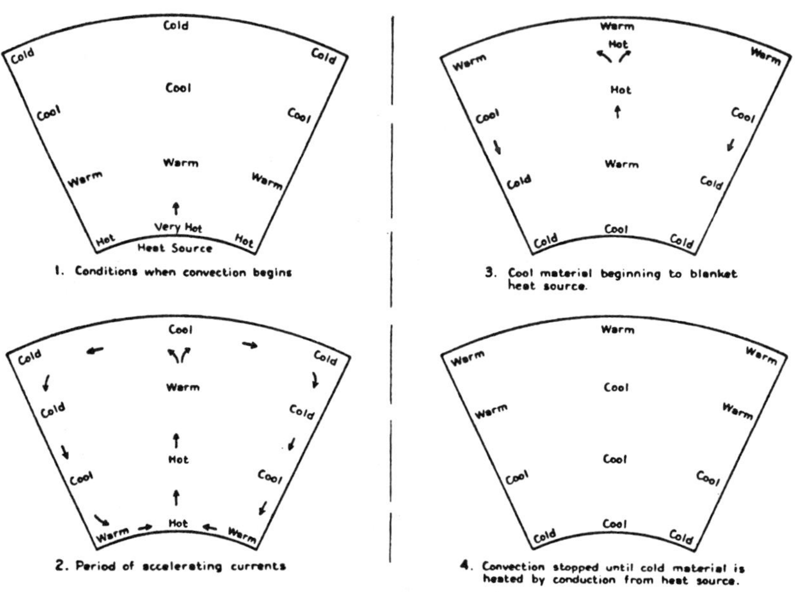

Fig. 12. Successive States in the Development of the Hypothetical Convection-Current Cycle in one Cell Extending from the Surface to the Core of the Earth.

as outlined above, this stress greatly accelerates the velocity of the flow, producing an increase in the tangential stress exerted on the crust by the current.

It will be seen that this increase in tangential stress is not proportional to the velocity of the currents, as would be the case in viscous flow, but is proportional to the increase in the driving force due to the greater density difference between the rising and sinking columns of material. Pekeris (1936, p. 357) shows that the tangential stresses are independent of the viscosity (or "equivalent viscosity" in this case—Griggs, 1939, p. 229).

As the hot material spreads over the surface, the cool material covers the bottom of the cell (Fig. 12-3). This decreases the density difference between the rising and sinking columns and slows down the currents. Finally, when the cool material rises in the central column, and the hot material begins to come down the sinking column, a stage is reached in which density equilibrium is again attained (Fig. 12-4). At this time the currents stop.

The temperature gradient is now the reverse of that before the convection began. No more convection will occur until the original unstable temperature gradient is re-established by conduction from the core and cooling from the surface. The time intervals of these four stages of the convection current cycle are estimated to be of the following order of magnitude:

First Phase—slowly accelerating currents—25 million years.
Second Phase—period of rapid currents—5 to 10 million years.
Third Phase—decelerating currents—25 million years.
Fourth Phase—quiescence—500 million years.

It would seem probable that during the period of quiescence when the substratum is regaining its unstable gradient, new currents would originate in some other cell in the substratum. For this reason, it seems impossible to set a definite period of time for recurrence of convection in some other cell.

A Dynamic Model of the Earth's Outer Shells.

In order to study the effect of sub-crustal convection currents on the continental crust, a dynamically similar scale model of the earth's outer shells was developed. By applying the laws of dimensional analysis and model theory (see Hubbert, 1937) it is possible to construct a model in which all the important properties of the earth's shells are accurately scaled down, and in which convection currents may be simulated.

Geologic models have progressed through several stages of approximation in the past, and no one of the various attempts has been subjected to a thorough dimensional analysis. The experiments most closely approaching dimensional correctness are those of Kuenen (1936). His apparatus (Fig. 13), utilized for the first time a fluid substratum, so that the crust was free to deform downward as well as upward. His choice of materials for the crust included a mixture of paraffin, vaseline, and oil

which had the strength appropriate to a crustal strength of
2,000 kg./cm.^{2}

Kuenen developed a compressive stress in the crust by pushing on it with a movable plunger. It was usually necessary to
localize deformation of the crust by artificially producing a
slight depression. The result of compressing the crust under
these conditions was to produce a downfold of the type shown
in Plate I. If, during the initial stages of an experiment, the broad geosynclinal depression was filled with thin
layers of material weaker than the crust itself, it was found
that these were intricately folded and thrust in a manner
resembling alpine deformations.

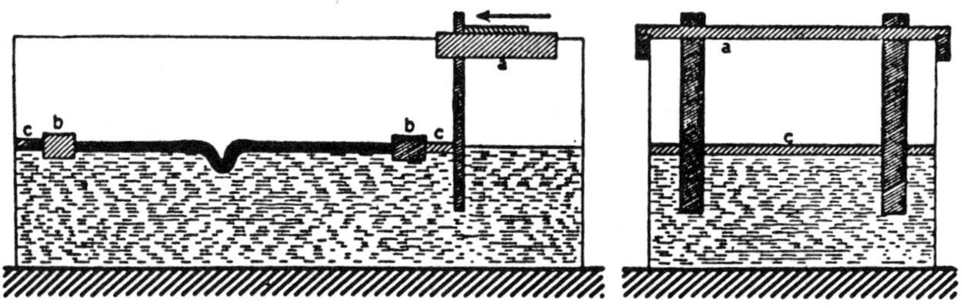

Side view and cross-section of apparatus.
a — pushing mechanism, b — floating beams, c — floating planks.

Fig. 13. Schematic Diagram of Kuenen's Model for Developing a Tectogene by Compression of a Crust Floating on a Liquid Substratum.

In this experiment of Kuenen's the strengths of the crust and
substratum were reproduced approximately to scale, and for
the first time in experimental geology the geometrical conditions
were favorable to the production of a Tectogene. One important factor was neglected, however—dynamical similarity.
Kuenen used water for his substratum, which because of its
extremely low viscosity did not provide sufficient viscous resistance to the over-riding crust to duplicate conditions in the
earth. This viscous resistance of the substratum exerts an
important drag on motion of the crust, as shown by the discussion on page 623 of this paper. It is instructive to make
a dimensional analysis of Kuenen's model to investigate this
point. The principles and symbols for this analysis are those
of Hubbert (1937), with Bridgman's (1931) concept of the
dimensionless product.

Dimensional Analysis of Kuenen's Model.

a. In order to have the time factor at all manageable, inertial forces must be vanishingly small as in the earth, so that body accelerations may be neglected.

b. The simplest way to avoid complications due to acceleration analysis in this derivation is to use weight (W) instead of mass as one of the fundamental units. The others are time (T) and length (L).

c. Dimensions of the important variables:

$$\begin{aligned} \text{Length} \quad &- \text{L} \\ \text{Time} \quad &- \text{T} \\ \text{Density} \quad &- \text{D} = \text{WL}^{-3} \\ \text{Viscosity} \quad &- \text{DLT} = \text{WL}^{-2}\text{T} \end{aligned}$$

d. Model Ratios (Ratio of the magnitude of the property in the model to the magnitude in the earth):

	Symbol	Dimensional Ratio	Value from Kuenen's model
Ratio of length	λ	1	2×10^{-7}
Ratio of density	δ	wl^{-3}	.35
Ratio of viscosity	ψ	$\mathrm{wl}^{-2}\mathrm{t}$	1×10^{-24}
Ratio of time	τ	t	to be determined

e. From these four ratios it is possible to choose a combination which will be dimensionless. The value of this "Dimensionless Constant" will then be the same in the model as in the earth, and will permit calculation of the unknown ratio.

$$\text{Dimensionless Product} = 1 = \frac{\delta\lambda\tau}{\psi} = \frac{\mathrm{wl}^{-3}\mathrm{lt}}{\mathrm{wl}^{-2}\mathrm{t}} \quad \dots\dots\dots\dots (3)$$

$$\text{or, } \tau = \frac{\psi}{\delta\lambda} = \frac{1 \times 10^{-24}}{7 \times 10^{-7}} = 1.4 \times 10^{-17}$$

This means that a process requiring 10 million years in the earth must be reproduced in the model in 1/300 second for dynamical similarity. Such rapidity would violate the basic assumption of this analysis—negligible inertial forces—and so it is impossible for this model to be dynamically similar.

If we are to design a model of the same size and density of materials to be dynamically similar, then we have one variable property—the viscosity—whose value is at our disposal. This analysis shows us that by correctly choosing the value of the viscosity we may make a dynamically similar model to operate at any speed we desire, limited only by the viscosity of the materials available. If, for example, we choose as convenient a model speed such that one minute in the model corresponds to a million years in the earth, we find:

$$\tau = 1.9 \text{ x } 10^{-12}$$

Solving equation (3) for the viscosity ratio:

$$\Psi = \delta\lambda\tau = .35 \text{ x } 2 \text{ x } 10^{-7} \text{ x } 1.9 \text{ x } 10^{-12} = 1.3 \text{ x } 10^{-19}$$

So that for the model, the substratum must have a viscosity of

$$\eta = 1.3 \text{ x } 10^{-19} \text{ x } 10^{22} = 1,300 \text{ c.g.s. units}$$

In setting up a model to illustrate the action of convection currents, the writer used two sizes—a large model in which an earth process requiring a million years could be reproduced in one minute, and a smaller one in which the same process could be reproduced in two seconds. The properties of the larger one are summarized in the following table:

Table I.

Properties of Dynamically Similar Model.

Property	*Model Ratio*	*Actual Value in Model*
Length	$1 \text{ x } 10^{-6}$	60 cm.
Density	.6	1.8 (substratum)
Viscosity	$.8 \text{ x } 10^{-18}$	6300 c.g.s. units
Strength	$.6 \text{ x } 10^{-6}$	.6 gm./cm.^{2} (ca.)
Time	$1.3 \text{ x } 10^{-12}$	1 min. = ca. 1 my.

It is interesting that when the viscosity of the substratum is chosen for dynamical similarity, the model does not behave in the same way as that of Kuenen. When the crust is compressed by a moving plunger in the same manner as in his experiments, it shows no tendency to develop a downfold, but instead the compression is taken up by thickening of the crust immediately in front of the advancing plunger. The viscous drag of the substratum seems sufficient to prevent the transmission of compressive stresses for long distances through the over-riding crust, and causes local thickening of the crust instead. This evidence from the behavior of a scale model provides some measure of confirmation of the conclusion reached on page 623 of this paper which would seem to be a substantial difficulty of the thermal-contraction theory of orogenesis.

The next problem is to produce the effects of subcrustal

convection currents in the model. Dimensional analysis shows that it is impossible to produce currents of sufficient velocity by thermal means. This would require a temperature difference between the bottom and top of the model of the same order of magnitude as that existing between the core and the crust of the earth—thousands of degrees. Since the effect on the crust is primarily that of a current passing under it and turning down at some point, however, this current can be simulated by the rotation of a paddle wheel strategically located in the substratum. In the actual model rotating drums were used, to avoid the irregularities of paddle motion. The friction at the surface of the drums sets up currents not greatly different from those observed in convection. This model corresponds to a vertical section across the boundaries of two adjacent cells. The third dimension is not reproduced.

THE DYNAMIC CONVECTION MODEL.

The principles of the dynamic model are best illustrated in Plate II, Fig. 1, which is a photograph of the small model. Here the small glass cell holds a clear liquid (glycerine) on which floats a black plastic crust (cylinder oil and fine sawdust). The drums are rotated by means of small pulleys belted to the larger pulleys which are turned by hand. When the rotation of the drums is slow, the plastic crust is thickened and pulled down into a slight trough (in this model the oil sticks to the glass walls so that it is impossible to see the surface configuration in the photograph). As the currents increase in velocity, the crust is thickened more and more until finally a narrow downfold is formed as shown in the photograph. At this time of rapid currents, the surface of the crust is irregularly folded, and its average level is just a little lower than before the currents began. If the currents are slowed down and stopped, the downward component of force disappears and the thickened mass rises to buoyant equilibrium, so that its surface is considerably above its original level. Thus the reaction of the crust to the three phases of the convection current cycle is demonstrated.

Because the inertial forces were not negligible in this small model, and because it was desired to study in more detail the structures produced, a larger model was built with the physical properties given in Table I. A photograph of this model in

41

operation is shown in Plate II, Fig. 1. Here the substratum
was very viscous waterglass and the continental crust was a
mixture of heavy oil and sand, which possessed a yield point
that could be varied by changing the proportion of oil used, and
so made to have the proper strength for the model scale. The
rotating drums show in the foreground. At the time the photograph was taken, the drums were in motion, and had developed
a downfold and the two thrust faults in the crust shown by the
dark lines. The actual development of these structures cannot
be satisfactorily depicted short of moving picture portrayal.

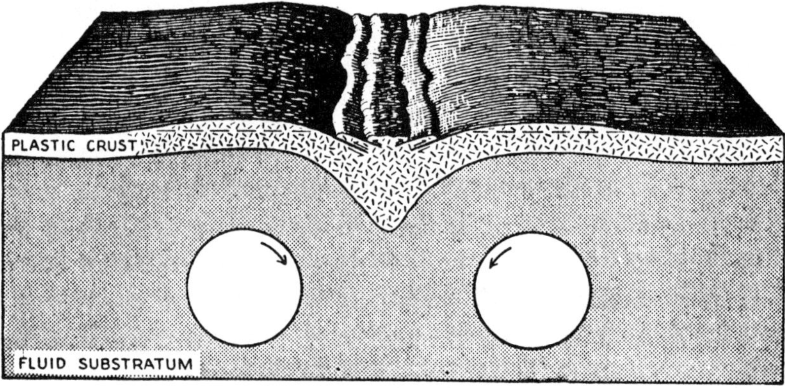

Fig. 14. Stereogram of Large Model with Both Drums Rotating, Showing Tectogene and Surface Thrust Masses with Relations Similar to
Kober's Orogen.

The writer has a series of movies of the apparatus and is
endeavoring to make arrangements so that anyone interested
can obtain them at a nominal cost.

The type of thrusts formed at the junction of two down-
currents is shown in Fig. 14. This sketch illustrates the relations during the height of current velocity. The similarity in
type of structure to that of Kober's Orogen is striking. An
interesting feature of these thrusts is that they are formed by
passive resistance of the overlying mass and active underthrusting of the foundation.

When only one drum is rotated (corresponding to the
development of a single convection cell), the effect on the crust
is different—the crustal downfold formed is not so narrow, and
is asymmetrical in that it is steeper on the side facing the cur-

rent.   The structures are correspondingly asymmetrical.   As the current velocity is increased, the crust is markedly thinned above it, and transported into the thickened part of the crust. Finally, the current sweeps all the crustal cover off and piles it up in a peripheral downfold.   This stage in the development is illustrated in Fig. 15, drawn from the model.

This indication that a singly active cell may sweep off the superjacent continental crust opens wide avenues for speculation as to the formation of the circum-Pacific mountains and indeed as to the primary segregation of the continental masses

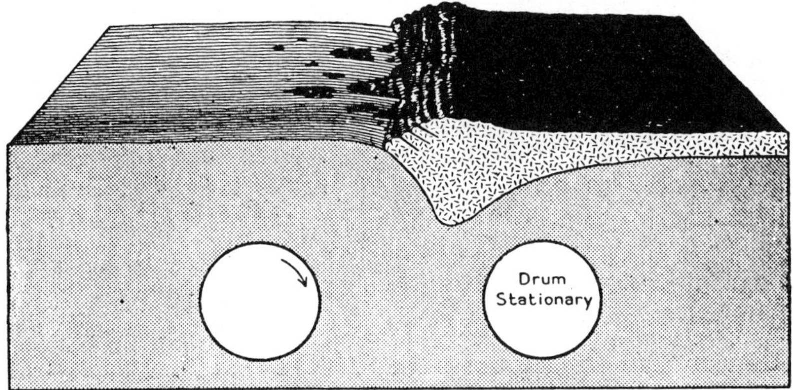

Fig. 15.   Stereogram of Large Model with only One Drum Rotating, Showing Development of Peripheral Tectogene.

themselves.   Here is a possible deformative force which could effectively counteract the tendency of erosion to distribute the continental material uniformly over the surface of the globe.

Correlation of the Convection and Orogenic Cycles.

We began with a generalized review of the mountain-building cycle; progressed through the hypothesis of thermal convection-current cycles in the earth; saw from model experimentation how subcrustal currents may deform the continental crust; and now we proceed to a synthesis of the facts and inferences gleaned from these varying modes of approach.

Fig. 16 shows the suggested correlation between the convection-current cycle and the mountain-building cycle.   During the first phase of the convection-current cycle, when the cur-

THE MOUNTAIN BUILDING CYCLE

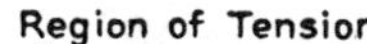

1. First stage in convection cycle — Period of slowly accelerating currents.

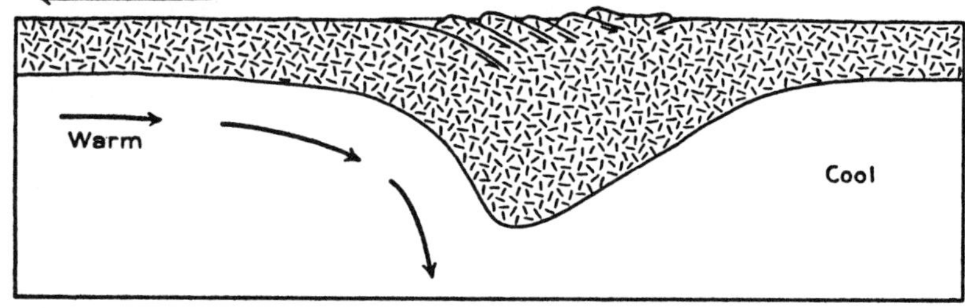

2. Period of fastest currents — Folding of geosynclinal region and formation of the mountain root.

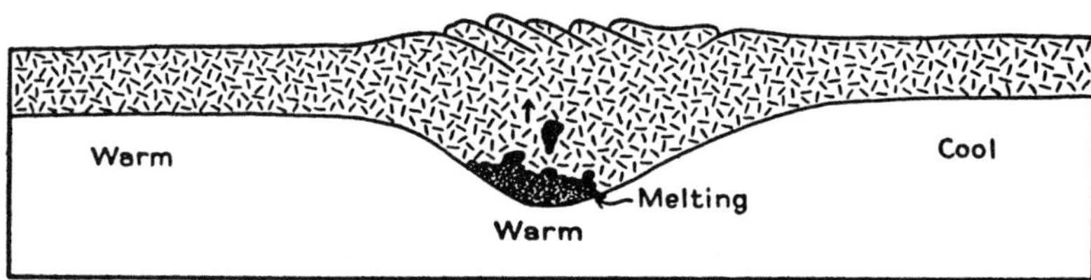

3. End of convection current cycle — Period of emergence. Buoyant rise of thickened crust aided by melting of mountain root.

Fig. 16. Hypothetical Correlation between Phases of the Convection-Current Cycle and Phases of the Mountain-Building Cycle. Structural Relations Drawn from the Model.

rents are slowly accelerating, they will exert an ever-increasing tendency to compress the crust. The compressive force in the crust reaches a maximum where the convection current dives down at the boundary of a cell. At this point there is a vertical component of stress as well as the tangential drag, and the combination of maximum compressive stress and the vertical stress localizes the crustal deformation which in the first stage of the mountain-building cycle takes the form of a geosyncline (Fig. 16-1). If the currents were constant in velocity, the geosyncline would reach an equilibrium position in which no further sinking would occur. With the slowly accelerating currents, however, the compressive stress is constantly increasing and causes continued depression of the geosynclinal trough, in agreement with the geological evidence for continued subsidence of geosynclines. According to the estimate given above, this period of acceleration would be of the order of 25 million years.

When the currents attain high enough velocity, the compressive stress on the crust is sufficient to deform it violently. This increase in compressive stress is accompanied by an increase in the vertical component at the cell boundary, and between the two the crust is downfolded as shown in Fig. 16-2. The thrusts shown in the diagram were added from a study of the model.

In these diagrams and in the discussion of the model, no attempt has been made to reproduce the sedimentary filling of the geosynclinal trough. The thrusts shown in Fig. 16-2 are foundation thrusts, corresponding to similar features in the Alps described by Swiss geologists, and it is to be expected that, as in the Alps, each foundation thrust would be connected with a nappe or thrust in the sedimentary rocks above. The folding and faulting in the sedimentary filling, however, must necessarily be very much more complicated than in the relatively homogeneous and massive granitic crust. This increase in complexity can certainly be better demonstrated in the region of the "roots of the nappes" in the Alps than in any model.

During this period of orogenic folding by the most rapid currents, which probably lasts from five to ten million years, the thickened mass of the crust is kept submerged by the downtow of the sinking current. The vertical component of stress developed by the current acts to prevent the swollen crust from rising to isostatic equilibrium. This persistent deviation from isostatic adjustment which has been shown in the model is corroborated by the common geological observation that mediterranean sedimentation occurred simultaneously with the peak of diastrophism in the mountain systems of the world.

As the currents decelerate in the third phase of the convection cycle, the compressive force in the crust decreases, the downtow decreases, and the thickened mass rises buoyantly, gradually attaining isostatic equilibrium as the currents slow and stop (Fig. 16-3). This produces the elevation which characterizes the third phase of the mountain-building cycle. Observation of the model shows that during this rise, the thickened part of the crust expands laterally and exhibits a tendency to flow away from the center of the folded portion of the crust under the influence of gravity. The absence of compression which characterizes this phase of the cycle, and the tendency toward lateral expansion are in agreement with geological observations that the period of elevation is not a period of dominant compression, but is one of dominant uplift and normal faulting. This change in character of deformation is difficult to explain by the hypothesis of thermal contraction, but can be readily demonstrated in the cyclic-convection theory.

The persistence of major isostatic disequilibria in the East Indies has puzzled Vening Meinesz, Umbgrove, and Kuenen (Kuenen, 1936, pp. 196-7). If the gravity deficiencies here observed are due to crustal downfolds, it is reasonable to suppose that they were formed at the time of the greatest folding of the sediments on the adjacent islands. This conclusion seems unequivocally established by the exact parallelism of regions of most intense folding and the negative anomaly bands. The age of this folding is Miocene, so that one must explain the persistence of the anomalies for something like 20 million years. This presents great difficulty to the thermal contraction theory of orogeny, but is to be expected from the cyclic-convection theory, since we have seen that the period of deceleration is estimated to last 25 million years.

*Application of the Cyclic-Convection Theory to
Earth Structures.*

ALTERNATIVE INTERPRETATIONS OF MAJOR STRUCTURE.

The primary purpose of this paper is to suggest a possible mountain-building mechanism. The application of that mechanism to details of earth structures is beyond the scope of this discussion. It may be of interest, however, to suggest in the most speculative way how this cyclic convection might operate to produce the major trends of the mountain systems.

The most fundamental question in applying this hypothesis to orogeny is the depth of the convective cells. In the absence of conclusive seismological evidence of the existence of a sufficiently sharp density discontinuity in the earth to prevent convective overturn between the crust and the core, it may be assumed for purely speculative purposes that the convection extends throughout this 2,900 km. depth. This would predict a size of the convection cells of the order of 7,000 to 10,000 km. in diameter (at the surface).

Holmes (1932) and Pekeris (1936) have suggested that the blanketing effect of the continents with their high radioactive content will cause sufficiently excess temperature under the continents to initiate rising currents there. These currents would act to spread the continents and to form mountains peripheral to them. Holmes published maps showing the hypothetical effect of this action on continental structures.

It seems conceivable to the writer that the temperature differences within the substratum inherited from the preceding convection cycle may be of more importance in localizing the cells than the blanketing effect of the continents. This opens the attractive possibility of a convection cell covering the whole Pacific basin, comprising sinking peripheral currents localizing the circum-Pacific mountains and rising currents in the center. Such an interpretation would partially explain the sweeping of the Pacific basin clear of continental material, in the manner demonstrated by the model. A minor cell might be suggested with its center in the southwest Indian Ocean, accounting for the Himalayan-Alpine bifurcation.

If this be assumed, then one may carry the speculation further and suppose that the previous cycle occurred as far from this location as possible—namely, about the central Atlantic Ocean. This location is nearly central to the Appalachians, Hercynian mountains, and the Post-Carboniferous mountains of Brazil and Africa.

In favor of the first alternative of Holmes and Pekeris we have the advantage of the thermal explanation for the localization of convection currents. The predominant development of thrusts inclined toward the ocean basins also indicates, by analogy with the model structures, currents rising under the continents. On the other hand, when slightly stronger crusts were used in the model, the direction of thrusting reversed. A tenuous argument in favor of oceanic rising currents is the distribution of deep-focus earthquakes. Visser, Leith and Sharpe, and Gutenberg and Richter all agree that foci of deep earthquakes in the circum-Pacific region seem to lie on planes inclined at about 45° toward the continents. It might be possible that these quakes were caused by slipping along the convection-current surfaces. These flow surfaces would be expected to dip toward the continents on the hypothesis of Pacific up-currents.

In contrast to the rest of the paper, this section is *purely* speculative.

Conclusion.

Of the five primary forces which have been invoked by various theorists to explain orogenic deformation of the earth's crust, three are totally inadequate in magnitude. In contrast, thermal contraction may produce abundant force, but is open to other objections as a mountain-building force. Chief among these are: (1) thermal contraction does not seem capable of providing sufficient shortening to produce the Tertiary mountain system; and (2) transmission of compressive stress through the earth's crust over long distances to provide the localized deformation of the mountain systems is greatly hindered by the viscous drag of the substratum on the overriding mass. Calculations indicate that this viscous drag would cause uniform thickening of the crust instead of localized downfolds. Kuenen produced localized downfolds in his model under conditions in which the crust transmitted the compressive stress as it would have to according to the theory of thermal contraction. Dimensional analysis of Kuenen's model shows that it is dynamically incorrect. A dynamically similar model was constructed and, under compression similar to Kuenen's, showed no formation of a downfold, but thickening of the crust at the point of stress application.

Pekeris', Vening Meinesz' and Hales' calculations show that convection currents can develop adequate compressive stress in the crust to cause orogenesis. A new mechanism of solid flow is suggested for the substratum, with experimental illustrations of this type of flow from two sets of creep tests. Consideration of what seem to be the first order factors leads to the hypothesis of a convection-current cycle. Experimentation with a dynamically similar model which simulates the action of the convection-current cycle produces effects resembling diastrophism of the earth's crust. The structures developed by the action of these currents in the model show striking resemblance to the structures developed in the earth. The phases of the convection-current cycle correlate with the phases of the orogenic cycle.

The hypothesis of orogeny developed from these observations is attractive because it seems to satisfy better than any other the three fundamental conditions of a mountain-building mechanism.

1. Provision of an adequate compressional force.
2. Local provision of sufficient contraction for orogenesis.
3. Explanation of the intermittent nature of orogenic processes and the threefold character of the mountain-building cycle.

Evidence sufficient to establish any theory of this kind can hardly be found within a short time, and has never been put forward in support of any previous theory. This difficulty seems inherent in the very nature of the problem. Accordingly the present theory, necessarily founded on insufficient evidence is here presented because it can be effectively tested only by use and the critical scrutiny of others.

Acknowledgments.

The writer wishes to express his deep appreciation to Mr. A. Lawrence Lowell and the Lowell Institute for providing incentive for the present work in the form of an invitation to deliver a series of lectures at the Lowell Institute. The Lowell Institute also financed the construction of the models. The suggestions of Professors L. J. Henderson, P. W. Bridgman, R. A. Daly, M. P. Billings, and Mr. R. W. Vose were most helpful.

REFERENCES.

Adams, F. D., and Bancroft, J. A.: "Internal Friction during Deformation and the Relative Plasticity of Different Types of Rocks," Jour. Geol., Vol. 25, 597-687, 1917.

Bridgman, P. W.: "Dimensional Analysis," Yale Univ. Press, 1931.

Bucher, W. H.: "The Deformation of the Earth's Crust," Princeton Univ. Press, 1933.

Bull, A. J.: "Further Aspects of Mountain Building," Proc. Geol. Assn. (England), Vol. 40, p. 105, 1929.

Daly, R. A.: "Strength of the Earth's Outer Shells," This Journal, Vol. 35, 401-425, 1938.

Griggs, D. T.: "Deformation of Rocks Under High Confining Pressures," Jour. Geol., Vol. 44, 541-577, 1936.

————: "Creep of Rocks," Jour. Geol., Vol. 47, 225-251, 1939.

Hales, A. L.: "Convection Currents in the Earth," Mon. Not. Roy. Ast. Soc., Geoph. Sup., Vol. 3, 372-379, 1936.

Hess, H. H.: "Gravity Anomalies and Island Arc Structure . . ." Proc. Am. Phil. Soc., Vol. 79, 71-95, 1938.

Holmes: "The Thermal History of the Earth," Jour. Wash. Acad. Sci., Vol. 23, pp. 169-195, 1932.

Hopkins, W.: "Researches in Physical Geology," Phil. Trans. Roy. Soc. London, 381-385, 1839.

Hubbert, M. K.: "Theory of Scale Models as Applied to the Study of Geologic Structures," Bull. G. S. A., Vol. 48, 1459-1520, 1937.

Ingersoll and Zobel: "Mathematical Theory of Heat Conduction," Ginn and Co., Boston, 1913.

Jeffreys, H.: "The Earth," Macmillan, New York, 2nd Ed., 1929.

————: "On the Materials and Density of the Earth's Crust," Monthly Notices of the Royal Ast. Soc., Vol. 4, 50-61, 1937.

Kuenen, P. H.: "The Negative Isostatic Anomalies in the East Indies (with Experiments)," Leidsche Geo. Mededeelingen, Vol. 8, 169-214, 1936.

Lambert, W. D.: "Mechanical Curiosities of the Earth's Field of Force," This Journal, Vol. 2, 129-158, 1921.

Lovering, T. S.: "Theory of Heat Conduction Applied to Geological Problems," Bull. G. S. A., Vol. 46, 69-94, 1935.

Pekeris, C. L.: "Thermal Convection in the Interior of the Earth," Mon. Not. Roy. Ast. Soc., Geophysical Sup., Vol. 3, 343-368, 1936.

Staub, R.: "Der Bewegungsmechanismus der Erde," Geb. Borntraeger, Berlin, 1928.

Vening Meinesz, F. A.: "Gravity Expeditions at Sea," Vol. II Netherlands Geodetic Commission, Delft, 1934.

von Karman, Th.: "Festigkeitversuche unter allseitigem Druck," Zeitschr. des Vereins deutscher Ingenieure, Vol. 55, 1749-1757, 1911.

SOCIETY OF FELLOWS,
HARVARD UNIVERSITY,
CAMBRIDGE, MASS.

2

Reprinted from pages 505–509 of *Principles of Physical Geology,* Ronald Press, New York, 1945, 532p.

PRINCIPLES OF PHYSICAL GEOLOGY

A. Holmes

[*Editor's Note:* In the original, material precedes this excerpt.]

THE SEARCH FOR A MECHANISM

It has been shown that in looking for a possible means of " engineering " continental drift we must confine ourselves to processes operating within the earth. To be appropriate, the process must be capable (*a*) of disrupting the ancestral Gondwanaland into gigantic fragments, and of carrying the latter radially outwards as indicated in Fig. 210 : Africa and India towards the Tethys ; Australasia, Antarctica, and South America out into the Pacific ; (*b*) of disrupting Laurasia, though much less drastically, and again with radially outward movements towards the Tethys and the Pacific, as indicated in Fig. 209. We have already seen that the peripheral orogenic belts probably mark the regions where opposing systems of sub-crustal currents came together and turned downwards. The movements required to account for the mountain structures are in the same directions as those required for continental drift, and it thus appears that the sub-crustal convection currents discussed on pages 408 to 413 may provide the sort of mechanism for which we are looking (Fig. 262).

To explain the peripheral orogenic belts three systems of convection currents are called for (or three co-ordinated groups of systems), with their ascending centres situated beneath Gondwanaland, Laurasia, and the Pacific respectively. Incidentally, it should be noticed that the coalescence of the usual chaotic or small convective systems into three gigantic ones involves a coincidence that can rarely have happened in the earth's history, and one that is just as likely to have come about during the Mesozoic era as at any other time. The often-asked question : How is it that Pangæa did not begin to break up and unfold until Mesozoic time ? thus ceases to have any significance. If continental drift could have been caused by the gravitational forces invoked by Wegener, then it should have occurred once and for all very early in the earth's history, since those forces have always been in operation. If convection currents are necessary, continental drift may have accompanied all the greater paroxysms of mountain building in former ages but, if so, it would usually have been on no more

51

than a limited scale. That there was a quite exceptional integration of effort in Mesozoic and Tertiary times is forcibly suggested by eruptions of plateau basalts and building of mountains on a scale for which it would be hard to find a parallel in any earlier age.

There are, therefore, good reasons for supposing that at this critical period of the earth's history the convective circulations became unusually powerful and well organised. Currents

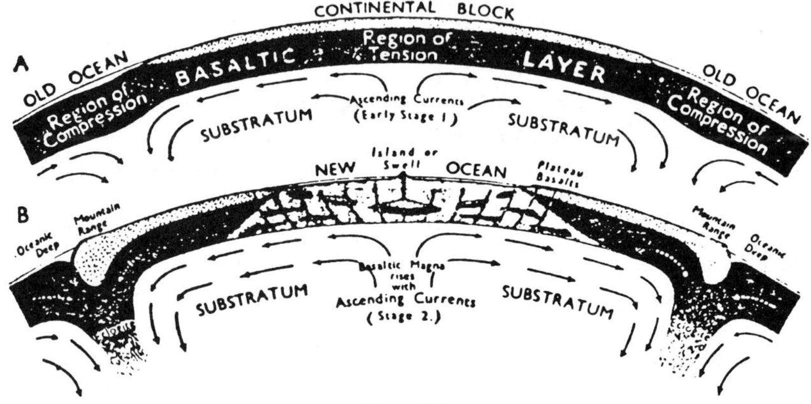

Fig. 262

Diagrams to illustrate a purely hypothetical mechanism for " engineering " continental drift. In **A** sub-crustal currents are in the early part of the convection cycle (Stage 1 of Fig. 215). In B the currents have become sufficiently vigorous (Stage 2 of Fig. 215) to drag the two halves of the original continent apart, with consequent mountain building in front where the currents are descending, and ocean floor development on the site of the gap, where the currents are ascending

flowing horizontally beneath the crust would inevitably carry the continents along with them, provided that the enormous frontal resistance could be overcome. The obstruction that stands in the way of continental advance is the basaltic layer, and obviously for advance to be possible the basaltic rocks must be continuously moved out of the way. In other words, they must founder into the depths, since there can be nowhere else for them to go (Fig. 262).

Now this is precisely what would be most likely to happen when two opposing currents come together and turn downwards beneath a cover of basaltic composition. The latter then suffers intense compression, and like the sial in similar circumstances it is eventually drawn in to form roots (*cf.* Figs. 215 and 216). On the ocean floor the expression of such a down-turning of the basaltic layer would be an oceanic deep. The great deeps bordering the island festoons of Asia and the Australasian arc (Tonga and Kermadec) probably represent the case where the sialic edge of a continent has turned down to form the inner flanks of a root, while the oceanic floor contributes the outer flanks.

It is not difficult to see that a purely basaltic root must have a very different history from one composed of sial. The density of sial is not significantly increased by compression. Consequently, when a sialic root is no longer being forcibly held down, it begins to rise in response to isostasy, heaving up a mountain range as it does so. But when rocks like basalt or gabbro (density 2·9 or 3·0) are subjected to intense dynamic metamorphism they are transformed into schists and granulites and finally into a highly compressed type of rock called *eclogite*, the density of which is about 3·4. Since this change is known to have happened to certain masses of basaltic rocks that have been involved in the stresses of mountain building, it may safely be inferred that basaltic roots would undergo a similar metamorphism into eclogite. Such roots could not, of course, exert any buoyancy, and for this reason it is impossible that tectonic mountains could ever arise from the ocean floor. On the contrary, a heavy root formed of eclogite would continue to develop downwards until it merged into and became part of the descending current, so gradually sinking out of the way, and providing room for the crust on either side to be drawn inwards by the horizontal currents beneath them (Fig. 262).

The eclogite that founders into the depths will gradually be heated up as it shares in the convective circulation. By the time it reaches the bottom of the substratum it will have begun to fuse, so forming pockets of magma which, being of low density, must sooner or later rise to the top. Thus an adequate source is provided for the unprecedented flows of plateau basalt that broke through the continents during Jurassic and Tertiary times. Most of the basaltic magma, however, would naturally rise with the ascending currents of the main convectional systems until it reached the torn and outstretched crust of the disruptive basins left behind the advancing continents or in the heart of the Pacific. There it would escape through innumerable fissures, spreading out as sheet-like intrusions within the crust, and as submarine lava flows over its surface. Thus, in a general way, it is possible to understand how the gaps rent in the crust come to be healed again ; and healed, moreover, with exactly the right sort of material to restore the basaltic layer. To sum up : during large-scale convective circulation the basaltic layer becomes a kind of endless travelling belt on the top of which a continent can be carried along, until it comes to rest (relative to the belt) when its advancing front reaches the place where the belt turns downwards and disappears into the earth.

To go beyond the above indication that a mechanism for continental drift is by no means inconceivable would at present be unwise. Many serious difficulties still remain unsolved. In particular, it must not be overlooked that a successful process must also provide for a general drift of the crust over the interior : a drift with a northerly component on the African side sufficient to carry Africa over the Equator, and Britain from the late Carboniferous tropics to its present position. The northward push of Africa and India, of which the Alpine system and the high plateau of Tibet are spectacular witnesses, could not have been sufficient by itself to shove Europe and Asia so far to the north. To achieve this the aid of exceptionally powerful sub-Laurasian currents directed towards the Pacific is required. The total northward components might then overbalance the southward components, and a general drift of the crust would be superimposed on the normal radial directions of drift.

It must be clearly realised, however, that purely speculative ideas of this kind, specially invented to match the requirements, can have no scientific value until they acquire support from independent evidence. The detailed complexity of convection systems, and the endless variety of their interactions and kaleidoscopic transformations, are so incalculable that many generations of work, geological, experimental, and mathematical, may well be necessary before the hypothesis can be adequately tested. Meanwhile it would be futile to indulge in the early expectation of an all-embracing theory which would satisfactorily correlate all the varied phenomena for which the earth's internal behaviour is responsible. The words of John Woodward, written in 1695 about ore deposits, are equally applicable to-day in relation to continental drift and convection currents : " Here," he declared, " is such a vast variety of phenomena and these many of them so delusive, that 'tis very hard to escape imposition and mistake."

SUGGESTIONS FOR FURTHER READING

A. WEGENER
The Origin of Continents and Oceans. Methuen, London, 1924.

A. L. DU TOIT
Our Wandering Continents. Oliver and Boyd, Edinburgh, 1937.

T. H. HOLLAND
The Evolution of Continents : A Possible Reconciliation of Conflicting Evidence. Proceedings of the Royal Society of Edinburgh, Vol. LXI., Part II., No. 13, 1941.

3

Reprinted from Nature **198**:925–929 (1963)

HYPOTHESIS OF EARTH'S BEHAVIOUR

By Prof. J. TUZO WILSON, O.B.E.

Institute of Earth Science, University of Toronto

FOR fifty years the question has been debated whether or not continents have moved and if they have, by what patterns[1-7]. Until an agreement is reached, no complete understanding is possible of geology, geophysics, evolution, or palæoclimatology.

The recent rapid advances in knowledge of the Earth's interior, ocean floors and rock magnetism combined with older information should soon enable a decision to be reached. This article selects and summarizes some of the conclusions of others in an endeavour to link them by a few original suggestions into an hypothesis of Earth's behaviour.

The uppermost layers of the Earth form a rigid lithosphere of strength equal to, or greater than, that of surface rocks to a depth of about 100 km[8,9]. A weaker asthenosphere lies beneath in which isostatic adjustment occurs by flow at rates of the order of a few cm a year[10]. The observed rates of post-glacial uplift indicate a viscosity of 5×10^{21} c.g.s. units[11]. Recent calculations[12] on the rate of uplift and perhaps satellite data[13] show that strength increases again below a few hundred km. The asthenosphere is thus the low-velocity layer long recognized by Gutenberg[14,15] and recently investigated by Ewing, Press *et al.*[16,17]. Isostatic adjustment shows that continents are free to move up and down vertically like ships held in a frozen surface. The outstanding question is whether the continents can be moved about in horizontal directions by the rifting of the ocean floor in some places and its compression and overriding at others, or whether they are immobile.

The crust is the upper part of the lithosphere above the Mohorovičić discontinuity where seismic velocities undergo a sudden increase to a value of 7·6–8·3 km/sec (ref. 18). This change may be due to a change in phase or in composition. The crustal thickness has been determined at about 300 places[19,20].

In the ocean basins, collection of samples[21], petrological investigations[22,23] and the uniformity of the crustal thickness at about 5 km (ref. 18) suggest that this crust, once considered basalt, may be altered mantle. Hess believes it is serpentinized peridotite. That the present ocean basins may also be young is suggested by the paucity of sediments[24,25], and the youth of all islands in the main ocean basins except Madagascar, Falkland and Seychelles Islands[26].

The composition, thickness and low velocities in the continental crust suggest, as Suess conjectured, that it is of different material (sial) from the mantle and oceanic crust (sima). The great age and zoned structure of continents suggest that they have grown[27,28], a view supported by isotopic investigations[29]. New evidence of three kinds suggests that great horizontal displacements of the crust have occurred. First, measurements by rock magnetism of latitude and azimuth are generally consistent for rocks of any one age and continent but have usually been interpreted to suggest that continents have moved independently relative to the poles[30]. Secondly, it is becoming realized that several dozen large faults, some of which, like the San Andreas and Great Glen faults, have long been known, have undergone horizontal displacements of tens or even hundreds of miles at different geological times[31,32]. Even more recent and striking is the recognition of great scarps on the floor of the eastern Pacific Ocean[33] and the demonstration that they offset magnetic anomalies by as much as 750 miles[34]. The only explanation so far offered is that the scarps are great transcurrent faults, now inactive. The age of the faults and why they stop abruptly at the coast of California have not yet been explained.

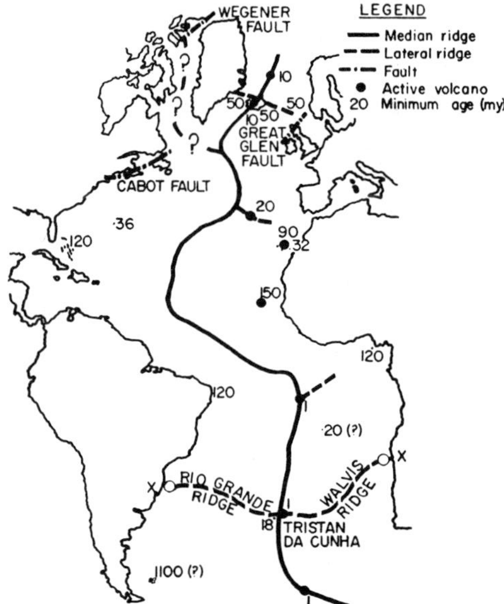

Fig. 1. Atlantic Ocean showing how islands appear to increase in age with distance from the mid-ocean ridge. Three ages are given for three zones in Iceland. Notice that the ends of pairs of lateral ridges define points which would match on the basis of fit of coastlines. A new branch of the mid-ocean ridge has been suggested in Baffin Bay

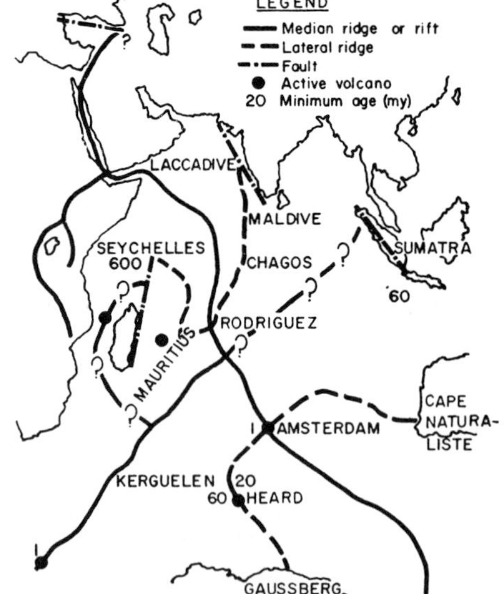

Fig. 2. Ridges of the Indian Ocean showing two postulated median ridges, between Africa and Madagascar and Rodriguez Island and Sumatra respectively. Four aseismic lateral ridges connect points believed to have been separated by continental motion

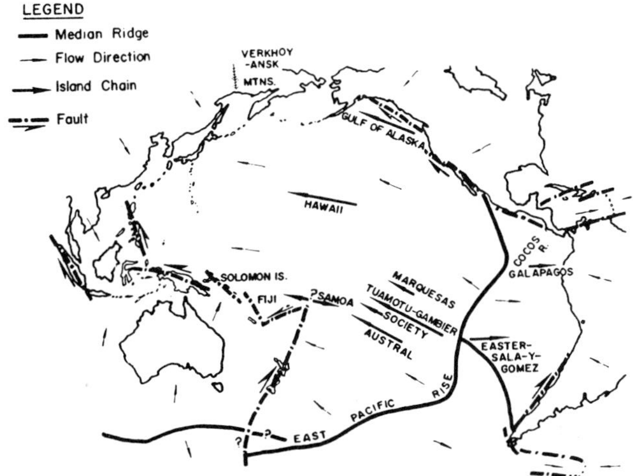

Fig. 3. Sketch of Pacific Ocean. Heavy arrows show nine linear chains of islands and seamounts which increase in age in direction of arrow. Single-headed arrows show direction of motion, where known, along large transcurrent faults. Small arrows show postulated direction of flow away from median ridges

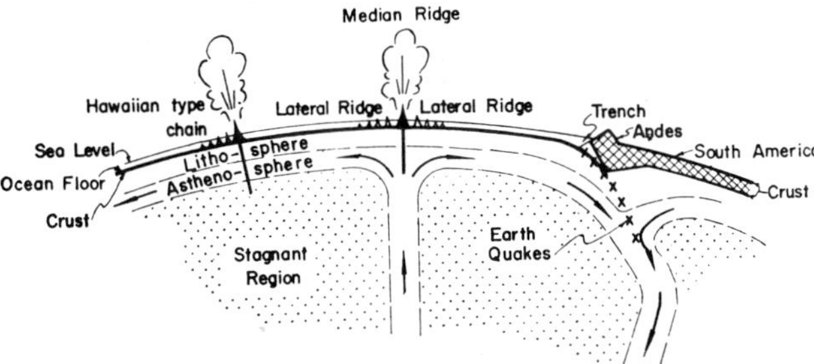

Fig. 4. Diagrammatic section showing a type of convection which might explain the origin of pairs of lateral ridges, median ridges, Hawaiian-type chains and mountain building

Thirdly, oceanic islands tend to increase in age with distance from mid-ocean ridges[26] (Figs. 1 and 2). In the Pacific Ocean the islands and seamounts of several chains have been reported to increase in age with distance from the East Pacific rise. These include the Hawaiian[35-37], Society[38], Austral[39], Galapagos Islands[40], the Pratt-Welker seamounts in the Gulf of Alaska[41] and probably the Marquesas[42]. The Tuamotus[43,44] are older than the Gambiers[45], and Sala-y-Gomez[46] than Easter Island[47] (Fig. 3).

Recently R. S. Dietz has pointed out to me that the Revilla Gigedo Islands off Mexico also increase in age with distance from the East Pacific rise, and I suspect that the same may also be true of the two Juan Fernandez Islands.

If parts of the crust have been shifted horizontally relative to other parts, it must have been stretched or rifted along some lines and compressed or overridden along others. In tension, rocks have little strength and break easily. If they have done so it is agreed that likely places are the mid-ocean ridges because Bullard *et al.*[48,49] have shown that they have high heat-flow, Ewing and Heezen that they are continuous and often rifted[50,51] and because islands increase in age away from them.

Suess pointed out that young continental mountains and island arcs form a continuous system with much evidence of compression. They have large gravity anomalies which Vening Meinesz[52] interpreted in terms of down-folding.

If the floor of the ocean basins, beneath an irregular veneer of sediment and lava, is altered mantle, Hess[53]

and Dietz[54] have pointed out that it can be overridden for unlimited distances either by continental crust or by other parts of the oceanic lithosphere. On descending it reverts to unaltered mantle. On the other hand, continental crust is not dragged down and two slabs on meeting pile up as in the Himalayas and Tibet. This may ultimately produce slippage beneath a continent, but no case has yet been identified. There is certainly a difference between mountains formed where two continental masses join and those formed where a continent overrides an ocean floor as in the Andes.

Although some hold that sections of the crust slip over the asthenosphere because of gravitational or rotational forces, Holmes[55], Griggs[56], and Vening Meinesz[10] have long advocated convection currents as the cause of motion. Currents rising under the mid-ocean ridges and sinking under the continental mountain system could explain respectively the tensional and compressional characteristics of those features in much the fashion illustrated by Bernal[57]. In spite of lack of direct evidence two arguments favour convection currents. Jeffreys[58] has used Rayleigh's theory to show that a viscosity of 10^{26} or 10^{27} would be necessary to stop convection in the mantle and, if currents in the mantle have rifted blocks of continental crust and carried them apart, it is apparent why the ridge where the currents rise is median. On the other hand, if the causes of motion have been forces acting on individual pieces of the crust this is not apparent.

Along the mid-Atlantic ridge are several active volcanoes, from most of which aseismic ridges extend laterally on one or on both sides. Thus from Tristan da Cunha a pair of lateral ridges, called Walvis and Rio Grande, extend to adjacent continents. Roughly speaking, each ridge is a mirror image of the other. If, due to the high heat-flow on the mid-ocean ridge, volcanoes have long been active in about the position of Tristan, horizontal currents would have carried volcanic piles successively off the ridge where they formed. They would have become detached from their source and hence inactive (Fig. 4). This is a possible origin of lateral ridges and can be checked, for if so, they should be progressively older from median ridge to continent.

It can be seen in Fig. 1 that the ends of the Rio Grande and Walvis lateral ridges are exactly opposite points on the coasts of South America and Africa which would fit on the basis of the match of the shape of the shorelines. What might otherwise be a remarkable coincidence becomes a natural consequence if these continents had once been together with a volcanic source between them. It is therefore suggested that submarine ridges provide a second precise method of fitting continents together in addition to that used by Wegener[1] and Du Toit[3].

In a similar manner the ridges which extend from Iceland to Greenland and from Iceland to the continental shelf of Europe serve to link them together in a way that provides a reasonable fit of opposite coasts[59,60]. Again submarine ridges which cross the Gulf of Aden[61,62] end at places which are opposite each other when the coastlines of the Gulf and of the Red Sea are fitted together.

These observations lead to two generalizations. First, where a mid-ocean ridge lies half-way between two continents, they were once in contact. Secondly, the ends of lateral ridges and the fit of shore lines may be used to re-assemble them in the positions in which they once lay.

The reverse case can also be argued. The rifting of the Atlantic Ocean not only involved separating South America from Africa and Europe from Greenland, but also the separation of Greenland from North America. Therefore a longitudinal, median ridge should extend up Baffin Bay. Six large and otherwise unexplained earthquakes have been recorded along this line[63]; but the bathymetry is as yet inadequately charted to prove that such a ridge exists. This can be checked.

Mid-ocean ridges appear to end either in large trans-current faults or by wedging out. One example of the first is the fault postulated by Wegener[1] and mapped by the Geological Survey of Canada[64,65] at the head of Baffin Bay.

The clearest example of a median ridge which terminates by wedging out appears to be in the Arctic Sea.

The Atlantic Ocean is well known to narrow to the north, and Jeffreys[6] has pointed out that this would require rotation through 15°. Fig. 5 shows that if Greenland is imagined to be moved back against Canada, the deep Siberian Basin is a triangular rift with the Lomonosov Ridge as one margin (for reasons unknown). The pivot of rotation lies at the apex of this basin near the New Siberian Islands. Beyond it rotation should produce compression and there the Verkhoyansk Mountains cross Eastern Siberia. According to A. P. Markovsky[66] these were folded and uplifted in Upper Jurassic to early Tertiary and again from Pliocene to Recent times. One might conclude from the Jurassic rocks on the coast of Norway and Greenland[59] and the Eocene[67] at the east and west ends of Iceland that that period was one of extensive rifting of the Atlantic. The Pliocene to Recent rocks in the central graben of Iceland indicate a second period of rifting. Both these periods exactly coincide with periods of compression of the Verkhoyansk Mountains.

The continents around the Indian Ocean have been re-assembled by various authors to form Gondwanaland. The patterns have been very varied[1,2,3]. Applying my new generalizations one can see that to create the Indian Ocean by rifting required the separation of four continents, Africa, India, Australia and Antarctica, and that in addition to the three known median ridges there should be a fourth extending north-easterly between India and Australia (Fig. 2). Several large earthquakes and some shallow soundings shown by Gutenberg and Richter[63] support the proposal that such a ridge may exist. Menard has included what may be an indication of it in his sketch of mid-ocean ridges. If it does exist it will no doubt be discovered during the International Indian Ocean Expeditions.

The opposing coasts of Australia and Antarctica lack large indentations by which they may be matched, but the Amsterdam-Cape Naturaliste ridge and the Amsterdam-Kerguelen-Gaussberg ridge can be seen to form a pair of lateral ridges on either side of the south-eastern median ridge. One is roughly a mirror image of the other. Our premise implies that Cape Naturaliste was once in contact with the Gaussberg coast and this gives a precise method of re-assembling Australia with Antarctica (which the ridge from Tasmania south to Cape Adare also supports). The known ages of Amsterdam[68], Kerguelen[69] and Heard[70] Islands get progressively greater. The Gaussberg volcanics and those at Bunbury near Cape Naturaliste should be older still and related in composition. Many such predictions enable this hypothesis to be checked.

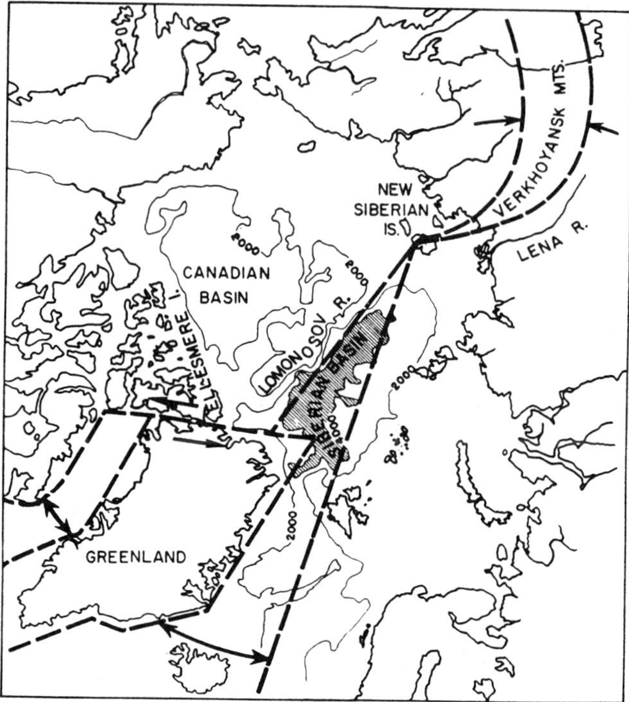

Fig. 5. Sketch map of arctic regions illustrating how continental drift could have moved Greenland, compressed the Verkhoyansk Mountains, and opened the deep Siberian basin, of which the boundary is Lomonosov Ridge

The Chagos-Maldive-Laccadive ridge forms another lateral ridge paired with the Amsterdam-Cape Naturaliste ridge across the postulated north-east median ridge. This suggests that southern India was once adjacent to western Australia and hence close to Gaussberg. If, when India was carried north and collided with the rest of Asia, the floor of the Arabian Gulf had continued to move northward, the fault long postulated along the west coast of India[71] would be explained.

Symmetry suggests that there should be a fourth lateral ridge. It must obviously be the Rodriguez-Mauritius Seychelles ridge, and again a large fault, known for fifty years[72], would have allowed the floor of the Arabian Sea to move northwards past Madagascar. If this fault be extended along a line of atolls, one of which, Providence[73], has a volcanic base, it would pass the Seychelles and may have carried this group away from Madagascar, thus explaining the unusual continental nature of those Islands[74]. A median ridge through the active volcano on Comores Island may exist and have separated Madagascar from Africa, and bathymetry suggests that such a ridge may exist and extend southwards.

Two median ridges apparently terminate in the Indian Ocean area. It may be more than a coincidence that there is a large transcurrent fault in Sumatra[75] where one ridge should end. The other ridge passes into the African and Jordan rift valleys. Again there is a large transcurrent fault in Turkey[76], but it seems to lie some distance from the end of the Jordan rift.

The observation that the rift, marked by a line of earthquakes, across the Arctic basin, and the East Pacific rise are not centrally located does not mean that these ridges are of a different nature from other mid-ocean ridges, but rather that they rifted old ocean floors, not old continents.

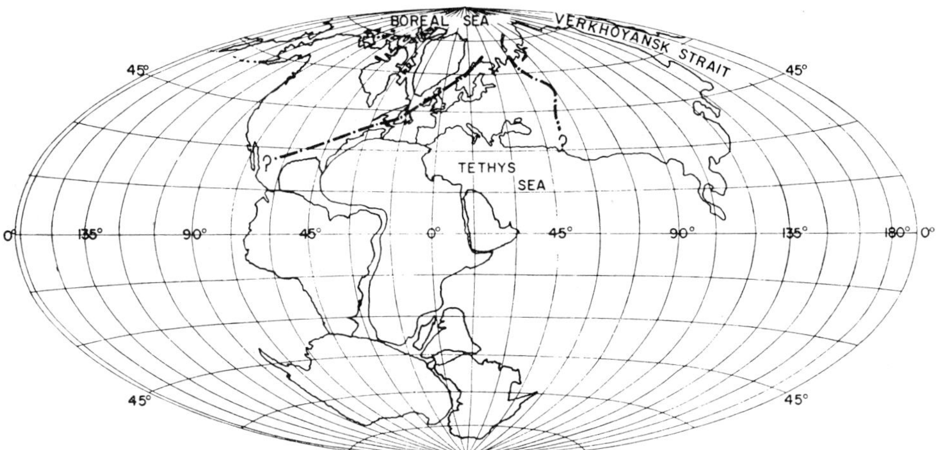

Fig. 6. Reconstruction of continents in mid-Mesozoic time assuming that continental drift occurred due to spreading from mid-ocean ridges along paths now marked by aseismic ocean ridges. The dashed lines represent some earlier mountains formed by unions of older continental blocks

According to our premise, to complete the rifting of the former continents there should also have been median ridges between South America and Antarctica (such a ridge has already been sketched)[77] and between North and South America. It is a striking fact that the two branches of the East Pacific rise head in these general directions. It is suggested that they once passed through Drake Strait and the Caribbean Sea. That they do not do so now is attributed to the growth in Tertiary time of part of Patagonia and Central America along the continental fracture system and to expansion of the ridges themselves.

If the continents stopped moving apart, but the ridges continued to expand by rifting, their growth could have caused them to move northwards, creating at their ends as they expanded the great and active right-hand shear faults off Chile[78] and through California[79].

This use of median and lateral ridges and the recognition of the Verkhoyansk Mountains as the proper hinge between Eurasia and North America instead of Alaska enables a unique and new reconstruction of continents to be made for Lower Jurassic and probably also Triassic time (Fig. 6).

This reconstruction has some features in common with those already proposed, but differs in others. Especially important is the fact that it has been assembled according to rules, and not merely by arbitrary matching of coasts. It has been given an approximate position from palæomagnetic data[30], but this could probably be improved.

It was not a primeval continent, and the inactive mountains marking junctions of older continental fragments can be plainly seen (Fig. 6).

Although the median ridges shown in Figs. 1–3 form a single system, the pattern of recent earthquakes shows that not all parts have been equally active simultaneously. There is evidence that in Upper Jurassic and Cretaceous time the Atlantic opened and joined the rift striking north-easterly across the Indian Ocean to move Africa and India northwards together[80].

In early Tertiary time the south-easterly rift opened in the Indian Ocean and extended south of Australia and across the Pacific to separate Antarctica and the Americas. Finally, in later Tertiary time the Atlantic rift has been active again, but this time has been connected through the western ridges of the Indian Ocean with the start of the Red Sea and the African rifts.

These rifts, now the most active, are meridional in the Atlantic, in Africa and in the Indian Ocean. The East Pacific rise trends north-easterly. Thus the greatest contraction and flow is in a generally west-north-westerly direction as indicated by the orientation of the Hawaiian and Society Islands. The ocean floors must therefore be disappearing in trenches lying at right angles to this direction. Such trenches are those of the East Asian Arcs, the Tonga-Kermadec Islands and the Andes. These are the most active seismic regions in the world with most of the deep earthquakes.

The large shears in Melanesia have been supposed to be left-handed in the Philippines[81], New Guinea[82] and between the New Hebrides and Tonga[83]. Thus all the directions of shearing in California, off British Columbia and in Melanesia, and all the directions of dip of the planes on which deep earthquakes lie in the Pacific, may possibly be explained by supposing that the floor of the Pacific Ocean is moving away from the East Pacific Rise, north-westerly to pass under East Asia and New Zealand and south-easterly beneath the Andes. Likewise, the lesser expansion in a northward direction due to the median ridge through the southern seas can explain the compression and activity of the Alpine-Himalayan-Indonesian region.

The right-hand shear through New Zealand[84], like those in Melanesia and the Western Americas, results from two directions of flow meeting at right angles. Because Macquarie Island[85] is the only island in the deep oceans to be well folded, prophyllitized and to have veins of sulphides, it seems to be partly continental in character. (So are the several groups of islands in shallow water south-east of New Zealand.) It is suggested that the Alpine Fault of New Zealand extends south past Macquarie Island and offsets the mid-ocean ridge near the Balleny Islands.

Where two horizontal currents meet at about 180° angles, arcs form, but where currents meet at about 90°, shears form, as Fig. 3 shows.

Much careful work by Gutenberg and Richter[63], by Benioff[79] and by Hodgson[86] has elucidated the distribution of earthquake foci and the direction of motion for many earthquakes.

Attempts to derive a simple interpretation of the results have been inconclusive. It is suggested that whereas Scheidegger's[87] explanation of the circular shape of island arcs by fracture is tenable for a brittle and uniform ocean floor in the shallow lithosphere, for deeper earthquakes an element of flow intervenes and that no consistent pattern exists. It is suggested that the reason why the planes marking the foci of deep earthquakes dip west under the East Asian area and the Tonga Islands, east under the Andes and nearly vertically under the Solomons and Bonin Islands is not decided by any special angles of fracture, but by plastic failure in places decided by the direction of flow. The dip is under continents where down currents have been overridden and it is near vertical where features strike parallel with the flow.

In the Pacific Ocean ten examples of another type of aseismic ridge have already been mentioned, of which the Hawaiian Islands are an example. These chains are straight and extend in one direction only from an active or recently active volcanic centre which is not on a median ridge. Betz and Hess[88] long ago made the sensible suggestion that these linear chains lie on large transcurrent faults, but later work has not supported this hypothesis. The faults afterwards mapped on the islands are not axial[36,89] but appear to be great landslides to which Fairbridge[90] directed attention and of which other large examples are recorded on Heard Island[70], the Marquesas[42] and Samoa[91]. The careful charting of Hamilton[92] has not disclosed a submarine fault east of Hawaii. The recent discovery[37] of Miocene fossils on Oahu suggests that the spread in age along the chain is greater than Stearns postulated[89]. Faulting is not a satisfactory explanation for these chains, but an explanation[93] is possible if the central parts of convection cells in the mantle move more slowly below about 200 km than above and if the sources of lava under the active ends of the chains lie so deep (Fig. 4). Observations have already marked the rise of lava from 60 km[36] and perhaps deeper[9].

The Hawaiian type of chains vary in length. The longer they are the higher the proportion of atolls. Darwin showed that atolls tend to be older than volcanic islands, which suggests that length is an indication of age of these chains.

This hypothesis of terrestrial behaviour is precise and can be tested in many places. It is based on much published work; but I would also like to acknowledge the help of many colleagues some of whom have been supported by grants to me from the National Research Council of Canada, the Geological Survey of Canada and the Vela Uniform Program of the United States Air Force. This is Contribution No. 22 to the Canadian Upper Mantle Project.

L. Smirnow and R. P. Morrison have been particularly helpful with advice, for which I am grateful.

1 Wegener, A., *The Origin of Continents and Oceans* (E. P. Dutton and Co., New York, 1924).

2 Van Waterschoot, van der Gracht, W. A., *Theory of Continental Drift* (Amer. Assoc. Petrol. Geol., 1928).

3 Du Toit, A. L., *Our Wandering Continents* (Oliver and Boyd, Edinburgh, 1937).

4 Carey, S. W., *Continental Drift, a Symposium* (Dept. Geol., Univ. Tasmania, Hobart, 1958).

5 King, L. C., *The Morphology of the Earth* (Hapner Publ. Co., New York, 1962).

6 Jeffreys, H., *The Earth*, fourth ed. (Camb. Univ. Press, 1958).

7 Runcorn, S. K. (editor), *Continental Drift* (Academic Press, New York and London, 1962).

8 Press, F., *Science*, 133, 1455 (1961).

9 Anderson, D. L., *Sci. Amer.*, 207, 52 (1962).

10 Heiskenen, W. A., and Vening Meinesz, F. A., *The Earth and its Gravity Field* (McGraw-Hill Book Co., Inc. New York, 1958).

11 Niskanen, E., *Publn. Isostatic Inst.*, *Helsinki*, 23, (1949).

12 McConnell, R. K., *The Viscoelastic Response of a Layered Earth to the Removal of the Fennoscandian Ice Sheet* (thesis, University of Toronto, 1963).

13 O'Keefe, J. A., Eckels, A., and Squires, R. K., *Science*, 129, 565 (1959).

14 Gutenberg, B., *Bull. Seis. Soc. Amer.*, 19, 11 (1926).

15 Gutenberg, B., *Science*, 131, 959 (1960).

16 Takeuchi, J., Press, F., and Kobayshi, N., *Bull. Seis. Soc. Amer.*, 49, 355 (1959).

17 Dorman, J., Ewing, M., and Oliver, J., *Bull. Seismol. Soc. Amer.*, 50, 87 (1960).

18 Worzel, J. L., and Shurbet, G. L., *Geol. Soc. Amer.*, Spec. Paper, 62, 87 (1955).

19 Deminitskaya, R. M., *Trans. Sci. Res. Inst. Arctic Geol.*, 115, (1961) (in Russian).

20 Steinhart, J. S., and Meyer, R. P., *Carnegie Inst. Wash. Publ.*, No. 622 (1961).

21 Hersey, J. B., *J. Geophys. Res.*, 67, 1109 (1962).

22 Hess, H. H., *Bull. Geol. Soc. Amer.*, 71, 235 (1960).

23 Ringwood, A. E., *J. Geophys. Res.*, 67, 857 (1962).

24 Kuenen, Ph. H., *Marine Geology* (Wiley, New York, 1950).

25 Hamilton, E. L., *J. Sediment. Petrol*, 30, 370 (1960).

26 Wilson, J. T., *Nature* 197, 536 (1963).

27 Gill, J. E., *Trans. Roy. Soc. Canada*, Sect. 4, Ser. 3, 43, 61 (1949).

28 Wilson, J. T., in Kuiper, G. P. (editor), *The Earth as a Planet*, 138 (Univ. Chicago Press, 1954).

29 Hurley, P. M., Hughes, H., Faure, G., Fairbairn, H. W., and Pinson, W. H., *J. Geophys. Res.*, 67, 5315 (1962).

30 Runcorn, S. K., *Continental Drift*, 1 (Academic Press, New York and London, 1962).

31 Benioff, H., in Runcorn, S. K. (editor), *Continental Drift*, 103 (Academic Press, New York and London, 1962).

32 Jacobs, J. A., Russell, R. D., and Wilson, J. T., *Physics and Geology* (McGraw Hill Book Co. Inc., New York, 1959).

33 Menard, H. W., *Bull. Geol. Soc. Amer.*, 70, 1491 (1959).

34 Vacquier, V., in Runcorn, S. K. (editor), *Continental Drift*, 135 (Academic Press, New York and London, 1962).

35 Dana, J. D., *U. S. Exploring Exped.* 1838–1842, 10 (1849).

36 Eaton, J. P., and Murata, K. J., *Science*, 132, 925 (1960)

37 Menard, H. W., Allison, E. C., and Durham, J. W., *Science*, 138, 896 (1962).

38 Williams, H., *B.P. Bishop Mus. Bull.*, 105, (1933).

39 Chubb, L. J., *Quart. J. Geol. Soc. Lond.*, 83, 291 (1927).

40 Banfield, A. F., Behre, jun., C. M., and St. Clair, D., *Bull. Geol. Soc. Amer.*, 67, 215 (1956).

41 Menard, H. W., and Dietz, R. S., *Bull. Geol. Soc. Amer.*, 62, 1263 (1951).

42 Chubb, L. J., *B.P. Bishop Mus. Bull.*, 68, (1930).

43 Aubert de la Rue, E., *Recherche géologique et Minérale en Polynésie Francaise* (Insp. Gén. Mines Géol., Paris, 1959).

44 Repelin, J., *C.R. Acad. Sci., Paris*, 168, 237 (1919).

45 Davis, W. M., *Amer. Geog. Soc. Spec. Publn.*, 9, 375 (1928).

46 Fisher, R. L., *Preliminary Report on Expedition Downwind*, IGY World Data Center A, IGY Gen. Rep. Ser., No. 2 (1958).

47 Bandy, M. C., *Bull. Geol. Soc. Amer.*, 48, 1589 (1937).

48 Bullard, E. C., Maxwell, A. E., and Revelle, R., *Adv. Geophys.*, 3, 153 (1956).

49 Menard, H. W., *Science*, 132, 1737 (1961).

50 Heezen, B. C., and Ewing, M., *Amer. Geophys. Union*, Monog. 1, 75 (1956).

51 Heezen, B. C., in Runcorn, S. K., *Continental Drift*, (Academic Press, New York and London, 1962).

52 Vening Meinesz, F. A., *Gravity Expeditions at Sea* 1923–1932, 2 (Netherlands Geod. Comm., Delft, 1934).

53 Hess, H. H., *Petrologic Studies: A Volume to Honor A. F. Buddington*, 599 (Geol. Soc. Amer., 1962).

54 Dietz, R. S., in Runcorn, S. K., *Continental Drift*, 289 (Academic Press, New York and London, 1962).

55 Holmes, A., *Trans. Geol. Soc. Glasgow*, 18, 559 (1928–29).

56 Griggs, D. T., *Amer. J. Sci.*, 237, 611 (1939).

57 Bernal, J. D., *Nature*, 192, 123 (1961).

58 Jeffreys, H., *The Earth*, second ed. (Camb. Univ. Press, 1929).

59 Donovan, D. T., *Meddel. om Grønland*, 155, No. 4 (1957).

60 Wilson, J. T., *Nature*, 195, 135 (1962).

61 Farquharson, W. I., *John Murray Exped.*, 1933–34, *Sci. Rep.*, 1, No. 2, Chart 1 (1935–36).

62 Loncarevic, B., and Matthews, D. H., *New Sci.*, 14, 513 (1962).

63 Gutenberg, B., and Richter, C. F., *Seismicity of the Earth*, second ed. (Princeton University Press, 1954).

64 Prest, V. K., *Geol. Surv. Canada*, Paper 52—32 (1952).

65 Christie, R. L., *Geol. Surv. Canada*, Paper 62—10 (1962).

66 Markovsky, A. P., translated by P. de Saint-Aubin, and Roget, J., *Structure Géologique de l'U.R.S.S.* (Centre Nat. Res. Sci., Paris, 1961).

67 Thorarinssen, S., *Mus. Nat. Hist. Reykjavik, Dep. Geol. Geog. Misc. Papers*, 25 (1960).

68 Davis, W. H., *Amer. Geog. Soc. Spec. Publ.*, 9, 154 (1928).

69 Aubert de la Rue, E., *Chronique Mines d'Outre-Mer, Paris*, 24, 162 (1956).

70 Lambeth, A. J., *Roy. Soc. New South Wales, J. and Proc.*, 86, 14 (1952).

71 Krishnan, M. S., *Geology of India and Burma*, fourth ed. (Higginbothams, Madras, 1960).

72 Lemoine, P., *Handbuch Reg. Geol.*, 6, Pt. 4 (1912).

73 Wiseman, J. D. H., *Trans. Linn. Soc. Lond.*, 19, Ser. 2, 437 (1926–36).

74 Miller, J. A., and Mudie, J. D., *Nature*, 192, 1174 (1961).

75 Westerveld, J., *Bull. Geol. Soc. Amer.*, 63, 561 (1952).

76 Richter, C. F., *Elementary Seismology* (W. H. Freeman and Co., San Francisco, 1958).

77 Runcorn, S. K., *Continental Drift*, 38 (Academic Press, New York and London, 1962).

78 St. Amand, P., *Los Terremotos de Mayo, Chile*, 1960 (Michelson Lab., U.S. Naval Ordnance Test Station NOTS TP2701, China Lake, California, 1961).

79 Benioff, H., in Runcorn, S. K., *Continental Drift*, 103 (Academic Press, New York and London, 1962).

80 Termier, H., and Termier, G., *Atlas de Paléogéographie* (Masson et Cie, Paris, 1960).

81 Allen, C. R., *J. Geophys. Res.*, 67, 4795 (1962).

82 Brouwer, H. A., *Quart. J. Geol. Soc. Lond.*, 106, 231 (1951).

83 Lensen, G. J., *Canada Dom. Obs. Publ.*, 24, No. 10, 391 (1960).

84 Wellman, H. W., *Geol. Rdsch.*, 43, 248 (1955).

85 Mawson, D., *Austral. Antarct. Exped.*, 1911–1914, *Sci. Reps.*, A, 5 (Sydney, 1943).

86 Hodgson, J. H., in Runcorn, S. K., *Continental Drift*, 67 (Academic Press, New York and London, 1962).

87 Scheidegger, A. E., and Wilson, J. T., *Proc. Geol. Soc. Canada*, 3, 167 (1950).

88 Betz, jun., F., and Hess, H. H., *Geog. Rev.*, 32, 99 (1942).

89 Stearns, H. T., *United States, Terr. Hawaii, Div. Hydrog. Bull.*, 8 (1946).

90 Fairbridge, R. W., *Geog. J.*, 115, 84 (1950).

91 Kear, D., and Wood, B. L., *New Zealand Geol. Surv. Bull.*, 63 (1959).

92 Hamilton, E. L., *Bull. Geol. Soc. Amer.*, 68, 1011 (1957).

93 Wilson, J. T., *Canadian J. Phys.* (in the press).

4

Continental Drift and Convection[1]

Leon Knopoff

BY NOW a considerable amount of information has been collected bearing on present ideas concerning the spreading of the sea floor and continental drift. Whether the ideas of sea-floor spreading and continental drift can be used to explain geologic processes in regions remote from oceanic ridges, trenches, and rift zones, and whether a process of convection must be invoked to explain the spreading and drifting observations, is as yet undetermined. The pertinent results will first be summarized briefly. By itself, none of these data is convincing; taken together, they raise the probabilities considerably that large-scale horizontal motions take place and have taken place in the past.

Continental Drift

The notion of continental drift was inspired in large part by the remarkable apparent fit of the present profiles of the facing coasts of

South America and Africa [*Wegener,* 1915]. The two continents were presumed to have been originally one protocontinent until about 150–200 million years ago; the date of the rupture was approximated by the ages of older common fossils on both shores. With the discovery of the Mid-Atlantic ridge, further geographic comparisons were possible; to a striking degree of accuracy, the central ridge is half-way between the shelves of the continents along the margins. *Bullard et al.* [1965] have tested the fit of the continents around the Atlantic at the 500-fathom contour and find the fit to be so good as to be outside the realm of chance. Support of this result is to be found in the observation that the contact between rocks of ages 2×10^9 years and 0.6×10^9, found on both the African and South American coasts, is in excellent accord with the geographical fit [*Hurley et al.,* 1967].

Vine and Matthews [1963] have shown that elongated magnetic anomalies in the Atlantic have a remarkable symmetry about the median ridge. Further, the distances to the anomalies

---

[1] Publication 736, Institute of Geophysics, University of California, Los Angeles.

61

from all the oceanic ridge axes are in excellent correlation [*Vine*, 1966] with a chronology for reversals of the earth's magnetic field [*Cox et al.*, 1964], with an apparent uniform 'velocity' for the distance-time relationship. Unfortunately, the chronology is available for only 4×10^6 years. Velocities range from about 1 cm/year for the Atlantic to more than 4 cm/year for the other oceans. Thus the correlation with the time scale is available for the central 100–350 km of the zone of magnetic anomalies. Alternations in the sign of magnetic stripes are found well beyond the distance corresponding to the chronologies; these can be used only as markers at present.

If the floor of the sea is spreading at oceanic ridges, the mechanism of earthquakes occurring on fracture zones should be one of transform faulting instead of transcurrent faulting [*Wilson*, 1965]. *Sykes* [1967], using Byerly's fault-plane method [*Stauder*, 1962], has shown that the sense of motion, if parallel to the principal geomorphic trend, is that for transform faulting and opposite to that for transcurrent faulting.

Oliver and Isacks [1967] showed that the deep ocean trenches are the loci where the surface area created at ridges is destroyed. The trenches are the places where plates of surface matter, perhaps 50–100 km thick, are being thrust into the mantle. Deep-focus earthquakes, usually associated with deep sea trenches, are thus possibly the consequence of thermal stresses set up by the plunging of cold, near-surface matter into a warm mantle. Below a depth of 600 to 700 km for the most rapidly moving blocks, the downward-thrusting block probably approaches a state of thermal equilibrium with its environment [*Griggs and Baker*, 1968]. The depth of deep-focus shocks is well-correlated with the velocity of the downward moving block [*Isacks et al.*, 1968] for the different geographic regions.

Paleomagnetic measurements of the remanent magnetic field in sediments serve to give the direction of the magnetic field of old rocks. Comparison of rocks of the same age from different continents give divergent magnetic poles. These can be brought into coincidence by invoking both continental drift and polar migration. The method involves assumptions regarding (1) the undisturbed nature of the samples since deposition, (2) the dipolar nature of the magnetic field, (3) the coupling of the magnetic and rotational poles, and (4) lack of scatter in the data. Under these assumptions, the magnetic data, as far back as 1.5×10^8 years ago, indicate consistency of continental drift with the geomorphic fit [*Creer*, 1965].

Finally, the work of *McKenzie and Parker* [1967], *Le Pichon* [1968], and *Morgan* [1968] has shown that the surface regions between ridges and trenches move as more-or-less rigid plates. The surface plates are bounded by sources of area at ridges, by sinks of area at trenches, and by transform faults where area is neither created nor destroyed. The edges of the plates are delineated by a map of earthquake epicenters [*Barazangi and Dorman*, 1969]; earthquakes occur where adjacent plates move relative to one another. At sources of area, the earthquakes are tensile in character; at sinks of area they are, in part, compressional; and at transform faults, they are strike-slip.

Note added in proof. A. E. Maxwell [*Trans. Am. Geophys. Union, 50,* 113, April 1969] reported results from the Deep Sea Sampling Program showing that the deepest sediments on the sea floor have ages proportional to their distance from the axis of the mid-ocean ridge in the South Atlantic. Results include sediments to ages about 90 million years, thus extending the previous chronology twentyfold and indicating that spreading has continued steadily at about 2 cm/yr for at least 90 m.y., and probably has been steady since the onset of spreading when the primeval protocontinent, Gondwanaland, ruptured.

Flow in the Mantle

A model has been described above of large-scale motions at the surface. Matter emerges **from the interior at rifts and** disappears into the interior at trenches. But as yet, this picture describes the motion of surface or near-surface matter only. No attempt has yet been made to describe the complete kinematics in terms of the return path of the flow, required by the continuity conditions for conservation of matter.

What is unknown is the rheological state of the mantle. It was formerly suspected that the deep mantle, below about 1000 km, had a viscosity of the order of 10^{26} cgs, a value sufficiently high to inhibit large-scale motions in the deep interior [*Munk and MacDonald*, 1960; *McKenzie*, 1968]. Recently it has been shown that this conclusion can be explained in an alternate way by renormalization of the second-order harmonic in the gravitational field as determined from the perturbation of satellite orbits [*Goldreich and Toomre*, 1969]. Thus we cannot say, at this time, whether there is significant motion in the deeper mantle.

The inference that the lower mantle had a very high viscosity was a consequence of the size of the second-order harmonic in the gravitational field; this was not appropriate to account for the observed polar flattening of the earth computed on the basis of the dynamics of a rotating fluid. A significant second-order term has always presented difficulties from the point of view of convection theory, since this must be produced by some process symmetric with respect to the axis of rotation; the other harmonics need not be so, and can, in principle, be generated by instabilities in the motion, as will be outlined below. *McKenzie* [1968] considered the possibility that the nonequilibrium flattening might be produced by mantle-wide poloidal currents caused by the failure of the isotherms to coincide with the gravitational equipotentials. If these do not coincide in a rotating fluid, flow must take place according to a theorem of von Zeipel. McKenzie concluded that this type of flow is not sufficient to account for the flattening and prefers the hypothesis that the flattening is associated with a deep-seated viscosity.

Inspection of the distribution of the perturbation gravitational fields as a function of spherical harmonic order number casts further light on the contribution of the second-order harmonic. The result shows that these coefficients, including the second-order term, vary as $1/n^2$ [*Anderle and Smith*, 1968]. This suggests, but not conclusively, that the process which produced the higher-order perturbations in the gravitational field also produced the second-order one.

The low-velocity channel appears to be a worldwide phenomenon occurring at different depths beneath the surface. Under more or less inactive areas, including shields, it is at a depth of about 115 km [*Brune and Dorman*, 1963]. Under more active regions, including mountains and oceans, it is at shallower depths [*Berry and Knopoff*, 1967; *Knopoff et al.*, 1966]. In addition, there appears to be a significant lateral variation of velocity in the channel. The reduction of velocity at these depths in some parts of the world may be due to partial melting. The increased fluidity would imply a somewhat reduced viscosity in this region. *Anderson* [1966] has shown that there is a correlation between viscosity and the specific attenuation factor $1/Q$ for seismic waves in about the same region as the low-velocity channel. The classical measurements of viscosity from elastic rebound of land masses due to the removal of a surficial load may in fact reflect the viscosity of the low-viscosity zones. If the low-velocity channel is a zone of reduced viscosity, it seems likely that the surficial plates described above move about using the low-velocity/low-viscosity zone as a lubricant. Thus the low-velocity channel might act as a zone of decoupling of large motions near the surface from those in the deeper interior where the motions are probably significantly reduced.

It has been shown [*Knopoff*, 1964, 1967a] that significant inhomogeneity in the form of phase transitions in the mantle might inhibit vertical motion across these boundaries if the transition is sharp enough. In addition to the boundaries of the low-velocity channel, significant discontinuities in the upper mantle are found at depths of about 360 and 600 km [*Johnson*, 1967]. These discontinuities, which seem to be well correlated with phase changes to high density polymorphs of magnesium silicates [*Anderson*, 1967], are imperfectly explored as yet; the degree of sharpness is not yet known, although the preliminary cross sections that are drawn through these regions do not appear to be so sharp as to inhibit vertical motions.

Anderson [1967] suggests that the mantle below 600 km has a higher iron-to-magnesium ratio compared with the region above this depth. This result depends significantly upon an equation of state derived from velocity-density relations for materials at laboratory pressures plus data derived from shock-wave studies. The laboratory data cannot be extrapolated accurately to give velocity-density relations at pressures corresponding to 600 km or more of depth; there are such significant uncertainties in the shock wave reductions [*Knopoff and Shapiro*, 1969] as to make the shock wave data unusable at this time. If the suggestion that the lower mantle is richer in iron than the upper mantle should prove to be correct, then vertical mass transport across the interface of the lower and upper mantle is forbidden, since convection is basically a homogenizing process; stirring would destroy the chemical stratification.

Scale of Mantle Motions

At this time, the evidence is extremely weak either for or against the idea that motions take place in a mantle-wide process. This point is raised because in several theories of convective processes the horizontal scale of convection is strongly dependent upon the vertical dimensions of the region involved in the calculations. If one imagines the surface plates to be the upper limbs of some large-scale convective process, with hot matter rising at the ridges and cold matter descending at the trenches, the distances between the trenches and ridges must be related to the vertical scale of the region of circulation. The horizontal distances between ridges and trenches seem more likely to be of the order of the depth to the core than, say, to the deepest major transition in the upper mantle, namely 600 km. Thus, if this point of view is taken, it is crucial that no barriers to the circulation be present anywhere in the mantle, whether they be of the type involving chemical stratification or of the type involving polymorphic phase changes over short ranges of depth. *Verhoogen* [1965] has shown that it is highly likely that most polymorphic transitions of the silicates probably do not present impenetrable barriers to vertical convection.

Driving Mechanism

It is of interest to see whether a model of a driving process that will generate these large horizontal motions at the surface can be constructed consistent with the observations mentioned thus far and consistent with the potential constraints on the motion. Three published models have been claimed, at various times, to be pertinent to the problem of continental drift. These are the models of (1) a layer of fluid heated from below (considered originally by Lord Rayleigh), (2) a fluid heated from within, and (3) a fluid with horizontal temperature gradients applied to it. The first model, that of a layer heated from below (summarized by *Knopoff* [1964]); involves marginal instability; circulation does not start until the vertical temperature gradient exceeds a certain critical value. At the condition of marginal instability,

the form of the convective flow is prismatic with a plan form of regular polygons. The horizontal scale factor is of the order of two to three times the vertical scale factor. If the superadiabatic vertical temperature gradient is significantly supercritical, the motion becomes irregular but, because of a high ratio of viscosity to thermal diffusivity, it probably retains approximately the ratio of dimension of plan form to vertical scale in some statistical sense.

The model of the fluid heated from within [*Tozer*, 1965] has certain similarities to the first model. It too involves marginal instability, but one governed by a different dimensional number than the first model. The scale factors appear to be of the same order as those for the first model.

In the presence of horizontal temperature gradients [*Allan et al.*, 1967], the fluid is always unstable, in contrast to the two foregoing models. The horizontal scale is once again of the order of the thickness of the layer for a delta function-like temperature distribution at the surface (Knopoff and Smith, unpublished results). Unfortunately, the nonlinear problem of heat transport has not yet been included in the calculations for this model, nor has it been taken into account adequately in other models.

Excluding the problems of the possible irregularity of the convection patterns in the mantle, which may itself be defeating to the argument of applying the above models, it would seem that the distance from the East Pacific rise to the Japan trench is significantly larger than the scale of the motion allowable by the above theories, even assuming that mantle-wide convection is possible. In addition, no mechanism is available to allow for the likelihood that the sources and sinks of area, the ridges and trenches, migrate. Migration is required to allow several sources of area to be adjacent to one another without the occurrence of intervening sinks. An example of this problem is to be found in the adjacent ridges of the southern Mid-Atlantic, the East African rift, and the Mid-Indian Ocean (Figure 1). No intervening trench is to be found between these three sources of area. These three rifts can all be sources of spreading if the distance between the rifts is continually increasing. Indeed, the four adjacent plates abutting on the three rifts mentioned above must all be increasing in net

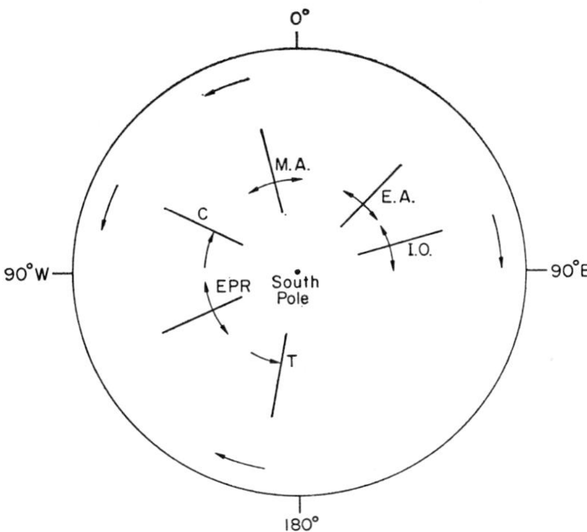

Fig. 1. Schematic diagram of the southern hemisphere showing major rift zones and trench structures. The inner arrows show motions relative to these features; the outer arrows show the motions of the features relative to one another. The East African rift has been taken as stationary in an inertial frame of reference, but this is arbitrary. *EA*, East African rift; *MA* Mid-Atlantic ridge; *IO*, Indian Ocean; *EPR,* East Pacific rise; *C,* Chilean trench; *T,* Tonga trench. The actual motions are not azimuthal as shown.

area, since the trenches at their outer edges, the Chile and the Tonga trenches, are oriented in the sense of disappearance of Pacific area. Hence these two trenches must be moving toward the East Pacific rise despite the spreading from the latter, since the net area of the Pacific will be reduced, to compensate for the increase in area of the four other plates.

A final consideration in the problem of mantle-wide convection is that the time constant to set up steady convection of this type is extremely long. Using rates of sea-floor spreading of the order of centimeters per year, the time for this process to become stable is of the order of 1 billion years [*Knopoff, 1967b*] or longer if the viscosity at depths is great; this value is approximately the time it takes a particle to complete one circuit of a regular convection cell. This time scale is long enough compared with the time for postulated rupture of the protocontinent of the southern hemisphere that it would seem unlikely that this model is appropriate.

It is therefore quite unlikely that mantle-wide convection is a pertinent model for the earth. The effects of possible chemical stratification,

the scale of distance, and especially the scale of time make this quite improbable. To restrict the circulation to the upper mantle only increases the difficulty with regard to the horizontal scale.

Episodic Movements

Additional geophysical evidence lends support to the point of view that convection is not a steady-state process. This is the observation that the spreading process may be episodic. Two independent sets of observations bear on this point. *Ewing and Ewing* [1967] have shown that there exists a rather abrupt change in the thickness of sediments of all the major oceans at a distance of the order of a few hundred kilometers from the axes of these ridges. If there has been continuous steady spreading, then an abrupt change in the rate of deposition of sediments in the oceans occurred several million years ago. On the other hand, it may be that the parts of the sea floor more remote from the axes of the ridges have been relatively stationary with respect to the location of the ridges for some time. Ewing and Ewing surmise that the latter explanation is the more likely and that the present episode of sea-floor spreading began about ten million years ago. This was preceded by a quiet interval in which no relative motion took place. An earlier episode of spreading may have terminated about forty million years ago. It is possible that the Atlantic Ocean opened up and spread to virtually its present shape in episodes ending about a million years ago and has only recently begun to widen again. The recent onset of spreading seems to have started more or less simultaneously in all oceans.

Grommé et al. [1967] find that earlier epochs of continental drift probably took place over quite short periods of time. Paleomagnetic pole positions for rocks from different continents seem to show consistent ages over long periods of time. Grommé et al. concluded that several spasmodic episodes of sea-floor spreading may have taken place in the lifespan of the earth and that the time during which the earth's surface is in a state of large-scale motion is probably small compared to the time during which it is in a stationary state.

If we assume that the motion is episodic, the transient process requires a somewhat different picture for the circulation. It is very likely that static stresses accumulate during the quiet period until they exceed certain critical stresses. When the critical state is reached, a major rupture takes place and the two adjacent plates are set in motion in opposite directions. Their momentum probably carries them a certain distance; ultimately the plates slow down because of viscous drag forces. A further quiet episode sets in once again. If this model is accepted, then it would seem likely that the source of the material to fill up the space between the diverging plates must be local. One possibility is that the material comes from the low-velocity channel itself.

The source of the instability is to be found in the static episode. The cause of the stresses that initiate the onset of spreading is a source of speculation at present, but it is undoubtedly thermal in nature. Once the surface parts of the earth begin to move, they readjust their positions to some new configuration, which is presumably in a lower energy state. The system then freezes and the static stresses begin to build up once again, much as some giant relaxation oscillator. A model somewhat along these lines was suggested by *Griggs* [1939], but in connection with a mantle-wide convection process. It seems useful to revive this picture in the context of the new information about sea-floor spreading.

REFERENCES

Allan, D. W., W. B. Thompson, and N. O. Weiss, Convection in the earth's mantle, in *Mantles of the Earth and Terrestrial Planets*, edited by S. K. Runcorn, pp. 507–512, Interscience Publishers, New York, 1967.

Anderle, R. J., and S. J. Smith, Observation of twenty-seventh and twenty-eighth order gravity coefficients based on Doppler observations, *J. Astronaut. Sci.*, *15*, 1–4, 1968.

Anderson, D. L., Earth's viscosity, *Science, 151*, 321–322, 1966.

Anderson, D. L., Phase changes in the upper mantle, *Science, 157*, 1165–1173, 1967.

Barazangi, M., and James Dorman, World seismicity map of ESSA Coast and Geodetic Survey epicenter data for 1961–1967, *Bull. Seismol. Soc. Am.*, *59*, 369–380, 1969.

Berry, M. J., and L. Knopoff, Structure of the upper mantle under the Western Mediterranean basin, *J. Geophys. Res.*, *72*, 3613–3626, 1967.

Brune, J. N., and J. Dorman, Seismic waves and earth structure in the Canadian shield, *Bull. Seismol. Soc. Am.*, *53*, 167–210, 1963.

Bullard, E. C., J. E. Everett, and A. G. Smith, The fit of the continents around the Atlantic, *Phil. Trans. Roy. Soc. London, A, 258*, 41–51, 1965.

Cox, A., R. R. Doell, and G. B. Dalrymple, Reversals of the earth's magnetic field, *Science, 144*, 1537–1543, 1964.

Creer, K. M., Palaeomagnetic data from the Gondwanic continents, *Phil. Trans. Roy. Soc. London, A, 258*, 27–40, 1965.

Ewing, J., and M. Ewing, Sediment distribution on the mid-ocean ridges with respect to spreading of the sea floor, *Science, 156*, 1590–1592, 1967.

Goldreich, P., and A. Toomre, Some remarks on polar wandering, *J. Geophys. Res.*, *74*(10), 1969.

Griggs, D., A theory of mountain building, *Am. J. Sci.*, *237*, 611–650, 1939.

Griggs, D. T., and D. W. Baker, Origin of deep focus earthquakes, in *Properties of Matter*, John Wiley & Sons, New York, in press, 1968.

Grommé, C. S., R. T. Merrill, and J. Verhoogen, Paleomagnetism of Jurassic and Cretaceous Plutonic rocks in the Sierra Nevada, California, and its significance for polar wandering and continental drift, *J. Geophys. Res.*, *72*, 5661–5684, 1967.

Hurley, P. M., F. F. M. deAlmeida, G. C. Melcher, U. G. Cordani, J. R. Rand, K. Kawashita, P. Vandoros, W. H. Pinson, Jr., and H. W. Fairbairn, Test of continental drift by comparison of radiometric ages, *Science, 157*, 495–500, 1967.

Isacks, B., J. Oliver, and L. R. Sykes, Seismology and the new global tectonics, *J. Geophys. Res.*, *73*, 5855–5899, 1968.

Johnson, L. R., Array measurements of *P* velocities in the upper mantle, *J. Geophys. Res.*, *72*, 6309–6325, 1967.

Knopoff, L., The convection current hypothesis, *Rev. Geophys.*, *2*, 89–122, 1964.

Knopoff, L., Thermal convection in the earth's mantle, in *The Earth's Mantle*, edited by T. F. Gaskell, pp. 171–196, Academic Press, New York, 1967*a*.

Knopoff, L., On convection in the upper mantle, *Geophys. J.*, *14*, 341–346, 1967*b*.

Knopoff, L., and J. N. Shapiro, Comments on the interrelationships between Grüneisen's parameter and shock and isothermal equations of state, *J. Geophys. Res.*, *74*, 1439–1450, 1969.

Knopoff, L., S. Mueller, and W. L. Pilant, Structure of the crust and upper mantle in the Alps from the phase velocity of Rayleigh waves, *Bull. Seismol. Soc. Am.*, *56*, 1009–1044, 1966.

Le Pichon, X., Sea-floor spreading and continental drift, *J. Geophys. Res.*, *73*, 3661–3697, 1968.

McKenzie, D. P., The influence of the boundary conditions and rotation on convection in the earth's mantle, *Geophys. J.*, *15*, 457–500, 1968.

McKenzie, D. P., and R. L. Parker, The North Pacific: An example of tectonics on a sphere, *Nature, 216*, 1276–1280, 1967.

Morgan, W. J., Rises, trenches, great faults, and

crustal blocks, *J. Geophys. Res.*, *73*, 1959–1982, 1968.

Munk, W. H., and G. J. F. MacDonald, *The Rotation of the Earth*, Cambridge University Press, Cambridge, 1960.

Oliver, J., and B. Isacks, Deep earthquake zones, anomalous structures in the upper mantle, and the lithosphere, *J. Geophys. Res.*, *72*, 4259–4275, 1967.

Stauder, W., The focal mechanism of earthquakes, *Advan. Geophys.*, *9*, 1–76, 1962.

Sykes, L. R., Mechanism of earthquakes and nature of faulting on the mid-oceanic ridges, *J. Geophys. Res.*, *72*, 2131–2153, 1967.

Tozer, D. C., Heat transfer and convection currents, *Phil. Trans. Roy. Soc. London, A*, 258, 252–271, 1965.

Verhoogen, J., Phase changes and convection in the earth's mantle, *Phil. Trans. Roy. Soc. London, A, 258*, 276–283, 1965.

Vine, F. J., Spreading of the ocean floor: New evidence, *Science*, *154*, 1405–1415, 1966.

Vine, F. J., and D. H. Matthews, Magnetic anomalies over oceanic ridges, *Nature*, *199*, 947–949, 1963.

Wegener, A., *Die Entstehung der Kontinente und Ozeane*, Vieweg u. Sohn, Braunschweig, 1915.

Wilson, J. T., A new class of faults and their bearing on continental drift, *Nature*, *207*, 343–347, 1965.

Part II

MODELS OF CONVECTION
IN THE MANTLE

Editor's Comments
on Papers 5 Through 8

Thermal convection in fluids placed in a gravitational field is driven by density differences resulting from temperature variations. Flow is maintained because viscous dissipation of kinetic energy is balanced by the potential energy released when relatively warm and, therefore, low-density material moves upward and cold, heavy material moves downward. In a homogeneous compressible fluid, convective flow will set in when its thermal gradient sufficiently exceeds the adiabatic gradient. In nature, convection is widespread on a variety of scales, and it is an important object of fluid dynamic studies. Convection is possible in the earth's mantle because over a long period of time it can creep significantly, and thus behaves like a fluid in response to stress.

Convection in a homogeneous fluid, having a Newtonian viscosity, that is heated from below (Rayleigh-Benard convection), was most extensively investigated, both theoretically and experimentally. Most important for determining the state of flow is the dimensionless Rayleigh number, which depends on the properties of the fluid, its geometry, and the superadiabatic thermal gradient. Convection begins when this number exceeds a critical value, which depends on the boundary conditions (Chandrasekhar, 1961). The critical Rayleigh number defines a state of marginal stability which has to be passed to maintain a balance between viscous dissipation and energy gained from bouyancy forces. Then small disturbances can grow into periodic convection cells with a finite amplitude. The stable form of flow close to the critical state is two dimensional and consists of steady-state elongated rolls, but in some situations hexagonal cells can also arise (Brindley, 1967). Above the critical state the flow depends also on the ratio between the kinematic viscosity and thermal diffusivity (the Prandtl number), which is very large in the mantle.

As the Rayleigh number increases, the flow changes in two important ways. First, transport of heat tends to become concentrated in progressively narrower layers along the cell boundaries. In these boundary layers the flow is considerably faster and the temperature gradients are higher than in the cores of the cells (Brindley, 1967). Second, the pattern of the flow undergoes several changes: The two-dimensional rolls change into three-dimensional forms, and at very high Rayleigh numbers the flow may become time dependent (Busse, 1967; Busse and Whitehead, 1971, 1974; Whitehead and Parsons, 1978). In all cases the aspect ratio of the cells, that is, the ratio of the distance between ascending and descending limbs of the cells and the depth of the convecting fluid layer, is about 1:2. This is true for regular rolls as well as for more complex flows at very high Rayleigh numbers.

Somewhat different flow regimes arise in internally heated fluids, or when the viscosity is variable. In all cases, convection sets in when the thermal gradient sufficiently exceeds the adiabatic gradient, and then a suitably defined dimensionless Rayleigh number exceeds some critical value (Roberts, 1967; Schubert, Turcotte, and Oxburgh, 1969).

The features of Rayleigh-Benard convection inspired the early models of convection in the mantle in which the flow was depicted as consisting of regular and periodic cells with material particles moving along closed circuits. As the evidence favoring a mobile earth accumulated, many fluid-dynamic studies were pursued to examine the possibility and nature of Rayleigh-Benard convection in systems having the gross features of the mantle. In the more refined models, the effects of various complications, such as variable viscosity, internal heat generation, externally applied forces, and so forth, were investigated. A very important early result (e.g., Knopoff, 1964; Tozer, 1965) was that the outer 700–1000 km of the mantle, and most likely also the entire mantle, if it is chemically homogeneous, seem to be well above the state of marginal stability, whether one envisages heating from below or from within. This is a very strong argument in favor of thermal convection within the earth's mantle. In fact, it seems that this is also true for the other terrestrial planets (Schubert, Turcotte, and Oxburgh, 1969).

Oxburgh and Turcotte (Paper 5) consider a simple model of steady-state, two-dimensional flow of a Newtonian fluid with constant properties, heated from below and having the material and thermal properties of the mantle. Since the viscosity is very much larger than the thermal diffusivity, a stable cellular flow will develop and turbulence will not arise. The Rayleigh number turns out to be much larger than the critical value, so that most of the flow and heat transfer take place in a relatively thin boundary layer, in which the thermal gradient is high. In the cores of the cells the thermal gradient is adiabatic and motion is slow. The oceanic lithosphere is the natural counterpart of the upper, cold, boundary layer. The model incorporates also an explanation of magma genesis under ridges which are supposed to overlie ascending currents, stressing the very important relation between magmatism and mass flow in the mantle.

Remarkably, this very simple model predicts horizontal velocities, heat flow and a lithospheric thickness of the right order of magnitude for plausible mantle properties. However, the symmetry of the flow and the aspect ratio of the cells are important shortcomings; the quantitative relations of the model show that for a lithosphere of constant thickness, variable surface velocities, such as characterize plate motions, require either considerable lateral variations of mantle properties (e.g., temperature) or different depths of convection. Neither of these features is considered. Nevertheless, the results justify the view that such simple models are fair first approximations.

McKenzie, Roberts, and Weiss (Paper 6) considered models in which, in addition to heating from below, volumetric heat generation results from decay of radionucleids in the mantle (an extended account of this study is given in McKenzie, Roberts, and Weiss, 1974). In this case, all parts of the cells must approach the surface to give off the heat generated within them. This makes the hot, rising, boundary layer less distinct when an increasing proportion of the heat flux at the surface is generated within the cell, while the cold boundary layer becomes more important. The highest temperatures will not be at the bottom of the cells unless heating from below is significant. Hence, internal heat generation has a profound influence on the flow. At very high Rayleigh numbers (exceeding 10^6) the flow becomes time dependent; descending blobs develop occasionally in the upper boundary layer and modify the flow. An important inference is that in all cases positive free air anomalies are expected over ascending currents, showing that gravity data can be used to help identify patterns of the deep flow. The numerical experiments also suggest that in order to achieve a steady-state flow, geologically long times may be required.

The necessity to consider both internal heating and heating from below, as done in this work, arises because the core must deliver heat to the base of the mantle. This is required in order to sustain the core's dynamo action which produced the magnetic field during most of the recorded history of the earth (Verhoogen, 1961; Braginsky, 1964; Gubbins, Masters, and Jacobs, 1979). Such a heat flux can produce a substantial boundary layer at the base of the mantle.

Houston and DeBremaecker (Paper 7) present models which allow both volumetric heat generation and variations of viscosity. The effect of variable viscosity is to accentuate the ascending boundary layer (also found by Torrence and Turcotte, 1971), which is opposite to the effect of volumetric heat production. In some cases, when the viscosity increases strongly downward, cells with large aspect ratios were obtained. Additional studies confirmed this conclusion (DeBremaecker, 1977; Davies, 1977). Understanding the conditions under which cells with large aspect ratios are stable is very important, because the length scale of plates is large when compared with the depth of convection, whether confined to the outer 700 km of the earth or the convection is mantle-wide. Thus, the return flow from subduction zones to ridges must occur on a horizontal scale that is much larger than the depth to which convection extends. Later studies showed that rigid surficial plates also stabilize long cells (Parmentier and Turcotte, 1978; Lux, Davies, and Thomas, 1979), as do some conditions of heat disposal (Hewitt, McKenzie, and Weiss, 1980).

The existence of inhomogeneities, and especially of phase changes, presents a problem for models of convection (e.g., Knopoff 1964). Schubert, Turcotte, and Oxburgh (Paper 8) investigated this problem in a simple case. When flow occurs, distortion of a phase boundary with a positive dT/dp, such as the olivine-spinel transition, is counteracted by the latent heat involved in the transition. It is shown that in some conditions the latter effect will be subordinate, and then such a phase change will enhance the flow. Schubert, Yuen, and Turcotte, (1975) made a more thorough analysis and again showed that a phase transition is not necessarily an obstacle to convection. Their study did not exclude, however, the possibility that in the earth's mantle separate convective flows will arise above and below the 650-km phase transition. However, if this is a compositional boundary (e.g., because of downward increase of the Fe/Mg ratio), then flow through this boundary is excluded (Richter, 1973; Richter and Johnson, 1974).

The works presented in this section demonstrate that thermal convection is to be expected in systems having the properties of the mantle or of its outer parts. They also indicate that the state of the mantle is far removed from the state of marginal stability, so that boundary layers should be important. As internal heat generation is important in the earth, the cold surficial boundary layer should have a dominant role. The oceanic lithosphere can actually be identified with the cold boundary layer and this approach predicts the correct order of magnitude for its flow velocity and the heat flux. On the other hand, the fluid-dynamic models differ from the earth's plate system in several important ways. The aspect ratios of the convection cells are generally found to be much smaller than is expected from the geometry of the earth's plates; the conditions for the development of cells having large aspect ratios are still incompletely known. Even more problematic is that in the models the convection cells have a uniform size and are arranged periodically, and sites of ascending motion alternate with sites of downward flow. Such a regular pattern contrasts with the variable sizes and shapes of the plates, their irregular motions, and with the complex arrangement of mid-ocean ridges and subduction zones.

These shortcomings are due to the simplicity of the theoretical models that consider two-dimensional, steady state, and generally constant viscosity flows in planar layers. To represent the mantle more realistically various complications must be considered, such as variable, temperature dependent viscosity, non-Newtonian rheology, and flow in spherical shells (Schubert, 1979), as well as time dependent three-dimensional flows. Such problems were addressed in many more recent works, but the fluid dynamic models do not yet provide a complete description of plate dynamics or of the scope, geometry, and history of mass motions in the mantle. Despite these problems, fluid-dynamic modeling is very valuable, because it allows to investigate well defined, though relatively simple, situations for which exact numerical solutions can be obtained. This helps clarify the basic physical processes and assess the influence of different parameters.

REFERENCES

Braginsky, S. I., 1964, Kinematic Models of the Earth's Hydromagnetic Dynamo, *Geomagn. Aeronomy* **4:**572–583 (English. transl.)

Brindley, J., 1967, Thermal convection in horizontal fluid layers, *Inst. of Math. Applications Jour.* **3:**313–343.

Busse, F. H., 1967, The Stability of Finite Amplitude Cellular Convection and Its Relation to an Extremum Principle, *Jour. Fluid Mechanics* **30:**625–649.

Busse, F. H., and J. A. Whitehead, 1971, Instabilities of Convection Rolls in High Prandtl Number Fluid, *Jour. Fluid Mechanics* **47:**305–320.

Busse, F. H., and J. A. Whitehead, 1974, Oscillatory and collective instabilities in large Prandtl number convection, *Jour. Fluid Mechanics* **66:**67–79.

Chandrasekhar, S., 1961, *Hydrodynamic and Hydromagnetic Stability,* Clarendon Press, Oxford, 652p.

Davies, G. F., 1977, Whole Mantle Convection and Plate Tectonics, *Royal Astron. Soc. Geophys. Jour.* **49:**459–486.

DeBremaecker, J. C., 1977, Convection in the Earth's Mantle, *Tectonophysics* **41:**195-208.

Gubbins, D., T. G. Masters, and J. A. Jacobs, 1979, Thermal Evolution of the Earth's Core, *Royal Astron. Soc. Geophys. Jour.* **59:**57-99.

Hewitt, J. M. D. P.McKenzie, and N. O. Weiss, 1980, Large Aspect Ratio Cells in Two-Dimensional Thermal Convection, *Earth and Planetary Sci. Letters* **51:**320-380.

Lux, R. A., G. F. Davies, and J. H. Thomas, 1979, Moving Lithospheric Plates and Mantle Convection, *Royal Astron. Soc. Geophys. Jour.* **57:**209-228.

Knopoff, L., 1964, The Convection Current Hypothesis, *Rev. Geophysics* **2:**89-122.

McKenzie, D. P., J. M. Roberts, and N. O. Weiss, 1974, Convection in the Earth's Mantle: Towards a Numerical Simulation, *Jour. Fluid Mechanics* **42:**465-538.

Parmentier, E. M., and D. L. Turcotte, 1978, Two Dimensional Mantle Flow beneath a Rigid, Accreting Lithosphere, *Physics Earth and Planetary Interiors* **17:**281-289.

Richter, F. M., 1973, Dynamical Models for Sea Floor Spreading, *Rev. Geophys. Space Phys.* **11:**223-288.

Richter, F. M., and C. E. Johnson, 1974, Stability of a Chemically Layered Mantle, *Jour. Geophys. Research* **79:**1635-1639.

Roberts, P. H., 1967, Convection in Horizontal Layers with Internal Heat Generation, *Jour. Fluid Mechanics* **30:**33-49.

Schubert, G., 1979, Subsolidus Convection in the Mantles of Terrestrial Planets, *Ann. Rev. Earth and Planetary Sci.* **7:**289-342.

Schubert, G., D. L. Turcotte, and E. R. Oxburgh, 1969, Stability of Planetary Interiors, *Royal Astron. Soc. Geophys. Jour.* **18:**441-460.

Torrance, K. E., and D. L. Turcotte, 1971, Structure of Convection Cells in the Mantle, *Jour. Geophys. Research* **76:**1154-1161.

Tozer, D. C., 1965, Heat Transfer and Convection Currents, *Royal Soc. London Philos. Trans.* ser. A, **258:**252-271.

Verhoogen, J., 1961, Heat Balance of the Earth's Core, *Royal Astron. Soc. Geophys. Jour.* **57:**209-228.

Whitehead, J. A., and B. Parsons, 1978, Observations of Convection at Rayleigh Numbers up to 760,000 in a Fluid with Large Prandtl Number, *Geophys. Astrophys. Fluid Dynamics* **9:**201-217.

5

Mid-Ocean Ridges and Geotherm Distribution during Mantle Convection

E. R. OXBURGH[1] AND D. L. TURCOTTE

Graduate School of Aerospace Engineering, Cornell University, Ithaca, New York 14850

INTRODUCTION

The geological evidence in favor of some kind of cellular movement in the mantle is strong but circumstantial [e.g., *Runcorn*, 1962; *Blackett*, 1965], but it has been considerably strengthened by the concept of ocean-floor spreading [*Hess*, 1962, 1965] and the associated interpretations of the linear magnetic anomaly patterns that characterize many parts of the ocean floors [*Vine and Matthews*, 1963; *Vine*, 1966]. On the other hand, objections or severe physical constraints on convective motion within the crystalline rock of the mantle have been advanced by others [*MacDonald*, 1965a; *McKenzie*, 1966], and the cellular character of any such motion has been questioned [e.g., *Elsasser*, 1963]. It is clearly desirable to seek further ways of testing the convection hypothesis.

Progress in testing the convection hypothesis has been hindered by the absence of any complete and internally consistent theoretical model for mantle convection relating the directions of movement and velocities of mantle material to its thermal behavior. The obstacles to the development of a rigorous model are formidable [*Knopoff*, 1964; *Tozer*, 1965], and in any case there is considerable uncertainty about many of the constants describing the physical behavior of the mantle at depth. In particular, it is not clear whether the mantle is better regarded as a plastic or a viscous body. *Gordon* [1965] has given strong arguments favoring a viscous model. The plastic model has been examined by *Orowan* [1965] and *Elsasser* [1963].

It is well known that cellular convection occurs when a confined fluid is heated from below. Solutions of the linearized stability problem that determine the onset of the convective motion were first obtained by *Rayleigh* [1916]. Convective motion is predicted if the dimensionless parameter $R = \alpha (T_{w1} - T_{w2}) d^3 g/\kappa\nu$ exceeds a critical value where α is the coefficient of thermal expansion, T_{w1} is the temperature

---

[1] Now at Division of Geological Sciences, California Institute of Technology, Pasadena, California. On leave from the Department of Geology and Mineralogy, University of Oxford, Oxford, England.

76

of the hot lower surface, T_{w2} is the temperature of the cold upper surface, d is the thickness of the layer of fluid, g is the acceleration of gravity, κ is the thermal diffusivity, and ν is the kinematic viscosity. The value of the critical Rayleigh number for free-surface boundary conditions is 657. The wavelength of the convective motion is $\lambda = 2^{3/2} d$. The extensive literature on the stability problem is summarized by *Chandrasekhar* [1961]. The predicted onset of convection is in good agreement with laboratory measurements.

For Rayleigh numbers that are large in comparison to the critical Rayleigh number, the linear theory is not applicable and it is necessary to solve a set of nonlinear partial differential equations. Extensions of the linear theory into the nonlinear regime have been given by *Malkus and Veronis* [1958], *Kuo* [1961], and *Platzman* [1965]. Numerical computations of cellular convection for Prandtl numbers of the order of 1 have been given by *Fromm* [1965].

For finite amplitude convection cells the Prandtl number, $Pr = \nu/\kappa$, enters the analysis as well as the Rayleigh number. Clearly the Prandtl number for mantle convection is large. A theoretical model for laminar cellular convection applicable for large values of the Prandtl and Rayleigh numbers has been given by *Turcotte and Oxburgh* [1967]. In this model temperature gradients are restricted to thin thermal boundary layers adjacent to the horizontal cell boundaries and to thin thermal plumes on the vertical boundaries between cells. The core of each convection cell is assumed to be highly viscous and isothermal. It has been shown by *Turcotte* [1967] that the boundary-layer theory for finite amplitude convection is in excellent agreement with laboratory measurements of heat flux. It is the purpose of this paper to examine the implications of the boundary-layer model for possible mantle convection.

The model adopted assumes that the mantle convection cells have the form of two-dimensional elongated rolls with a horizontal dimension in the direction of movement of thousands of kilometers and a persistence along the axis of the rolls of many times this distance. Oceanic ridges, such as the East Pacific rise and the mid-Atlantic ridge, are regarded as the sites of ascending limbs, and the ocean floor is regarded as the upper surface of the cells, as in the ocean-floor spreading hypothesis of continental drift [*Hess*, 1962]. A possible distribution of ascending and descending limbs is illustrated by Figure 1 [after *Girdler*, 1965]. That convection cells can be expected to have this form has been verified in the laboratory. A photograph of finite amplitude laminar convection cells, as obtained by *Somerscales and Dropkin* [1966], is given in Figure 2. It is seen that the convection cells have the form of elongated two-dimensional rolls.

Theoretical Model

To determine the steady structure of Benard cells, the conservation equations for mass, momentum, and energy are required. It has been shown by *Jeffreys* [1928] that the influence of the earth's rotation on convection in the mantle is negligible. *Jeffreys* [1930] has also shown that the Boussinesq approximation may be used if the temperature in the Boussinesq equations is taken to be the difference between the actual temperature and the adiabatic temperature at that point, provided that this temperature difference is not too large. The temperature T is defined to be the difference between the actual temperature and the adiabatic temperature at that point. The viscosity, thermal diffusivity, and coefficient of thermal expansion are taken to be constant; in the body force term of the momentum equation a linear relation is assumed between the variations of temperature and density:

$$\rho - \rho_0 = -\rho_0 \alpha (T - T_0) \qquad (1)$$

where T_0 is the temperature at which the density ρ is equal to the reference density ρ_0. Introducing $\Theta = T - T_0$ and $P = p + \rho_0 py$ we can write the equations for conservation of mass, momentum, and energy for a steady, laminar flow as

$$\nabla \cdot \mathbf{u} = 0 \qquad (2)$$

$$(\mathbf{u} \cdot \nabla)\mathbf{u} = -\frac{1}{\rho_0} \nabla P + \nu \nabla^2 \mathbf{u} + \alpha \Theta g \mathbf{j} \qquad (3)$$

$$(\mathbf{u} \cdot \nabla)\Theta = \kappa \nabla^2 \Theta \qquad (4)$$

Chandrasekhar [1961] considered the effect of a spherical geometry on the stability and did not find a strong geometrical influence. Therefore, we will consider only the two-dimen-

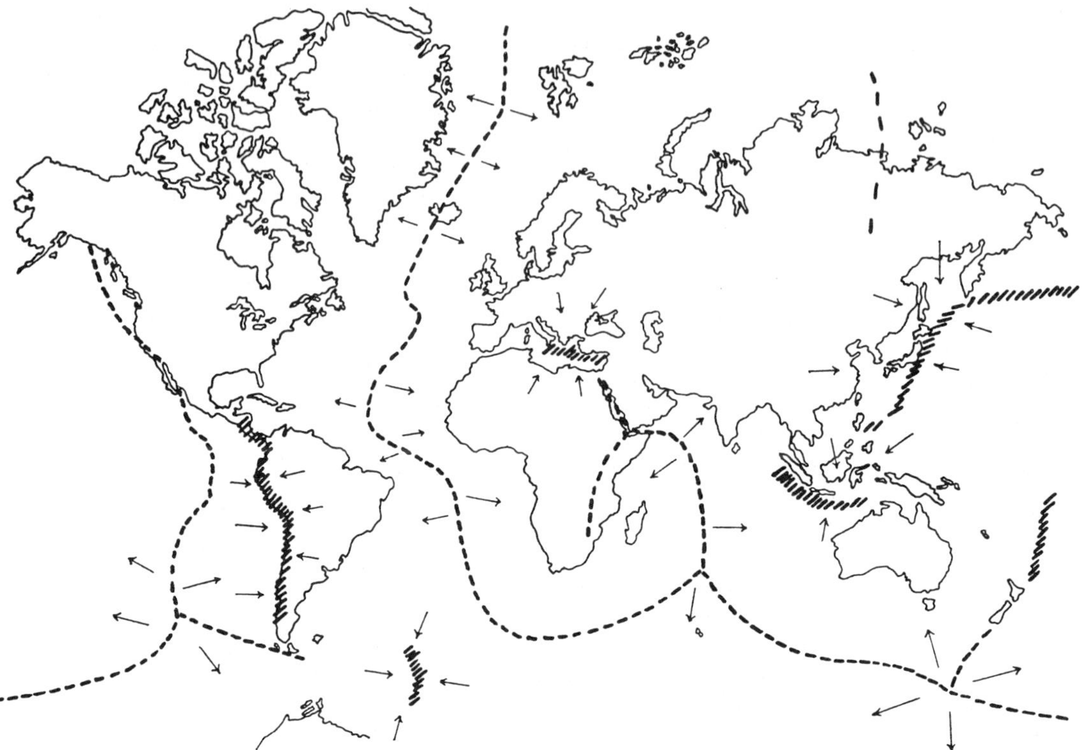

Fig. 1. The distribution of ascending and descending limbs over the earth's surface. The zones of shallow seismicity associated with mid-oceanic ridges and attributed to ascending limbs are denoted by a dashed line. The zones of deep-focus earthquakes associated with ocean trenches and attributed to descending limbs are denoted by slanted lines [after *Girdler*, 1965].

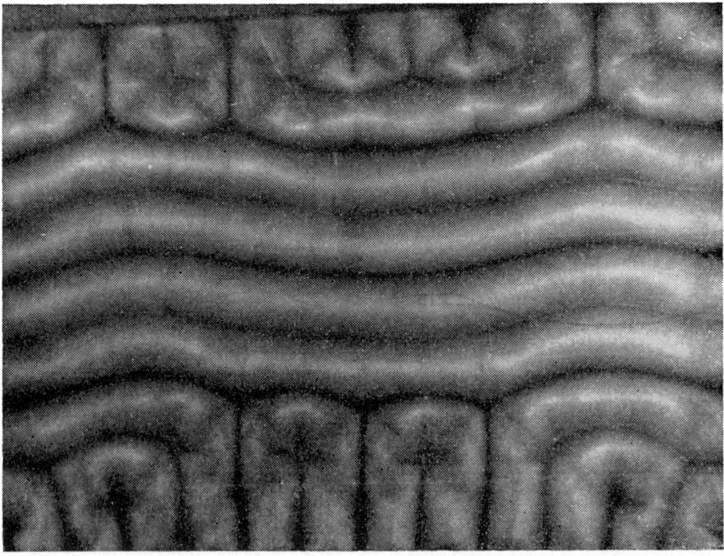

Fig. 2. Top view of laboratory cellular convection of a silicon oil obtained by *Sommerscales and Dropkin* [1966] in an 8-inch by 6-inch container, $d = 0.551$ inch; $R = 3040$, $Pr = 522$. The dark lines correspond to the surface traces of ascending and descending limbs.

sional planar problem. We neglect the production of heat by radioactivity in the convecting layer and assume that the gravitational acceleration is a constant. We solve for the flow in convection cells confined between two horizontal boundaries. It is convenient to let $T_0 = \frac{1}{2}(T_{w1} + T_{w2})$, so that $\Theta_{w1} = \frac{1}{2}(T_{w1} - T_{w2})$ on the lower boundary at $y = 0$ and $\Theta_{w2} = \frac{1}{2}(T_{w2} - T_{w1}) = -\Theta_{w1}$ on the upper boundary at $y = d$; T_{w1} and T_{w2} are constants. The condition that there be no flow through the boundaries requires that $v = 0$ at $y = 0$ and d. A second velocity boundary condition is required on the horizontal boundaries. The standard alternatives are the fixed wall boundary condition, $u = 0$ at $y = 0$ and d, or the free surface boundary condition that the tangential component of the shear stress vanish at the horizontal boundaries. The second condition would be applicable at the ocean floor if the oceanic crust is being transported by the cellular flow. Under a continental mass the fixed wall boundary condition might be applicable. The appropriate boundary condition on the lower cell boundary is not well defined since the boundary itself is not well defined. In this paper free surface boundary conditions are applied to both boundaries. Since our analysis is restricted to two-dimensional flows, the free surface boundary conditions require that $\partial u/\partial y = 0$ at $y = 0$ and d.

For a fluid with large Prandtl and Rayleigh numbers it is appropriate to use a boundary-layer theory to solve for the steady, finite amplitude cellular convection. The fluid is divided into two-dimensional rolls with rectangular cross sections; alternate rolls rotate in the clockwise and counterclockwise directions. Each cell has a height d and a width w. It is assumed that the aspect ratio of each cell is that given by the linear stability theory; that is, $w = 2^{1/2}d$. Since thermal gradients are assumed to be restricted to thin layers on the sides of each cell, it is appropriate to solve for the isothermal core flow in a rectangle with dimensions d and $2^{1/2}d$. For constant fluid properties an analytic solution for the flow field is easily obtained.

Thin thermal boundary layers are formed on the horizontal boundaries of each cell. With the prescribed boundary conditions the hot and cold boundary layers have similar temperature

profiles. A solution for the boundary-layer temperature profiles is obtained by using the mean horizontal velocity at the horizontal boundaries and by setting the vertical velocity equal to zero in the boundary layers. The solution is a good approximation except in the vicinity of the vertical boundaries between cells. The local heat flux to the horizontal boundaries is also found.

When the two hot thermal boundary layers from adjacent cells meet, they separate from the lower horizontal boundary and form a hot thermal plume that rises to the upper surface. Similarly, when the two cold boundary layers from adjacent cells meet, they separate from the upper horizontal boundary and form a cold thermal plume that descends to the lower surface. The center line of each thermal plume is the dividing line between adjacent cells. A solution for the temperature distribution in the vertical plumes is obtained by using the mean vertical velocity at the vertical boundaries between cells and by setting the horizontal velocity equal to zero in the plumes. It is assumed that the initial temperature distribution at the base of the plume is the same as the temperature distribution in the thermal boundary layers adjacent to the base of the plume. The decrease in the maximum (minimum) temperature and the growth of the plume are obtained. The buoyancy force in the plumes drives the viscous flow. The magnitude of the velocities associated with the cellular convection is obtained by matching the integral of the temperature excess (deficit) within the plume to the velocity gradient at the edge of the plume.

When the convective plumes impinge on the horizontal surfaces, stagnation-point thermal boundary layers are formed. Using the velocities obtained from the core flow, we can obtain temperature profiles for the stagnation-point boundary layers. The stagnation-point thermal-boundary layers merge smoothly into the thermal-boundary layers on the horizontal boundaries. The stagnation-point heat flux is also determined.

The boundary-layer model for finite amplitude, steady cellular convection is illustrated in Figure 3. The analysis that leads to a solution has been given by *Turcotte and Oxburgh* [1967]. The results that will be applied to mantle convection are reproduced here. The horizontal

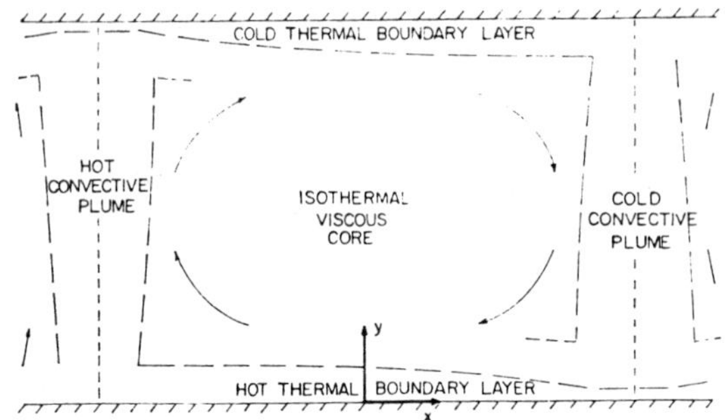

Fig. 3. Illustration of the boundary-layer model for cellular convection.

velocity component within each cell is given by

$$\frac{ud}{\kappa} = \sum_{n=1,3,\ldots} \left[\frac{3.56R^{2/3}\cos{(n\pi y/d)}}{n^2\pi^2\cosh{(n\pi/2^{1/2})}}\right]$$

$$\cdot\left[\frac{n\pi}{2^{1/2}}\tanh\frac{n\pi}{2^{1/2}}\cosh\frac{n\pi x}{d} - \frac{n\pi x}{d}\sinh\frac{n\pi x}{d}\right] \quad (5)$$

and the vertical velocity component is given by

$$\frac{vd}{\kappa} = \sum_{n=1,3,\ldots} \left[\frac{3.56R^{2/3}\sin{(n\pi y/d)}}{n^2\pi^2\cosh{(n\pi/2^{1/2})}}\right]$$

$$\cdot\left[\left(1 - \frac{n\pi}{2^{1/2}}\tanh\frac{n\pi}{2^{1/2}}\right)\sinh\frac{n\pi x}{d}\right.$$

$$\left. + \frac{n\pi x}{d}\cosh\frac{n\pi x}{d}\right] \quad (6)$$

where R is the Rayleigh number. The coordinates x and y are measured from the center of the lower boundary as shown in Figure 3. The mean value of the horizontal velocity on the horizontal boundaries is given by

$$u_m d/\kappa = 0.142R^{2/3} \quad (7)$$

and the mean value of the vertical velocity on the boundaries between cells is given by

$$v_m d/\kappa = 0.250R^{2/3} \quad (8)$$

The temperature profiles in the horizontal boundary layers are given by

$$\Theta = \Theta_w\left[1 - \int_0^{0.188(y_1/d)(d/x_1)^{1/2}R^{1/3}}\right.$$

$$\left. \cdot \exp{(-\eta)}\, d\eta\right] \quad (9)$$

where x_1 and y_1 are the distances from the corners of the cell, as shown in Figure 3. This solution is valid for $x_1 > 0.0266d$. The local

heat flux to the horizontal boundaries due to the horizontal boundary layers is

$$\frac{q_w d}{2\Theta_{w1}k} = 0.106\left(\frac{d}{x_1}\right)^{1/2}R^{1/3} \quad (10)$$

This relation is valid for $x_1 > 0.0266d$. The thickness of the thermal boundary layer, $y_{1\delta}$, is defined as the value of y_1 where $\Theta/\Theta_w = 0.1$. The maximum boundary layer thickness is at $x_1 = 2^{1/2}d$ and is given by

$$y_{1\delta}/d = 7.38R^{-1/3} \quad (11)$$

This result verifies the boundary layer approximation. As long as the Rayleigh number is sufficiently large, the horizontal boundary layers are thin in comparison to the cell dimensions. Since the thickness of the plumes is of the same order as the thickness of the boundary layers, the plumes are also thin for large Rayleigh numbers. The temperature distribution in the thermal plumes is given by

$$\Theta = \Theta_w\left(\frac{d}{\pi y_1}\right)^{1/2}$$

$$\cdot\int_{-\infty}^{\infty}\left[1 - \frac{2}{\pi^{1/2}}\int_0^{0.635|\xi|}\exp{(-\eta^2)}\,d\eta\right]$$

$$\cdot\exp\left[-\left(0.50\frac{x_1}{d}R^{1/3} - \xi\right)^2\frac{d}{y_1}\right]d\xi \quad (12)$$

The temperature distribution in the stagnation-point thermal-boundary layer is given by

$$\Theta = \Theta_w\left[1 - \frac{2}{\pi^{1/2}}\right.$$

$$\left.\cdot\int_0^{1.155(y_1/d)R^{1/3}}\exp{(-\eta^2)}\,d\eta\right] \quad (13)$$

This solution is valid for $x_1 < 0.0266d$. Comparing (13) with (9) shows that the structure of the stagnation-point thermal-boundary layer is identical to the structure of the horizontal thermal-boundary layer at $x_1 = 0.0266d$. The local heat flux to the horizontal boundaries due to the stagnation-point thermal-boundary layers is

$$q_w d/2\Theta_{w1}k = 0.652R^{1/3} \quad (14)$$

This relation is valid for $x_1 < 0.0266d$. The thickness of the stagnation point thermal boundary layer is given by

$$y_{1\delta}/d = 1.01R^{-1/3} \quad (15)$$

The mean heat flux to the horizontal boundaries Q_w is obtained by averaging q_w over the width of a cell using (10) and (15) with the result

$$Q_w d/2\Theta_w k = 0.165 R^{1/3} \qquad (16)$$

APPLICATION TO MANTLE CONVECTION

Before the boundary-layer theory for finite amplitude cellular convection can be applied to mantle convection, the properties of the mantle must be specified. Some of the properties are not well known, and some vary with depth in an unknown manner. Obviously, a theory that assumes constant values for the properties can be only approximately valid. For these reasons and others, close agreement between observed phenomena and theoretical predictions must be regarded as somewhat fortuitous.

The physical properties of the mantle have been studied by many authors including *Verhoogen* [1958], *Knopoff* [1964], *Clark and Ringwood* [1964], *McConnell* [1965], *Jaeger* [1965], *Tozer* [1965], and *MacDonald* [1965b]. Probably the most serious uncertainty relates to the kinematic viscosity ν. When the theoretical analysis of diffusion creep given by *Gordon* [1965] is compared with the analysis of the unlift of Fennoscandia by *Haskell* [1935] and the uplift of Lake Bonneville by *Crittenden* [1963], however, we conclude that for the upper mantle a mean value of $\nu = 10^{22}$ cm²/sec should have an uncertainty of less than an order of magnitude [see also *McKenzie*, 1966].

The thermal state of the upper mantle is subject to considerable controversy. Some authors believe that the earth is cooling [see *Jeffreys*, 1929] and others believe that the earth is heating up because of radioactivity [see *Lubimova*, 1958]. There appears, however, to be no earth thermal history model in the literature that takes into account the possible effect of continued transfer of heat through the mantle by convection. The heat fluxes associated with a convecting mantle are so high that a steady-state thermal condition is not impossible. To explain the measured values of oceanic heat flux, it has been proposed that under oceanic areas the bulk of the radioactive components of the mantle are concentrated in the upper few hundred kilometers. With mantle convection such an assumption is no longer necessary, and it is appropriate to assume a uniformly low distribution of radioactivity over the mantle. For a uniform distribution of radioactivity and a steady thermal state, the average thermal gradient in excess of the adiabatic across the convecting layer is taken to be $\beta = 1.5$ °K/km, and this value should have an uncertainty of less than a factor of 2. The thermal conductivity is taken to be $k = 2 \times 10^{-2}$ cal/cm sec °K, which is a factor of 3 higher than the thermal conductivity at the upper edge of the mantle but should be representative of an average value of the thermal conductivity through the upper mantle. The coefficient of thermal expansion is taken to be $\alpha = 2 \times 10^{-5}$ °K⁻¹; the thermal diffusivity, $\kappa = 3 \times 10^{-2}$ cm²/sec; the acceleration of gravity, $g = 10^{3}$ cm/sec². The depth of the convecting layer d is assumed to be 1500 km. This value is uncertain by about a factor of 2. Since the width of each cell is $2^{1/2}d$ the corresponding cell width is 2100 km. Using the above values, the Rayleigh number is found to be $R = 5 \times 10^{5}$, which is near the center of the range of uncertainty. A reasonable estimate for the possible error in the Rayleigh number would be two orders of magnitude. The Prandtl number is $Pr = 3 \times 10^{23}$.

The mean horizontal velocity on the surface is found from (7) to be 1.8×10^{-7} cm/sec (5.68 cm/yr). Considering the range of uncertainty, this value is in good agreement with the velocities of continental drift obtained from paleomagnetic and other evidence; 10^{-7} cm/sec [*Tozer*, 1965], 1.25 cm/yr [*Orowan*, 1965], 3–5 cm/yr [*Allen*, 1965], and 1–4.4 cm/yr [*Vine*, 1966]. The mean vertical velocity on the centerline of the convective plumes from (8) is 3.2×10^{-7} cm/sec. The dependence of the horizontal component of velocity on the distance from the vertical rise is obtained from (5) and is given in Figure 4 for several depths. The dependence of the vertical component of velocity on the distance from the vertical rise is obtained from (6) and is given in Figure 5 for several depths. It is seen that vertical motion is concentrated in the regions adjacent to the boundaries between cells. The dependence of the temperature on the depth in the thermal boundary layer is obtained from (9) and is given in Figure 6 for several distances from the vertical rise. The dependence of the temper-

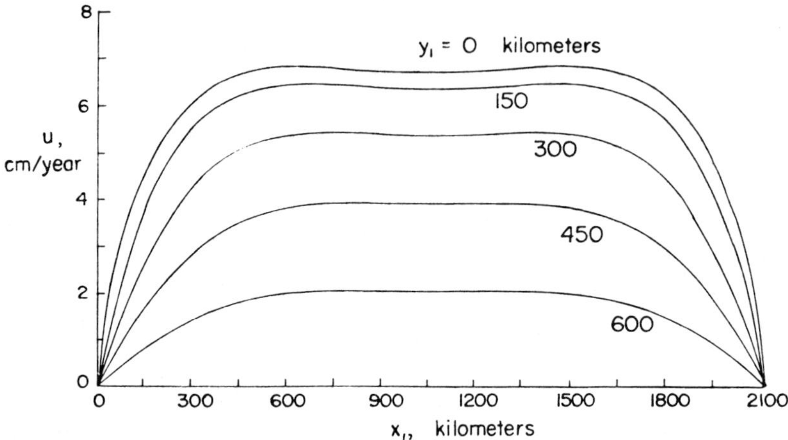

Fig. 4. Dependence of the horizontal component of velocity on the distance from the vertical rise at several depths.

ature on depth in the vicinity of the vertical rise is given in Figure 7. Included are profiles obtained from the stagnation-point thermal-boundary layer solution, (13), which are matched to the plume temperature outside the boundary layer. The stagnation-point solution is valid at distances of less than 40 km from the vertical rise. The temperature distribution in the rising convective plume, as obtained from (12), is given in Figure 8. The temperatures plotted are the difference between the local temperature and the adiabatic temperature at that depth. The maximum thickness of the thermal-boundary layer obtained from (11) is 139 km. The thickness of the stagnation-point boundary layer as obtained from (15) is 19.1 km.

From (10) and (14) the local heat flux to the surface can be obtained. The predicted values are given in Figure 9. Included for comparison are the measured values obtained on the East Pacific rise. The experimental points are taken directly from a figure published by *Lee and Uyeda* [1965]. The mean value of the heat flux as obtained from (16) is 3.93 $\times$ 10^{-6} cal/cm^2 sec. This value compares with the value 1.58 $\times$ 10^{-6} cal/cm^2 sec given by *Lee and Uyeda* [1965]. The more recent analysis of heat flow measurements by *Langseth et al.* [1966] gives similar values. Once again the agreement between theory and experiment is well within the range of possible error in the choice of mantle properties. The agreement between predicted and observed heat fluxes could be made closer

by the adoption of a more realistic thermal conductivity for the upper boundary layer, in place of the mean upper mantle conductivity used at present. To adopt this more realistic representation, however, would scarcely be justified in view of the other simplifying assumptions made.

ISOTHERM DISTRIBUTION

The steady-state distribution of isotherms predicted by the model is shown in Figure 10; the ascending and descending limbs are symmetrically related, as are the upper and lower boundary layers. Strong horizontal thermal

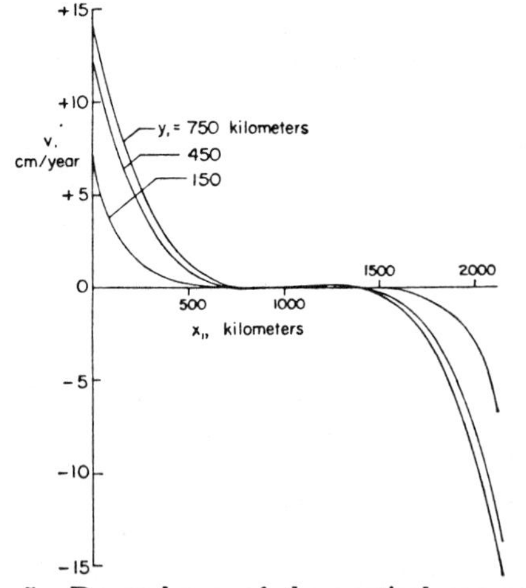

Fig. 5. Dependence of the vertical component of velocity on the distance from the vertical rise at several depths.

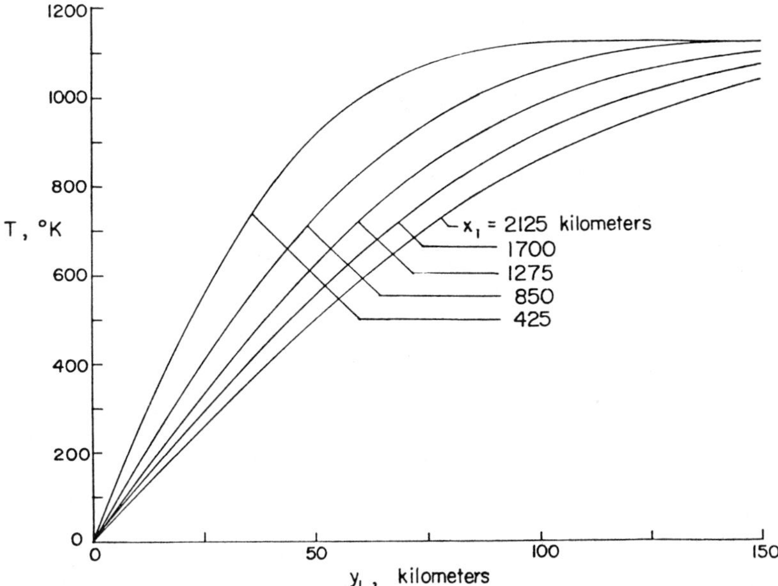

Fig. 6. Dependence of the temperature excess on depth at various distances from the vertical rise.

gradients exist only near the vertical plumes. The isotherm distribution shown for the uppermost part of the ascending limb is, however, unlikely to be realized in nature. As discussed in more detail below, partial fusion of the mantle takes place in this zone, and temperatures are lowered by latent heats of fusion and by the rapid transfer of heat to the surface by magmas.

It is of interest to note the extent of isotherm depression within the descending plume. The tectonic hypotheses that depend on the assumption that the base of the crust above the de-

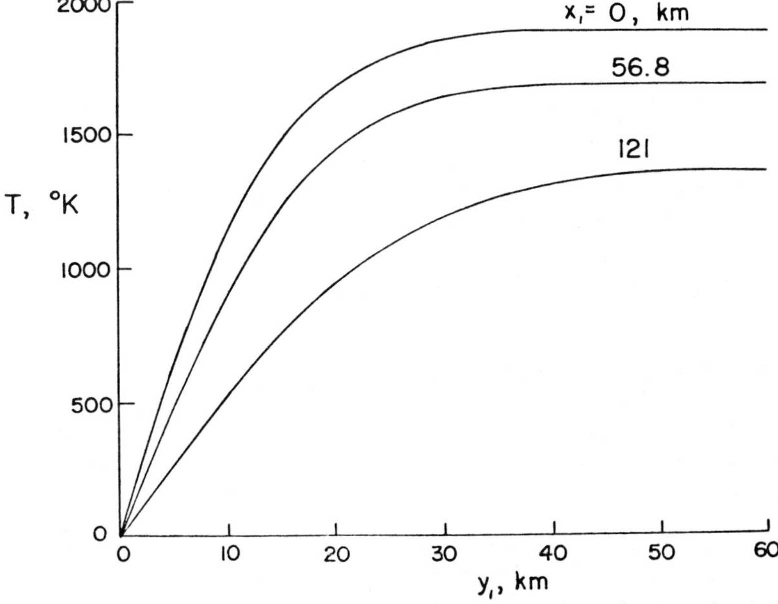

Fig. 7. Dependence of the temperature excess on depth at various distances from the vertical rise.

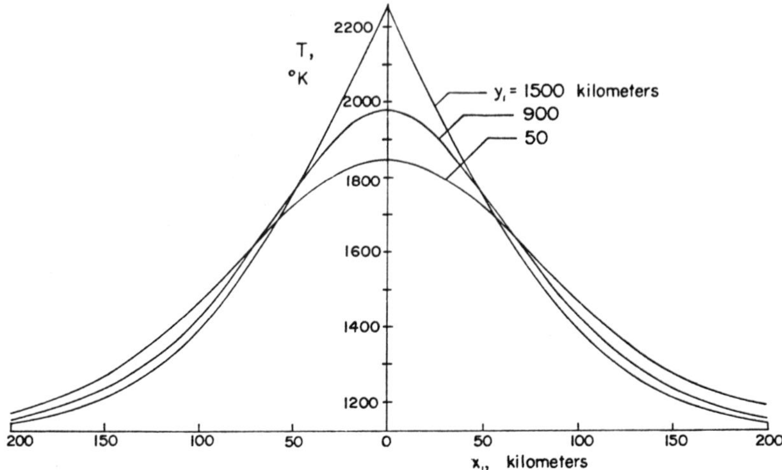

Fig. 8. Dependence of the temperature excess on the distance from the vertical rise at several depths.

scending limb of a mantle convection current is dragged downward and heated may be seen to be somewhat improbable or, at any rate, dependent on convective theory very different from that presented here. It is, however, difficult to visualize any convection model in which the depression of the isotherms in the descending plume is not greater than any reasonable amount of depression of the base of the crust.

UPPER PART OF ASCENDING PLUME

Mantle material traveling upward in the ascending plume undergoes a reduction in pressure that is large in comparison to its reduction in temperature (Figure 10). Whether this material melts depends on whether the geotherms displaced by convection intersect the isotherms for the beginning of mantle melting; in mantle material that does not have large-scale lateral compositional inhomogeneities, these isotherms will be approximately horizontal and parallel to the earth's surface.

It will be accepted here that the upper mantle is of broadly peridotitic composition and that basalt represents a low melting fraction derived from it. The evidence for these assumptions will not be reviewed here, but it is fully treated in the literature [e.g., *Yoder and Tilley*, 1962; *Oxburgh*, 1964; *Green and Ringwood*, 1966]. The melting behavior of basalts at high pressures should, therefore, provide a useful guide to the early melting behavior of the mantle. There are, however, considerable differences in melting

behavior of basalts having different chemical compositions, and there is no general agreement as to which basaltic types represent the unmodified first melting products of the mantle and which may have been contaminated by the solution of extraneous material or may have undergone differentiation. *Engel et al.* [1965] have found that olivine tholeiite basalt of extremely uniform composition is the most abundant volcanic rock of the ocean ridges. Its abundance, uniformity of composition, and occurrence in the deep oceans, where there is little opportunity for contamination with silicous crustal material,

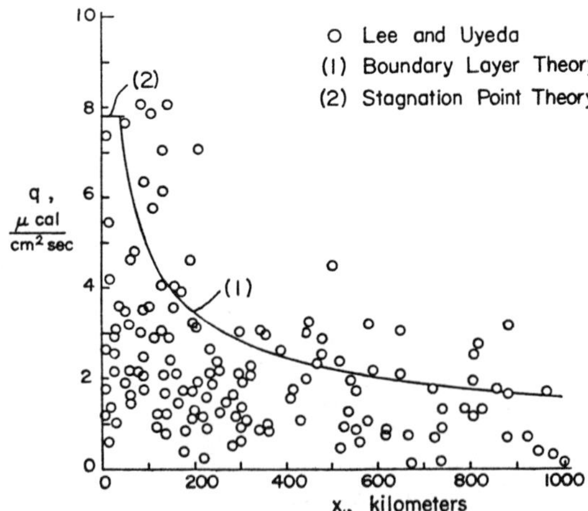

Fig. 9. Comparison of measured values of the heat flux as a function of the distance from the East Pacific rise with the boundary layer theory.

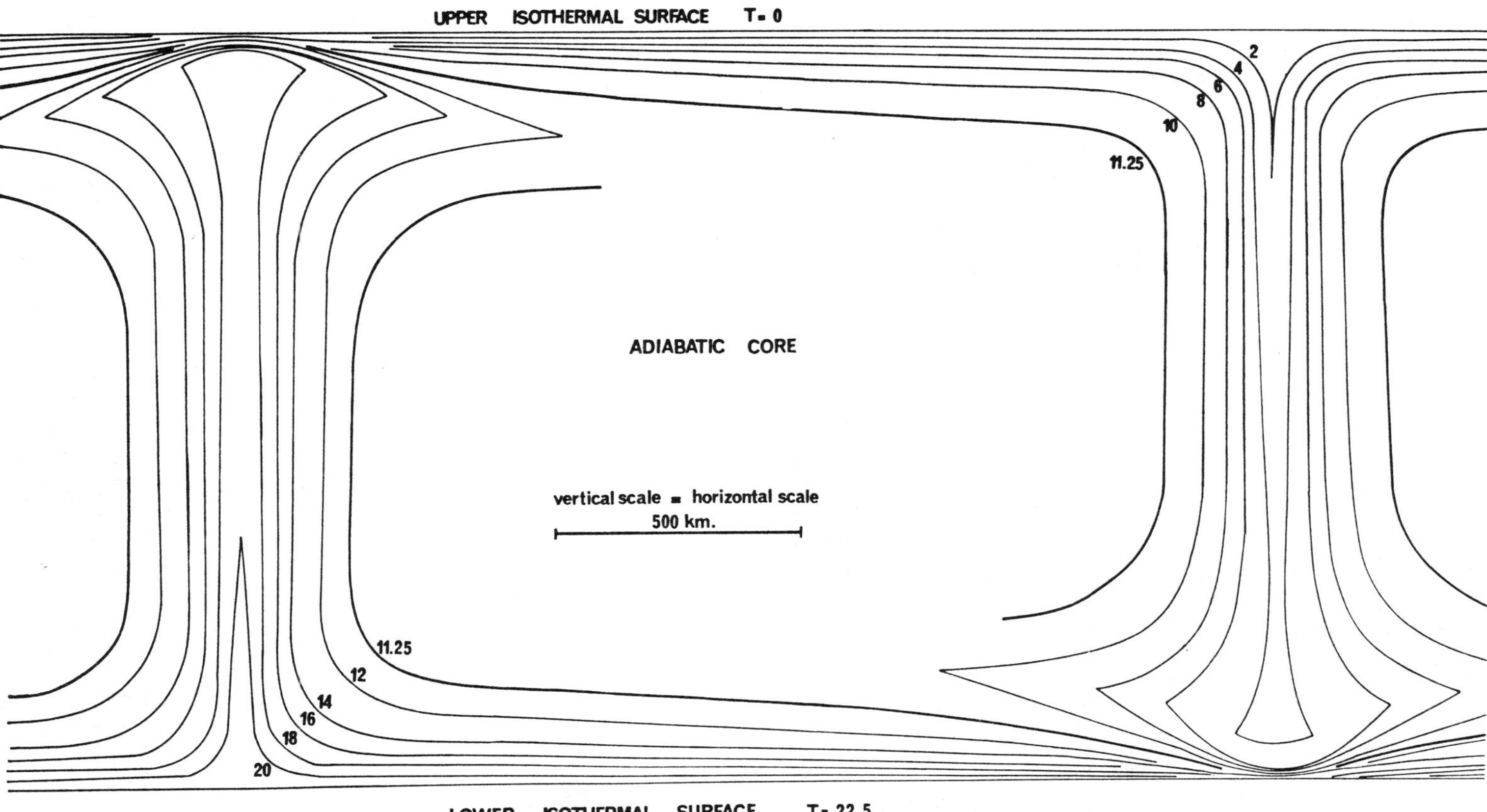

Fig. 10.   The steady-state isotherm distribution predicted by the boundary-layer model; the values of the isotherms are given as superadiabatic temperatures $\times 10^{-2}$ °C.

all suggest that it may well be a relatively unmodified first melting fraction of the mantle.

Figure 11 shows the relationship between the predicted geotherm distribution within the upper part of the ascending plume in the absence of partial melting and the isotherms for the melting of olivine tholeiite. A zone of partial melting is defined by the intersection of the geotherms with melting isotherms of equal or lesser value. (The cross-sectional shape of this zone is shown in Figure 12.) The zone is somewhat more than 400 km wide and 200 km deep.

The amount of fusion occurring within this zone and the actual temperature distribution within it cannot be established without more precise information on the composition of the upper mantle than is available at present. These problems may be approached qualitatively, however, by considering the passage of a small reference volume of mantle through the zone of fusion; on the one hand the liquidus for the reference volume tends to be lowered by the progressive reduction in confining pressure during ascent, while on the other it is relatively elevated by the melting of the lower tempera-

ture fractions. The situation is represented diagramatically in Figure 13, where a family of liquidus curves corresponding to different confining pressures are plotted on a temperature-composition diagram. Consider the case of a particle that is following a streamline such that it enters the zone of fusion at a point where the conditions correspond to those at A. The diagram shows various possibilities for the subsequent changes in temperature, pressure, and liquid composition in the system represented by the reference particle. The limiting cases are paths A–X and A–W. The path A–X represents the situation in which the relative elevation of the liquidus through progressive melting is negligible in comparison with the depression through pressure reduction; this could happen only in a system with a bulk composition close to that of olivine tholeiite and would result in a temperature distribution within the zone of fusion that, for every depth, was close to the tholeiite liquidus. The other limiting case, path A–W, represents the converse situation in which a very small amount of melting is accompanied by a rapid relative

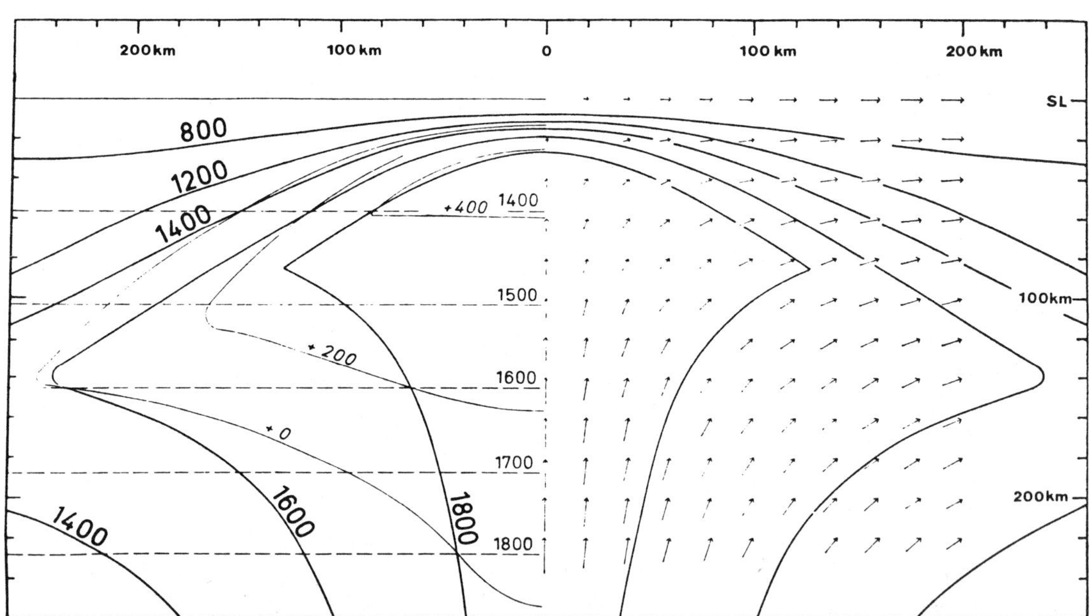

Fig. 11. Detail of the top of the ascending plume. Isotherms are shown as heavy lines (values in degrees Celsius in heavy numerals); liquidus temperatures for olivine tholeiite indicated by horizontal dashed lines, temperatures in degrees Celsius. The outline of the left half of the zone of fusion is indicated by the light line +0, and the zone within which the predicted temperature exceeds the liquidus temperature by 200°C or more by the light line +200. Small arrows are velocity vectors for particle motion.

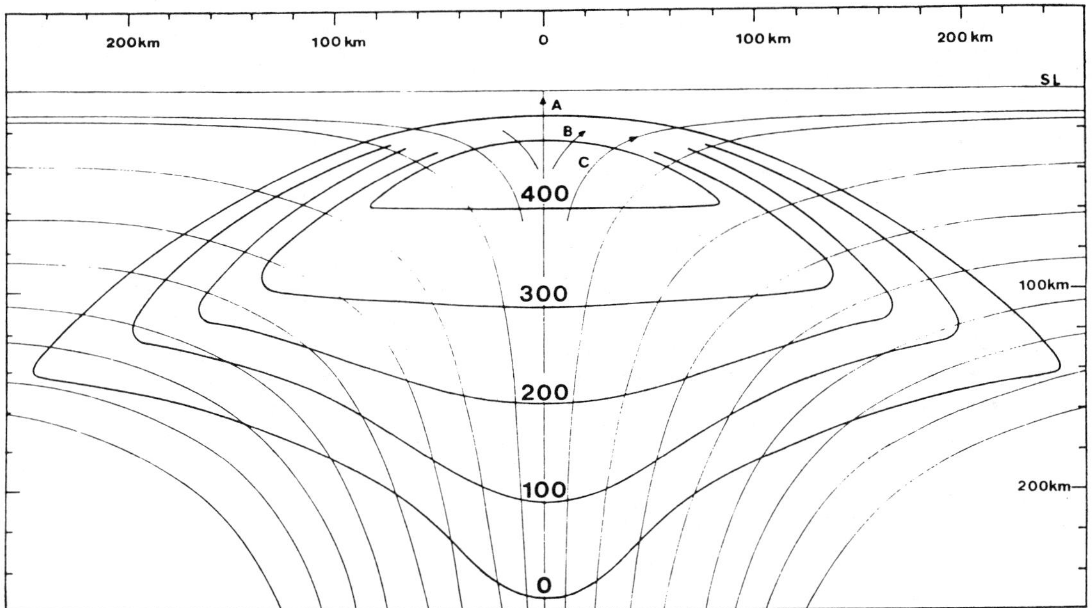

Fig. 12. Top of the ascending limb; heavy lines indicate zone of fusion and zones within it where the predicted temperature exceeds the olivine tholeiite fusion temperature by 100, 200°C, etc. Lighter lines indicate particle paths. For A, B, and C see Figure 14.

elevation in the liquidus temperature, so that the total amount of liquid produced in the fusion zone is very small and the temperatures within it correspond closely to the temperatures expected in the absence of any melting. The actual temperature distribution must lie between these extremes, e.g. path A–Y. Path B–Z would be followed by a particle that is on a path farther from the center of the rising column than A and that consequently enters the fusion zone at a lower temperature and pressure.

The compositional variations displayed by the melting of a peridotite-basalt system clearly cannot be properly represented in terms of two components, as is done in Figure 13, and the successive liquid compositions developed during passage through the fusion zone would vary in a complex fashion. In addition, the shape of the liquidus curves would probably vary with pressure; i.e., the P_8 curve would not have the same shape as the P_1 curve (Figure 13). This would reflect changes in the number and composition of the phases present in the system at different pressures.

Figure 12 shows a series of 'excess temperature' contours within the zone of fusion; the

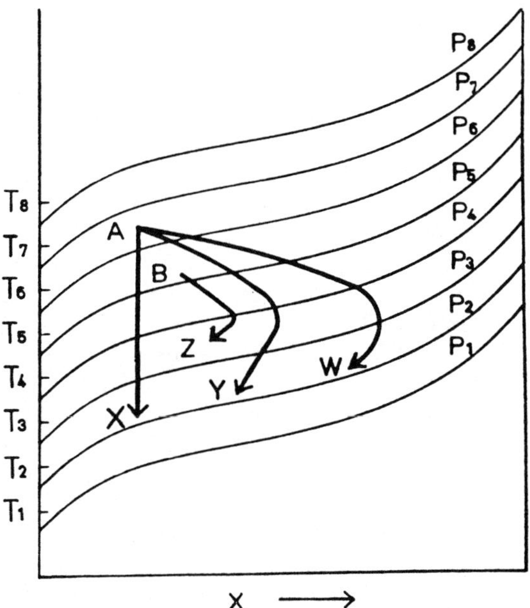

Fig. 13. Schematic relations between temperature, $T_1 < T_8$, pressure, $P_1 < P_8$, and liquid composition, x, within the zone of fusion. For discussion see text.

contours are drawn for every hundred degrees by which the predicted isotherms exceed the basalt fusion temperature. The 'excess heat' within the fusion zone would be sufficient to fuse olivine tholeiite equivalent to about 60% of the zone. This would correspond to fusion along path A–X in Figure 13. This path is the upper limiting case for the degree of fusion and, as discussed above, the degree of fusion that actually occurs is much less.

Figure 11 shows the velocity vectors for particle motion near the top of the hot plume computed for points on a 20-km square grid (see also Figures 4 and 5). From the velocity vectors streamlines may be derived, as shown in Figure 12. The streamlines diverge as the velocity decreases in the area of the stagnation point at the top of the hot plume, but they converge again after the corner is rounded and the flow becomes horizontal. The streamline separation is greater on the horizontal limb than in the rising plume, corresponding to horizontal velocities that are lower than the velocities in the plume. It can be seen from Figure 12 that only mantle material ascending along streamlines 100 km or less from the plume center line will pass through the fusion zone; mantle that has passed through the fusion zone forms the upper 130 km of the horizontal limbs. This would result in a compositional layering within the upper mantle. The uppermost mantle would be depleted in basalt and would pass downward into mantle material that had not passed through the fusion zone and was therefore unaffected [*Oxburgh*, 1965]. The depleted layer could be the source of some olivine-nodule xenoliths in basalts.

The model described above has several other petrogenetic implications. Within each half of the zone of fusion the combination of temperature and pressure conditions at every point is unique. This means that the liquids produced by the partial melting of a complex silicate system should vary continuously in composition throughout the fusion zone. In the parts of the fusion zone where the degree of melting is sufficiently great to allow their escape, these liquids should migrate upward to the surface to form hyperbyssal intrusions and surface eruptions. During their upward passage such liquids could mix with other liquids produced elsewhere within the fusion zone and having different compositions. It is evident that magmas erupted at the surface comprise a mixture of different liquids and that gradational variations in magma compositions are to be expected. It may be that such gradational variation in composition as has commonly been observed in volcanic sequences is partly to be explained in terms of the processes outlined above, rather than as entirely due to differentiation. If mantle material were of a composition such that its melting behavior followed a path similar to A–W in Figure 13, i.e. in the lower parts of the olivine tholeiite zone of fusion, little or no melting occurred, temperatures in the upper part of the zone should be sufficiently high to allow the formation of ultrabasic magmas within a narrow zone in the center of the ascending plume.

It should also be pointed out that, if ascending plumes are located under mid-ocean ridges, the main volcanic activity of the ridge should be concentrated within 200 km of the ridge crest, the lateral extent of the zone of fusion. Although this may be generally true, oceanic volcanic activity is certainly not confined to such a narrow belt. Active volcanos are found 500 km or more from the ridge; this may mean that the zone of fusion is wider than predicted by the present model.

NATURE OF LAYER 3 IN THE OCEANIC CRUST

Seismic refraction measurements at sea have shown that in oceanic areas the M discontinuity is generally at little more than 10 km below sea level. Taking M as the base of the oceanic crust, the upper mantle is characterized by compressional wave velocities of 8.0–8.2 km/sec, whereas the oceanic crust comprises three horizontal layers, in descending order, layers 1, 2, and 3. Typical thicknesses and velocities are as follows [*Hill*, 1957; *Raitt*, 1963]:

	V_p, km/sec	Thickness, km
Layer 1	1.45 to 2.0	0.45, but variable
Layer 2	5.07 ± 0.63	1.71 ± 0.75
Layer 3	6.69 ± 0.26	4.86 ± 1.42

Layer 1 makes the floor of the oceans and may be sampled directly; it is known to be predominantly unconsolidated sedimentary material. Layers 2 and 3 present more of a problem; their character has been discussed by a number of authors, notably *Hess* [1962, 1965]. Briefly, however, layer 2 has a velocity compatible with a composition of indurated sediment, possibly containing intercalated volcanics. The velocity of layer 3 is compatible with basalt or serpentinite; a few other known crustal rocks have similar velocities but basalt and serpentinite are the only compositions that have been seriously proposed for layer 3. The work of *Vine and Matthews* [1963] and *Vine* [1966] has shown that the steep local gradients in the magnetic field on the flanks of mid-ocean ridges require that the magnetic material responsible for the anomalies lies predominantly in layer 2. This substantiates the idea that layer 2 has a significant basaltic component. The anomalies also suggest that layer 3 is either nonmagnetic or weakly so. This might appear to eliminate the possibility of a basaltic layer 3, in that basalt is generally rich in the magnetic oxides. Serpentinite, however, as proposed by Hess, both has a suitable velocity and is very weakly magnetic.

The model for mantle convection presented in this paper suggests some difficulties in the formation of a serpentinite crust [see also *Menard*, 1965]. In the system $MgO-SiO_2-H_2O$ the upper temperature limit for the stability of serpentinite is known to be about 450°C at a pressure of 1.5 kb [*Scarfe and Wyllie*, 1967], i.e. the pressure at the base of the oceanic crust. It has been proposed [*Hess*, 1962, 1965] that as the peridotite of the mantle ascends in the rising limb of a convection cell and changes direction to flow away laterally, the olivine and enstatite of mantle material that is outermost cool sufficiently to react with interstitial water to produce serpentinite. Thus, layer 3 of what is recognized seismically as the crust would in fact have been produced by a change of phase within material that had previously been part of the upper mantle.

The principal difficulty with this model is that it is not clear how a seismic discontinuity sharp to a few tens of meters, as it appears to be in oceanic areas away from the ridges, could develop between the serpentinized mantle and the mantle proper; a more gradational contact would be expected. In addition, surface heat flow measurements indicate that, away from oceanic ridges the 450°C isotherm at which the serpentinization reaction proceeds, must be

considerably deeper than the M discontinuity.

Hess recognized this and proposed that the oceanic M discontinuity represents the 'fossilized' isotherm for the temperature of the serpentinization reaction [*Hess*, 1962]; this requires that at some temperature less than 450°C the serpentinization reaction rate drops sharply, effectively restricting the reaction to a rather narrow temperature interval, here called the reaction interval. The peridotite immediately below M and now at temperatures of less than 450°C must be assumed to have cooled through the reaction interval too rapidly for serpentinization to have taken place, whereas the peridotite immediately adjacent to it and, on the horizontal limb, above it, cooled sufficiently slowly for the reaction to proceed.

The model of mantle convection presented in this paper allows the cooling history of particles of mantle material following various flow paths to be established. This was done for the particles moving along three selected paths shown in Figure 12 as A, B, and C; the path A corresponds to an ascent along the center line of the plume ($x = 0$) and subsequent lateral motion along the upper surface ($y = 0$), and B and C are slightly deeper paths. The results are plotted in Figure 14. Although velocities are very low near the stagnation point ($x = 0$, $y = 0$), this does not prevent very rapid cooling on path A as it nears the upper surface. In fact path A gives more rapid cooling than either B or C; streamlines farther away from the surface pass more slowly through the reaction interval the deeper they are. The reaction interval is here arbitrarily designated

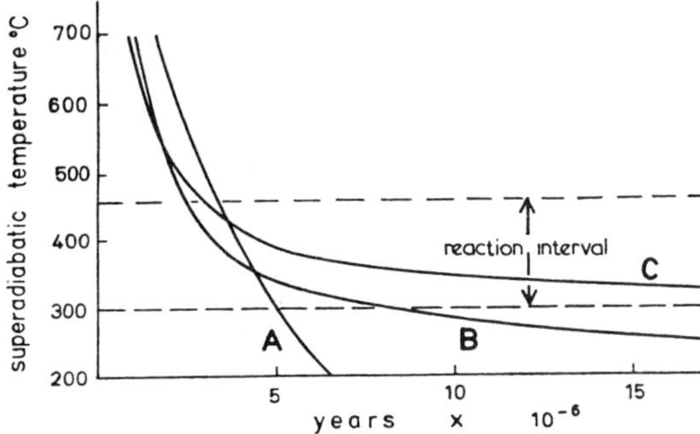

Fig. 14. Calculated cooling curves for particles following paths A, B, and C shown in Figure 12. For discussion see text.

as 450°–300°C, but the result would be similar for any other reasonable lower limit.

It is necessary to conclude that, if layer 3 is composed of serpentinite produced by hydration of the upper mantle, the M discontinuity is unlikely to have developed because the material below it did not have time to react; the deeper the streamline is, the more time there is available for the reaction. It is possible that a serpentinite layer 3 could develop if only the outer few kilometers of the mantle contained sufficient water for serpentinization to occur, but there does not seem to be any obvious way in which this could come about. It is proposed below that layer 3 comprises predominantly rocks of basaltic composition but that the magnetic properties of their oxide minerals may have been destroyed during low-grade metamorphism near the mid-oceanic ridge [*Cann and Vine*, 1966].

CRUSTAL STRUCTURE OF MID-OCEANIC RIDGES

The crustal structure of mid-oceanic ridges has been reviewed in great detail by workers from the Lamont Geological Observatory [*Le Pichon et al.*, 1965; *Talwani et al.*, 1965; *Heirtzler and Le Pichon*, 1965; *Ewing et al.*, 1966; *Langseth et al.*, 1966]. These workers were primarily concerned with the East Pacific rise and part of the mid-Atlantic ridge; although these ridges differ in a number of ways, gravity and seismic observations indicate that they have several significant features in common:

1. Although on the flanks of both ridges typically layered oceanic crust is found, beneath a zone extending several hundreds of kilometers away from the ridge crest on either side, no clear M discontinuity has been observed, and under a layer that is continuous with layer 2 on the flanks, a variety of P-wave velocities ranging from 6.4 to 8.2 km/sec are measured; particularly notable are velocities in the range 7.2 to 7.6 km/sec, which are absent on the flanks.

2. Both ridges are characterized by negative Bouguer gravity anomalies centered over their axial zones. These anomalies require the presence of a substantial mass deficiency immediately below the axial zone and somewhat smaller deficiencies on the flanks.

Talwani et al. [1965] have proposed solutions for the East Pacific rise and mid-Atlantic ridge structures that satisfy both gravity and seismic observations, and a regional crust and upper mantle structure has been proposed for the North Atlantic by *Tryggvason* [1961].

The model for mantle convection proposed in this paper has a bearing on the interpretation of both gravity and seismic observations. It is known in a general way that the effect of increase in temperature is to reduce seismic velocities in rocks, whereas increase in confining pressure is associated with an increase in velocity. A considerable amount of experimental work has been done in this field at moderate temperatures and pressures (reviewed by *Press* [1966]); relatively little is known about the temperature and pressure dependence of velocity at pressures up to 100 kb and at the high temperatures proposed here for the top of the ascending plume. In particular, the effects of partial melting on velocities in crystalline aggregates do not appear to have been investigated. It is to be expected, however, that the physical parameter most drastically affected by partial melting would be the rigidity; by comparison, the decrease in density could be relatively small. Taking the usual equations for the velocities of propagation of compressional waves V_p and shear waves V_s,

$$V_p = [(\tfrac{4}{3}\mu + \kappa)/\rho]^{1/2} \qquad (17)$$

$$V_s = (\mu/\rho)^{1/2} \qquad (18)$$

where k is the bulk modulus, μ is the rigidity, and ρ is the density of the transmitting medium, it is evident that a large reduction in rigidity accompanied by a small reduction in density might bring about a 10% reduction in V_p, but a more marked reduction in V_s. A similar conclusion is reached by *Shimozuru* [1963].

In addition to direct effects on elastic wave velocity, the temperature distribution proposed here should have strong effects on Q, the anelasticity of the upper mantle [*Anderson et al.*, 1965; *Anderson*, 1967]. Although lateral variations in Q for the upper mantle have been observed in some places [*Oliver and Isacks*, 1967], there appear to be no studies bearing directly on mid-oceanic ridge structure.

In the absence of supporting experimental evidence it is proposed that compressional wave velocities in the range 7–8 km/sec correspond to peridotite ($V_p = 8.0$–8.2 km/sec) of the upper mantle that has begun to melt and that consists of an aggregate of weakly connected crystals the interstices between which are occupied by basaltic liquid. It should be possible to verify this suggestion by measurements of shear wave velocities and wave attenuation phenomena under oceanic ridges. It is seen from Figure 12 that tongues of completely crystalline upper mantle with normal mantle velocities are to be expected above the zone of partial fusion on either side. Partially serpentinized peridotite is almost the only other rock known to have compressional wave velocities in the range 7–8 km/sec (Birch in *Hess* [1962]), and it has been shown earlier that temperatures at the top of the plume are too high for the existence of serpentinite at the depths required. An alternative suggestion has been made by *Hess* [1962] that these anomalous velocities are the result of microfracturing in the peridotite where the direction of flow changes rapidly. This proposal is difficult to evaluate.

A detailed discussion of gravity profiles computed for the convection model proposed here will be presented elsewhere. It may be noted, however, that the geometry of the isotherms and thus roughly the densities, as proposed, is consistent with the solutions of *Talwani et al.* [1965] and *Tryggvason* [1961]. The present model, however, would require the anomalous masses to be rather larger than the values proposed by *Talwani et al.* [1965] and to be situated at somewhat greater depth. It is also possible that the density contrasts employed by the authors cited are too great. The authors employed the Nafe-Drake empirical relationship between seismic velocity and density. Whereas this relationship is known in many circumstances to give reasonable solutions, it is possible that in situations such as the one described here, where rocks are close to or above their temperatures for the beginning of fusion, the relationship breaks down as rigidity decreases more rapidly than density.

KINETIC IMPLICATIONS OF THE PROPOSED MODEL

It remains to consider some of the larger-scale and long-term implications of the convection model proposed here. For the cell dimensions and rates of movement assumed here, one complete convective overturn would require approximately 10^9 years, i.e. nearly a quarter of the age of the earth. Isotherms would, however, have reached an approximately steady-state configuration after 5×10^8 years. All these values would be somewhat reduced, however, if the aspect ratio of the model cell were modified as discussed below. Thus, the earliest stages of ocean spreading or continental drift by the spreading mechanism might be expected to be devoid of volcanic activity.

As briefly mentioned in an earlier section, the mantle material that passes through the zone of fusion subsequently changes direction and moves horizontally away from the ridge, forming the upper 130 km of mantle on the ridge flanks. The removal of the low melting fraction during passage through the fusion zone renders the uppermost mantle different in composition from the material that it overlies. The basaltic lava formed in the fusion zone should rise to the surface above and there be erupted upon earlier lavas, or, near the stagnation point, upon upper mantle material itself; in either case, the basalt would be expected to form a passive sheet resting on the upper mantle which is continuously transported away laterally upon it, in this way forming layer 3 of the oceanic crust. Thus, after the early stages, as long as the convective motion persists, new layer 3 should be generated at or near the mid-oceanic ridges.

If layer 3 is formed in the way suggested, its thickness gives some idea of the degree of fusion in the fusion zone. For a materials balance, a 4-km thickness of layer 3 (possibly with some of layer 2) added to a 130-km thickness of depleted upper mantle immediately below M, should have the same bulk composition as the upper mantle below 130 km which has not been melted. This suggests that the average degree of fusion within the fusion zone may be as little as 3%, although locally it would be rather higher, particularly if not all the fused material finds its way to the surface.

Figures 4, 5, and 11 show that particles on the upper surface of the convection cell undergo a rapid acceleration for about 300 km away from the stagnation point at the beginning of their horizontal paths. The widths of the magnetic anomaly bands [*Vine*, 1966] at different distances from the ridge indicate, however, that spreading rates at the surface do not vary in this fashion. It is evident that here the assumption of uniform viscosity has failed and that the cold upper surface layer of the convecting medium has such a high viscosity that it effectively moves as a rigid plate. The thickness of the layer behaving in this fashion would be related to the depth of the cold thermal boundary layer and would thus increase away from the ridge. A rigid surface layer of this kind should have a marked effect on the geometry of the descending flow, possibly along the lines suggested by *Oliver and Isacks* [1967].

Conclusions

A quantitative model for upper mantle convection has been proposed that satisfies reasonably well the requirements of heat transfer, velocity, and scale imposed by the hypothesis of ocean-floor spreading. It appears also that the model may be consistent with seismic observations on mid-oceanic ridges, although the experimental data necessary to demonstrate this are not available. Geotherm distributions predicted by the model have been used to explain mid-oceanic volcanic activity and the generation of layer 3 of the oceanic crust. A serpentinite layer 3 is not compatible with the thermal model used in this paper unless the position of the discontinuity is governed primarily by the availability of water for the serpentinization reaction.

A large number of simplifying assumptions have been made in setting up the model. The effects of radioactivity have been ignored; although many of the arguments for the strong concentration of radioactive constituents in the upper part of the mantle are invalid if mantle convection occurs, the currents proposed here take so long to complete a cycle that relatively low concentrations of radioactivity could have significant effects. One such effect could be the modification of the cell aspect ratio [*Tritten and Zarraga*, 1967; *Roberts*, 1967]. This could explain the existence of convection currents with the horizontal dimensions required for large continental displacements but at the same time sufficiently shallow to be unaffected by the high viscosities that appear to characterize the deeper parts of the mantle [*McKenzie*, 1966; *Gordon*, 1965]. The effects of more realistic boundary conditions on mantle convection must also be investigated.

It has also been assumed that the mantle behaves in a viscous fashion and that viscosity does not vary with temperature; the former assumption has been seriously questioned [e.g., *Orowan*, 1965] and the latter assumption is certainly wrong, but it is not yet known how this might affect convective circulation. The provisional agreement, however, between the model proposed here and the natural phenomena that it purports to describe is sufficiently good as to suggest that the simplifying assumptions either have effects that cancel each other or are less important than they first appear.

Acknowledgments. The research was partially supported by the Air Force Office of Scientific Research of the Office of Aerospace Research under contract AF 49(638)-1346. The research was undertaken when one of the authors (D. L. Turcotte) held a National Science postdoctoral research fellowship at the Department of Engineering Sciences, University of Oxford.

References

Allen, C. R., Transcurrent faults in continental areas, *Phil. Trans. Roy. Soc. London, A, 258*, 82, 1965.

Anderson, D. L., Latest information from seismic observations, in *The Earth's Mantle*, edited by T. F. Gaskell, pp. 355–420, Academic Press, New York, 1967.

Anderson, D. L., A. Ben-Menahem, and C. B. Archambeau, Attenuation of seismic energy in the upper mantle, *J. Geophys. Res., 70*, 1441, 1965.

Blackett, P. M. S., Introduction; A symposium on continental drift, *Phil. Trans. Roy. Soc. London, A, 258*, vii, 1965.

Cann, J. R., and F. J. Vine, An area on the crest of the Carlsberg ridge: Petrology and magnetic survey, *Phil. Trans. Roy. Soc. London, A, 259*, 198, 1966.

Chandrasekhar, S., *Hydrodynamic and Hydromagnetic Stability*, Oxford University Press, Oxford, 1961.

Clark, S. P., and A. E. Ringwood, Density distribution and constitution of the mantle, *Rev. Geophys., 2*, 35, 1964.

Crittenden, M. D., Effective viscosity of the earth derived from isostatic loading of Pleistocene Lake Bonneville, *J. Geophys. Res., 68*, 5517, 1963.

Elsasser, W. M., Early history of the earth, in *Earth Science and Meteoritics*, pp. 1–30, North Holland, Amsterdam, 1963.

Engel, A. E. J., C. G. Engel, and R. G. Havens, Chemical characteristics of oceanic basalts and the upper mantle, *Bull. Geol. Soc. Am., 76*, 719, 1965.

Ewing, M., X. Le Pichon, and J. Ewing, Crustal

structure of the mid-ocean ridges, 4, Sediment distribution in the South Atlantic Ocean and Cenozoic history of the mid-Atlantic ridge, *J. Geophys. Res., 71*, 1611, 1966.

Fromm, J. E., Numerical solutions of the nonlinear equations for a heated fluid layer, *Phys. Fluids, 8*, 1757, 1965.

Girdler, R. W., The formulation of new oceanic crust, *Phil. Trans. Roy. Soc. London, A, 258*, 252, 1965.

Gordon, R. B., Diffusion creep in the earth's mantle, *J. Geophys. Res., 70*, 2413, 1965.

Green, D. H., and A. E. Ringwood, The genesis of basaltic magmas, in *Petrology of the Upper Mantle, Publ. 444*, Department of Geophysics and Geochemistry, Australian National University, 1966.

Haskell, N. A., The motion of a viscous fluid under a surface load, *Physics, 6*, 265, 1935.

Heirtzler, J., and X. Le Pichon, Crustal structure of the mid-ocean ridges, 3, Magnetic anomalies over the mid-Atlantic ridge, *J. Geophys. Res., 70*, 4013, 1965.

Hess, H. H., History of ocean basins, in *Petrologic Studies*, pp. 599–620, Geological Society of America, New York, 1962.

Hess, H. H., Mid-oceanic ridges and tectonics of the sea floor, in *Submarine Geology and Geophysics, Colston Papers*, vol. 17, pp. 317–334, Butterworths, London, 1965.

Hill, M. N., Recent geophysical exploration of the ocean floor, *Phys. Chem. Earth, 2*, 129, 1957.

Jaeger, J. C., Application of the theory of heat conduction to geothermal measurements, in *Terrestrial Heat Flow Geophys. Monograph 18*, pp. 7-23, American Geophysical Union, Washington, D. C., 1965.

Jeffreys, H., Some cases of instability in fluid motion, *Proc. Roy. Soc. London, A, 118*, 195, 1928.

Jeffreys, H., *The Earth*, 2nd ed., Cambridge University, Cambridge, 1929.

Jeffreys, H., The instability of a compressible fluid heated below, *Proc. Cambridge Phil. Soc., 26*, 170, 1930.

Knopoff, L., The convection current hypothesis, *Rev. Geophys., 2*, 89, 1964.

Kuo, H. L., Solution of the non-linear equations of cellular convection and heat transport, *J. Fluid Mech., 10*, 611, 1961.

Langseth, M. G., X. Le Pichon, and M. Ewing, Crustal structure of the mid-ocean ridges, 5, Heat flow through the Atlantic ocean floor and convection currents, *J. Geophys. Res., 71*, 5321, 1966.

Le Pichon, X., R. E. Houtz, C. L. Drake, and J. E. Nafe, Crustal structure of the mid-ocean ridges, 1, Seismic refraction measurements, *J. Geophys. Res., 70*, 319, 1965.

Lee, W. H. K., and S. Uyeda, Review of heat flow data, in *Terrestrial Heat Flow, Geophys. Monograph 8*, pp. 87-190, American Geophysical Union, Washington, D. C., 1965.

Lubimova, H. A., Thermal history of the earth with consideration of the variable thermal conductivity of the mantle, *Geophys. J., 1*, 115, 1958.

MacDonald, G. J. F., Continental structure and drift, *Phil. Trans. Roy. Soc. London, A, 258*, 215, 1965a.

MacDonald, G. J. F., Geophysical deductions from observations of heat flow, in *Terrestrial Heat Flow, Geophys. Monograph 8*, pp. 191-210, American Geophysical Union, Washington, D. C., 1965b.

Malkus, W. V. R., and G. Veronis, Finite amplitude cellular convection, *J. Fluid Mech., 4*, 225, 1958.

McConnell, R. K., Isostatic adjustment in a layered earth, *J. Geophys. Res., 70*, 5171, 1965.

McKenzie, D. P., The viscosity of the lower mantle, *J. Geophys. Res., 71*, 3995, 1966.

Menard, H. W., Sea floor relief and mantle convection, in *Physics and Chemistry of the Earth, 6*, pp. 315–364, Pergamon, London, 1965.

Oliver, J., and B. Isacks, Deep earthquake zones, anomalous structures in the upper mantle and the lithosphere, *J. Geophys. Res., 72*, 4259, 1967.

Orowan, E., Convection in a non-Newtonian mantle, continental drift and mountain building, *Phil. Trans. Roy. Soc. London, A, 258*, 284, 1965.

Oxburgh, E. R., Petrological evidence for the presence of amphibole in the upper mantle and its petrogenetic and geophysical implications, *Geol. Mag., 101*, 1, 1964.

Oxburgh, E. R., Volcanism and mantle convection, *Phil. Trans. Roy. Soc. London, A, 258*, 142, 1965.

Platzman, G. W., The spectral dynamics of laminar convection, *J. Fluid Mech., 23*, 481, 1965.

Press, F., Seismic velocities, in *Handbook of Physical Constants, Mem. 97*, pp. 195–218, Geological Society of America, New York, 1966.

Raitt, R. W., The crustal rocks, *The Sea*, vol. 3, 85, Interscience, New York, 1963.

Rayleigh, J. W. S., On convective currents in a horizontal layer of fluid when the higher temperature is on the under side, *Phil. Mag., 32*, 529, 1916.

Roberts, P. H., Convection in horizontal layers with internal heat generation: Theory, *J. Fluid Mech., 30*, 33, 1967.

Runcorn, S. K., Palaeomagnetic evidence for continental drift and its geophysical cause, in *Continental Drift*, pp. 1–40, Academic Press, New York, 1962.

Scarfe, C. M., and P. J. Wyllie, Experimental redetermination of the upper stability limit of Serpentine up to 3-kb pressure (abstract), *Trans. Am. Geophys. Union, 48*, 225, 1967.

Shimozuru, D., Geophysical evidence for suggesting the existence of molten pockets in the earth's upper mantle, *Bull. Volcanol., 26*, 181, 1963.

Somerscales, E. F. C., and D. Dropkin, Experimental investigation of the temperature distribution in a horizontal layer of fluid heated from below, *Intern. J. Heat Mass Transfer, 9,* 1189, 1966.

Talwani, M., X. Le Pichon, and M. Ewing Crustal structure of the mid-ocean ridges, 2, Computed model from gravity and seismic refraction data, *J. Geophys. Res., 70,* 341, 1965.

Tozer, D. C., Heat transfer and convection currents, *Phil. Trans. Roy. Soc. London, A, 258,* 252, 1965.

Tritton, D. J., and M. N. Zarraga, Convection in horizontal layers with internal heat generation: Experiments, *J. Fluid Mech., 30,* 21, 1967.

Tryggvason, E., Wave velocity in the upper mantle below the Arctic-Atlantic Ocean and northwest Europe, *Ann. Geofis., 14,* 380, 1961.

Turcotte, D. L., A boundary-layer theory for cellular convection, *Intern. J. Heat Mass Transfer, 10,* 1065, 1967.

Turcotte, D. L., and E. R. Oxburgh, Finite amplitude convection cells and continental drift, *J. Fluid Mech., 28,* 29, 1967.

Verhoogen, J., Temperatures within the earth, *Phys. Chem. Earth, 1,* 17, 1958.

Vine, F. J., Spreading of the ocean floor: New evidence, *Science, 154,* 1405, 1966.

Vine, F. J., and D. H. Matthews, Magnetic anomalies over oceanic ridges, *Nature, 199,* 947, 1963.

Yoder, H. S., and C. E. Tilley, Origin of basalt magmas: An experimental study of natural and synthetic rock systems, *J. Petrol., 3,* 342, 1962.

(Received June 30, 1967;
revised December 1, 1967.)

6

NUMERICAL MODELS OF CONVECTION IN THE EARTH'S MANTLE

DAN McKENZIE[*], JEAN ROBERTS[*] and NIGEL WEISS[**]

[*] *Department of Geodesy and Geophysics, University of Cambridge, Cambridge (Great Britain)*
[**] *Department of Applied Mathematics and Theoretical Physics, University of Cambridge, Cambridge (Great Britain)*

(Accepted for publication January 25, 1973)

ABSTRACT

McKenzie, D., Roberts, J. and Weiss, N., 1973. Numerical models of convection in the earth's mantle. In: E. Irving (Editor), *Mechanisms of Plate Tectonics. Tectonophysics,* 19(2): 89–103.

Calculations on convection in Newtonian fluids at large Rayleigh numbers show behaviour in general agreement with geophysical observations. The calculated values of surface velocities, regional gravity and topographic anomalies all lie within the range of values found at the earth's surface. The sign of the gravity anomaly over rising fluid is positive because the gravitational effect of the surface deformation is greater than that of the density deficit. This result appears to agree with geophysical observations and should permit the flow within the mantle to be followed in some detail.

INTRODUCTION

This paper contains a condensed account of some numerical experiments and their geophysical implications carried out at Cambridge over the last two years. A full account will be published elsewhere (McKenzie et al., 1973). The work began as a direct consequence of the success of plate tectonics. The original theory formed a kinematic description of relative motions between large aseismic spherical caps. Though attempts have recently been made to relate these motions to a frame fixed to the lower mantle, it is not yet clear whether it is possible to do so. Moreover the success or otherwise of this attempt in no way affects the success of plate tectonics.

Before the kinematic theory was understood many features of the sea floor were believed to be the surface expressions of convection in the mantle. The elevation and high heat flow of ridges were thought to be caused by a rising convection current beneath them, and the depression and large negative gravity anomalies of trenches were believed to be features of a sinking region. Early attempts at producing a convective model for the mantle concentrated on modelling these features of the sea floor (Langseth et al., 1966; Turcotte and Oxburgh, 1967). This approach lead to many difficulties, especially on ridges. A different approach was to argue that ridges were structures associated with the production of a cold strong boundary layer, or plate, whose motion was maintained by convective forces acting on its lower surface. The motion of this layer need not be closely related to the motions of

95

the mantle below. In this model trenches were sites where boundary layers became detached from the surface (McKenzie, 1969). This model allowed detailed calculations to be carried out because the resulting equations were linear, and has since been shown to agree remarkably well with the relevant geophysical observations (Sclater and Franchetau, 1970; Sclater et al., 1971). In view of this success it is important to determine which surface observations cannot be explained by the behaviour of the boundary layer. This same question was considered four years ago (McKenzie, 1968), but advances in plate tectonics have since then considerably reduced the list.

The most obvious and striking effect which must be accounted for is the plate motions themselves, with velocities between 10 and 200 mm/year. Elsasser's (1969) attempts to maintain these motions by the traction exerted by the sinking slabs cannot account for the motions between Africa, America and Antarctica, which are not especially slow (~ 40 mm/ year in the South Atlantic). None of these plates has a sinking slab of any appreciable size attached to it. A more recent and more sophisticated attempt of similar type has been made by Howard et al. (1970), who argued that the high concentration of radioactive elements in continental rocks could maintain horizontal temperature gradients in the mantle and hence drive convective motions. Their argument neglects the fact that three plates in rapid motion: the Pacific, Nasca and Phillipine Sea plates, have as the only continents to maintain their motion the South Island of New Zealand and the southern coast ranges of California. Therefore, though Howard et al.'s mechanism may exist so must some other effective mechanism. Since this other mechanism can move plates apart at 200 mm/year it can clearly maintain all plate motions. The simplest hypothesis is that it does so, and this view will be adopted here and continental radioactivity will be neglected.

Though the heat lost from the mantle through the formation of the plates accounts for about 30% of the heat lost by conduction through the floor of the oceans, the remainder must be transported to the base of the plates by some form of convection. Whether there is a direct relation between this convection and the plate motions is not yet clear.

The external gravity field determined from the motion of artificial satellites has long been believed to be the consequence of a nonhydrostatic density distribution maintained by convection in the mantle (Runcorn, 1965). The relationship between the two is complicated by the surface deformation which accompanies the convection, and it is not obvious whether the force due to gravity should be greater or less above a rising region of hot material. It is, however, obvious that the surface of the earth must be elevated above a rising convection current. The calculations below show that the gravity anomaly is dominated by the surface deformation and is therefore positive over rising regions. The correlation between these two effects promises to be a useful method of determining the form of the flow (see below).

The calculations we have carried out have all been two-dimensional, and have used a flat layer of fluid 700 km thick. Recently there has been considerable discussion as to whether the lower mantle below a depth of 700 km is also convecting. The detached boundary layer beneath island arcs does not penetrate below 700 km and appears to meet with strong resistance to its motion at this depth (Isacks and Molnar, 1971). Furthermore, the atomic arrangements within possible phases below this depth is likely to make deformation more difficult than in the upper mantle. Despite these arguments, Morgan (1971) has recently proposed that convective plumes rising from the lower mantle are the main driving mechanism for plate motions. It is difficult to understand how such plumes can arise unless the radioactive elements within the mantle are concentrated in a layer at some depth below 700 km. Such a distribution does not seem likely, and a uniform distribution cannot produce plumes for geometric reasons (see below). It is also difficult to understand how a hot rising plume with no strength can penetrate the boundary at 700 km when the strong cold plumes beneath island arcs cannot do so. We therefore believe that convection in the upper mantle is the main mechanism by which the plate motions are maintained, and the calculations described below are directed towards gaining an understanding of this problem.

THE GOVERNING EQUATIONS

Any convection calculations require an expression which relates the shearing stress to the strain and strain rate. The simplest such relationship is linear and the fluid flow can then be described by a Newtonian viscosity. There are theoretical reasons why solids should deform in this way under sufficiently small stresses. Whether the stresses involved in the mantle convection are sufficiently small is not yet clear, and no laboratory experiments have yet been carried out on relevant materials under conditions comparable to those within the mantle. Such experiments are necessary because dislocations may well be in motion under conditions where the deformation is rate limited by diffusion. Whatever the outcome of such experiments may be we do not believe calculations with complicated relations between stress and strain rate are useful until the behaviour of simple Newtonian fluids is known and understood, and therefore we used a linear relation between stress and strain rate described by a constant viscosity. The calculations described below were all carried out with a temperature independent viscosity. Since the cold boundary layers then possess no mechanical strength, these calculations will not be able to model the plates. The movement of the fluid may therefore differ in important ways from that of the earth's mantle, and may not be directly relevant to geophysics.

The equations governing the conservation of mass, momentum and energy of a convecting fluid can be simplified when applied to mantle convection (McKenzie, 1968):

$$\nabla \cdot v = 0$$

$$\frac{1}{\rho}\left[(\nabla \cdot \eta\nabla)v + \frac{\partial \eta}{\partial x_j}\frac{\partial v_j}{\partial x_i}\right] + \nabla U - \frac{1}{\rho}\nabla P = 0 \tag{1}$$

$$\rho C_p\left[\frac{\partial T}{\partial t} + v\cdot\left\{\nabla T - (\nabla T)_s\right\}\right] = \nabla \cdot k\nabla T + H + \frac{\eta}{2}\left(\frac{\partial v_i}{\partial x_j} + \frac{\partial v_j}{\partial x_i}\right)^2$$

where v is the fluid velocity, ρ its density, η its viscosity, C_p its specific heat and k its thermal conductivity. U is the gravitational potential, P the fluid pressure, $(\nabla T)_s$ the adiabatic temperature gradient and H the rate of internal heat generation. These equations must be solved with appropriate boundary conditions. If η and k are taken to be constant, and the temperature and density scale heights are large compared to 700 km, we may write:

$$\mathbf{v} = \frac{g\alpha T_0}{\nu}\, l^2 \left(\frac{\partial \psi'}{\partial y'},\, -\frac{\partial \psi'}{\partial x'},\, 0 \right)$$
$$x = lx' \qquad T = T_0 T'$$
$$y = ly' \qquad t = \frac{\nu t'}{g\alpha T_0 l} \tag{2}$$

where primes denote dimensionless quantities, g is the acceleration due to gravity, α the thermal expansion coefficient, l the thickness of the convecting layer and T_0 an appropriate scaling temperature. x, y are Cartesian co-ordinates with y vertical. Eq. 1 then reduce to:

$$\nabla^4 \psi = \frac{\partial T}{\partial x}$$
$$\frac{\partial T}{\partial t} + \frac{\partial \psi}{\partial y}\frac{\partial T}{\partial x} - \frac{\partial \psi}{\partial x}\frac{\partial T}{\partial y} = \frac{1}{R}\left(\nabla^2 T + \epsilon_1 \right) \tag{3}$$

where:

$$R = \frac{g\alpha T_0 \rho C_p l^3}{k\nu}, \quad \epsilon_1 = \frac{Hl^2}{k}$$

where the primes on the dimensionless variables have been omitted. The boundary conditions used here are that:

$$\psi = \nabla^2 \psi = 0 \tag{4}$$

on all boundaries of a rectangular box, and that:

$$T = 0 \tag{5}$$

on the upper surface:

$$\frac{\partial T}{\partial x} = 0 \tag{6}$$

on the two sides and:

$$\frac{\partial T}{\partial y} = \text{constant} \tag{7}$$

on the lower boundary. The heat flux through the bottom and H were varied to keep the mean flux through the top of the box constant. The values used for the parameters are given in Notation I.

Eq. 3 contain two non-linear terms governing the horizontal and vertical convection of heat. They were solved by a conservative difference scheme on a staggered mesh (Roberts and Weiss, 1966). Moore et al. (in preparation) and McKenzie et al. (1973) give further de-

NOTATION I

Values used in calculations

$$K = \frac{k}{\rho C_p} \qquad = 1.5 \cdot 10^{-6}\ \mathrm{m^2/sec}$$

$$\rho \qquad = 3.7 \cdot 10^3\ \mathrm{kg/m^3}$$

$$C_p \qquad = 1.2 \cdot 10^3\ \mathrm{J\ kg^{-1}\,^\circ C^{-1}}$$

$$g \qquad = 10\ \mathrm{m/sec^2}$$

$$\nu \qquad = 2 \cdot 10^{17}\ \mathrm{m^2/sec}$$

$$\alpha \qquad = 2 \cdot 10^{-5}/^\circ C$$

$$\text{Surface heat flux } f \qquad = 5.85 \cdot 10^{-2}\ \mathrm{W/m^2}$$

$$l \qquad = 7 \cdot 10^2\ \mathrm{km}$$

tails of the method used. The results below were all obtained using a square mesh with 24 vertical and 24 horizontal divisions. When the flow reached steady state the solution was stretched onto a 48 × 48 mesh and allowed to reach a steady state. The difference between the two solutions was minor in all cases, though this is not a sufficient test of the accuracy of the solution (McKenzie et al., 1973).

A long series of experiments was carried out with the numerical scheme outlined above, in order to understand the development of boundary layer structure and to follow various types of time dependent behaviour. Since full description of these results will be found in McKenzie et al (1973), only those experiments with possible geophysical relevance will be described below.

GEOPHYSICAL APPLICATIONS

Fig.1 shows the form of flow when all the heat is conducted through the lower boundary, Fig.2 when half the heat is generated within and half conducted through the bottom, and Fig.3 when all the heat is generated within the fluid. Fig.2 most closely resembles the best estimates of the behaviour of the mantle. In all three calculations the mean heat flow through the upper surface of the box was $5.85 \cdot 10^{-2}\ \mathrm{W/m^2}$. Also shown are plots of the heat flow through the upper surface of the boxes, calculated from the temperature gradient, the surface deformation and the gravity anomaly obtained by integrating the attraction of the density perturbations resulting from the flow. The horizontal velocity and the horizontal mean of the temperature are also shown. One of the major advantages of numerical, rather than laboratory, experiments is that all these quantities of geophysical interest can be calculated easily from the stream function, the vorticity and temperature distribution once the relevant subroutines have been properly tested. There is no corresponding method of observing these quantities in a laboratory experiment.

The flow and temperature fields in Fig.1–3 show various interesting features. When the heat enters only from below it can be transported to the upper boundary by the rapid motion of a thin hot sheet of fluid, leaving the temperature of the main mass of the fluid unchanged. However, when the heat is generated within the fluid the heat cannot be transported by the motion of a boundary layer, and all the fluid must therefore move close to the upper boundary to lose heat. All experiments were started with a single roll in a square

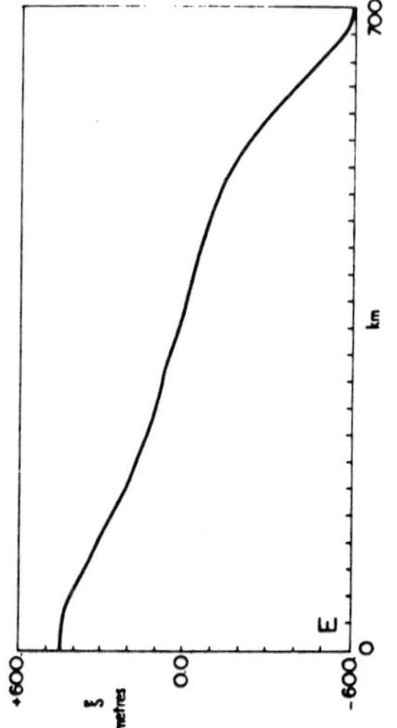
+600
00
-600
metres
km
E
700
0

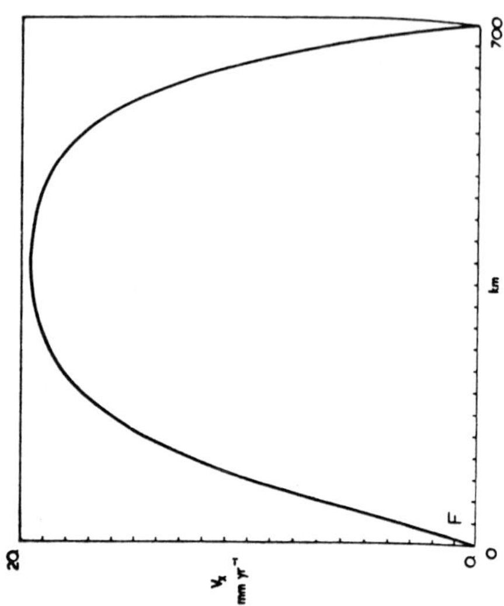
20
0
V_x
mm yr^{-1}
km
F
700
0

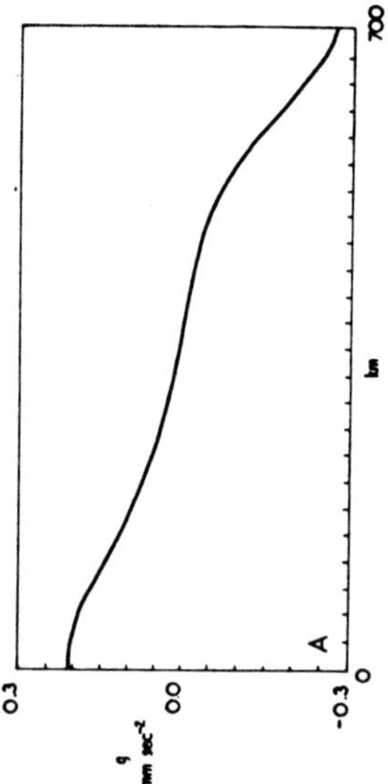
0.3
00
-0.3
q
mm sec^{-2}
km
A
700
0

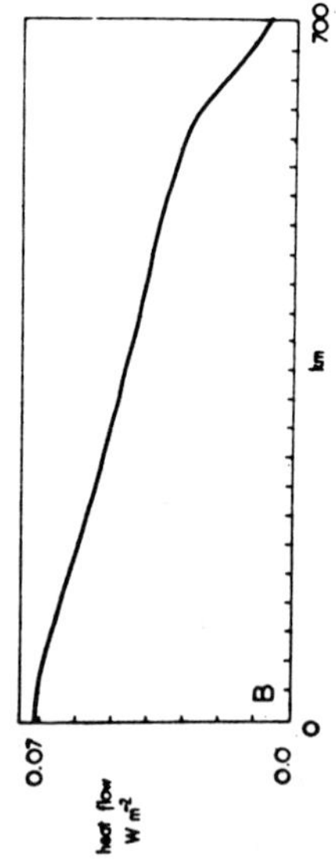
0.07
00
heat flow
W m^{-2}
km
B
700
0

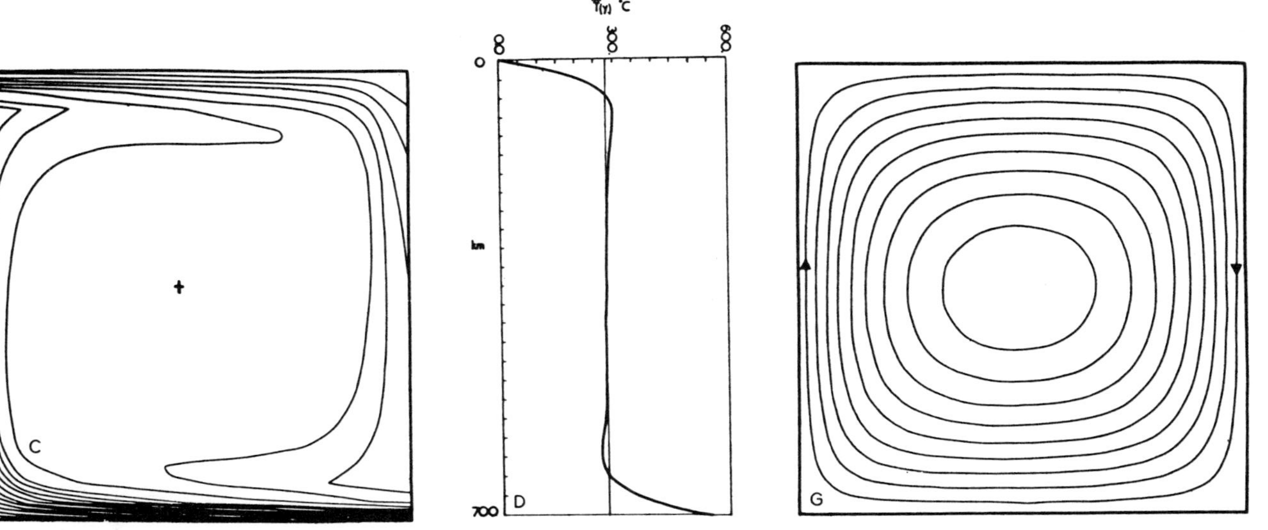

Fig. 1. Flow in a square box heated from below, and no heat generated within. A. The gravity anomaly ($1\,mm/sec^2 \equiv 100$ mGal). B. The heat flow ($0.0418\,W/m^2 \equiv 1\,\mu cal.cm^{-2}sec^{-1}$). C. The isotherms. D. The vertical variation of the mean horizontal temperature. E. The deformation of the upper surface. F. The horizontal velocity at the upper surface. G. The stream lines. The boundary conditions on the upper and lower surfaces are specified in the text: those on the vertical boundaries are satisfied if the convection pattern is extended horizontally by reflection. The flow is steady state and the calculation was started with a temperature perturbation to produce a single cell. The time taken to reach a steady state was about 10^9 years. Notice that both the surface elevation and the gravity anomaly are positive over the rising part of the cell. The heat flow through the bottom of the box is given in Notation I. The marks on the horizontal axes in A, B, E and F and on the vertical axis in D show the spacing of the mesh points for a 24 × 24 mesh. The calculations shown in this figure are those for a 48 × 48 mesh.

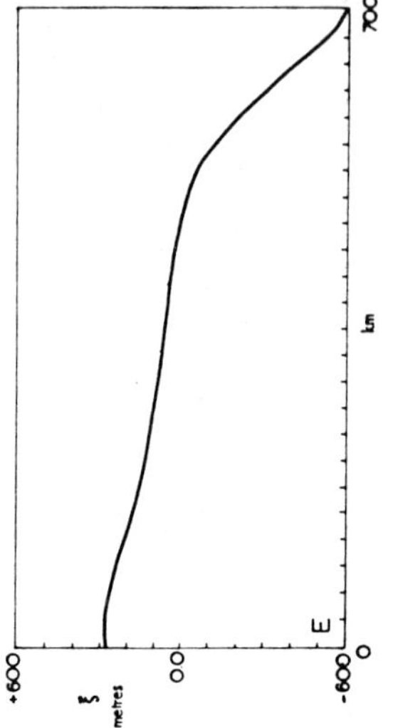
+600
ξ
metres
00
-600
km
E

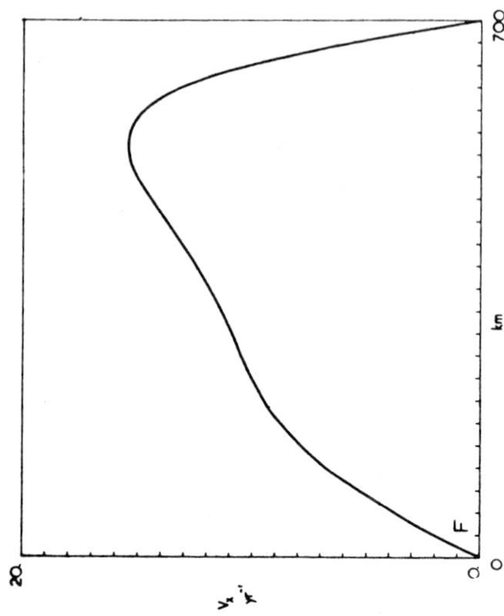
20
v_r °k⁻¹
0
km
F

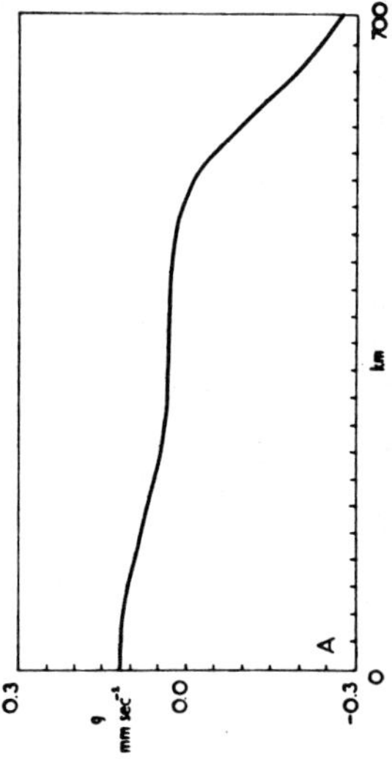
0.3
g
mm sec⁻¹
00
-0.3
km
A
700

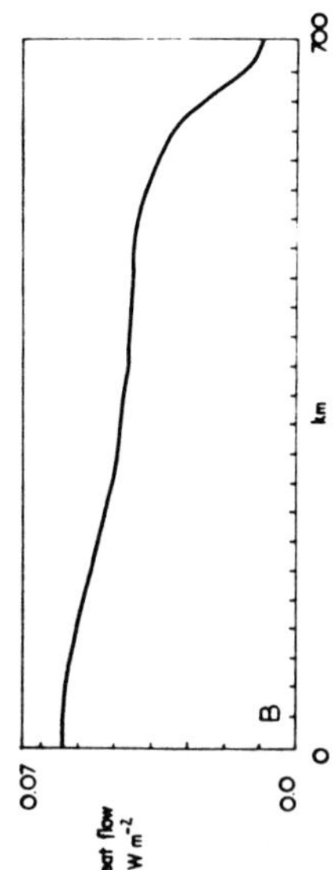
0.07
heat flow
W m⁻²
00
km
B
700

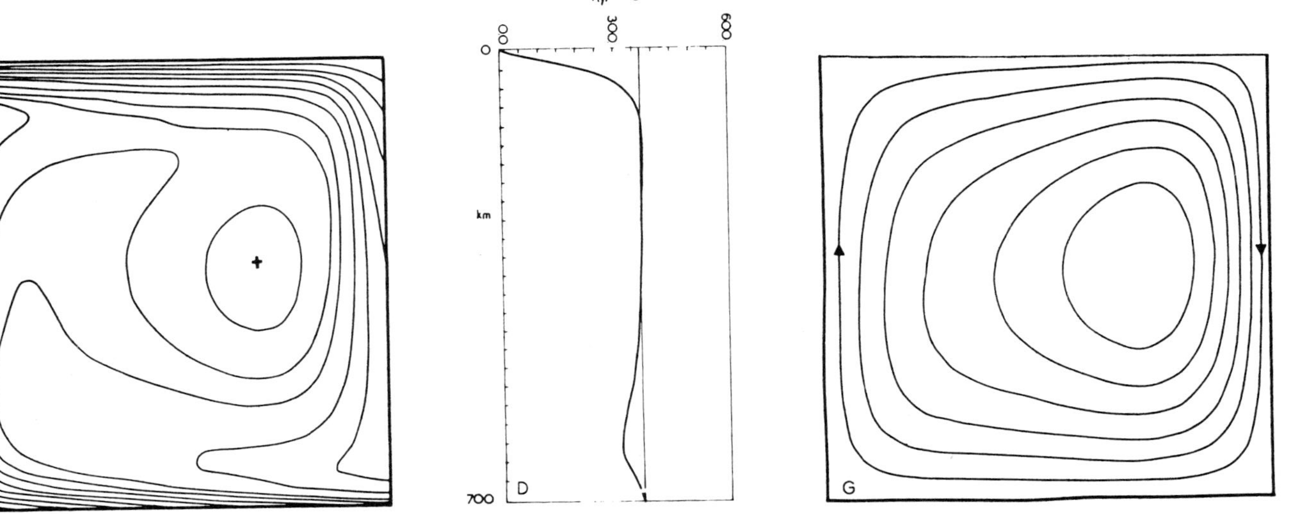

Fig. 2. As for Fig. 1, except that the heat flux through the base of the box is $3 \cdot 10^{-2}$ W/m^2. The remainder of the mean heat flux through the top of the box of $5.85 \cdot 10^{-2}$ is generated uniformly within the box.

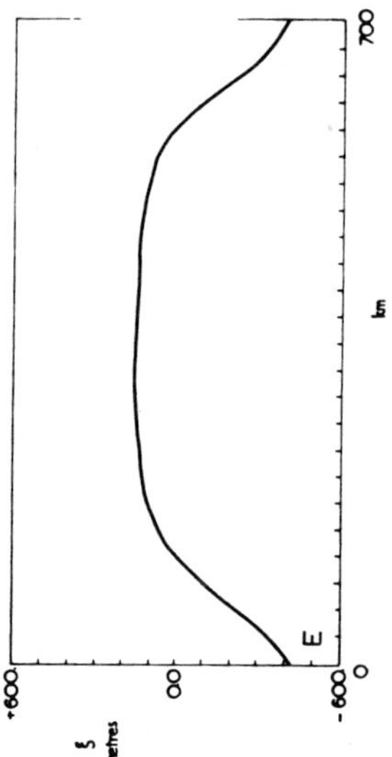
E
ξ
metres
+600
0
-600
km

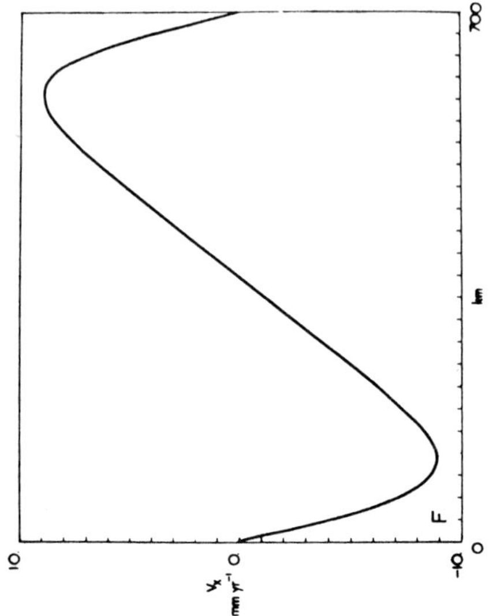
F
10
0
-10
v_x
mm yr⁻¹
km

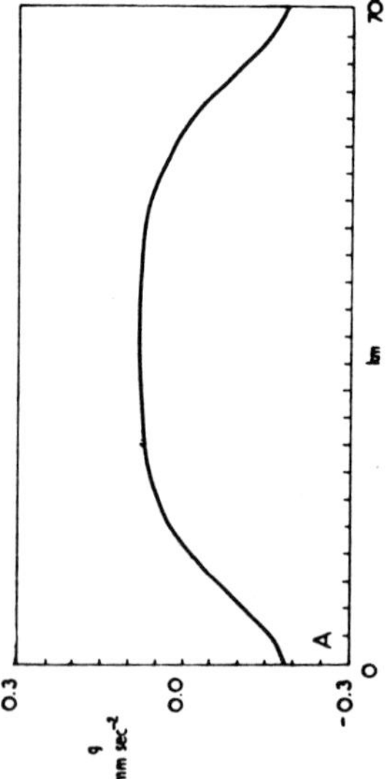
A
0.3
0.0
-0.3
q
mm sec⁻²
km

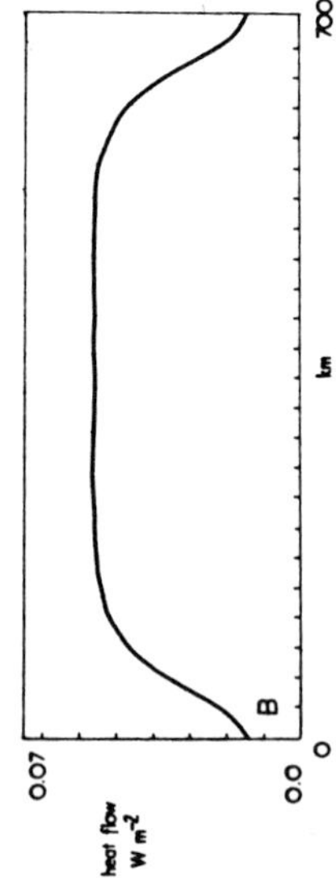
B
0.07
0.0
heat flow
W m⁻²
km

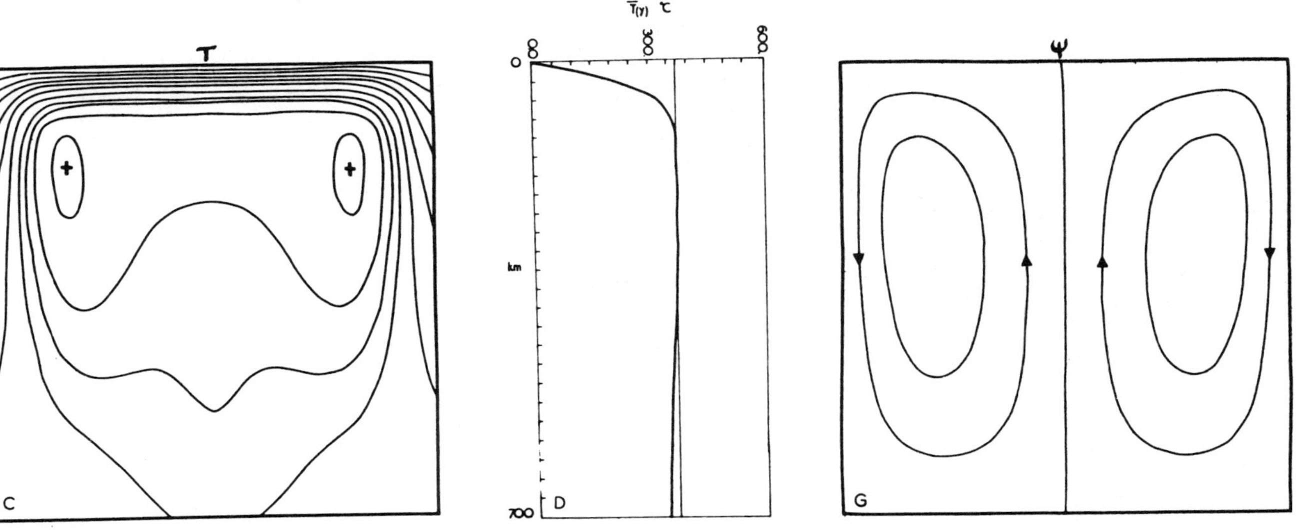

Fig. 3. As for Fig. 1, except that all the heat lost from the box is generated within it, and none enters from below. The mean surface heat flux is that given in Notation I. This numerical experiment was started with a single cell, but the upper boundary layer became unstable and two cells were formed.

box. In the two cases in which heat came from below a steady-state solution resulted which also had one roll in the box. However, when all the heat was generated within, the boundary layer became unstable and produced a sinking blob of cold fluid at the left-hand side of the box. This instability set up two rolls in the square box and this geometry remained. The times taken to reach an approximate steady state varied but were about 3,000 million years for the three cases. It is therefore unlikely that steady-state flow is ever achieved within the earth. Because of plate evolution boundary conditions are unlikely to remain unchanged for such long periods. Furthermore even the development of instabilities takes times of the order of 40 m.y., therefore rapid changes in plate-motion directions or rates are not likely to be due to changes in the convection pattern beneath them.

Various other features of the flow are of geophysical interest. The heat flow, gravity field, and surface deformation are all relatively constant over the rising region when the heat comes from within the fluid, and have narrow minima over the sinking sheets. When, however, the heat all enters from below all quantities vary over the upper surface. The uniformity of the heat flow over most of the box when the heating is from within may explain the uniformity of the heat flow through the deep ocean basins.

The sign of the gravity anomalies and their correlation with surface elevation is particularly important, since it suggests a simple test of these results which if successful will permit the flow within the earth to be mapped in considerable detail. As explained in the previous section most authors previously argued that the gravity anomaly should be negative over hot rising regions because their density was lower than that of the main body of the fluid. This argument neglected the surface deformation above the rising region, and simple

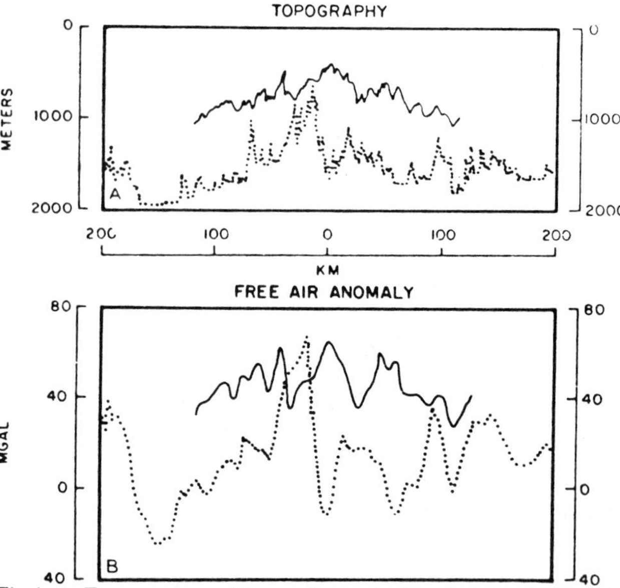

Fig.4. A. Topography and B. gravity in the North Atlantic (Talwani et al., 1971). The solid lines show the values observed across the Reykjanes Ridge, the dotted lines at 30°N. Notice that the sign of the correlation between gravity anomaly and depth variations is the same as in the theoretical models.

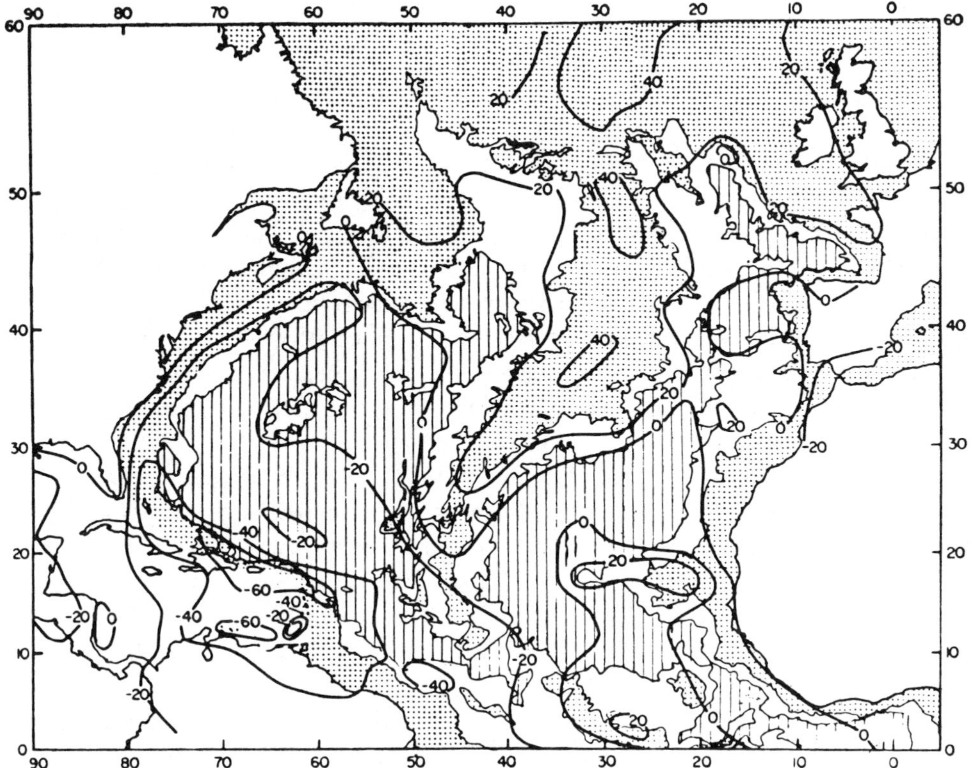

Fig.5. Regional variation of free-air gravity anomaly and depth in the North Atlantic (Talwani and Le Pichon, 1969).

calculations carried out by Pekeris (1935) and McKenzie (1968) showed that the influence of surface deformation was more important than the decreased density when convection of heat could be neglected. Though McKenzie (1968) supposed that the sign of the anomaly would change when the temperature was governed by convection of heat, the present calculations show that this is not the case. The gravity anomaly in all experiments reported by McKenzie et al. (1973) was positive over the rising region.

A geophysical test of this result is possible only if care is taken to exclude the effects which variation of plate temperature and crustal thickness have on surface elevation. The simplest part of the earth's surface is undoubtedly the ocean basins, and provided the influence of the time-dependent thermal structure of the plates is corrected for, a comparison can easily be made between the free-air gravity anomaly and the regional ocean depth. Fig.4, taken from Talwani et al. (1971), shows two profiles of gravity and bathymetry across two different parts of the Mid-Atlantic Ridge axis. Since the plate has only just been formed at the Ridge axis, any difference between the bathymetric profiles cannot be due to cooling of the crust. Furthermore there is a striking correlation between topography and gravity throughout the North Atlantic (Talwani and Le Pichon, 1969, Fig.5). The sign of this correlation and the magnitudes of the elevation and gravity anomalies agree well with those calculated in Fig.2. Since the theoretical values for the elevation were obtained for a free surface with a vacuum above they must be multiplied by 1.5 to take account of the density of sea water. A more systematic comparison between gravity and topography at sea has recently been made by Anderson et al. (1973), and they find similar agreement between the observations and the theory. It therefore appears possible to use these results to discover the flow pattern within the mantle. Since the sign of the regional gravity anomaly must agree with the regional bathymetric anomaly when the age of the sea floor has been taken into account, this argument provides two independent methods of mapping the flow.

Also plotted in Fig.1—3 are the horizontal velocities at the upper surface of the box. These should be compared with the half spreading rates of ridges, which vary from 8 to 120 mm/year. The calculated values lie in the middle of this range, but unlike the observed velocities vary continuously across the upper surface of the box. This difference is the consequence of using a constant viscosity in the calculations.

Another difference between geophysical observations and the model calculations is the vertical variation of the horizontally meaned temperature. Though the calculated mean temperature shows a thin surface boundary layer, the temperature change across it is only between 250°C and 400°C. The temperature of basalts erupted on ridge axes and from volcanoes is between 1100°C and 1200°C, and since the lavas come from beneath the boundary layer the calculated temperature difference is too small. This difference is also the result of using a constant viscosity. Torrance and Turcotte (1971) have carried out similar calculations with a temperature dependent viscosity and have shown that the thickness of and temperature difference across the boundary layer both increase.

These calculations show that the main features of simple models agree reasonably well with geophysical observations.

CONCLUSION

The model calculations described above were all carried out on simple models. They show that convection in the upper mantle can account for those geophysical observations which are believed to be maintained by mass motions within the mantle. It is our intention to investigate the stability of our solutions to three-dimensional disturbances, and also to increase the complexity of our models in three ways. A more realistic boundary condition to apply to the velocity field on the lower surface of the box is that the horizontal velocity, not the shear stress, should vanish. Since this difference will alter the structure of the boundary layer it may have an important influence on the flow. The term due to shear-stress heating is believed to have important local effects within the earth. It is easily included in the calculations and again may be important because of its influence on the boundary layers. Finally the variation of viscosity with temperature must be included if the cold boundary layers are to behave like plates.

ACKNOWLEDGEMENT

This research was supported by a grant from the Natural Environmental Research Council.

REFERENCES

Anderson, R.N., McKenzie, D.P. and Sclater, J.G., 1973. Gravity, bathymetry and convection within the earth. *Earth Planet. Sci. Lett.*, 18: 391–407.

Elsasser, W.M., 1969. Convection and stress propagation in the upper mantle. In: S.K. Runcorn (Editor) *The Applications of Modern Physics to the Earth and Planetary Interiors.* Interscience, New York, N.Y.

Howard, L.N., Malkus, W.V.R. and Whitehead, J.A., 1970. Self-convection of floating heat sources: A model for continental drift. *Geophys. Fluid Dyn.*, 1: 123–142.

Isacks, B. and Molnar, P., 1971. Distribution of stresses in the descending lithosphere from a global survey of focal-mechanism solutions of mantle earthquakes. *Rev. Geophys.*, 9: 103–174.

Langseth, M.G., Le Pichon, X. and Ewing, M., 1966. Crustal structure of the mid-ocean ridges, 5: Heat flow through the Atlantic Ocean floor and convection currents. *J. Geophys. Res.*, 7: 5321–5355.

McKenzie, D.P., 1968. The influence of the boundary conditions and rotation on convection in the earth's mantle. *Geophys. J.*, 15: 457–500.

McKenzie, D.P., 1969. Speculations on the causes and consequences of plate motions. *Geophys. J.*, 18: 1.

McKenzie, D., Roberts, J. and Weiss, N.O., 1973. Toward the numerical simulation of convection in the earth's mantle. *J. Fluid Mech.* (in press).

Moore, D.R., Peckover, R.S. and Weiss, N.O., in preparation. Difference methods for time-dependent two-dimensional convection.

Morgan, W.J., 1971. Convection plumes in the lower mantle. *Nature*, 230: 42.

Pekeris, C.L., 1935. Thermal convection in the interior of the earth. *Mon. Not. R. Astron. Soc. Geophys. Suppl.*, 3: 343.

Roberts, K.V. and Weiss, N.O., 1966. Convective difference schemes. *Math. Comput.*, 20: 272.

Runcorn, S.K., 1965. Changes in the convection pattern in the earth's mantle and continental drift: evidence for a cold origin of the earth. *Philos. Trans. R. Soc., Lond., Ser. A*, 258: 228.

Sclater, J.G. and Francheteau, J., 1970. The implications of terrestrial heat-flow observations on current tectonic and geochemical models of the crust and upper mantle of the earth. *Geophys. J.*, 20: 509.

Sclater, J.G., Anderson, R.N. and Bell, M.L., 1971. The elevation of ridges and the evolution of the Central Eastern Pacific. *J. Geophys. Res.*, 76: 7888–7915.

Talwani, M. and Le Pichon, X., 1969. Gravity field over the Atlantic Ocean. In: P.J. Hart (Editor), *The Earth's Crust and Upper Mantle.* Am. Geophys. Union, Washington, D.C.

Talwani, M., Windisch, C.C. and Langseth, M.G., 1971. Reykjanes ridge crest: a detailed geophysical study. *J. Geophys. Res.*, 76: 473–517.

Torrance, K.E. and Turcotte, D.L., 1971. Thermal convection with large viscosity variations. *J. Fluid Mech.*, 47: 113.

Turcotte, D.L. and Oxburgh, E.R., 1967. Finite amplitude convection cells and continental drift. *J. Fluid Mech.*, 28: 29–42.

7

Numerical Models of Convection in the Upper Mantle

M. H. HOUSTON, JR. AND J.-CL. DE BREMAECKER

Two-dimensional numerical models of steady state convection show that convection cells of aspect ratio as large as 8.6 are possible for variable viscosity convection in the upper mantle. Our models include the effects of variable viscosity, viscous dissipation, internal heating, heat flow through the bottom, and the adiabatic gradient. The large aspect ratio of the convection cells is primarily due to the large viscosity contrast between the lithosphere and the asthenosphere. It appears possible for multiple convection cells to occur in a low-viscosity zone while the surface velocities give the appearance of a single cell. The details of the viscosity law relevant to mantle materials and conditions are presently uncertain but are of crucial importance; temperature, viscosity, and flow patterns are inextricably entwined. Convection decreases the overall temperature gradient; consequently, generally accepted temperatures for most of the mantle are too high. The controversies over plate-mantle decoupling and passive versus active plates are probably due to oversimplifications that disregard hydrodynamic concepts.

The concept of 'plate tectonics,' which has evolved from earlier theories of continental drift and sea floor spreading, has been remarkably successful in explaining the global distribution of seismicity [*Isacks et al.*, 1968] and the linear magnetic anomalies symmetric with respect to the mid-ocean ridges [*Vine and Matthews*, 1963].

Although the relative motions of the major plates are fairly well known, the driving mechanism for plate motion is not clearly understood. A consideration of the energy necessary to maintain plate motion, as derived from the average worldwide seismic energy [*Gutenberg*, 1956], suggests that the most plausible energy source for plate motion is some type of thermal convection within the earth's mantle [*McKenzie*, 1972]. The details of the convection are little known and probably cannot be derived directly from surface features, since the thick lithospheric plate effectively integrates the thermal and stress variations. We are faced with a geophysical inverse problem, and we can expect that there will be no unique dynamic model consistent with all the geologic and geophysical constraints. Nevertheless, by constructing a plausible family of models that incorporate what appear to be the most important physical mechanisms, at least we can eliminate those models whose geologic and geophysical consequences are inconsistent with observation. In addition, these models can point out critical areas for observation and experiment.

It would appear that for the earth, internal heating is almost certainly very important, that variable viscosity greatly influences the convection pattern, and that viscous dissipation may not always be negligible. For these reasons we have included these factors in the present model. Finally, the convection solutions are essential to obtain the temperature distribution within the upper mantle: the energy equation (heat transfer equation) and the momentum equation (Navier-Stokes equation) form a system of strongly coupled partial differential equations and cannot be considered separately.

It would be desirable to investigate the problem in three dimensions. This method is not practical on existing computers. On the other hand, there is no definitive experimental evidence to help decide whether the flow is essentially two- or three-dimensional: physical experiments cannot be run at Rayleigh numbers of 10^6, Prandtl numbers of the order of 10^{23}, and negligibly small Reynolds numbers [*Knopoff*, 1967].

The nature of the plate motions themselves suggests that two-dimensional models may incorporate the dominant features of upper mantle flow. Because of the magnitude of the calculations we are forced to restrict ourselves to two-dimensional computations. This fact should be borne in mind when the geophysical significance of our results is assessed.

EQUATIONS TO BE SOLVED

In two dimensions the momentum equation of an incompressible fluid of variable viscosity may be written as a generalized biharmonic equation [*Andrews*, 1972] in terms of the stream function:

$$\frac{\partial^2}{\partial x \, \partial y}\left(4\eta \frac{\partial^2 S}{\partial x \, \partial y}\right) + \left(\frac{\partial^2}{\partial y^2} - \frac{\partial^2}{\partial x^2}\right)$$

$$\cdot \left[\eta\left(\frac{\partial^2}{\partial y^2} - \frac{\partial^2}{\partial x^2}\right)S\right] - \frac{\partial}{\partial x}(\rho_0 g\alpha T) = 0 \qquad (1)$$

The energy equation is

$$\frac{\partial T}{\partial t} = -u\frac{\partial T}{\partial x} - v\frac{\partial T}{\partial y}$$

$$+ \frac{1}{\rho_0 C}\left[\frac{\partial}{\partial x}\left(k\frac{\partial T}{\partial x}\right) + \frac{\partial}{\partial y}\left(k\frac{\partial T}{\partial y}\right) + 4\eta\left(\frac{\partial^2 S}{\partial x \, \partial y}\right)^2\right.$$

$$\left. + \eta\left(\frac{\partial^2 S}{\partial y^2} - \frac{\partial^2 S}{\partial x^2}\right)^2 + gv\rho_0\alpha T + H\right] \qquad (2)$$

where

α thermal expansivity;
η viscosity;
ρ_0 density at 0°C;
C specific heat;
g gravitational acceleration;
H radioactive heat generated per unit volume;
k thermal conductivity;
S stream function;
t time;
T temperature;
u horizontal velocity, $\partial S/\partial y$;
v vertical velocity, $-\partial S/\partial x$;
x horizontal coordinate, pointing east;
y vertical coordinate, pointing down.

110

We wish to solve these equations in a rectangular enclosure whose top is horizontal and with the following boundary conditions.

Thermal boundary conditions. The sides of the enclosure have zero horizontal temperature derivative. The top is at 0°C. The heat flux through the bottom is specified.

Velocity boundary conditions. The horizontal derivative of the vertical velocity is zero along the sides; i.e., they are free-slip (zero shear stress) surfaces. The top surface is also free slip with zero vertical velocity. The bottom is a no-slip surface, and both vertical and horizontal velocities vanish. All velocities normal to the boundary vanish at that boundary. The conditions on the stream function immediately follow from the above. Free-slip and no-slip boundaries are often referred to as free and rigid boundaries, respectively. In common usage, though, both the shear stress and the normal stress vanish at a 'free' boundary, whereas the wording 'rigid' boundary conveys only that the velocity normal to the boundary vanishes there. We thus prefer the terms free slip and no slip.

Turcotte et al. [1973] show how equations equivalent to (1) and (2) may be derived from the exact equations of convection at the limit $\alpha\Delta T = 0$. It may be noted that the influence of the compressibility χ results only in a second-order correction because the pressure P in the earth is almost purely hydrostatic. This means that the term in $\chi\,\partial P/\partial x$ that has been neglected in (1) is negligibly small. This problem was first discussed by *Jeffreys* [1930] and later generalized by *Spiegel and Veronis* [1960] and by *Knopoff* [1964].

NUMERICAL SOLUTION OF THE EQUATIONS

Details of the method of solution are given in another paper [M. H. Houston and J. Cl. De Bremaecker, manuscript in preparation, 1974; see also *Houston*, 1973]. We use the offset grid taken from the MAC (marker and cell) method of *Harlow and Welch* [1965] [see also *Welch et al.*, 1965] and recently used by *Andrews* [1972]. In order to obtain a better resolution near the surface where the temperature varies most rapidly, we use a special mesh with the variable vertical spacing first suggested by Samarskii [*Saul'yev*, 1964, p. 149] and rediscovered by *Sundqvist and Veronis* [1970].

In order to obtain a steady state solution we start with an assumed temperature distribution and solve the stream function equation by an extension of the alternating direction implicit (ADI) method of *Conte and Dames* [1958, 1960]; the energy equation is then solved by the ADI method [*Douglas and Rachford*, 1956] using 'upwind' differencing [*Forsythe and Wasow*, 1960, p. 397] and thus yielding new temperatures to find a new stream function. This process is continued until the temperatures and the stream function remain essentially stationary. The convergence criterion for the stream function and the temperature equations is that the normalized root mean square change between successive iterations be less than 2×10^{-6} and 1×10^{-5}, respectively. The consistency of the solutions has been established by using both a finer and a coarser grid.

PARAMETERS OF THE MODELS

We will discuss successively models in which the viscosity and diffusivity are constant, are a function of depth only, or are a function of temperature and depth. The values of the parameters used are given in Table 1. They are taken from the literature.

Depth. We have chosen the depth of the cell to be 700 km. This assumption has often been made [e.g., *Tozer*, 1967; *Torrance and Turcotte*, 1971b]. It rests partly on the absence of earthquake foci at greater depth and partly on considerations of viscosity and geochemistry. If the aspect ratio of the cell remains constant, an increase in the cell depth will have little effect on convection patterns in a fluid with either a constant viscosity or the depth-dependent viscosity that we have used. For a fluid in which the viscosity varies with both depth and temperature an increase in the allowable convection depth may affect the convection patterns. However, such changes are expected to be minor because the high-viscosity lithosphere dominates the general circulation.

Conductivity. For the relationship of conductivity to temperature and pressure we use the experimental fit for olivine given by *Schatz and Simmons* [1972, equations 9 and 10]. If the conductivity is constant, K_0 (Table 1) is used; if it varies only with depth, that variation is used together with a classical temperature profile. In any case, the influence of variable conductivity appears slight.

Viscosity. If the viscosity is constant, ν_0 (Table 1) is used; if it varies with depth only, we use the profile given in Figure 1, which is based on the nonlinear profile of *Weertman* [1970].

Viscosity as a function of pressure and temperature is modeled as a Newtonian viscosity due to diffusion creep [*Heering*, 1950; *Nabarro*, 1967]:

$$\eta = \frac{kTR^2}{10\,D_0 V_a}\exp\left(\frac{E^* + pV^*}{kT}\right) \qquad (3)$$

where T is in degrees Kelvin and p is in dynes per square centimeter. Consistent with deduced values of mantle viscosity and with experimentally determined crystalline parameters, we

TABLE 1. Parameters of the Models

	cgs or Other Units	SI Units
Depth	$7 \cdot 10^7 = 700$ km	$7 \cdot 10^5$
Width*	$1.96 \cdot 10^8 \cong 2000$ km	$1.96 \cdot 10^6$
Density ρ_0	3.5 g/cm^3	$3.5 \cdot 10^3$
Specific heat C	$1.3 \cdot 10^7$ ergs/g °C	$1.3 \cdot 10^3$
Thermal expansivity α	$3.7 \cdot 10^{-5}/$°C	$3.7 \cdot 10^{-5}$
Conductivity K_0	$1.1 \cdot 10^{-2}$ cal/cm °C s	4.6
Kinematic viscosity ν_0	$5 \cdot 10^{21}$ cm^2/s	$5 \cdot 10^{17}$
Radiogenic heating*† H	$1.9 \cdot 10^{-7}$ ergs/g s	$6.7 \cdot 10^{-8}$ W/m^3
Heat flux through bottom*	0.3 HFU	$1.25 \cdot 10^{-2}$

*Unless stated otherwise in Table 2.
†Distributed over the depth of the cell, radiogenic heating corresponds to a surface heat flow of 1.1 HFU = $4.6 \cdot 10^{-2}$ SI units. Therefore it is called 'equivalent radioactive heating.'

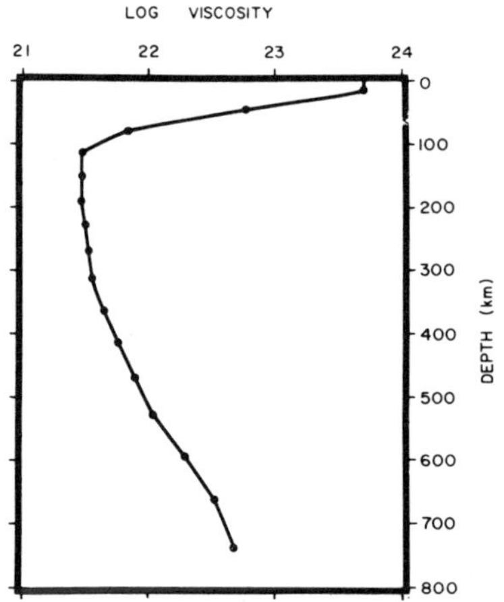

Fig. 1. Assumed kinematic viscosity profile as a function of depth only, in units of square centimeters per second.

choose $V^* = V_a = 15 \times 10^{-24}$ cm³, $D_0 = 20$ cm²/s, $R = 0.05$ cm, and $k = 1.38 \times 10^{-16}$ ergs/°K is Boltzmann's constant [*Torrance and Turcotte*, 1971*a, b*].

We choose two values for the activation energy E^*, namely, $E^* = 7.21 \times 10^{-12}$ ergs (4.5 eV) and $E^* = 5.52 \times 10^{-12}$ ergs (3.45 eV). These correspond to viscosities derived from creep experiments by using, respectively, 'dry' and 'wet' mantlelike materials [*Carter and Avé Lallemant*, 1970].

The presence of water decreases the effective viscosity (a water content of 1% is petrologically very 'wet'). Decreasing the activation energy has the effect of decreasing the viscosity for the same temperature. The resulting expression reduces to

$$\eta = 115T \exp\left(\frac{E^*/k + 0.3804d}{T}\right) \qquad (4)$$

where d is in kilometers and T is in degrees Kelvin.

RESULTS

Table 2 shows the differences among the models corresponding to the various numerical experiments illustrated in the figures. In these figures the heat flow is expressed in heat flow units (1 HFU = μcal cm⁻² s⁻¹), and the temperatures in degrees Celsius. The stream function is normalized and expressed in units of $\eta_0 = k_0 \rho^{-1} c^{-1} = 1$ mm²/s; thus a change in the stream function of one unit per kilometer corresponds to a velocity of 10^{-9} m/s = 3.15 cm/yr.

Constant Viscosity

Figure 2 shows the results obtained for a model with con-stant viscosity and the parameters given in Table 1 (the width is 2,000 km). The dots show the points at which the computations were performed. The Rayleigh number, calculated from internal heating alone, exceeds 10⁶.

The stream function and the isotherms are symmetric with respect to the middle of the enclosure. The stream function maximums (Figure 2*c*) are displaced toward the descending limb of the cells. Correspondingly, the vertical velocity in the descending limb is greater than that in the ascending limb [*McKenzie et al.*, 1973, 1974]. The maximum horizontal velocity (1.6 cm/yr) occurs at the surface, approximately above the stream function maximums.

The isotherms (Figure 2*b*) show that a temperature profile versus depth taken through the circulation maximums exhibits a surface boundary layer about 100 km thick, a nearly isothermal region with minor temperature inversions from 100 to 500 km, and a weak second boundary layer near the bottom, caused mainly by the bottom heat flux.

The surface heat flow (Figure 2*a*) decreases from 2.07 HFU above the rising limb of the cell to 0.87 HFU at the sides. The average heat flow is within 4.5% of the theoretical heat flow as computed from the addition of the equivalent radioactive heating (Table 1) plus the bottom heat flow.

This discrepancy is attributable to an imperfect estimate of the surface temperature gradient due to the small number of temperature nodes within the surface thermal boundary layer [*McKenzie et al.*, 1974].

The center line symmetry of the two-cell circulation depicted in Figure 2 is further proof of the validity of the solution. The results are similar to those of *McKenzie et al.* [1973, 1974], but these simple constant viscosity solutions probably bear little resemblance to convection patterns within the earth. For example, the temperatures are too low to be a realistic representation of the earth's upper mantle. Increasing the viscosity would increase the overall temperature but would also decrease the velocities.

Viscosity Varying With Depth Only

Distributed heating and bottom heat flux. The results obtained for the next model are shown in Figure 3; all the parameters are identical to those of Table 1 except that the viscosity varies with depth only, as is shown in Figure 1.

The most striking feature (Figure 3*c*) is that this system turns as one cell with the stream function maximum within the low-viscosity zone. The maximum horizontal velocity, 1.6 cm/yr, occurs at the surface, above the stream function maximum. Just as in the constant velocity case, the velocity in the descending limb is greater than that in the ascending limb, and the stream function maximum is slightly displaced toward the descending limb.

The isotherms (Figures 3*b* and 11) show that the upper ther-

TABLE 2. Differences Among the Models

Figure	Viscosity	Equivalent Radioactive Heat, HFU	Bottom Heat Flux, HFU	Width, km	Aspect Ratio
2	constant	1.1	0.3	2000	2.8
3, 9	varies with depth	1.1	0.3	2000	2.8
4, 10	varies with depth	0	1.4	2000	2.8
5	varies with depth	1.4	0	2000	2.8
6	varies with depth	1.1	0.3	6000	8.6
7, 11	'Herring-Nabarro dry'	1.1	0.3	2000	2.8
8, 12	'Herring-Nabarro wet'	1.1	0.3	2000	2.8

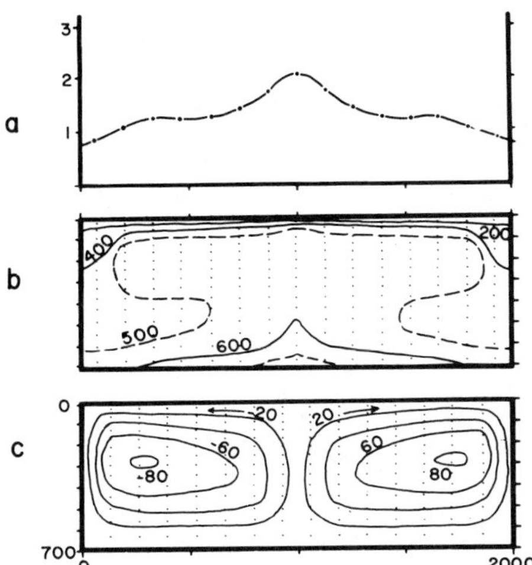

Fig. 2. Constant viscosity case with internal heating and bottom heat flux; parameters are as given in Tables 1 and 2. (*a*) Heat flow (heat flow units). (*b*) Temperatures (degrees Celsius). (*c*) Stream function (square millimeters per second), arrows indicating direction of flow.

mal boundary layer increases in thickness from about 50 km above the hotter, rising limb of the convection cell to several hundred kilometers at the colder, descending limb. The isothermal region, which extends from the surface boundary layer to nearly 500 km, exhibits a tongue of advecting hotter fluid. A temperature profile through the center of the cell (Figure 11) shows a slight temperature inversion supported by convection. The bottom boundary layer is thinnest near the descending limb and thickest near the rising limb. The maximum temperature within the enclosure has increased to over 1200°C at 700 km, but the overall temperature still seems somewhat low from a geologic point of view [*Green*, 1973*a*, *b*].

A plot of the surface heat flux versus distance (Figure 3*a*) shows an essentially linear trend in which the heat flux falls from a maximum value of 2.3 HFU above the rising portion of the cell to a minimum value of 0.6 HFU above the descending limb.

It is worth noting that the solution shown in Figure 3 was obtained from an initial temperature distribution purposely chosen to give rise to two equal convection cells turning in opposite directions. As the solution progressed toward steady state, one cell gradually increased, and eventually the pattern of Figure 3 resulted.

The high aspect ratio of the present case confirms in a general way the prediction of *McFadden and Smylie* [1968] [see also *McFadden*, 1969] based on marginal stability analysis. These authors investigated Rayleigh-Bénard convection in a fluid in which the viscosity has a minimum at middepth. They found that for nonslip boundaries the aspect ratio remains close to unity but for free-slip boundaries the aspect ratio increases rapidly as the viscosity contrast increases. They noted that this might imply horizontal wavelengths well in excess of five for the mantle.

On the other hand, the marginal stability analysis of *Peltier* [1972] deals with convection in a medium consisting of an elastic plate at the top, which constitutes a nonslip boundary for the fluid below. The bottom boundary is also nonslip. The viscosity is minimum under the plate. In this case, the aspect ratio does not exceed 1.5. Since this case is physically close to

the nonslip case of *McFadden and Smylie* [1968], the similarity of the conclusions is encouraging.

We further note that if a convection cell was entirely under the ocean, its upper boundary would be free slip; if it was entirely under the continents, approximately no-slip conditions might prevail, the result being a different aspect ratio for the two cases. If the viscosity profile passes through a pronounced minimum in the fluid, the physical reason for the difference between the free-slip and the nonslip cases can be explained qualitatively by the following argument: in the nonslip case the high viscosity near the top effectively prevents motion near the boundary; the motion will thus be more or less concentrated in the low-viscosity zone, in which roughly equidimensional cells will form. In the free-slip case, on the other hand, the highly viscous top 'plate' participates in the motion. Since it would require more energy for it to bend at several places, it moves as a unit, and a general single circulation results.

Bottom heat flux only and distributed heating only. Figures 4 and 5 present models in which the driving energy source is, respectively, bottom heat flux alone or internal heating alone. These two cases may be compared with the case of Figure 3, in which both types of energy sources are present. In all cases the total heating is equivalent to a surficial heat flux of 1.4 HFU.

A comparison of the surficial features of the two contrasting models reveals a surprising similarity. The surface heat flux and the thermal boundary layer are nearly indistinguishable. It is only below the thermal (and viscous) boundary layer that the models are significantly different.

The model driven by bottom heat flux (Figure 4) shows a nearly isothermal region between the surface boundary layer and the well-developed bottom boundary layer. In this case the temperature distribution is similar to that suggested by *Tozer* [1967] and *Turcotte and Oxburgh* [1967]. The model driven by internal heat sources (Figure 5) also shows a nearly isothermal region below the surface boundary layer but no bottom boundary layer. Temperatures within the isothermal region are 100°–200°C hotter for the internally heated model.

The circulation patterns are generally similar, but important

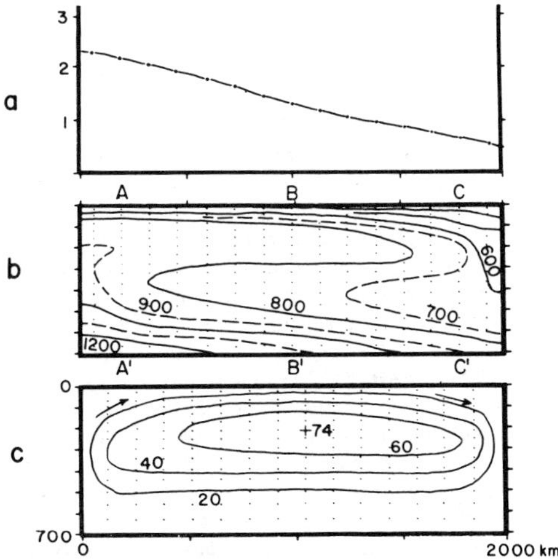

Fig. 3. Depth-dependent viscosity (as in Figure 1) with internal heating and bottom heat flux; other parameters are as given in Tables 1 and 2. (*a*) Heat flow (heat flow units). (*b*) Temperatures (degrees Celsius). (*c*) Stream function (square millimeters per second).

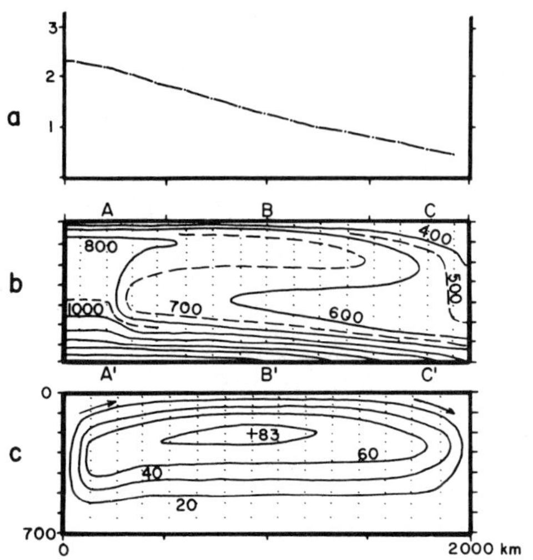

Fig. 4. Depth-dependent viscosity (as in Figure 1) with bottom heat flux alone; other parameters are as given in Tables 1 and 2. (*a*) Heat flow (heat flow units). (*b*) Temperatures (degrees Celsius). (*c*) Stream function (square millimeters per second).

differences that reflect the energy source exist between the two models. The model driven by internal heat sources (Figure 5) shows that the largest horizontal temperature gradients and hence the highest vertical velocities exist within the downgoing limb of the convection cell. In contrast, the bottom-heated model shows nearly equal vertical velocities in the two limbs, although those in the rising limb are slightly greater. This increase in the velocities in the descending limb for an internally heated fluid is also present in our constant viscosity model (Figure 2) and has previously been noted by *McKenzie et al.* [1973, 1974]. It is essentially due to the fact that heating is here a diffuse, i.e., volumetric, process, whereas cooling is a localized, i.e., surface, one (D. L. Turcotte, personal communication, 1973).

In both cases the maximum horizontal velocities occur at the surface, approximately above the stream function maximum; they are 1.9 cm/yr in Figure 4 and 1.5 cm/yr in Figure 5.

The model shown in Figure 3 is driven by both boundary and volumetric heat sources and is intermediate in characteristics between the two extreme cases of Figures 4 and 5. We infer from a comparison of these three models that the nonlinearity of the convection process does not prevent a limited interpolation. This conclusion enhances the usefulness of selected models and encourages the cautious extrapolation of simple models to more general circumstances.

Cell with large aspect ratio. It has often been suggested [*McKenzie*, 1969; *Richter*, 1973] that in a convective system, cells with a high aspect ratio should have a smaller surface velocity than those with a low one. In contrast, the average temperature should increase as the aspect ratio increases, a situation that might increase the speed of convection. The following model seeks to investigate these questions.

In Figure 6 the width of the cell is 6,000 km, whereas the depth remains 700 km. The viscosity is that shown in Figure 1. All the other parameters are given in Table 1. The computation grid is indicated by dashes on the inside of the enclosure. Again (Figure 6c) the system turns as a single cell despite the

fact that the initial temperature distribution was on purpose such that two cells initially formed.

The circulation and the surface velocity are 45% higher in the present case than in Figure 3. The maximum horizontal velocity, 2.35 cm/yr, occurs at the surface, approximately above the stream function maximum.

The temperatures (Figure 6*b*) in the ascending limb are approximately 400°C higher than those in the comparable previous case (Figure 3*b*; 2000 × 700 km); those in the descending limb are essentially unchanged, and those in the middle are correspondingly 200°C higher. The 1000°C isotherm is now within 45 km of the surface at the ascending limb and slopes only very gently.

The surface heat flow (Figure 6*a*) decreases roughly exponentially from 2.6 HFU at the rising limb to 0.5 HFU at the descending limb.

Viscosity Varying With Temperature and Depth

In the remaining experiments the viscosity and the conductivity are functions of both temperature and depth (Herring-Nabarro viscosity). The key parameter is the activation energy E^*, which is not well known for the materials or the conditions within the mantle. We limited the viscosity variation to four orders of magnitude by setting limits within the program.

In all previous models a clear steady state solution was obtained. This situation did not occur in the present case. Instead, the solutions appeared to oscillate between states differing only slightly in their temperatures and stream functions.

Although these oscillations may be caused by phase distortion [*Houston and De Bremaecker*, 1974], we prefer to interpret these iterative transients as 'timelike' transients in a nonlinear system that has no true steady state solution. Justification of this interpretation is provided by the work of *Garabedian* [1956] and others who have noted a close similarity between transients of sequential iterations of steady state solutions and actual time-dependent solutions. Furthermore, previous workers using a variety of numerical techniques to investigate

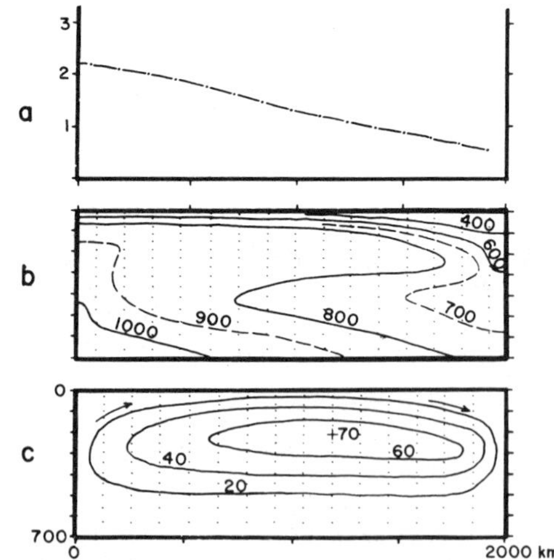

Fig. 5. Depth-dependent viscosity (as in Figure 1) with internal heating alone; other parameters are as given in Tables 1 and 2. (*a*) Heat flow (heat flow units). (*b*) Temperatures (degrees Celsius). (*c*) Stream function (square millimeters per second).

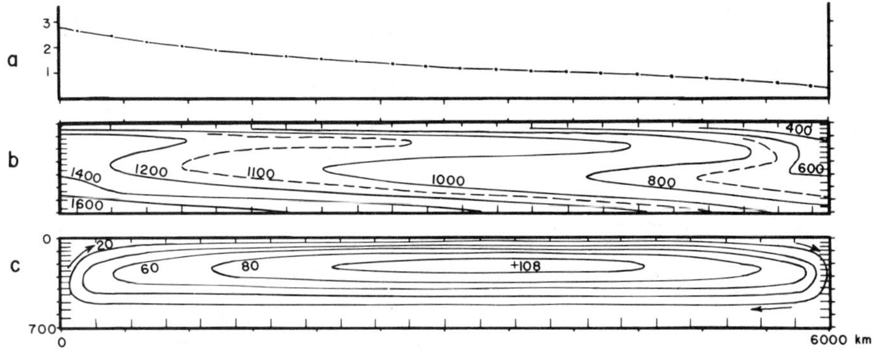

Fig. 6. Depth-dependent viscosity (as in Figure 1) with internal heating and bottom heat flux. Width of the cell is 6000 km; other parameters are as given in Tables 1 and 2. (*a*) Heat flow (heat flow units). (*b*) Temperatures (degrees Celsius). (*c*) Stream function (square millimeters per second).

two-dimensional convection at Rayleigh numbers greater than 10^6 have reported observing such oscillations in 'steady state' solutions [*Fromm*, 1971; *Torrance and Turcotte*, 1971*a*; *Andrews*, 1973; *McKenzie et al.*, 1974].

Thus although a numerical cause cannot be ruled out absolutely for the oscillations shown in the following cases, we believe that they are real. They can be qualitatively explained as follows. All things being equal, if cooling is a surface effect and heating mostly volumetric, the velocity of the descending limb is higher than that of the ascending limb (see above). Correspondingly, for a clockwise circulation the stream function maximum is displaced to the right of the middle of the enclosure. On the other hand, if the viscosity decreases as the temperature increases, the velocity of the ascending limb becomes greater than that of the descending limb, and the stream function maximum migrates to the left. The latter case is conspicuous in the results of *Torrance and Turcotte* [1971*b*]. The oscillations, then, correspond to the tendency of the system to satisfy these two conflicting requirements.

Since the number of steady state iterations may be viewed as a distorted time scale, we first present the solution that exists for the largest percentage (60–80%) of steady state iterations. We then present the corresponding solution for the 'transient' state.

'Dry' mantle. Figure 7 shows the results of a model in which the activation energy (4.5 eV) corresponds to a 'dry' mantle. It is clear that the introduction of the 'Herring-Nabarro' viscosity function considerably alters the convection pattern. The major portion of the circulation is limited to the left half of the cell, and the maximum circulation has increased. The isotherm structure has also changed (Figure 7*b*). These results are similar to those of *Torrance and Turcotte* [1971*b*], but there exist important differences.

The distinct surface thermal boundary layer is now limited to the left 1200 km of the computation cell. The 1400°C isotherm, which forms the approximate lower limit of the surface boundary layer in this region, rises to within 80 km of the surface (see also Figure 13). The surface boundary layer gradually thickens away from the rising limb until near 1200 km the boundary layer structure gradually disappears. This region near 1200 km is also the limit of vigorous circulation (Figure 7*c*). Thus a distinct boundary layer complements vigorous advective heat transport. Near the extreme right-hand side of the cell the temperature increases nearly uniformly with depth (Figure 13).

The two small arrows on Figure 7*c* show the maximum horizontal velocity (1.5 cm/yr at 150 km depth) and the maximum surface velocity (0.5 cm/yr). The upward motion is confined to a width of 450 km, compared with a total cell width of 2000 km.

The kinematic viscosity (Figure 7*d*) was constrained between a minimum of $5 \cdot 10^{20}$ (-1 contour) and $5 \cdot 10^{24}$ cm²/s ($+3$ contour). The viscosity distribution is determined by the model itself as being consistent with the temperatures

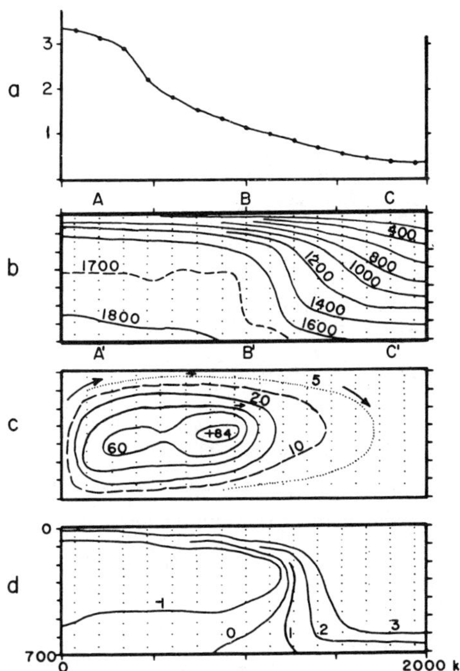

Fig. 7. 'Steady state' solution for a model with Herring-Nabarro viscosity corresponding to a 'dry' mantle with internal heating and bottom heat flux; other parameters are as given in Tables 1 and 2. (*a*) Heat flow (heat flow units). (*b*) Temperatures (degrees Celsius). (*c*) Stream function (square millimeters per second). (*d*) Scaled kinematic viscosity. Contours indicate integer values of $\log_{10} (\nu/\nu_0)$ ($\nu_0 = 5 \times 10^{21}$ cm²/s).

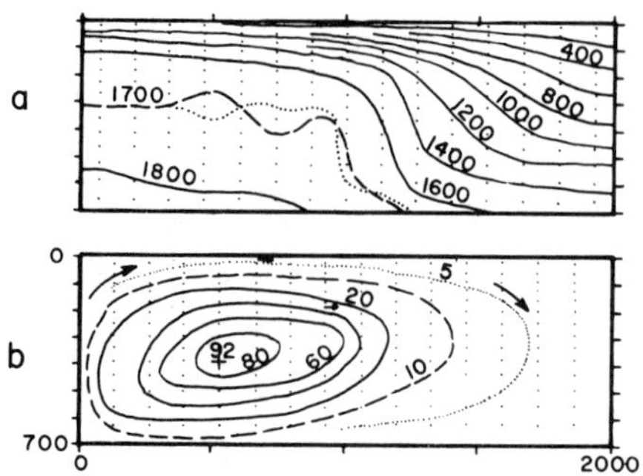

Fig. 8. 'Transient' conditions for a model with Herring-Nabarro viscosity corresponding to a 'dry' mantle with internal heating and bottom heat flux; other parameters are as given in Tables 1 and 2. (*a*) Temperatures (degrees Celsius); the dotted 1700°C isotherm is reproduced from Figure 7. (*b*) Stream function (square millimeters per second).

and flow. In the left 1200 km of the enclosure a low-viscosity region is present. centered at a depth of roughly 300 km. The upper boundary of the low-viscosity layer is abrupt and coincides with the lower boundary of the surface thermal boundary layer. Below the viscosity minimum the viscosity increases very slightly with depth. The region of minimum viscosity tapers from a maximum thickness of ~500 km on the extreme left of the cell to zero near 1200 km from the rising limb. The right third of the enclosure is a large region of high viscosity.

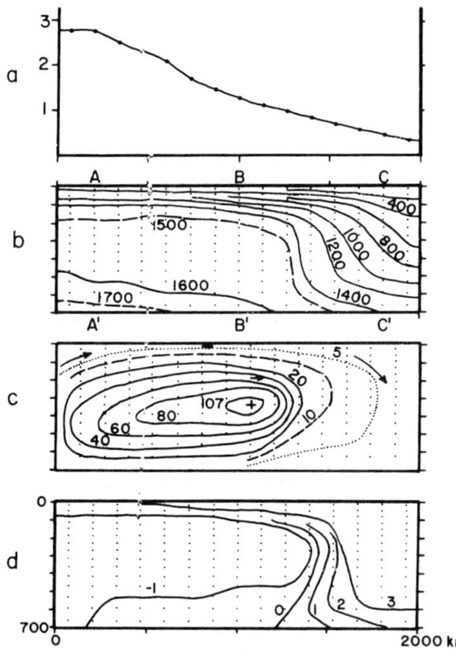

Fig. 9. 'Steady state' solution for a model with Herring-Nabarro viscosity corresponding to a 'wet' mantle with internal heating and bottom heat flux; other parameters are as given in Tables 1 and 2. (*a*) Heat flow (heat flow units). (*b*) Temperatures (degrees Celsius). (*c*) Stream function (square millimeters per second). (*d*) Scaled kinematic viscosity. Contours indicate integer values of $\log_{10}(\nu/\nu_0)$ ($\nu_0 = 5 \times 10^{21}$ cm²/s).

The heat flux (Figure 7*a*) varies from a maximum of 3.3 HFU at the rising limb to a minimum of 0.34 HFU at the descending limb.

Figure 8 illustrates the 'transient' solution for convection in a dry mantle. As is apparent, the isotherms (Figures 7*b* and 8*a*) for these two figures are practically identical except for the 1700°C isotherm. (The surface heat flow curves are indistinguishable.) Figures 7*c* and 8*b* show that the stream function varies only within the 20-mm²/s contour: the first one shows a single maximum 500 km from the rising limb, the second one a double maximum, one near 800 km and one near 400 km. In the first case the rising limb is fairly broad (~500 km); it is only half as broad in the second case.

Finally, it should be noted that the stream function maximum occurs at a depth of 350–400 km in the present case, whereas it is a depth of 200 km in Figures 3, 4, and 5. This difference is related to the extension of the low viscosity to a greater depth in the present case.

'Wet' mantle. The last case to be discussed is shown in Figures 9 and 10. It differs from the previous one by having the activation energy decreased by about 25%, i.e., $E^* = 3.45$ eV.

Because of the great similarity between the results in these two cases, Figure 9 will not be discussed in detail. It is only necessary to point out that the circulation (Figures 9*c* and 10*b*) has increased by about 20% and that the maximum has moved 500 km to the right. The maximum horizontal velocity (arrows) within the fluid and at the surface are, respectively, 2.0 and 0.5 cm/yr. The isotherms (Figures 9*b* and 10*a*) change very little as the system oscillates. The temperature in most of the enclosure is about 150°C lower than that in the previous case (see also Figure 14). The heat flow maximum is also lower (2.75 HFU), and the low-viscosity zone extends farther down and toward the descending limb.

A previous, slightly less exact, computation with a 'wet' mantle and a kinematic viscosity varying from $5 \cdot 10^{19}$ to $5 \cdot 10^{23}$ cm²/s rather than $5 \cdot 10^{20}$ to $5 \cdot 10^{24}$ cm²/s showed very similar results, except that most temperatures decreased by ~200°C, as was expected from the lower overall viscosities.

It may be remarked that the difference between the two Herring-Nabarro models supports *Tozer*'s [1970] conclusion that the temperatures and viscosities in the mantle are 'self-regulating.' What he implies is simply that a steady state solution, if it exists, will reflect the variation of viscosity with temperature and pressure. As our results confirm, there is an

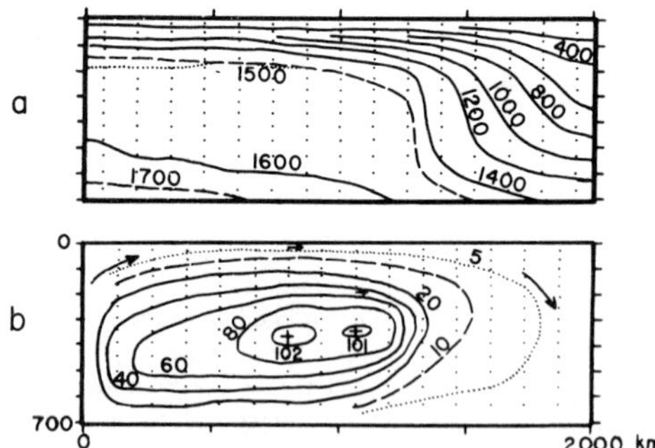

Fig. 10. 'Transient' conditions for a model with Herring-Nabarro viscosity corresponding to a 'wet' mantle with internal heating and bottom heat flux; other parameters are as given in Tables 1 and 2. (*a*) Temperatures (degrees Celsius); the dotted 1500°C isotherm is reproduced from Figure 9. (*b*) Stream function (square millimeters per second).

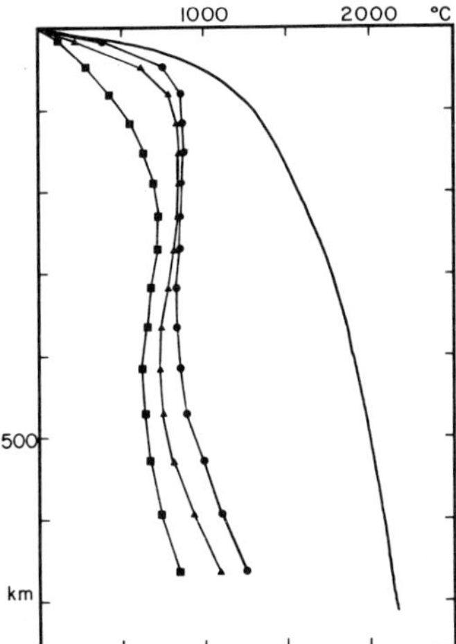

Fig. 11. Temperature versus depth corresponding to Figure 3, i.e., depth-dependent viscosity, internal heating, and bottom heat flux. Circles represent ascending limb (AA' on Figure 3); triangles, center of the cell (BB'); and squares, descending limb (CC'). Solid line shows geotherm after *Clark and Ringwood* [1964].

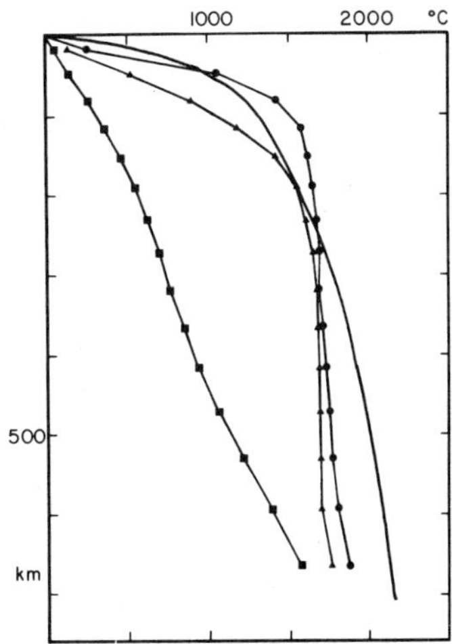

Fig. 13. Temperature versus depth corresponding to Figure 7, i.e., 'Herring-Nabarro dry' viscosity, internal heating, and bottom heat flux. Circles represent ascending limb (AA' on Figure 9); triangles, center of cell (BB'); and squares, descending limb (CC'). Solid line shows geotherm after *Clark and Ringwood* [1964].

intimate relationship among temperature, viscosity, and flow within a convecting mantle.

RESTRICTIONS AND IMPLICATIONS OF OUR RESULTS

Temperatures in the earth. Temperature versus depth in four representative cases is shown in Figures 11 to 14. These strongly suggest that the commonly accepted geotherm [*Clark and Ringwood*, 1964] is too high, at least below approximately 200 km. Moreover, large lateral temperature differences occur.

Within the upper 200 km the lateral temperature differences in all models are associated with the downstream growth of a thermal viscous boundary layer. Within the nearly isothermal layer these differences are as much as 300°C at 300 km for the depth-dependent viscosity (Figures 11 and 12); for the Herring-Nabarro viscosity these differences are much smaller within the region of vigorous circulation (Figures 13 and 14) but are as much as 700°C across the whole width of the cell. Such temperature differences should be reflected in seismic and gravity surface measurements. *McKenzie et al.* [1974] have shown that the exact gravity signature over a convection cell depends critically on the surface distortion. The surface displacement, in turn, is sensitive to the boundary conditions and to the near-surface distortion of viscosity.

Relief, heat flow, and surface velocity. Various methods

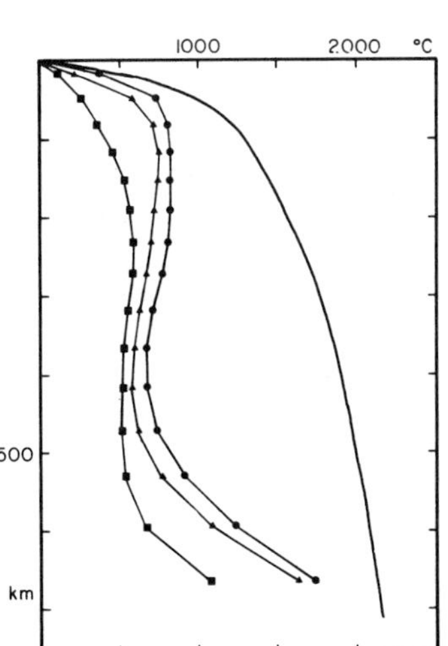

Fig. 12. Temperature versus depth corresponding to Figure 4, i.e., depth-dependent viscosity and bottom heat flux alone. Circles represent ascending limb (AA' on Figure 3); triangles, center of the cell (BB'); and squares, descending limb (CC'). Solid line shows geotherm after *Clark and Ringwood* [1964].

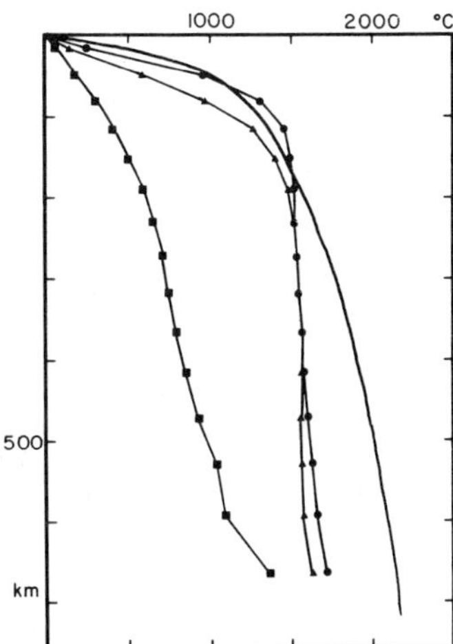

Fig. 14. Temperature versus depth corresponding to Figure 9, i.e., 'Herring-Nabarro wet' viscosity, internal heating, and bottom heat flux. Circles represent ascending limb (AA' on Figure 7); triangles, center of cell (BB'); and squares, descending limb (CC'). Solid line shows geotherm after *Clark and Ringwood* [1964].

have been proposed to compute the relief of the free upper boundary [*Torrance and Turcotte*, 1971b; *McKenzie et al.*, 1974]. In essence these methods compute the elevation that would cancel the normal stress on the boundary. *McKenzie et al.* [1974] include an additional term, which computes the effect of the divergence of the near-surface flow on the local pressure. In all computations to date, though, the local velocities are not altered by the computed relief.

We observe that in the real earth the upper boundary of the convection system is essentially flat (negligible curvature), that it moves with constant horizontal velocity, and that the heat flow decreases roughly exponentially away from the ridges. From the constancy of the horizontal velocity we have

$$\frac{\partial u}{\partial x} = \frac{\partial^2 S}{\partial x\, \partial y} = 0$$

and since $S = 0$ on the boundary, it follows that in the real earth the upper streamline should be essentially parallel to the surface.

This is not the case in any of our models or any other published model, except of course those that impose $u = \text{const}$ at the surface. This difference in the near-surface streamline leads to an appreciable difference in the heat flow: in all models to date (see exception stated above), heat is advected to the surface (along the streamlines) in approximately half the cell and away from the surface in the other half. When the upward and downward flows are concentrated at either end of the cell, the streamline being parallel to the surface elsewhere, the heat flow will be appreciably different. It is therefore encouraging to note that Figure 6, in which the upper streamline comes closest to being parallel to the upper surface, also has the most nearly correct heat flow pattern.

Because of the preceding considerations we believe that a realistic computation of the relief and gravity must await a better approximation of the free surface condition.

In this connection it may be pointed out that since the bottom of the ocean slopes fairly uniformly, a tilted rectangular cell might be a better approximation to the real situation than a horizontal one. Tilting the axes correspondingly would, of course, result in an additional term in $\partial/\partial y\,(\rho_0 g\alpha Ts)$ in (1) (s is the slope of the surface). Despite the smallness of the slope ($\sim 10^{-3}$) this term will be important in the thermal boundary layer, where it may be of the same order as the driving function presently considered, i.e., $\partial/\partial x\,(\rho_0 g\alpha T)$. The importance of the slope of the surface has been pointed out in a slightly different context by *Hales* [1969] and *Jacoby* [1970].

'Decoupling' of the plates. Figure 8 shows that the local motions in the low-viscosity zone (asthenosphere) can be 'decoupled' from the motions in the high-viscosity zone (lithosphere), but they also show that the overall circulation pattern is always dominated by the plate motion. The argument about decoupling thus appears to rest on an 'either-or' conception, which disregards hydrodynamic concepts. The same may be said about the discussion concerning 'active' and 'passive' plates.

Conclusions

The choice of particular values for the various physical parameters that go into a numerical model of convection may appear somewhat arbitrary. For example, the depth of the convection cell, the amount and distribution of radioactivity, the boundary conditions, the effects of phase changes, and especially the viscosity functions are all parameters whose choice

is not completely circumscribed by relevant data. However, the convection solution must satisfy strong geologic and geophysical constraints. Hence numerical experiments can explore in a quantitative way the effect of particular hypotheses.

Although substantial difficulties remain to be overcome before realistic numerical models of mantle convection can be obtained, important conclusions about the driving mechanism and the temperatures have emerged. The two critical remaining problems are an adequate and practical modeling of the boundary conditions at the free surface and the three-dimensional nature of the motions in the mantle.

Acknowledgments. The help of H. H. Rachford of the Mathematics and Mathematical Sciences Department, Rice University, is gratefully acknowledged. We wish to thank D. L. Turcotte of Cornell University for very stimulating exchanges of views and the Geophysical Laboratory, Marine Science Institute, University of Texas, Galveston, Texas, for providing computer time. Acknowledgment is made to the donors of the Petroleum Research Fund administered by the American Chemical Society for the major support of this research. One of us (M.H.H.) was supported by the National Defense Education Act during this period. Contribution 29, Geophysical Laboratory, Marine Science Institute, University of Texas, Galveston, Texas.

References

Andrews, D. J., Numerical simulation of sea floor spreading, *J. Geophys. Res., 77,* 6470–6481, 1972.

Andrews, D. J., Temperature and viscosity in the mantle (abstract), *EOS Trans. AGU, 54,* 469, 1973.

Carter, N. L., and H. G. Avé Lallemant, High temperature flow of dunite and periodotite, *Geol. Soc. Amer. Bull., 81,* 2181–2202, 1970.

Clark, S. P., Jr., and A. E. Ringwood, Density distribution and constitution of the mantle, *Rev. Geophys. Space Phys., 2,* 35–88, 1964.

Conte, S. D., and R. T. Dames, An alternating direction method for solving the biharmonic equation, *Math. Tables Aids Comput., 12,* 198–205, 1958.

Conte, S. D., and R. T. Dames, On an alternating direction method for solving the plate problem with mixed boundary conditions, *J. Ass. Comput. Mach., 7,* 264–273, 1960.

Douglas, J., Jr., and H. H. Rachford, On the numerical solution of heat conduction problems in two and three space variables, *Trans. Amer. Math. Soc., 82,* 421–439, 1956.

Forsythe, G. E., and W. R. Wasow, *Finite Difference Methods for Partial Differential Equations,* p. 397, J. Wiley, New York, 1960.

Fromm, J. E., Numerical method for computing nonlinear, time-dependent, buoyant circulation of air in rooms, *IBM J. Res. Dev., 15,* 186–196, 1971.

Garabedian, P. R., Estimation of the relaxation factor for small mesh size, *Math. Tables Aids Comput., 10,* 183–186, 1956.

Green, D. H., Contrasted melting relations in a pyrolite upper mantle under mid-oceanic ridge, stable crust and island arc environments, *Tectonophysics, 17,* 284–297, 1973a.

Green, D. H., Experimental melting studies on a model upper mantle composition at high pressure under water-saturated and water-undersaturated conditions, *Earth Planet. Sci. Lett., 19,* 37–53, 1973b.

Gutenberg, B., Great earthquakes, 1896–1903, *EOS Trans. AGU, 37,* 608–614, 1956.

Hales, A. L., Gravitational sliding and continental drift, *Earth Planet. Sci. Lett. 6,* 31–34, 1969.

Harlow, F. H., and J. E. Welch, Numerical calculation of time-dependent viscous incompressible flow of fluid with free surface, *Phys. Fluids, 8,* 2182–2183, 1965.

Herring, C., Diffusional viscosity of a polycrystalline solid, *J. Appl. Phys., 21,* 437–445, 1950.

Houston, M. H., Jr., Numerical models of free convection in the earth's upper mantle with radiogenic heating and variable viscosity, Ph.D. thesis, Rice Univ., Houston, Tex., 1973.

Houston, M. H., Jr., and J. Cl. De Bremaecker, ADI solution of free convection in a variable viscosity fluid, *J. Comput. Phys.,* in press, 1974.

Isacks, B. L., J. Oliver, and L. R. Sykes, Seismology and the new global tectonics, *J. Geophys. Res., 73,* 5855–5899, 1968.

Jacoby, W. R., Instability in the upper mantle and global plate movements, *J. Geophys. Res.*, *75*, 5671–5680, 1970.

Jeffreys, H., The instability of a compressible fluid heated below, *Proc. Cambridge Phil. Soc.*, *26*, 170–172, 1930.

Knopoff, L., The convection current hypothesis, *Rev. Geophys. Space Phys.*, *2*, 89–122, 1964.

Knopoff, L., Thermal convection in the earth's mantle, in *The Earth's Mantle*, edited by T. F. Gaskell, chap. 8, pp. 171–196, Academic, New York, 1967.

McFadden, C. P., The effect of a region of low viscosity on thermal convection in the earth's mantle, Ph.D. thesis, Univ. of West. Ont., London, Canada, 1969.

McFadden, C. P., and D. E. Smylie, Effect of a region of low viscosity on thermal convection in the mantle, *Nature, 220*, 468–469, 1968.

McKenzie, D. P., Speculations on the consequences and causes of plate motions, *Geophys. J., 18*, 1–32, 1969.

McKenzie, D. P., Plate tectonics, in *The Nature of the Solid Earth*, edited by E. C. Robertson, pp. 323–360, McGraw-Hill, New York, 1972.

McKenzie, D. P., J. Roberts, and N. Weiss, Numerical models of convection in the earth's mantle, *Tectonophysics, 19*, 89–103, 1973.

McKenzie, D. P., J. M. Roberts, and N. O. Weiss, Convection in the earth's mantle: Towards a numerical simulation, *J. Fluid Mech., 62*, 465–538, 1974.

Nabarro, F. R. N., Steady-state diffusional creep, *Phil. Mag., Ser. 8, 16*, 231–237, 1967.

Peltier, W. R., Penetrative convection in the planetary mantle, *Geophys. Fluid Dyn., 5*, 47–88, 1972.

Richter, F. M., Convection and the large-scale circulation of the mantle, *J. Geophys. Res., 78*, 8735–8745, 1973.

Saul'yev, V. K., *Integration of Equations of Parabolic Type by the Method of Nets*, p. 149, Macmillan, New York, 1964.

Schatz, J. F., and G. Simmons, Thermal conductivity of earth materials at high temperatures, *J. Geophys. Res., 77*, 6966–6983, 1972.

Spiegel, E. A., and G. Veronis, On the Boussinesq approximation for a compressible fluid, *Astrophys. J., 131*, 442–447, 1960.

Sundqvist, H., and G. Veronis, A simple finite-difference grid with non-constant intervals, *Tellus, 22*, 26–31, 1970.

Torrance, K. E., and D. L. Turcotte, Thermal convection with large viscosity variations, *J. Fluid Mech., 47*, 113–125, 1971a.

Torrance, K. E., and D. L. Turcotte, Structure of convection cells in the mantle, *J. Geophys. Res., 76*, 1154–1161, 1971b.

Tozer, D. C., Heat transfer and convection currents, *Phil. Trans. Roy. Soc. London, Ser. A, 258*, 252–271, 1965.

Tozer, D. C., Towards a theory of thermal convection in the mantle, in *The Earth's Mantle*, edited by T. F. Gaskell, pp. 325–353, Academic, New York, 1967.

Tozer, D. C., Factors determining the temperature evolution of thermally convecting earth models, *Phys. Earth Planet. Interiors, 2*, 393–398, 1970.

Turcotte, D. L., and E. R. Oxburgh, Finite amplitude convection cells and continental drift, *J. Fluid Mech., 28*, 29–42, 1967.

Turcotte, D. L., K. E. Torrance, and A. T. Hsui, Convection in the earth's mantle, in *Methods in Computational Physics*, vol. 13, *Geophysics*, edited by B. A. Bolt, pp. 403–406, Academic, New York, 1973.

Vine, F. J., and D. H. Matthews, Magnetic anomalies over ocean ridges, *Nature, 199*, 947–949, 1963.

Weertman, J., The creep strength of the mantle, *Rev. Geophys. Space Phys., 8*, 145–168, 1970.

Welch, J. E., F. H. Harlow, J. P. Shannon, and B. J. Daly, The MAC method; a computing technique for solving viscous, incompressible, transient fluid-flow problems involving free surfaces, *LA-3425, VC-32*, 146 pp., Los Alamos Sci. Lab., Los Alamos, N. Mex., 1966.

(Received December 4, 1973;
revised October 10, 1974;
accepted October 10, 1974.)

8

PHASE CHANGE INSTABILITY IN THE MANTLE

Gerald Schubert, D. L. Turcotte, and E. R. Oxburgh

Abstract. *In the presence of a temperature gradient, phase changes of the type believed to exist in the upper mantle, in which the less dense phase lies above the dense phase, may be unstable. Approximate calculations show such phase change instabilities are possible for both the 400-kilometer olivine-spinel phase transition and also for partial melting at shallower depths. The resulting flow patterns may provide a driving mechanism for the new global tectonics.*

Thermal convection within the earth's mantle has been proposed to explain continental drift (*1*). The Rayleigh number for the mantle, based on the value of viscosity inferred from postglacial uplift of Scandinavia (*2*), is several orders of magnitude larger than the critical Rayleigh number for the onset of convection (*3, 4*). Also, estimates of surface velocity and heat flux from constant property theories of thermal convection are in good agreement with observations (*5*).

Seismological (*6*) and geochemical (*7, 8*) evidence indicates that one or more phase changes occur in the upper mantle at a depth of about 400 km. The most important is likely to be the olivine-spinel phase transition. It is of interest to consider the influence of such a phase change on mantle convection. The volume and the entropy changes have the same sign for this phase transition (*7, 8*), so that heat is evolved when going from the olivine (light) to the spinel (heavy) phase and is absorbed when going from the heavy to the light. In this case the Clapeyron curve has a positive slope. Seismic evidence shows that the dense phase lies beneath the light phase; under isothermal conditions such a phase change is stable. If the fluid moves upward, the dense phase transforms to the light one and heat is absorbed. Thus the fluid is cooled and there is a downward stabilizing body force. On this basis it has been argued that the mantle phase change would act as a barrier to thermal convection (*4*). However, others (*9*) have argued qualitatively

that large-scale convection could penetrate the phase change interface. A quantitative analysis of the influence of phase transformations on the stability of a fluid has recently been made (*10*). In this report we apply this theory to phase changes in the mantle.

We consider a simple model in which a two-phase fluid is confined between horizontal planes separated by the distance $2d$. The phases are assumed to be in thermodynamic equilibrium, so that the location of the phase boundary is determined by the intersection of the Clapeyron curve with the pressure-temperature curve for the fluid. In order for the light phase to lie above the heavy phase, it is necessary that the slope of the pressure-temperature curve exceed the slope (assumed positive) of the Clapeyron curve. The univariant phase boundary is initially midway between the planes. Both phases are assumed to have the same values of absolute viscosity μ, thermal conductivity k, and specific heat at constant pressure c_p (μ, k, and c_p are constants). A constant negative temperature gradient of absolute magnitude β, and a constant pressure gradient $-\rho g$ (g is the gravitational acceleration and ρ is the density) are present in this static state. The change in density $\Delta\rho$ at the phase boundary is an essential feature of the model. However, elsewhere each phase can be assumed to be an incompressible fluid of density ρ (for the olivine-spinel phase change the fractional density change $\Delta\rho/\rho$ is only about 0.1); the thermal expansion of the fluid is not considered. Since the coefficients

120

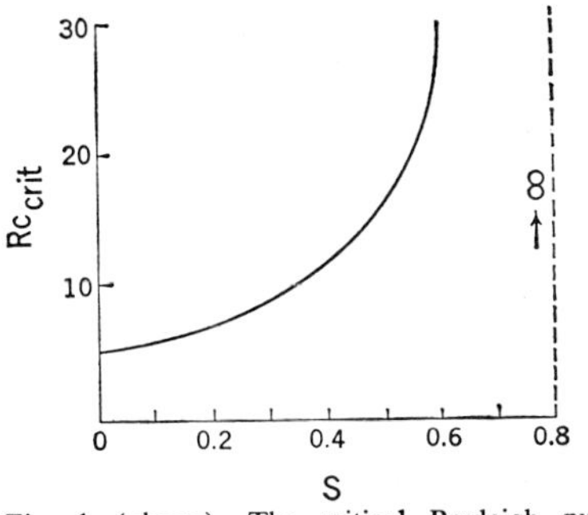

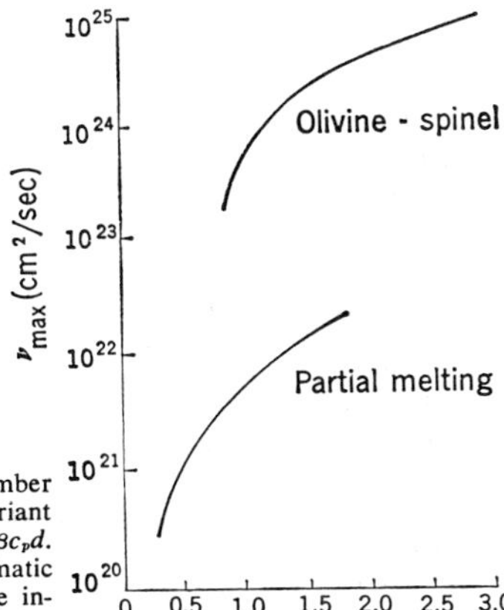

Fig. 1 (above). The critical Rayleigh number Rc_{crit} for the onset of instability of a univariant phase boundary as a function of $S = Q/\beta c_p d$. Fig. 2 (right). The maximum value of kinematic viscosity ν which will lead to a phase change instability versus the magnitude of the negative temperature gradient for the olivine-spinel phase transition and for partial melting.

of thermal expansion for both phases are taken to be zero, the Rayleigh instability is *not* present.

It is found that this basic state may be unstable to infinitesimal disturbances. The instability mechanism may be understood as follows. In a region of downward flow there is an inflow of relatively cold material from above the interface (due to the zero-order temperature gradient). Since the interface must remain on the Clapeyron curve, the lower temperature forces the interface to a region of lower hydrostatic pressure—that is, upward. With the interface displaced upward, the heavier material below the interface gives a hydrostatic pressure head tending to drive the flow downward, which leads to instability. However, the downward flow of fluid through the interface releases heat, thus tending to warm the fluid and return the phase boundary to its unperturbed location. The inflow of cold material tends to promote instability, whereas release of heat by the phase change promotes stability. The necessity of overcoming viscous dissipation in the fluid provides a further stabilization mechanism.

A quantitative analysis of this instability has been given elsewhere (*10*). The parameters determining the stability of the unperturbed state against infinitesimal perturbations are

$$Rc = \frac{gd^3 \Delta\rho/\rho}{8\bar{\nu}\kappa(\rho g/\gamma\beta - 1)}$$

and

$$S = Q/\beta c_p d$$

where Q is the energy required per unit mass to change the dense into the less

dense phase, T is the temperature at which the phase transition occurs, $\nu = \mu/\rho$ is the kinematic viscosity, $\kappa = k/\rho c_p$ is the thermal diffusivity, and $\gamma = Q\rho^2/T\Delta\rho$ is the slope of the Clapeyron curve in a pressure-temperature diagram evaluated at the point on the curve that represents the unperturbed state. The parameter Rc plays the role of a Rayleigh number. The parameter S is the ratio of the temperature increase due to the energy released in the phase change Q/c_p to the temperature difference between the boundaries and the interface βd. This parameter provides a quantitative measure of the stabilizing influence of the latent heat release versus the destabilizing influence of the inflow of cold material from above. The static state is unstable if $0 < S < 4/5$ and $Rc > Rc_{\text{crit}}$, where Rc_{crit} is given as a function of S in Fig. 1.

We assume that a breakdown of olivine to a spinel structure occurs in the mantle near a depth of 400 km. After approximating this phase change by a simple univariant system, we apply the results discussed above to determine whether such a phase change would be stable or unstable to small perturbations. First we will determine the value of the temperature gradient necessary for the parameter S to be sufficiently small for instability. For the olivine-spinel phase change, we take $Q = 9.8$ kcal mole^{-1} (*7*), and $c_p = 0.3$ cal g^{-1} °K^{-1}. Since the phase change occurs at a depth of 400 km, we take $d = 400$ km. If $S = 4/5$, it is necessary to have $\beta = 0.73$°K km^{-1}. Values of $\beta > 0.73$°K km^{-1} could lead to instability. Conduction gradients in the

mantle are 10°K km^{-1} or higher and the adiabatic gradient in the mantle is near 0.5°K km^{-1}. Therefore it is likely that β will be sufficiently large for instability to occur.

It is also necessary that the equivalent Rayleigh number Rc must exceed its critical value if an instability is to occur. We take $\gamma = 0.0625$ kb °K^{-1}, $\rho = 4$ g cm^{-3}, $g = 10^3$ cm sec^{-2}, $T = 1800$°K, and $\kappa = 10^{-2}$ cm^2 sec^{-1}. All the quantities necessary for evaluation of the Rayleigh number can be estimated with fair accuracy except the viscosity. Therefore we find the maximum value of the viscosity which leads to instability. This maximum value of the kinematic viscosity is plotted as a function of β in Fig. 2. It is seen that the maximum viscosity is of the order of 10^{24} cm^2 sec^{-1} for reasonable values of the temperature gradient β. If the viscosity is less than this value, an instability can be expected. Uplift data suggest that, at least over some range of depths, the viscosity of the mantle is in the range $\nu = 10^{21}$ to 10^{22} cm^2 sec^{-1}, so that this requirement on the viscosity is not unreasonable.

Although the analysis presented above is only approximately valid for the mantle, the results obtained indicate that the phase change instability may occur. It is doubtful that a more accurate analysis can be made until more information is available on the mantle phase change.

A second possible application of the phase change instability is to the problem of migrating mid-oceanic ridge systems. The causes of the motions of the large lithospheric plates which are generated at mid-oceanic ridges are at present rather uncertain, and a number of alternative hypotheses must be entertained (*11*). If, however, the plates are driven by a diverging upward flow centered beneath the ridge and extending deep into the mantle for 500 km or more, it is difficult to understand how the ridges themselves are able to migrate relatively rapidly (in some cases, at many times the spreading rate) because the deep flow structure would have to migrate with them.

It is possible that the phase change instability provides a mechanism for the migration of flow beneath ridges. One of the essential features of mid-oceanic ridges is that they are the loci of intense volcanic activity and that sufficient basaltic magma is continuously extruded close to the crest for a surface layer some 4 to 5 km thick to be built up, continuously generating new oceanic crust. We therefore consider the phase change instability which might exist between the crystalline peridotite of the upper mantle and partially molten peridotite (the partial melt fraction has the composition of basalt).

If the mantle at, say, a depth of 100 km were just above its temperature of beginning of melting (as has been suggested for the seismic low-velocity zone), the introduction of some pressure perturbation (conceivably arching of the lithosphere through tectonic activity, or some kind of surface rifting) might be sufficient to initiate a self-sustaining upward flow beneath the locus of the perturbation. If fragmentation of the lithosphere ensued, sea-floor spreading could take place with ridge formation, eruption of magmas, and generation of new crust. If the ridge remained stationary or nearly so, a relatively deep flow might ultimately become established. The boundary conditions for the motion of the new plates generated at the ridge might, however, be such that the ridge itself was constrained to migrate; in that the ridge is the locus of rifting, this amounts to the migration of a pressure perturbation. In turn, this should cause the phase change instability to propagate laterally at the level of the low-velocity zone; the instability should be able to sustain the horizontal surface flows as it migrated.

This suggestion is advanced somewhat tentatively, since the theory of the instability is only approximately valid for this phase change (it is at least divariant). It would, however, imply that slowly moving or static ridges, if they could be recognized, should be characterized by deep flows, and rapidly migrating ones by shallow flows. The perturbation of mantle isotherms associated with these two kinds of flow might ultimately be distinguishable by seismic methods. It is also possible that the two kinds of flow might be characterized by slightly different compositions of magmas extruded at the surface.

Finally, we test this melting instability quantitatively. We assume one phase to contain no liquid and the other phase to contain a partial melt fraction of 5 percent. We take the partial melt frac-

tion to be basalt. The denser (no partial melt) phase lies beneath the lighter phase, and heat is absorbed in going from the denser to the lighter phase. For the 5 percent partial melt fraction, we take $Q = 5.4$ cal g^{-1}, $c_p = 0.3$ cal g^{-1} $°K^{-1}$, $d = 100$ km, $\gamma = 0.118$ kb $°K^{-1}$, $\rho = 3$ g cm^{-3}, $g = 10^3$ cm sec^{-2}, $T = 1500°K$, and $\kappa = 10^{-2}$ cm^2 sec^{-1}. We can determine the maximum value of the kinematic viscosity which will lead to an instability as a function of β. The result is given in Fig. 2. The maximum value of the kinematic viscosity for instability is of the order of $v = 10^{21}$ cm^2 sec^{-1}. This value is not unreasonable, particularly since the presence of the partial melt is likely to reduce the viscosity significantly.

References and Notes

1. A. Holmes, *Trans. Geol. Soc. Glasgow* **18**, 559 (1931).
2. N. A. Haskell, *Physics New York* **6**, 265 (1935).
3. C. L. Pekeris, *Mon. Notic. Roy. Astron. Soc. Geophys. Suppl.* **3**, 343 (1935); D. C. Tozer, *Phil. Trans. Roy. Soc. Ser. A* **258**, 252 (1965).
4. L. Knopoff, *Rev. Geophys.* **2**, 89 (1964).
5. D. L. Turcotte and E. R. Oxburgh, *J. Fluid Mech.* **28**, 29 (1967); *J. Geophys. Res.* **74**, 1458 (1969); E. R. Oxburgh and D. L. Turcotte, *ibid.* **73**, 2643 (1968).
6. K. E. Bullen, *Bull. Seismol. Soc. Amer.* **30**, 235 (1940); H. Jeffreys, *Mon. Notic. Roy. Astron. Soc. Geophys. Suppl.* **4**, 498 (1939).
7. S. Akimoto and H. Fujisawa, *J. Geophys. Res.* **73**, 1467 (1968).
8. S. P. Clark and A. E. Ringwood, *Rev. Geophys.* **2**, 35 (1964); A. E. Ringwood, *Earth Planet. Sci. Lett.* **5**, 401 (1969).
9. F. A. Vening Meinesz, in *Continental Drift*, S. K. Runcorn, Ed. (Academic Press, New York, 1962), p. 145; J. Verhoogen, *Phil. Trans. Roy. Soc. Ser. A* **258**, 276 (1965).
10. F. H. Busse and G. Schubert, in preparation.
11. E. R. Oxburgh and D. L. Turcotte, in preparation.
12. Supported in part by NASA under NGL 05-007-002. One of us (G.S.) would like to thank the Alexander von Humboldt Foundation for a fellowship while visiting the Max-Planck-Institut für Physik and Astrophysik, München. We thank Dr. F. H. Busse for helpful discussions.

Part III

MANTLE DIAPIRS AND HOT SPOTS

Editor's Comments
on Papers 9 Through 12

In the previous section, mass motions in the mantle, and especially ascending motions, were envisaged as distributed in large volumes. However, there is also evidence for areally restricted rising masses. These are associated with igneous activity. Melting does not normally occur in the mantle, so that some type of disturbance is called for under all sites of magmatic activity. Fast (adiabatic) uprising, with little loss of heat, is the most reasonable model for such a disturbance (Verhoogen, 1954; Green and Ringwood, 1967); the pressure acting on the rising material is reduced, so that on reaching a low enough pressure this material will cross the solidus temperature, and melting will begin. This may occur either on the limbs of convection cells, as envisaged by Oxburgh and Turcotte (Paper 5), or in isolated plumes or mantle diapirs under areally restricted igneous manifestations, especially in the interior of plates.

Wyllie discussed the problems of magma genesis in the mantle (Paper 9). He stressed the role of diapirism, that is localized upwelling, in the mantle, and the role of water. Wyllie showed how the presence of water may enhance upward movement by allowing small degrees of melting. These arguments justify the assertion that sites of igneous activity can generally be interpreted as surface expressions of upward movements in the underlying mantle. The only exception may be island-arc magmatism, which is perhaps related to frictional heating or to local release of volatiles during descent of subducted lithospheric slabs. Heating by depleted mantle material, as suggested by Yoder (1978), also requires local upwelling.

Wilson (Paper 3) suggested that chains of volcanic islands becoming younger in one direction (e.g., Hawaii-Emperor chain) were the result of motion of lithospheric plates over magma sources originating at depth and not participating in plate motions. This profound observation had to wait for almost a decade until Morgan (Paper 10) showed that the directions of such chains were indeed compatible with the known geometry of plate motions. These localized magma sources, or "hot spots," which originate in plumes rising from below the lithosphere, provide a frame of reference that is independent of the plates, but is tied to the underlying mantle. Motions relative to the hot spots or relative to the mantle became known as "absolute motions." Morgan also observed that localized magmatism occurs along future plate boundaries just prior to continental breakup. He therefore suggested that initiation of the underlying plumes could split continents and thus control production of new ridges. He also suggested that the plumes underneath the hot spots originated in the lower mantle and then spread out under the lithosphere, providing the forces that drive plates. These latter two assumptions do not, however, depend on the previous ones, and their validity is not required to confirm the other assumptions about the hot spots.

Later analysis confirmed that the relative positions of the hot spots have not changed appreciably in the last few million years (Minster and Jordan, 1978; Chase, 1978). However, this was probably not so during longer periods (Burke, Kidd, and Wilson, 1973; Molnar and Atwater, 1973). Molnar and Francheteau (1975) suggested that in the Cenozoic the oceanic hot spots moved one relative to the others, but much slower than the relative motions between fast-moving plates.

Wilson (Paper 11) made an inventory of sites of active intraplate magmatism, which he thinks are expressions of underlying rising plumes in the mantle. He also related several geologic observations to the absolute motion of the plates. It is noteworthy that this list does not include any example of a continental linear volcanic chain with a well documented progression of the age of volcanism, as is the case with oceanic island chains, like Hawaii. In fact some linear arrays of igneous bodies in continents do not display any regular temporal migration of ages of emplacement (e.g., Wright, 1973; Cantagrel, Jamond, and Lasserre 1978). Most of the continental hot spot magmatism occurs in restricted areas and does not define any linear trends at all.

Further dating of oceanic islands and seamounts confirmed that in some chains in the Pacific Ocean the site of magmatic activity migrated as envisaged by Wilson and Morgan, but in some chains such a migration is not very clear (Bonatti et al., 1977; Jarrard and Clague, 1977; McDougall and Duncan, 1980). However, the directions of the chains are as expected from plate kinematics (cf. Paper 10). Thus it seems that there are significant differences among oceanic hot spots, and not just between continental and oceanic ones, in terms of duration of igneous activity and its spatial migration relative to the plates.

The hot spots generally have two other important characteristics. First, they generally occur over young topographic highs on land as well as in the oceans, which also implies a local source of heat (Le Bas, 1971; Crough, 1979). Second, the basaltic rocks of hot spots have a characteristic geochemical fingerprint, and are distinct from the mid-ocean ridge basalts (Schilling, 1975a, 1975b; O'Nions and Pankhurst, 1974; Sun and Hanson, 1975; O'Nions, Hamilton, and Evensen, 1977; DePaolo, 1979). This shows that the source of the hot spots differs from the reservoir that supplies material to the mid-oceanic ridges. The location of the reservoirs is not well constrained, but one possibility is that the circulation involved in overturn of oceanic lithosphere is confined to the outer 650 km of the mantle, whereas the lower mantle is a distinct geochemical reservoir from which the hot spots arise (Sun and Hanson, 1975; Wasserburg and DePaolo, 1979). Such a layered mantle precludes mantle-wide convection and indicates separate flow systems. In any case the diapirs are superimposed on the large-scale flow involved in plate motion and overturn, but the interaction between them is not well understood. The relation of some hot spots to continental breakup and to formation of new oceanic ridges suggests that occasionally the mantle diapirs interfere with, and modify, the large-scale flow. Together, these observations on hot spots support the Wilson-Morgan hypothesis of plumes arising below, and independent of, the moving plates. The major difficulty is posed by the cases where migration of magmatism is not appatent. This led to the alternative suggestion that lithospheric fractures, "hot lines," are the main control of intraplate magmatism (e.g. Turcotte and Oxburgh, 1978). This explanation appears to be less satisfactory than the plume hypothesis as it does not explain the formation of swells and the geochemical observations. Fracturing of the lithosphere is important for the transport of magmas, but probably does not control their formation.

Diapirs are generally thought to form when a fluid layer overlies another fluid with a smaller density. In that case disturbances having a particular wavelength, which depends on the viscosities of the fluids, will grow faster than others (Chandrasekhar, 1961). Ramberg (1972a, 1972b), discussed this phenomenon further and showed that, in the mantle, diapiric upwellings having the appropriate wavelength can grow to a significant size within geologically short times, and are thus likely sources of magmas. Whitehead and Luther (1975) confirmed these conclusions and also showed how the diapirs grow and become localized upwellings. Yuen and Schubert (1976) studied the flow in rising diapirs, assuming different rheologies, and showed that it can be fast. The problem with such simple models is that they do not explain how the density inversion arises and why hot spots appear intermittently. A possible solution is that the hot spots are nucleated in hydrodynamic instabilities, perhaps in the boundary layer at the base of the lower mantle (Yuen and Peltier, 1980).

While evidence for motion of lithospheric plates is lacking from the Moon, Mercury, and Mars, the latter planet has some features analogous with terrestrial hot spots. Carr (Paper 12) described the largest fault system on Mars—in the Tharsis region—which is associated with huge shield volcanoes and a major topographic high. He explained all these features as resulting from a mantle plume that has been active for a long time. The volcanoes are much higher than any known on Earth, which seems to indicate that the Martian lithosphere is thicker and stronger than the Earth's. Some major faults form large rift valleys—the Valles Marineris—but unlike on Earth, these are a part of a radial system and are not directly associated with volcanoes. Nevertheless, the analogy with the African rifts was repeatedly mentioned (Wood and Head, 1977; Frey 1979). Other volcanic features on Mars were also compared with a terrestrial hot spot (Malin, 1977). Courtillot, Allegre, and Mattauer (1975) interpreted the Valles Marineris in terms of plate boundaries and lateral slip of unspecified magnitude. However, they could not identify displaced features or zones of plate convergence, which their model requires. Thus, a comparison with terrestrial rifts is still the most viable approach. Perhaps the most noteworthy difference from Earth is that in the Tharsis region the history of igneous activity and faulting lasted much longer than in any analogous feature on Earth, and several fault systems were formed successively (Mutch et al., 1976). This could have resulted from the absence of plate motion on Mars. On the other hand, convection in the Martian mantle is most likely (Schubert, 1979), so its possible interaction with plumes should be studied. The Martian surficial tectonics and internal dynamics, particularly the history and origin of the Tharsis region are important objects of continuing research. In particular, the presence of Martian "hot spot"-like features emphasizes the role of the terrestrial equivalents.

REFERENCES

Bonatti, E., C. G. A. Harrison, D. E. Fisher, J. Honnorez, J. G. Schilling, J. J. Stipp, and M. Zentilli, 1977, Easter Volcanic Chain (Southeast Pacific) Mantle Hot Line, *Jour. Geophys. Research* **82:**2457-2478.
Burke, K., W. S. F. Kidd, and J. T. Wilson, 1973, Relative and Latitudinal Motion of Atlantic Hot Spots, *Nature* **245:**133-137.
Cantagrel, J. M., C. Jamond, and M. Lasserre, 1978, La magmatisme alcalin de

la ligne du Cameroun au Tertiaire inferieur; données géochronologiques K/Ar, *Soc. Géol France Compte Rendu* 6, p. 300–303.

Chandrasekhar, S., 1961, *Hydrodynamic and Hydromagnetic Stability*, Clarendon Press, Oxford, 652p.

Chase, C. G., 1978, Plate kinematics: The Americas, East Africa, and the Rest of the World, *Earth and Planetary Sci. Letters* v **37**:355–368.

Courtillot, V. E., Allegre, C. J., and M. Mattauer, 1975, On the Existence of Lateral Relative Mations on Mars, *Earth and Planetary Sci. Letters* v **25**:279–285.

Crough, T. S., 1979, Hotspot Epeirogeny, *Tectonophysics* **61**:321–333.

DePaolo, D. J., 1979, Implications of Correlated Nd and Sr Isotopic Variations for the Chemical Evolution of the Crust and Mantle, *Earth and Planetary Sci. Letters* **43**:201–211.

Frey, H., 1979, Martian Canyons and African Rifts: Structural Comparisons and Implications, *Icarus* **37**:142–155.

Green, D. H., and A. E. Ringwood, 1967, The Genesis of Basaltic Magmas, Contr. *Mineralogy and Petrology* **15**:103–190.

Jarrard, R. D., and D. A. Clague, 1977, Implications of Pacific island and Seamount Ages for the Origin of Volcanic Chains, *Rev. Geophys. Space Phys.* **15**:57–75.

Le Bas, M. J., 1971, Per-alkaline Volcanism, Crustal Swelling, and Rifting, *Nature, Phys. Sci.* **230**:85–87.

Malin, M. C., 1977, Comparison of Volcanic Features of Elysium (Mars) and Tibesti (Earth), *Geol. Soc. American Bull.* **88**:908–919.

McDougall, I., and R. A. Duncan, 1980, Linear Volcanic Chains—Recording Plate Motions? *Tectonophysics* **63**:275–295.

Minster, J. B., and T. H. Jordan, 1978, Present Day Plate Motions, *Jour. Geophys. Research* **83**:5331–5354.

Molnar, P. and T. Atwater, 1973, The Relative Motion of Hot-Spots in the Mantle, *Nature* **296**:288–291.

Molnar, P., and J. Francheteau, 1975, The Relative Motion of "Hot Spots" in the Atlantic and Indian Oceans during the Cenozoic, *Royal Astron. Soc. Geophys. Jour.* **43**:763–774.

Mutch, T. A., Arvidson, R. E., Head, J. W., Jones, K. L., and Saunders, R. S. 1976, *The Geology of Mars*, Princeton University Press, Princeton, N.J., 400p.

O'Nions, R. K., and R. J. Pankhurst, 1974, Petrogenetic Significance of Isotope Trace Element Variations in Volcanic Rocks from the Mid-Atlantic, *Jour. Petrology* **15**:603–634.

O'Nions, R. K., Hamilton, P. J., and Evensen, N. M., 1977, Variations in ^{143}Nd/^{144}Nd and ^{87}Sr/^{86}Sr Ratios in Oceanic basalts, *Earth and Planetary Sci. Letters* **34**:13–22.

Ramberg, H., 1972a, Theoretical Models of Density Stratification and Diapirism in the Earth, *Jour. Geophys. Research* **77**:877–889.

Ramberg, H., 1972b, Mantle Diapirism and its Tectonic Magmagenic Consequences, *Physics Earth and Planetary Interiors* **5**:45–60.

Schilling, J. G., 1975a, Rare-Earth Variations across 'normal segments' of the Reykjanes Ridge, 60°–53°N, Mid-Atlantic Ridge, 29°S, and East Pacific Rise, 2°–19°S, and Evidence on the Composition of the Underlying Low-Velocity Layer, *Jour. Geophys. Research* **80**:1459–1473.

Schilling, J. G., 1975b, Azores Mantle Blob: Rare Earth Evidence, *Earth and Planetary Sci. Letters* **25**:103–115.

Schubert, G., 1979, Subsolidus Convection in the Mantles of Terrestrial Planets, *Ann Rev. Earth and Planetary Sci.* **7**:289–342.

Sun, S. S., and Hanson, G. H., 1975, Evolution of the Mantle: Geochemical Evidence from Alkali Basalts, *Geology,* **3**:297–302.

Turcotte, D. L. and Oxburgh, E. R., 1978, Intra-Plate Volcanism, *Royal Soc. London Philos. Trans.,* ser. A, **288**:561–579.

Verhoogen, J., 1954, Petrological Evidence on Temperature Distribution in the Mantle of the Earth, *Am. Geophys. Union Trans.* **35**:85–92.

Whitehead, J. A., and D. S. Luther, 1975, Dynamics of Laboratory Diapir and Plume Models, *Jour. Geophys. Research* **80**:705–717.

Wood, C. A. and J. W. Head, 1977, Rift Valleys on Earth, Mars and Venus, *Paleorift Symp. Oslo, Proc.* D. Reidel Publishing Company, Dordrecht, pp. 401–408.

Wright, J. B., 1973, Continental Drift, Magmatic Provinces and Mantle Plumes, *Nature* **244**:565–567.

Wasserburg, G. J., and D. J. DePaolo, 1979, Models of Earth Structure Inferred from Neodymium and Strontium Isotopic Abundances, *Nat. Acad. Sci. Proc.* **76**:3594–3598.

Yuen, D. A., and W. R. Peltier, 1980, Mantle Plumes and the Stability of the D″ layer, *Geophys. Research Lett.* **7**:625–628.

Yuen, D. A., and G. Schubert, 1976, Mantle Plumes: A Boundary Layer Approach for Newtonian and Non-Newtonian Temperature Dependent Rheologies, *Jour. Geophys. Research* **81**:2499–2510.

Yoder, H. S., Jr., 1978, Basic Magma Generation and Aggregation, *Bull. Volcanol.* **41**:301–316.

9

Role of Water in Magma Generation and Initiation of Diapiric Uprise in the Mantle

PETER J. WYLLIE

Department of Geophysical Sciences, University of Chicago, Chicago, Illinois   60637

The content and distribution of water is a critical factor in determining mantle properties, especially in the low-velocity and Benioff zones. Estimated temperature distributions vary widely, depending on assumptions regarding the relative significance for heat transfer of conduction, radiation, and convection. Comparison of estimated geotherms with known or inferred phase relationships in the system peridotite-eclogite-water provides information about the possible physical state of the mantle at various depths for comparison with geophysical measurements. Water is stabilized in minerals such as amphibole and phlogopite in the upper mantle; at greater depths, it may exist as intergranular fluid probably adsorbed on mineral surfaces, or it is dissolved in interstitial silicate magma. The most satisfactory explanation for the low-velocity zone involves incipient melting due to traces of water. Water rising from the deep mantle in preferred zones would augment the interstitial silicate magma at the base of the low-velocity zone, initiating the diapiric uprise necessary for basaltic magma generation. A petrological model for the suboceanic mantle suggests that a lens of gabbro-rich material may account for the gravity anomaly and seismic refraction measurements at mid-oceanic ridges. A petrological model for the area beneath island arcs suggests that the seismic low-velocity zone should not be continuous across it. Magma may be generated beneath island arcs by partial fusion of down-going oceanic crust by frictional heating, by migration of water from the lithosphere into overlying mantle, or by diapiric uprise of material from the downgoing lithosphere. Diapiric uprise could be initiated by influx of water into the layer including the mantle–upper lithosphere boundary over a down-going slab. The composition of the liquid generated depends on many variables, including the water content and the stability of hydrous minerals. Magma generation may depend on the regime of water, rising from the deep mantle in preferred zones and carried down with the lithosphere in Benioff zones.

For consideration of magma generation in the mantle, we need to know the temperature distribution, the mantle composition, and the properties of the mantle material. These parameters remain uncertain.

Figure 1 shows three estimated temperature distributions; these vary, depending on assumptions as to the relevant significance for heat transfer of conduction, radiation, and convection. At 500 km depth, there is a difference of about 1000°C between the uppermost and lowest curves. According to recent reviews [*Ringwood,* 1969a; *Wyllie,* 1970], the upper mantle is composed of peridotite and eclogite, possibly with traces of water and carbon dioxide. The relevant mantle property for magma generation is the melting temperature. Solidus curves have been measured experimentally up to pressures corresponding to about 150 km depth; solid lines in Figure 1 to 100-km depth

show the range of reported curves. These limits are extrapolated linearly to 500 km, and an indication of the area of uncertainty is shown by the shaded band.

According to these estimates of temperature distribution and melting temperatures, magma does not form in dry mantle under normal conditions. Of the many mechanisms proposed to raise the temperature above the solidus, the most reasonable model is that involving diapiric uprise of solid mantle [*Green and Ringwood,* 1967]. I examine here the effect of water on the conditions for melting in the mantle, and on the model of diapiric uprise. To illustrate patterns of behavior of mantle materials, I shall arbitrarily adopt the oceanic geotherm from Figure 1, and the line drawn within the shaded band for the solidus.

The amphibole and phlogopite in mantle-derived eclogite and peridotite nodules in kimber-

130

lite has been cited as evidence for water in the mantle (see review by *Wyllie* [1970]). G. C. Kennedy (personal communication, 1970) contends that this evidence has been misinterpreted because the hydrous minerals occur only in the outer parts of nodules, as late alteration products. Nevertheless, the effects of even traces of water in the mantle are so significant that we must evaluate them. Water was certainly present in the mantle at some earlier stage of its history, and according to the new global tectonics it is probably being carried down into the mantle beneath island arcs [*Isacks et al.*, 1968].

Petrology of the Mantle

Phase diagrams for mantle materials show the mineral facies, and the mantle mineralogy at any depth is given by following geotherms through the facies.

Phase diagrams for peridotite-eclogite-water. Figures 2 and 3 illustrate schematically the phase relationships in peridotite and gabbro up to pressures corresponding to 600 km depth; these diagrams were obtained by combination of and extrapolation from experimental results. Each diagram has the solidus from Figure 1; peridotite has a wide melting interval, but gabbro and eclogite have a narrow melting interval. Two subsolidus transformation intervals are shown for each rock.

With increasing pressure, feldspathic peridotite transforms to spinel peridotite and then to garnet peridotite; this transformation is accomplished through the low-pressure shaded band. Garnet peridotite persists to the high-pressure shaded band, where olivine is transformed into a spinel-like phase and the pyroxene dissolves in the garnet [*Ringwood*, 1969*b*]. Phase transitions occurring at greater pressures are not shown. In Figure 3, the transformation of gabbro to eclogite is shown by the low-pressure shaded band, and by analogy with Figure 2 it is assumed that the pyroxene dissolves in the garnet through the high-pressure shaded transformation interval.

Figure 4 compares the phase relationships in dry gabbroic material with those in the presence of excess water to pressures corresponding to 100-km depth. The effect of adding excess water is to: (1) lower (initially) the temperature of beginning of melting, (2) increase the melting interval, and (3) introduce a hydrous mineral,

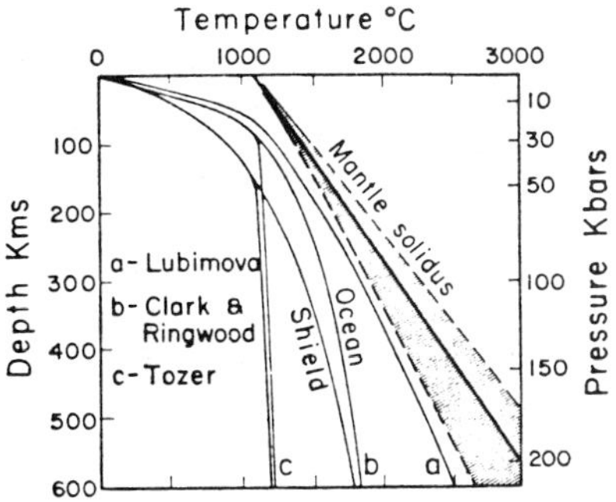

Fig. 1. Estimated geotherms for a conduction model [after *Lubimova*, 1967], a conduction-radiation model [*Clark and Ringwood*, 1964], and a convection model [*Tozer*, 1967]. Curves *b* were calculated to 400 km by Clark and Ringwood, and extended to 1400 km by Tozer. Experimentally determined solidus curves for peridotites and eclogites lie within the two solid lines extending to about 100-km depth [*Green and Ringwood*, 1967, 1968; *Cohen et al.*, 1967; *Ito and Kennedy*, 1967, 1968; *Kushiro et al.*, 1968]. The shaded band gives an idea of the range of uncertainty in extrapolating the mantle solidus to 600 km depth. The line within this band is adopted arbitrarily in following diagrams.

amphibole, which becomes stable with liquid. At about 15-kb pressure, the plagioclase feldspar breaks down, and the slope of the solidus (dP/dT) changes from negative to positive. Its slope becomes similar to that of the solidus for the dry eclogite. Similarly, as amphibole breaks down to yield garnet at pressures above about 15 kb, the slope of its breakdown curve reverses so that it passes below the solidus curve at about 25 kb. The gabbro-eclogite phase transition is masked within the stability field of amphibole.

The general pattern of the phase relationships for gabbro in the presence of a trace of water can be deduced from Figure 4 [*Lambert and Wyllie*, 1970*b*; *Wyllie*, 1970], and the result is shown in Figure 5. This isoplethal section for a gabbro-water mixture with insufficient water present to saturate the rock shows that all the water is stored in the amphibole within its stability field. The solidus between 1 and 25 kb is given by the curve where amphibole begins to break down and yield water for solution in a H_2O-undersaturated liquid.

Combining Figure 5 with the extrapolation in Figure 3 gives Figure 6 as an isopleth for gabbroic material in the presence of 0.1% wa-

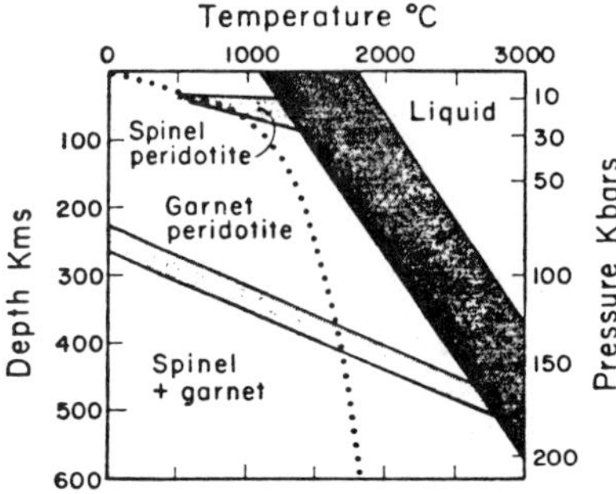

Fig. 2. Schematic phase diagram for peridotite. The melting interval is extrapolated from *Ito and Kennedy* [1967]. Subsolidus phase transformations are based on reviews by *Ringwood* [1969b] and *Wyllie* [1970]. Dotted ocean geotherm is from Figure 1.

ter. A similar isopleth is given in Figure 7 for peridotite-water based on preliminary data for the peridotite-water solidus and the synthesis field of amphibole in the peridotite-water system [*Kushiro, 1970*].

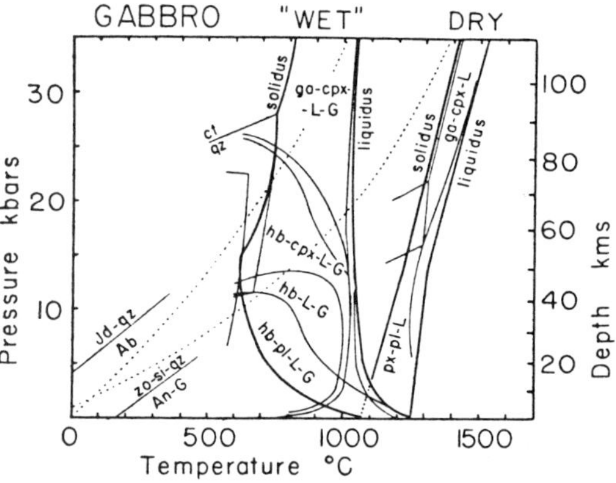

Fig. 4. Experimentally based phase relationships in gabbro [based on *Cohen et al.*, 1967, and *Green and Ringwood*, 1967] and gabbro in the presence of excess water (unpublished data by Lambert and Wyllie; see *Lambert and Wyllie* [1968, 1970a, b], *Hill and Boettcher* [1970]). The interval for the coexistence of amphibole (hb) and garnet (ga) is probably too narrow, because of the reluctance of garnet to nucleate; with seeded runs, the garnet curve would probably be determined at lower temperatures. Dotted lines are the geotherms *b* from Figure 1. Abbreviations: ga, garnet; cpx, clinopyroxene; px, pyroxene; pl, plagioclase; hb, amphibole; qz, quartz; ct, coesite; Jd, jadeite; Ab, albite; zo, zoisite; si, sillimanite; An, anorthite; L, liquid; G, gas or vapor phase, aqueous. The presence of epidote or zoisite, and sillimanite or kyanite, was suspected in high-pressure runs, but not proved.

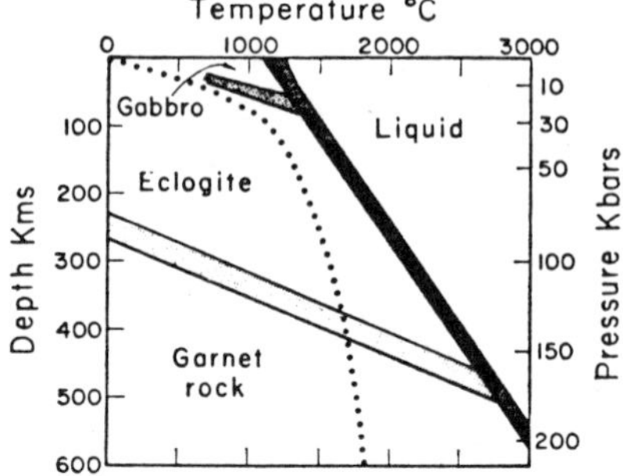

Fig. 3. Schematic phase diagram for gabbroic material. The melting interval is extrapolated from *Cohen et al.* [1967] and *Green and Ringwood* [1967]. Gabbro-eclogite transformation is from *Green and Ringwood* [1967] and *Ito and Kennedy* [1970]. Eclogite-garnet rock transformation is assumed to be in the same position as the corresponding transformation in Figure 2.

In Figures 6 and 7, the subsolidus phase transformations are the same as in Figures 3 and 2, respectively. The melting intervals are in two parts; the dark-shaded bands correspond closely to the dry melting intervals, and the light-shaded bands show where there is a trace of silicate liquid. These latter bands show where incipient melting occurs in the presence of traces of water. Notice the curve labeled Hb, for am-

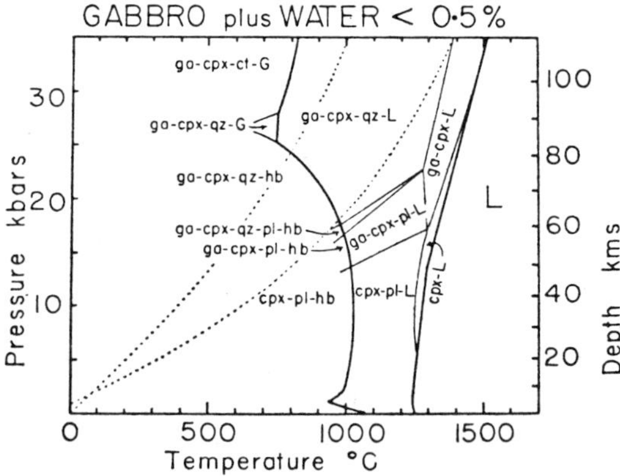

Fig. 5. Estimated isopleth for gabbro in the presence of a small proportion of water, <0.5% by weight, based on results in Figure 4 [see *Lambert and Wyllie,* 1970b; *Wyllie,* 1970]. The solidus is the breakdown curve for amphibole, which should be at somewhat higher temperatures under these water-deficient conditions than in Figure 4. Notice the subsolidus transition from amphibole-gabbro to amphibole-quartz eclogite. Dotted lines are geotherms *b* from Figure 1. Abbreviations are as in Figure 4. Compare Figure 6.

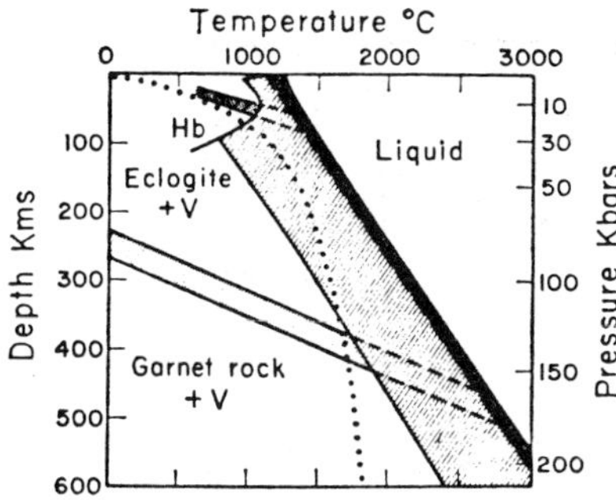

Fig. 6. Schematic isopleth for gabbro in the presence of 0.1% water, based on Figure 3 and extrapolation from Figure 5. The melting interval consists of two parts: the light-shaded band represents incipient melting, and the dark-shaded band is equivalent to normal melting as in Figure 3.

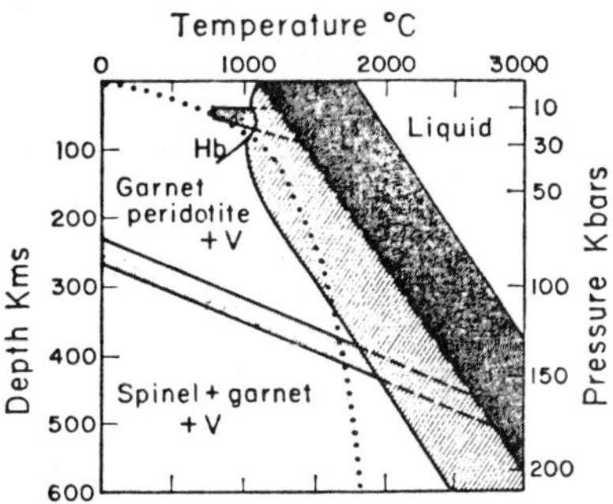

Fig. 7. Schematic isopleth for peridotite in the presence of 0.1% water, based on Figure 2, experimental data by *Kushiro* [1970] for excess water studies, and by analogy with Figure 6. If phlogopite is present, its stability limit would extend to higher temperatures than the hb line (remaining below the dry peridotite solidus [*Modreski and Boettcher*, 1970] before curving back to intersect the wet solidus at a presssure of 50 kb or more.

phibole breakdown; within these curves, all the water is contained within the minerals. I assume arbitrarily in Figures 6 and 7 that the solidus with water present becomes parallel with the dry solidus at pressures corresponding to depths greater than about 200 km. Figures 2, 3, 6, and 7 are schematic, but there is little variation possible in the general pattern of phase relationships depicted.

The occurrence of phlogopite in peridotite mantle could extend the vapor-absent region to depths probably in the range 150 to 200 kms.

Mineral facies in the upper mantle. Figure 8 shows patterns for the petrology of hypothetical mantle sections composed of either peridotite or eclogite with a trace of water present, in two different environments. These are derived by following the shield and oceanic geotherms (*b* in Figure 1) through the schematic phase diagrams in Figures 6 and 7. The geotherms pass through the intervals of incipient melting. Therefore, a trace of interstitial liquid exists in the depth zones where the geotherms exceed the solidus curves, as indicated by the shaded bands in Figure 8. The shield geotherm is at a lower temperature, so the zone of incipient melting is narrower. If the mantle is dry, then no melting occurs and only the subsolidus sequence of mineral facies is developed (compare Figures 2 and 3).

Lambert and Wyllie [1968, 1970a and b], using results in Figures 4 and 5, interpreted the low-velocity zone of the upper mantle in terms of incipient melting in the presence of traces of water. They listed several possibilities for downward termination of the low-velocity zone; it is here accomplished by passage of the geotherm through the solidus at depth. The amount of interstitial liquid produced is very small, and almost a direct function of water content (see Figure 9B); it does not vary significantly through wide variations in temperature and depth. For a given water content, the amount of liquid in peridotite is much less than in eclogite. The thickness of the shaded zones with interstitial liquid in Figure 8 vary with the geotherm, and thus with the tectonic environment, and the distribution of peridotite and eclogite in the mantle also affects its position. The versatility of this model for the low-velocity zone is one of its attractive features.

The production of liquid by the presence of water is likely to change the position of the geotherm used in Figures 6 and 7, which in turn would change the sequence of mineral facies in Figure 8. This procedure of using extrapolated experimental data and arbitrarily selected geotherms is intended only to show patterns for the petrology of the mantle; specific temperatures and depths can be modified as required by the acquisition of extended and improved theoretical and experimental data.

Magmas in the System Peridotite-Water

The most reasonable model for magma generation involves the diapiric uprise of mantle

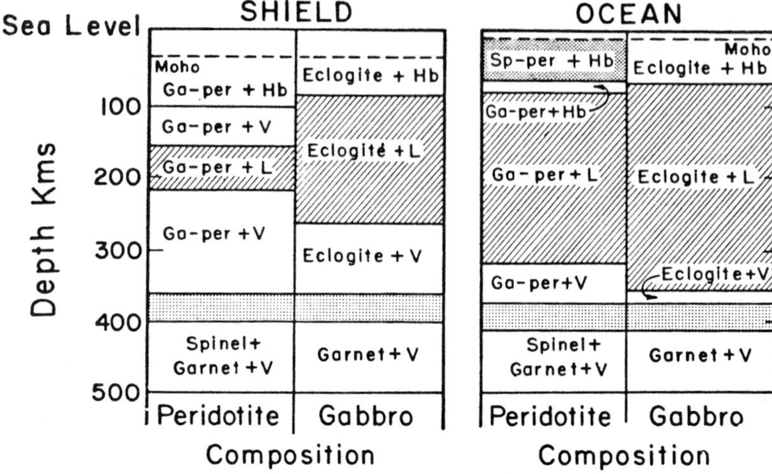

Fig. 8. Schematic sections through the upper mantle in two different tectonic environments, for mantle material composed of either peridotite or eclogite, in the presence of traces of water. These were determined by following geotherms through the mineral facies in Figures 6 and 7. Shading has same meaning as in Figures 6 and 7. The thickness of the zone of incipient melting depends on both environment (geotherm) and composition. This zone is equated with the seismic low-velocity zone.

material [*Green and Ringwood*, 1967]. The physical behavior of partially melted mantle depends on the amount of liquid present, and the variation in liquid composition through the melting interval is a significant factor in petrogenesis (Figure 9). The process that causes gravitational instability is a matter for speculation. One possibility is that upward migration of juvenile water from the deep mantle could operate as shown in Figures 10 and 11.

Figure 9*A* shows a linear pattern of melting for dry mantle peridotite (pyrolite) assumed by Green and Ringwood for at least 40% fusion. This is an idealized situation. In rock systems where the percentage of melting has actually been measured, the pattern is step-like rather than linear. The dry melting curve in Figure 9*B* is my estimate based on the results of *Ito and Kennedy* [1967], who reported little increase in liquid content in a natural peridotite between 1320°C and 1600°C at 20 kb. This curve is similar to that suggested by *O'Hara* [1968], who discussed the partition of elements between crystals and liquid in five stages of fusion, without giving temperatures. These stages are shown in Figure 9*B*. The changes in slope of the dry fusion curve are related to exchanges in the mineral assemblages coexisting with the liquid, as shown by the column between Figures 9*A* and *B*.

Green [1970] presented a comprehensive scheme for peridotite-water in the form of a petrogenetic grid that provides an internally consistent working model for igneous petro-

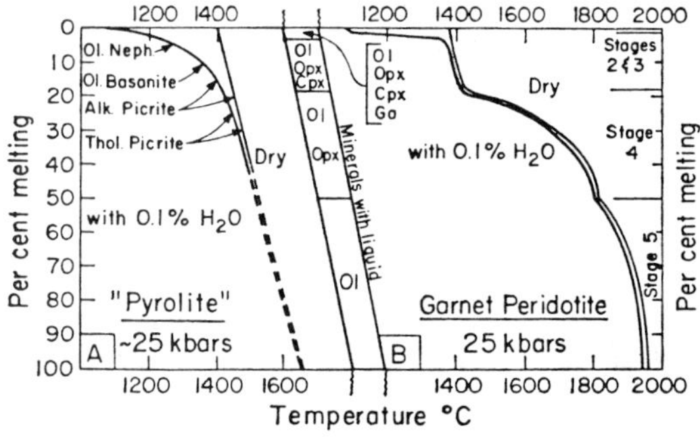

Fig. 9. Fusion curves for peridotite dry, and in the presence of 0.1% water, showing the percentage liquid produced at successive temperatures above the solidus, and the minerals coexisting with the liquid in different parts of the melting interval. (A) Dry curve from *Green and Ringwood* [1967]; curve with H_2O and estimated liquid compositions after *Green* [1970]. (B) Dry curve estimated from partial experimental data by *Ito and Kennedy* [1967], with stages according to *O'Hara* [1968]; curve with H_2O contrasting with the curve in *A*, and corresponding closely to *Ringwood*'s [1969a] estimate, except for the formation of a finite amount of liquid just above the solidus, consequent on dehydration of amphibole.

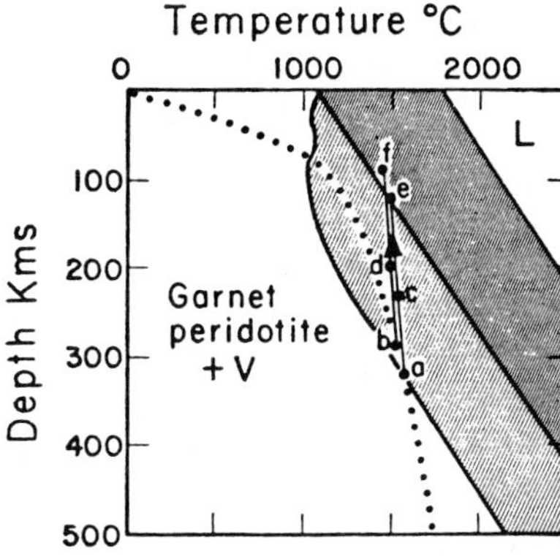

Fig. 10. Peridotite-water isopleth from Figure 7, and dotted oceanic geotherm from Figure 1. Diapiric uprise of layer *ab* (Figure 11*A*) to successive positions *cd* (Figure 11*B*) and *ef* (Figure 11*C*) is shown.

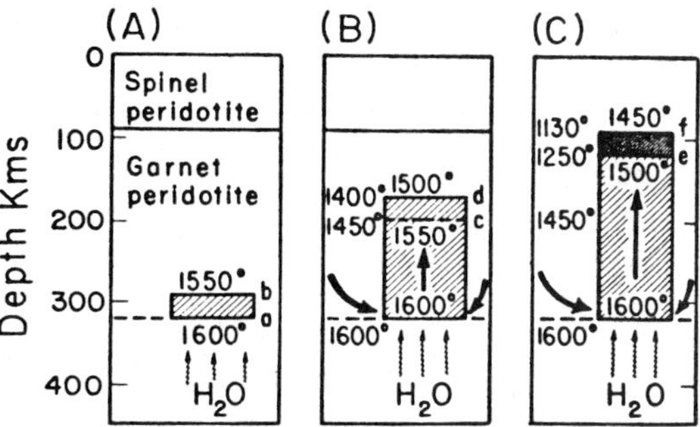

Fig. 11. Schematic mantle sections showing the diapiric uprise of layer *ab* at the base of the low-velocity zone into successive positions *cd* (*B*) and *ef* (*C*) on the adiabatic curves *ace* and *bdf* (*B*) Figure 10. Water migrating from the deep mantle increases liquid present in layer *ab*, causing diapiric uprise from the base of the low-velocity zone, and basaltic magma generation toward the top of the zone.

genesis. This is based on reconnaissance experiments in water-deficient regions with peridotites and basaltic compositions. Part of the model is shown in Figure 9*A* by the curve showing the percentage of melting produced with 0.1% water at each temperature above the solidus, at a pressure of 25 kb. The solidus is at 1080°C, where amphibole breaks down (compare Figure 7). The position of this curve is estimated on the basis of experimental results, but the data and method have not been published as yet.

Green's curve for melting of peridotite with 0.1% H_2O indicates surprisingly high percentages of liquid at subsolidus temperatures. According to the curve in Figure 9*A*, 0.1% H_2O is sufficient to produce about 17% melting at the dry solidus temperature, and to dissolve all of the garnet and most of the clinopyroxene. Another value indicates 25% liquid at the dry solidus. Until this curve can be considered as securely based on experimental data, I prefer the alternative pattern shown in Figure 9*B*, which is similar to that proposed by *Ringwood* [1969*a*].

In Figure 9*B*, a finite amount of liquid is produced within a few degrees of the solidus where hornblende breaks down (contrast Figure 9*A* and *Ringwood* [1969*a*]), and very little additional liquid is produced until the temperature closely approaches that of the dry solidus; at higher temperatures it follows the dry curve, rather than approaching it asymptotically as

indicated by Green's curve between about 30 and 40% melting. I have not attempted to estimate actual percentages of melting because I do not have adequate data, and the percentages guessed are simply to illustrate a pattern that contrasts with Green's pattern. Distinction between these two patterns is important, because the amount of interstitial liquid produced by traces of water below the dry melting temperature affects markedly the physical properties of the mantle and the prospects that the liquid can escape from its host for independent uprise as a magma.

The composition of the liquid produced is in dispute. From experimental studies, *Green* [1970] inferred that highly alkaline, undersaturated liquids could form by partial melting of mantle peridotite in the presence of water, and Figure 9*A* shows the compositions of the liquids developed at successive stages of partial fusion. From experiments in synthetic systems, *Kushiro* [1970] concluded that the liquids are andesitic or tholeiitic in composition.

Kushiro [1970] studied the composition of liquids coexisting with synthetic forsterite, orthopyroxene, clinopyroxene, and garnet at 20-kb pressure both dry and with excess water. The dry liquid is nepheline-normative, and the H_2O-saturated liquid is quartz-normative. Probe measurement of the glass confirms that the latter liquid composition was andesitic (Kushiro, personal communication, 1970).

Green [1970] concluded that both sets of experimental data were correct, but he described preliminary experiments suggesting that at 22.5 kb the conclusions from Kushiro's synthetic systems cannot be extrapolated to a typical basalt. He doubted that a quartz-normative tholeiitic or andesitic liquid could coexist with olivine at water pressure greater than 10 kb. Kushiro's results show that in the absence of hydrous minerals (e.g. amphibole) the first liquid produced is andesitic, and H_2O-saturated. The amount of andesitic liquid produced depends on the water content; with continued fusion in the vapor-absent region, the liquid compositions trend toward nepheline-normative.

The apparent conflict of interpretation between Green and Kushiro may be resolved when all the variables are considered and their effects adequately determined. In different parts of a petrogenetic grid such as that proposed by Green, we have partial fusion of the following assemblages, each of which may involve a liquid of different composition: (1) anhydrous minerals only, with vapor; (2) anhydrous and hydrous minerals, with vapor; (3) anhydrous and hydrous minerals, with no vapor. In compositions 1 and 2, the first liquid is H_2O-saturated, but in 3 the first liquid is H_2O-undersaturated. Figure 7 shows that a given fixed composition in the system peridotite-H_2O may change from condition 3 to condition 1 as a simple function of depth. *Robertson and Wyllie* [1971] have reviewed these factors in the water-deficient region of rock-water systems.

Diapiric Uprise from the Base of the Low-Velocity Zone

It has often been suggested that magma generation begins with diapiric uprise from the upper level of the low-velocity zone. I propose that uprise may begin at the base of the low-velocity zone, at depths of the order of 300 km, and that uprise may be triggered by the outward migration of water from within the deep mantle. The process is illustrated in Figures 10 and 11.

Figure 10 shows the isopleth for peridotite with 0.1% H_2O, from Figure 7. The dotted curve is a standard suboceanic geotherm from Figure 1. Figure 11 shows three schematic sections through the mantle representing successive stages in time that can be correlated with

Figure 10. The heavy dashed line at about 320 km is the bottom of the low-velocity zone, marked by *a* in Figure 10. The low-velocity zone between *a* and a depth of about 90 km contains interstitial liquid (not represented in Figure 11).

If water migrates upward along the geotherm in Figure 10, it reaches the low-velocity zone at *a*, and the water content of a layer such as *ab* increases; the increase of water causes more melting in this layer, approximately in proportion to the amount of water added. The increased melting lowers the density and viscosity of the layer relative to the surrounding mantle, and it tends to rise as shown in Figure 11*B* [*Ramberg*, 1967]. It rises adiabatically (Figure 10), and when the layer *ab* reaches the position *cd* it is about 100°C hotter than the surrounding mantle. Continued uprise carries it through the level near the dry solidus, where significant melting occurs (see Figure 9*B*), and at *ef* magma generation occurs in the normal way and subsequent events follow the dry model of *Green and Ringwood* [1967].

Petrology of the Mantle Beneath the Oceans

Figure 8 shows that the petrology of the mantle varies with temperature distribution as well as with composition. Horizontal temperature gradients produce horizontal petrological variations. *Oxburgh and Turcotte* [1968] considered the problem of mantle convection, and presented a steady-state distribution of isotherms beneath an ocean ridge, computed on the basis of a boundary layer theory, and assuming a constant viscosity, independent of temperature. They concluded that, although their assumed viscosity relations were certainly wrong, the provisional agreement between the model and natural phenomena suggests that the simplifying assumptions had not too great an effect. There is a mushrooming isotherm distribution associated with the upper ascending limb beneath the mid-oceanic ridge. Actual temperatures here would be decreased by upward transfer of heat by basaltic magmas.

The general pattern of the petrology of the upper mantle beneath the ridges and oceanic lithosphere can be determined from this model for temperature distribution and the schematic phase diagrams in Figures 6 and 7; the results are shown in Figures 12 and 13. Each point in the mantle section is defined by a specific pressure (depth) and temperature, and the phase assemblage for a given material at each point can be determined from the appropriate phase diagram, Figure 6 or 7.

Figure 12 shows the distribution of phase assemblages for a mantle composed of peridotite with a trace of water, and Figure 13 shows the phase assemblages produced in gabbroic material with a trace of water. The layered sequence for mantle some distance from the ridge is similar to that summarized in Figure 8. The presence of a trace of water produces a zone of incipient melting, which is equated with the low-velocity zone. Material of gabbroic composition exists as eclogite. The low-velocity zone increases in thickness considerably near the ridge crest, extending downward to the level of the olivine-spinel transition.

There is a large zone of partial melting where the temperature exceeds the dry solidus of peridotite, and gabbro in most of the corresponding zone in Figure 13 is completely melted. The zone is more than 400 km wide and 200 km thick. Oxburgh and Turcotte noted that only mantle material ascending along stream lines

within 100 km of the plume center would pass through this fusion zone, and that this material then forms the upper 130 km of the horizontal limb. If all the basaltic magma generated within the fusion zone escapes to the surface at or near the ridge, then this 130-km layer would be composed of residual peridotite. On the other hand, if some of the interstitial magma crystallizes as it is transported laterally through the upper boundary of the fusion zone, we have to consider phase assemblages in the upper mantle for material of basaltic composition.

Figure 13 shows that basaltic magma emerging from the fusion zone above about 80-km depth crystallizes as gabbro, and that below 100 km it crystallizes as eclogite. Lateral transportation of the lithosphere causes cooling of the gabbro, and if equilibrium is maintained the gabbro is transformed into eclogite through a wide zone of garnet granulite [*Ito and Kennedy*, 1970]. This illustrates the dynamic model for the suboceanic mantle proposed by *Press* [1969]. The low density layer of gabbro and garnet granulite exhibits a general pattern similar to that required by *Talwani et al.* [1965] for

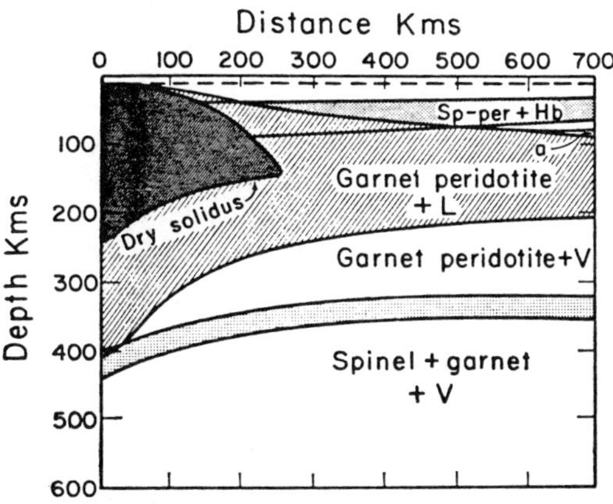

Fig. 12. Schematic section showing the petrology of peridotite with traces of water in the suboceanic mantle, extending from the crest of a mid-oceanic ridge. Far from the ridge, the section is similar to one in Figure 8. The area *a* is for garnet peridotite plus vapor. The shading has the same significance as in Figure 7 and 8. Each point in the mantle section is defined by a specific pressure (depth) and temperature, using *Oxburgh and Turcotte's* [1968] isotherms as basis, and the appropriate phase assemblages are plotted accordingly from Figure 7.

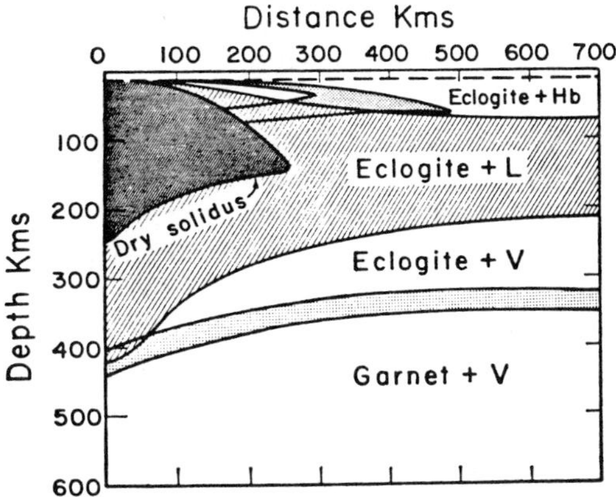

Fig. 13. Schematic section showing the petrology of a hypothetical suboceanic mantle composed of eclogite with traces of water, extending from the crest of a mid-oceanic ridge. Far from the ridge, the section is similar to one in Figure 8. This shows mineral facies for mantle layers or pockets of eclogite, with traces of water. The shading has the same significance as in Figure 6 and 8. This is based on *Oxburgh and Turcotte's* [1968] isotherms. Notice the shallow (less than 100 km) lens of gabbro and garnet granulite extending out to about 500 km from the ridge crest before the temperature cools sufficiently to produce eclogite (under equilibrium conditions).

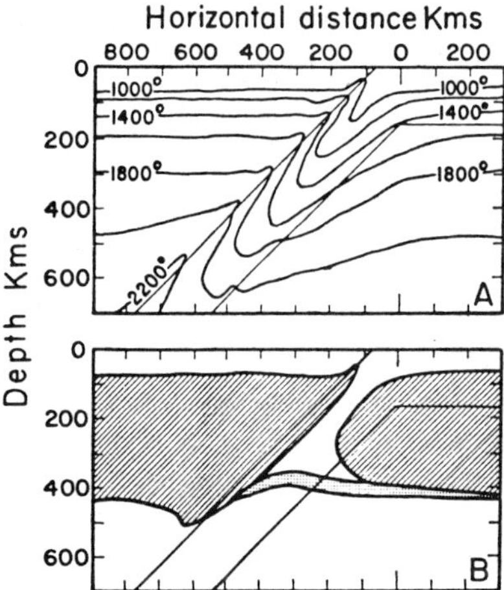

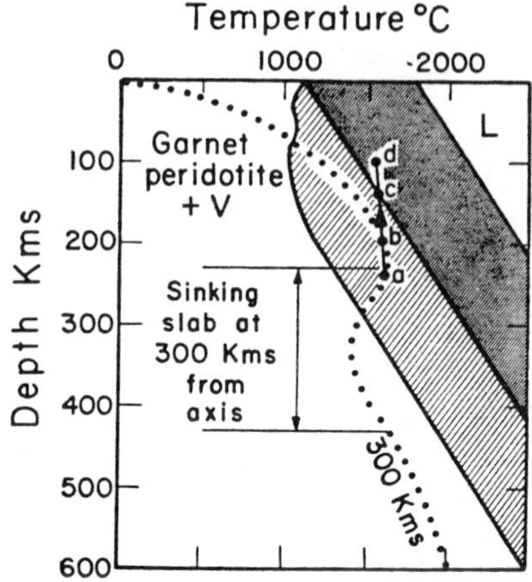

Fig. 14. Schematic sections through the mantle and slab of lithosphere in mantle beneath an island arc. The lithosphere, 160-km thick, dips at 45° from the axis at 0 km. (A) Isotherms computed by *Minear and Toksöz* [1970], according to their Figure 9, which takes into account shear-strain heating along the edges of the lithosphere slab. (B) Petrology of the mantle section, assuming peridotite with traces of water (Figure 7) and the temperature distribution in A. The shaded zone shows incipient melting, interrupted by the slab. Uprise of the olivine-spinel transformation zone within the cold slab is shown. The corresponding diagram for gabbroic material is very similar. Normal fusion temperatures for the generation of basaltic magmas are not reached anywhere according to this temperature distribution, and according to the extrapolated solidus from Figure 1.

Fig. 15. Peridotite-water isopleth from Figure 7, and the position of the sinking slab in the mantle at 300 km from the axis in Figure 14A. The dotted line is the geotherm for this position, taken from the isotherms in Figure 14A. Temperature decreases within the slab. This shows the diapiric uprise of layer *ab* (see Figure 16B) along an adiabat to the level *cd* (Figure 16C), where basaltic magma is generated.

the structure of mid-oceanic ridges. Although modification of the temperature distribution and experimental determination of the phase diagrams will change the details in Figures 12 and 13, the general pattern is probably realistic.

PETROLOGY OF THE MANTLE BENEATH ISLAND ARCS

The concept of plate tectonics requires that the lithosphere extend down into the mantle beneath island arcs [*Isacks et al.*, 1968], as shown schematically in Figure 14. *Minear and Toksöz* [1970] used a quasi-dynamic scheme and a finite difference solution of the conservation of energy equation to determine the effects of several factors on the temperatures in a downgoing slab. Thermal regimes were calculated for a 160-km-thick slab downwarping at 45° to the horizontal, with the complexity of the physical

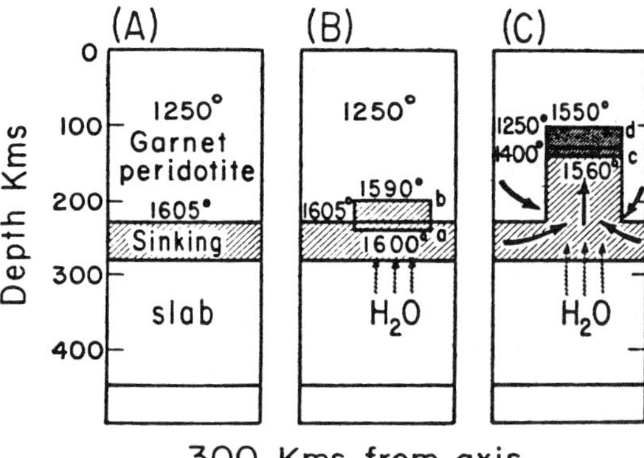

Fig. 16. Schematic mantle sections at a distance of 300 km from the axis in Figure 14 showing diapiric uprise of layer *ab* (B) to level *cd* (C) along the adiabat shown in Figure 15: (A) The sinking slab with interstitial water in a dry mantle of garnet peridotite. Only the upper part of the slab has a temperature high enough even for incipient melting. (B) Migration of water from the slab into layer *ab*, across the upper boundary of the slab, increases the extent of partial melting, leading to diapiric uprise of slab and adjacent mantle as in C. (C) Basaltic magma is generated in layer *cd*.

model being increased in a series of steps to demonstrate the effects of the various heat-generating processes and the spreading rate. Figure 14A shows the distribution of isotherms taken from their Figure 9, which assumes a 1-cm/year spreading velocity, shear-strain heating of 1.6×10^{-4} erg/cm³ sec along the top edge of the slab, and 1.0×10^{-5} erg/cm³ sec along the ends and bottom, and no contribution from phase changes or adiabatic compression. The dominant pattern is one of isotherms depressed deeply into the mantle. The zone of shear-strain heating with temperatures greater than those in normal mantle is less than 30-km wide above 200 km, and the greatest temperature increases occur at depths less than 100 km.

Each point in Figure 14A is specified by a temperature and pressure, and assuming a composition of peridotite with 0.1% water, the appropriate phase assemblage can be assigned to each point from Figure 7; the results are shown in Figure 14B. No melting occurs in peridotite (or in gabbro or eclogite) in this mantle section in the absence of water, but, given a trace of water, incipient melting occurs in the shaded zone. If this layer does correspond to the low-velocity layer, as was suggested in connection with Figure 8, then Figure 14B suggests that the layer is not continuous beneath island arcs. Upward migration of the olivine-spinel transition zone within the cool slab is shown.

Magma may be generated beneath an island arc system in three ways: (1) The oceanic crust at the surface of the lithosphere, including siliceous and water-bearing sediments, could be partially fused either by thermal conduction from the mantle, or by frictional heating, producing the calc-alkaline series [*Hamilton*, 1969; *Oxburgh and Turcotte*, 1970]. (2) Dehydration of the lithosphere and upward migration of water into hotter mantle overlying the slab could produce partial melting in the mantle, yielding intermediate and acid magmas for andesites and batholiths [*McBirney*, 1969]. (3) Diapiric uprise of solid peridotite or eclogite under adiabatic conditions may occur from the zone of frictional heating, leading to magma generation according to the *Green and Ringwood* [1968] model.

Figure 15 shows a geotherm at a distance of 300 km from the axis in Figure 14, and the depth interval occupied by the sinking slab at this distance. The temperature decreases within the slab, and there is a temperature maximum at or within a small layer *ab* across its upper boundary. Figure 15 also shows the isopleth for peridotite in the presence of a trace of water, transferred from Figure 7. If we assume a mantle containing no water, the slab is bounded by dry garnet peridotite, with the potential for incipient melting above the slab if water is added (compare Figures 14B and 15). If we assume that water is carried down into the mantle in the lithosphere slab, we have the situation depicted in Figure 16A for a vertical profile at a distance of 300 km from the axis. The position of the sinking slab is shown, and the upper part of this slab contains interstitial silicate liquid (compare Figures 14B, 15, and 16A). If water migrates from the sinking slab into the overlying mantle, this causes incipient melting in the layer *ab*, as is shown in Figure 16B (compare Figure 15). Influx of water from the slab into this layer increases the percentage of liquid present, decreases the density and viscosity, and facilitates diapiric uprise as shown in Figures 15 and 16C by the path *abcd*. This process would produce basaltic magmas from peridotite, or andesite from eclogite, at depths considerably above the Benioff zones. If water is abundant, liquids of intermediate composition may be generated in similar fashion from peridotite; we have seen that the compositions of liquids depend on a number of variables.

Discussion

Future correlation of the observed products of volcanoes with the compositions of liquids from various crystalline materials under known experimental conditions should permit us to locate the sources of specific magmas beneath island arcs and beneath ocean ridges. This would provide invaluable fixed points in terms of depths (pressures) and temperatures for many geophysical calculations and models, such as the computed temperature distribution in Figure 14A. This in turn would place limits on assumed physical properties of mantle materials. The way in which petrology, geophysics, and experimental petrology and geophysics are coming to be mutually dependent is very heartening. Until more precise data are available, we will have to be satisfied with general patterns such as Figures 8, 12, 13, and 14B, based on extrapolated experimental data (Figures 6 and 7) and possible temperature distributions such as Figure 14*A*. These provide insight into processes, and guidelines for further experiments.

Acknowledgments. I thank the National Science Foundation for grant GA-15718 and the Advanced Research Projects Agency for support from grant SD-89.

References

Clark, S. P., and A. E. Ringwood, Density distribution and constitution of the mantle, *Rev. Geophys., 2*, 35, 1964.

Cohen, L. H., K. Ito, and G. C. Kennedy, Melting and phase relations in an anhydrous basalt to 40 kilobars, *Amer. J. Sci., 265,* 475, 1967.

Green, D. H., Compositions of basaltic magmas as indicators of conditions of origin: Application to oceanic volcanism, *Phil. Trans. Roy. Soc. London,* in press, 1970.

Green, D. H., and A. E. Ringwood, The stability fields of aluminous pyroxene peridotite and garnet peridotite and their relevance in upper mantle structure, *Earth Planet. Sci. Lett., 3,* 151, 1967.

Green, T. H., and A. E. Ringwood, Genesis of the calc-alkaline igneous rock suite, *Contrib. Mineral. Petrol., 18,* 105, 1968.

Hamilton, W., Mesozoic California and the underflow of Pacific mantle, *Geol. Soc. Amer. Bull., 80,* 2409, 1969.

Hill, R. E. T., and A. L. Boettcher, Water in the earth's mantle: Melting curves of basalt-water and basalt-water-carbon dioxide, *Science, 167,* 980, 1970.

Isacks, B., J. Oliver, and L. R. Sykes, Seismology and the new global tectonics, *J. Geophys. Res., 73,* 5855, 1968.

Ito, K., and G. C. Kennedy, Melting and phase relations in a natural peridotite to 40 kilobars, *Amer. J. Sci., 265,* 519, 1967.

Ito, K., and G. C. Kennedy, Melting and phase relations in the plane tholeiite-lherzolite-nepheline basanite to 40 kilobars with geological implications, *Contrib. Mineral. Petrol., 19,* 177, 1968.

Ito, K., and G. C. Kennedy, The fine structure of the basalt-eclogite transition, *Mineral. Soc. Amer. Spec. Pap. 3,* 77, 1970.

Kushiro, I., Systems bearing on melting of the upper mantle under hydrous conditions, *Carnegie Inst. Wash. Yr. Book 68,* 240, 1970.

Kushiro, I., Y. Syono, and S. Akimoto, Melting of a peridotite nodule at high pressures and high water pressures, *J. Geophys. Res., 73,* 6023, 1968.

Lambert, I. B., and P. J. Wyllie, Stability of hornblende and a model for the low velocity zone, *Nature, 219,* 1240, 1968.

Lambert, I. B., and P. J. Wyllie, Melting in the deep crust and upper mantle and the nature of the low velocity layer, *Phys. Earth Planet. Int., 3,* 316, 1970a.

Lambert, I. B., and P. J. Wyllie, Low-velocity zone of the earth's mantle: Incipient melting caused by water, *Science, 764,* 1970b.

Lubimova, E. A., Theory of thermal state of the earth's mantle, in *The Earth's Mantle,* edited by T. F. Gaskell, p. 232, Academic, New York, 1967.

McBirney, A. R., Compositional variations in Cenozoic calc-alkaline suites of Central America, *Oreg. Dep. Geol. Miner. Ind. Bull., 65,* 185, 1969.

Minear, J. W., and M. N. Toksöz, Thermal regime of a downgoing slab and new global tectonics, *J. Geophys. Res., 75,* 1397, 1970.

Modreski, P. J., and A. L. Boettcher, The stability of phlogopite in the earth's mantle. Abstracts with programs, *Geol. Soc. Amer., 2,* 626, Denver, 1970.

O'Hara, M. J., The bearing of phase equilibria studies in synthetic and natural systems on the origin and evolution of basic and ultrabasic rocks, *Earth Sci. Rev., 4,* 69, 1968.

Oxburgh, E. R., and D. L. Turcotte, Mid-ocean ridges and geotherm distribution during mantle convection, *J. Geophys. Res., 73,* 2643, 1968.

Oxburgh, E. R., and D. L. Turcotte, Thermal structure of island arcs, *Geol. Soc. Amer. Bull., 81,* 1665, 1970.

Press, F., The suboceanic mantle, *Science, 165,* 174, 1969.

Ramberg, H., *Gravity, Deformation and the Earth's Crust,* Academic, London, 214 pp., 1967.

Ringwood, A. E., Composition and evolution of the upper mantle, in *The Earth's Crust and Upper Mantle, Geophys. Monograph 13,* edited by P. J. Hart, p.1, AGU, Washington, D. C., 1969a.

Ringwood, A. E., Phase transformations in the mantle, *Earth Planet. Sci. Lett., 5,* 401, 1969b.

Robertson, J. K., and P. J. Wyllie, Magma generation and crystallization: Rock-water systems, with special reference to the water-deficient region, submitted to *Amer. J. Sci.,* 1971.

Talwani, M., X. Le Pichon, and M. Ewing, Crustal structure of the mid-ocean ridges, *J. Geophys. Res., 70,* 341, 1965.

Tozer, D. C., Towards a theory of thermal convection in the mantle, in *The Earth's Mantle,* edited by T. F. Gaskell, p. 327, Academic, New York, 1967.

Wyllie, P. J., Ultramafic rocks and the upper mantle, *Mineral. Soc. Amer. Spec. Pap. 3,* 3, 1970.

(Received August 11, 1970.)

10

Deep Mantle Convection Plumes and Plate Motions[1]

W. JASON MORGAN[2]
Princeton, New Jersey 08540

INTRODUCTION

We may account for the main features of the Hawaiian Islands (the long linear chain, the uniform progression of ages toward the northwest, the transition from the tholeiitic main stage to more alkalic later stages of volcano growth) by assuming that the Pacific plate is moving northwestward over a fixed-mantle "hot-spot." Likewise the Greenland-Iceland and Iceland-Faeroe ridges emanating from Iceland, and the Rio Grande and Walvis ridges emanating from Tristan da Cunha and Gough Island, may be interpreted as the result of plates moving away from fixed hot-spots located on the crest of a spreading mid-ocean rise. Wilson (1963a, b; 1965a, b) advanced this hypothesis for the origin of island chains and aseismic ridges, and in a sequence of papers developed how these features may be used to determine the present motion of each plate, how the aseismic ridges are important guides in reconstructing pre-drift continental configurations, and how aseismic ridges and transform faults interrelate. Morgan (in press) has presented three additional lines of evidence supporting the concept of rigid plates moving over fixed-mantle hot-spots; this evidence is summarized in Figures 1–3.

In Figure 1, I quantify observations of the parallelism of the Pacific island chains and the continuity of the Hawaiian Islands and Em-

peror Seamount chains. Four sites of present volcanism are noted: (1) the Juan de Fuca Rise near Cobb Seamount, (2) Hawaii, (3) MacDonald Seamount (for a report on its discovery see Johnson, 1970), and (4) the Pacific-Nazca rise near Easter Island. The four heavy lines in Figure 1 were generated by rotating the Pacific plate backward in time over these four fixed hot-spots, first 34° about a pole at 67°N, 73°W (0–40 m.y.), then 45° about a pole at 23°N, 110°W (40–100 m.y.). The close agreement of these predicted lines with the trends of the Gulf of Alaska seamount chains, the Hawaiian-Emperor chain, the Austral-Gilbert-Marshall chain, and the Taumotu-Line chain substantiates this hypothesis. Even more exact agreement can be obtained by removing the constraint of fixed hot-spots and allowing the hot-spots to migrate at about ½ cm/year—a small fraction of the roughly 7-cm/year motion of the Pacific plate.

Figure 2 shows paleomagnetic pole positions determined from seamounts in the Pacific. The letter shown with each circle of confidence identifies the seamount group used by Francheteau *et al.* (1970); the number shows the age of the pole in millions of years. The heavy line is the Pacific polar-wandering curve predicted by the motion of the Pacific plate shown in Figure 1. (It is implicitly assumed that, at least during the past 100 m.y., the geomagnetic pole has not wandered relative to the hot-spots fixed in the lower mantle.) The paleomagnetic evidence thus verifies the plate motion predicted by the hot-spot trajectories. This test should be repeated for each of the other major plate units.

Figure 3 shows the present plate motion over the fixed hot-spots. This figure was constructed by finding the relative plate motions deduced from fault strikes and spreading rates on the

[1] Manuscript received, March 4, 1971; accepted, May 24, 1971.

[2] Department of Geological and Geophysical Sciences, Princeton University.

Kenneth Deffeyes first made several of the order-of-magnitude estimates shown herein. I thank him and many others at Princeton for their contributions to this problem. This work was partly supported by the National Science Foundation and the Office of Naval Research.

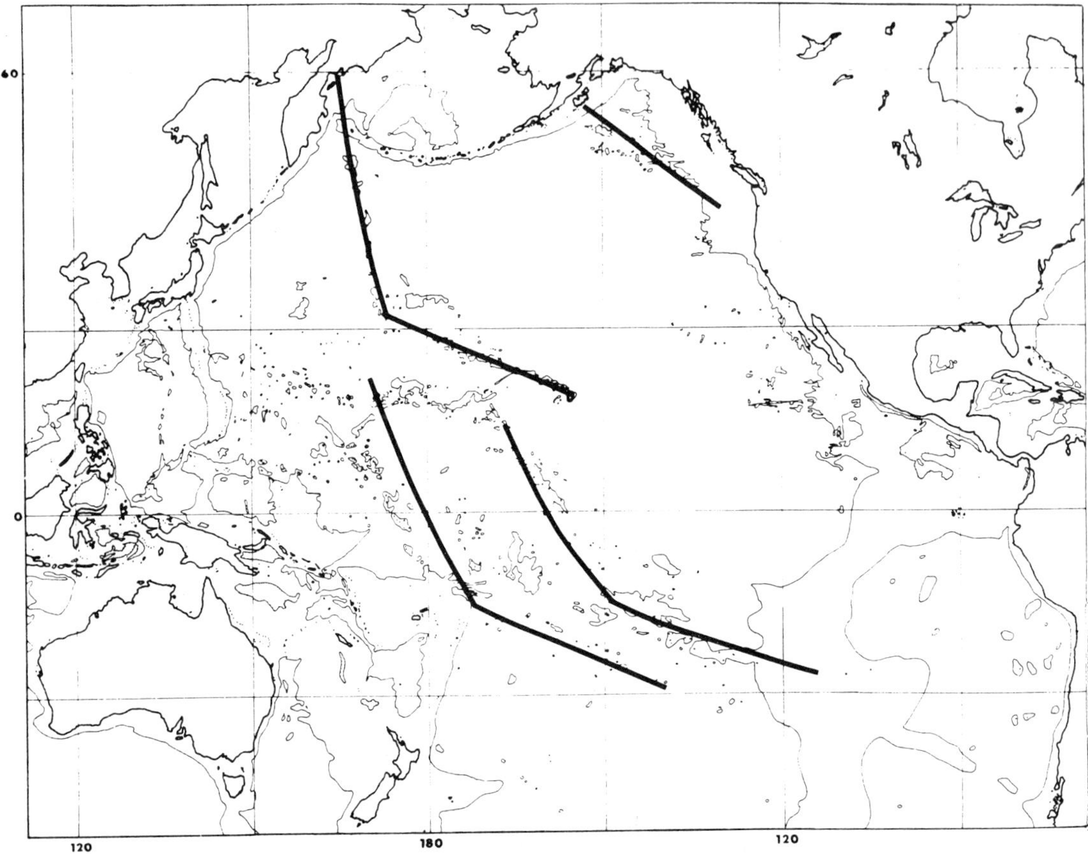

Fig. 1—Hot-spot trajectories constructed by rotating Pacific plate over four fixed hot-spots.

rise boundaries, and then adding a constant rotation to make the Pacific plate rotate properly over its hot-spots. If the synthesis tabulated by Morgan (in press) is correct, and if all the hot-spots are fixed in the mantle, then the velocity vectors shown in Figure 3 should predict accurately the trends of the island chains/aseismic ridges away from hot-spots.

Morgan (1971; in press) proposed that these hot-spots are surface manifestations of lower mantle convection which provides the motive force for continental drift. Assume that about 20 deep mantle plumes bring heat and relatively primordial material up to the asthenosphere, producing horizontal currents in the asthenosphere which flow radially away from each plume. The points of upwelling have unique petrologic and kinematic properties, but I assume there are no corresponding unique points of downwelling—the return flow is uni-

formly distributed throughout the mantle. The deep convection thus has a thunderhead character, whereas the shallow convection, constrained by the rigid plates at the top surface, has a roll or two-dimensional character.

Some of the consequences of the interactions of rigid plates with localized upwellings and an interpretation of the observed petrologic differences of oceanic island type basalt and oceanic ridge type basalt were presented by Morgan (1971; in press). In this paper I shall amplify the arguments supporting the claim that the hot-spots provide the driving force for continental drift. These arguments fall into three categories: (1) the observation that most hot-spots are near rise crests and evidence that hot-spots become active before continents split apart; (2) an interpretation of the gravity and topography around each hot-spot, showing that the mantle plumes generate moderately large stresses; and (3) estimates comparing the magnitude of stresses generated by plumes to the magnitude of rise and trench stresses.

LOCATION OF HOT-SPOTS

The primary criterion for selection of the hot-spots in Figure 3 was recent volcanic islands not associated with andesitic trench-type activity. Several volcanic islands were removed from this list based on the assumption that some volcanic activity is delayed by a magma-storage mechanism somewhere in the lithosphere. For example, the oldest rocks on Heard Island on the Kerguelen Ridge are 40 m.y. old, but there has been some activity in historic times. I assume that the entire Kerguelen Ridge is a hotspot feature and that the recent activity on Heard Island is a delayed action of hot-spot material placed there in the lithosphere 40 m.y. ago; thus Heard is eliminated from the present hot-spot list. A more puzzling case is the Cameroon Trend, which lines up with the ridge that heads northeast from St. Helena. I have tentatively assumed that the Cameroons are a delayed action of the St. Helena hot-spot, and that the Cretaceous volcanic rocks northeast of Mt. Cameroon are the expression of the St. Helena hot-spot prior to the breakup of South America and Africa.

Four oceanic centers of volcanism are not near mid-ocean rises: Hawaii and MacDonald in the Pacific plate, the Canary Islands in the

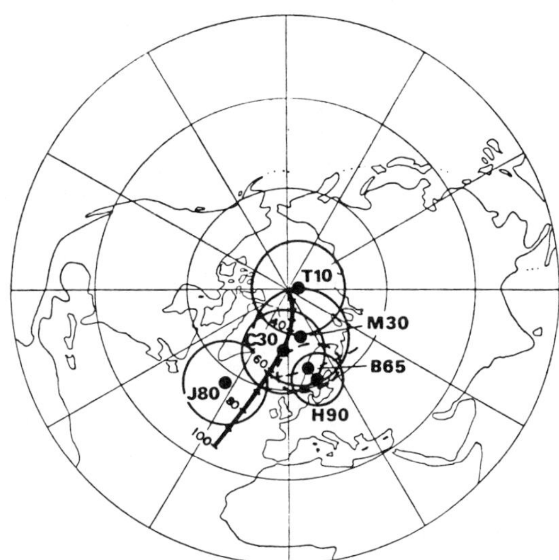

Fig. 2—Pacific paleomagnetic pole positions (adapted from Francheteau *et al.,* 1970) and polar-wander curve predicted by motion of Pacific plate shown in Figure 1.

African plate, and the Comores Islands in the Somalian plate. In addition, Yellowstone (and the Snake River basalts), Tibesti (in central Sahara), and Mount Kenya have characteristics suggestive of continental hot-spots. The Yellowstone, Kenya, and Comores hot-spots are near present-day breakups in the western

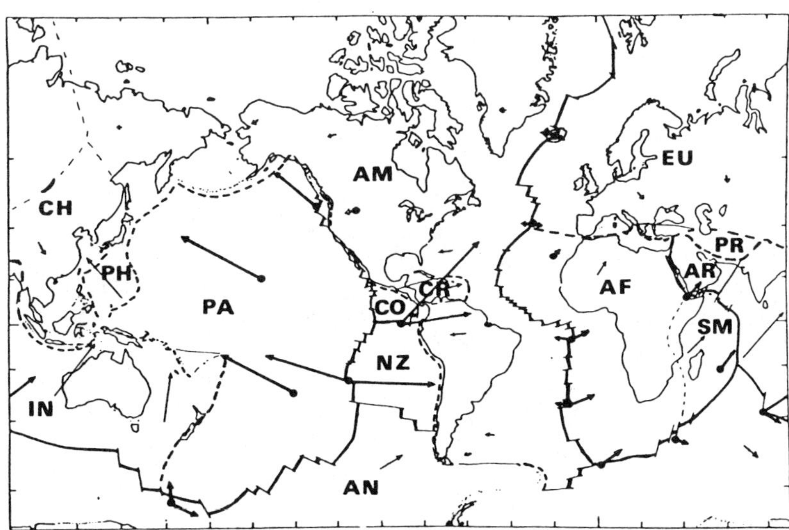

Fig. 3—Present motions of plates over hot-spots. Relative plate motions were determined from fault strikes and spreading rates on rise boundaries; with an appropriate constant rotation added, absolute motions of each plate over mantle were determined. Lengths of arrows are proportional to plate speed.

United States and East Africa, therefore only four of the locations listed are far from present spreading axes. In contrast, 14 hot-spots are near rise axes; four are in the Pacific (Juan de Fuca, Galápagos, Easter, and Balleny) and two lesser hot-spots suggested by seamount chains are near the mouth of the Gulf of California and the Eltanin fracture zone. There are four in the Indian Ocean (Amsterdam, Reunion, Afar, and Prince Edward), and six in the Atlantic (Bouvet, Tristan de Cunha-Gough, St. Helena, Ascension, Azores, and Iceland). One may argue that the lack of identifiable hot-spots on continents results from continental complexities camouflaging their presence. However, the reverse argument is no less valid; the hot-spots are mostly in the open ocean because they have pushed the crust away.

More dramatic than the location of the present hot-spots near the present rises is the evidence that the same hot-spots became active *before* the rises were formed. This evidence is best displayed in the lands bordering the Atlantic. The Jurassic volcanics in Patagonia may be regarded as the early expression of the Bouvet plume. (The even earlier Cape Volcanics in South Africa may be an expression of this plume. This interpretation depends on how Gondwanaland moved over this plume.) The flood basalts in the Parana basin and the ring dike complex of Southwest Africa may be due to the Tristan da Cunha plume. The White Mountain Magma Series in New Hampshire can be associated with the same hot-spot that produced the New England Seamount chain (probably the Azores plume). The Skaergaard

and the Scottish Tertiary volcanic province are associated with the Iceland plume. I claim this line of plumes produced currents in the asthenosphere which led to the continental breakup creating the Atlantic. Likewise the Deccan Traps (Reunion plume) were symptomatic of the forthcoming Indian Ocean rifting, and, if my premise is accepted as proved, the Snake River basalts (Yellowstone plume) foretell a breakup of North America.

Gravity and Topographic Highs

Figure 4 shows a worldwide gravity map computed for spherical harmonics up to order 16 (Kaula, 1970). Isolated gravity highs are apparent over Iceland, Hawaii, and most of the other hot-spots (Galápagos is a notable exception). Such gravity highs are symptomatic of rising currents in the mantle—the less dense material in the rising plume produces a broad negative gravity anomaly; but the satellite passes closer to the excess mass in the elevated surface pushed up by this current, and the net gravity anomaly in the area over the rising current is positive. The mid-ocean rises are exceptionally shallow near the hot-spots; note particularly the 10^6 sq km areas surrounding the Iceland, Juan de Fuca, and Galápagos plumes. This regional high topography is another manifestation of the rising plume, and I shall now use the magnitude of the high topography and gravity to estimate roughly the size of the rising current.

The formulas following are adapted from derivations shown by Morgan (1965). A spherical ball of mass deficiency M, located a distance D

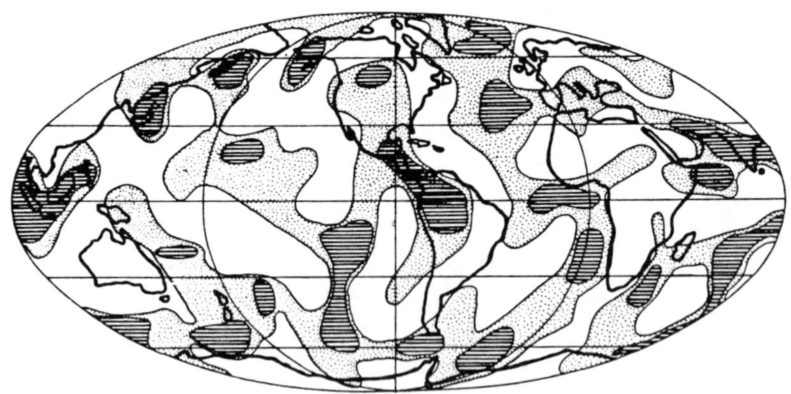

Fig. 4—Isostatic gravity map of earth constructed from spherical harmonic coefficients of degree 6 through 16. Shaded areas are regions of positive anomalies; heavier shaded areas are regions where anomalies are greater than +10 mgal. Note correlations of gravity highs with Iceland, Hawaii, and most other hot-spots (adapted from Kaula, 1970).

below the surface, is rising in a fluid of uniform viscosity. The Navier-Stokes equations were solved for a uniform half-space with rigid-plate boundary conditions for the top surface. The formulas give the normal stress σ_n, the shear stress σ_s, and the gravity anomaly δg at the rigid-plate boundary. The height the surface is pushed up is related to the normal stress by $\sigma_n = \rho g h$, and δg is the gravity anomaly produced by adding the effect of the raised surface to the gravity effect of the ball M. In these formulas, g is the gravitational field strength, G is the Newtonian Gravitational Constant, and r is the horizontal distance from the point directly above the ball.

$$\delta g = GMD \frac{(2D^2 - r^2)}{(D^2 + r^2)^{5/2}} \tag{1}$$

$$\sigma_n = \frac{Mg}{2\pi} \frac{3D^3}{(D^2 + r^2)^{5/2}} \tag{2}$$

$$\sigma_s = \frac{Mg}{2\pi} \frac{2r}{(D^2 + r^2)^{3/2}} \tag{3}$$

From Figure 4, we estimate that δg over a plume is typically $+20$ mgal and that the gravity anomaly falls off with distance, so that it is zero at 1,000 km away. Using equation (1) with these values, I find $D = 700$ km and $M = 0.8 \times 10^{21}$ g. If from topographic maps we estimate 1 km extra height as typical of the region near a plume, then equation (2) yields a value of $M = 3 \times 10^{21}$ g. These estimates are based on a uniform viscosity model; what effects would a more complicated viscosity pattern have? The uniform viscosity case yields the result that the total mass excess of the elevated surface equals the mass deficiency of the rising ball; i.e., the stresses produced by the rising ball are balanced by the stresses produced by the increased surface load. If plumes deep in the earth are rising in pipes surrounded by a very viscous mantle (much more viscous than the asthenosphere), then much of the stress of the rising material will be distributed throughout the more rigid mantle and will not produce a local elevated surface. Thus, a larger mass deficiency may be present than that estimated from the uniform viscosity formulas. We therefore estimate $M = 10^{21} - 10^{22}$ g as a typical mass deficiency of a single plume.

Using plume dimensions to be discussed later, the density of a plume can be calculated from its total mass deficiency ($M = 3 \times 10^{21}$ g). Assuming a cylindrical shape 150 km in diameter and assuming that only the top 1,000 km of the cylinder contributes to the M estimated by the surface gravity and topography, we find $\delta \rho = -0.2$ g/cu cm, or about a 5 percent density deficiency. This density change could be produced by a migration of "400–600"-km phase change boundary. This magnitude density deficiency is ideal; if the mass and dimensions yielded a density difference 10 times larger or 10 times smaller, the result would be respectively unreasonable or uninteresting.

Formula (3) may be used to estimate the shear stresses acting on the plate bottom. On the assumption that $M = 3 \times 10^{21}$ g and $D = 700$ km, then at $r = 500$ km the shear stress is 80 bars. How the shear stress falls off with distance away from the plume is very sensitive to the exact viscosity pattern; however I shall use 100 bars as a rough estimate of the shear stress on a plate near a plume.

STRESSES AT RISES AND TRENCHES

If the stresses produced by plume currents were clearly larger than the push of a rise or the pull of a trench, then the problem of finding the stresses acting on the plates would be greatly simplified. I have constructed the following mathematical model with three sources of stress: (1) stresses on plate bottoms falling off as $1/r$ away from each hot-spot, (2) a drag stress on the bottom of each plate proportional to the plate's velocity over the lower mantle and (3) stresses generated by plate-to-plate interactions of rises, trenches, and faults. The last category would have a moderately complicated set of equations predicting the stress generated by a specified closing rate at a trench or slip rate at a fault, etc., but it would have Newton's Third Law of action and reaction as a simplifying feature. With this model the torques on plates with the present boundary locations can be determined, and the direction and rate of motion of each plate predicted. The parameters specifying plume size and plate-to-plate interactions then could be accurately found by adjusting them until the present observed plate motions were predicted. If the plate-to-plate interactions were smaller than the stresses produced by plumes, very elementary assumptions could be made about rises and trenches, as small errors in this specification will not be important. If the push of rises and pull of trenches are stronger than the plume-generated stresses (as appears to be the case), the equations relating stress and strain rate at boundaries must be known accurately. Thus I consider the evidence relating to the magnitude of stresses at rises and trenches, and in particular reexamine the argument that the symmetry of rises shows that the rises exert no push on the plates.

A simple calculation places an upper limit on the amount of stress that can be generated by a spreading rise. Equate the work done pushing the plates apart (the total force on a plate times the rate of displacement) with the gravitational energy available in the light material rising into the spreading area (the buoyant force times the rate of upward movement). This calculation is performed most easily in a triangular geometry with a wedge-shaped unit rising and pushing two lithospheric plates apart. The calculation is independent of the shape of the triangle and the spreading rate; it depends only on the thickness of the lithosphere and the density deficiency of the material entering at the bottom compared to the average density of the lithosphere. If it is assumed that $L = 70$ km and $\delta\rho/\rho = 3$ percent, the horizontal compressive stress in the lithosphere would be 300 bars. Any of the gravitational energy of the rising wedge that is dissipated in viscous flow will not be available to do the work required to push the plates apart; thus 300 bars is an upper limit. A 1 percent density deficiency may more accurately describe the material entering the region below the rise; in this case the horizontal compressive stress averaged over the thickness of the lithosphere would be less than 100 bars.

Focal mechanism studies of earthquakes along rise axes have shown that the lithosphere at a rise is under tension in the direction of spreading. Wyss (1970b) has concluded that this tension has a magnitude of roughly 200 bars; this result is based on studies of the seismic moments of rise earthquakes showing an "apparent stress release" of 20 bars, combined with an estimate of 10 percent for the seismic efficiency of stress release. This tension of 200 bars can be interpreted several ways. In one interpretation, we assume that some distant forces are pulling the plates apart and that the asthenosphere is rising passively to fill the void that would be created by plate separation. The lithosphere is very thin beneath the rise crest, and this "necking" of the lithosphere acts as a stress concentrator. That is, the tensional stress may be 200 bars in a region 5 km thick at the rise crest, but only 50 bars spread out over the entire 70 km thick lithosphere at some distance away from the rise. The amount of stress concentration in the thin lithosphere depends on how the stress load is distributed between the cool, strong lithosphere and the hotter, weaker asthenosphere flowing into the broad "gap" between the plates. In an alternate interpretation, the rising asthenosphere is pushing the plates apart, causing horizontal compression in the plates except in the small thin section of lithosphere at the rise which resists the separation. The average compression in the plates thus will be reduced by a factor which depends on the effective viscosity and thickness of the lithosphere at the rise crest. The dissipation in this thin section of lithosphere may form the major part of the viscous dissipation mentioned in the preceding paragraph. Therefore, the stresses at rises may be compressive or tensile, but in any event have a magnitude less than a few hundred bars.

Why are the mid-ocean rises "mid-ocean", and why is the seafloor magnetic pattern symmetrical about the rise crest? It would be easy to imagine that a rise creates new sea floor on one side only, analogous to the one-sided consumption of crust in a trench system, and yet new sea floor is created in equal amounts on the two sides of a rise. As a consequence, rise crests cannot be fixed with respect to the mantle; they must migrate over the mantle to maintain their position midway between continents. An example of such rise migration is seen for the rise boundaries that enclose Africa on three sides. As the Mid-Atlantic Rise spreads symmetrically, there is ever more sea floor between the rise crest and the African coastline. With a similar increase in the distance from Africa to the crest of the Mid-Indian Rise, the distance from the crest of the Mid-Atlantic Rise to the crest of the Mid-Indian Rise must be increasing. Thus both rises cannot be fixed with respect to the mantle—one or both must be migrating over the mantle.

The Mid-Atlantic Rise apparently is fixed to the mantle, as all the Atlantic plumes are near the present crest. Thus it is the Mid-Indian Rise that is migrating east at a rate faster than the African plate is moving northeastward. The Mid-Indian Rise has migrated over the Reunion plume—this plume was once in the Indian plate (Deccan Traps, Laccadive-Maldive island chain), but is now on the African side of the rise. Similarly, the growth of the Afar Triangle on the southwest may be regarded as the plume staying fixed while the Red Sea and Aden rifts migrate northeastward.

It has been argued that the symmetrical spreading and the migrating rise crest indicate

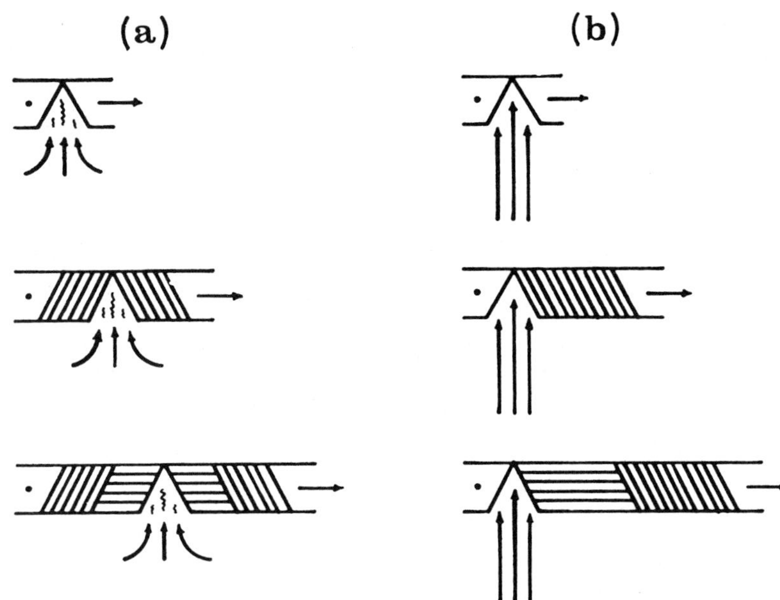

Fig. 5—In these symmetrical and asymmetric models of sea-floor spreading, left lithospheric plate is constrained to be fixed and right plate to move away at constant velocity. (a) If asthenosphere near rise fills gap made as plates move apart, symmetrical sea-floor spreading results. (b) If location of rising current is influenced strongly by conditions near bottom of asthenosphere, one-sided sea-floor spreading results.

that the rises do not drive the plates. If the two sides of a rise are moving away from a fixed rise crest with equal velocities, there is little problem; in contrast in Figure 5 I postulate a case in which the left plate has zero velocity over the mantle and the right plate is moving away with constant velocity. I do not inquire why the plates have these velocities—there may be a trench nearby on the right and another rise off left, or whatever is needed to produce the motions depicted in Figure 5. As the plates move apart at the rise crest, material from the asthenosphere rises to fill the void that would otherwise develop. The exact center of the most recently injected "dike" is hotter than any other part of the lithosphere, and because strength is extremely temperature dependent, this is where the plates will tear apart and another dike be inserted. Thus if the temperature pattern about the rise crest is symmetrical, the temperature dependence of strength will assure a symmetrical pattern of sea-floor spreading.

Any arguments about symmetrical or asymmetric spreading must thus concentrate on those factors which will make a symmetrical or asymmetric temperature pattern within a rise. It was concluded that the important factor is to have a soft asthenosphere below the plates, and that material flowing into the "gap" should be

drawn from very shallow depths to avoid any kind of coupling with conditions at the bottom of the asthenosphere. It appears that passive pulling of asthenosphere into the gap or active driving of the asthenosphere upward into the gap is not related to the question of symmetry. The tensional or compressive nature of rises is related to whether or not there is a density inversion between the lithosphere and asthenosphere; the symmetrical spreading is related to the laws of heat conduction and the temperature dependence of strength. Thus it is thought that earlier conclusions of the writer (Morgan, 1971) and of Elsasser (1969) relating symmetry to passiveness are in error. A two-dimensional numerical rise model with viscosity and density varying with temperature could aid in answering this question.

A related problem concerns the existence of fracture zones. Objectors to sea-floor spreading have used the pattern of transform faults as an argument against spreading; namely, it is inconceivable that convection currents beneath the surface could have the numerous offsets of the surface pattern. The notion of a crustal plate removes the objection, as the deeper flow may have a smoother, more fluid pattern and only the "rigid" lithosphere need be broken into the irregular pattern observed at the surface. How-

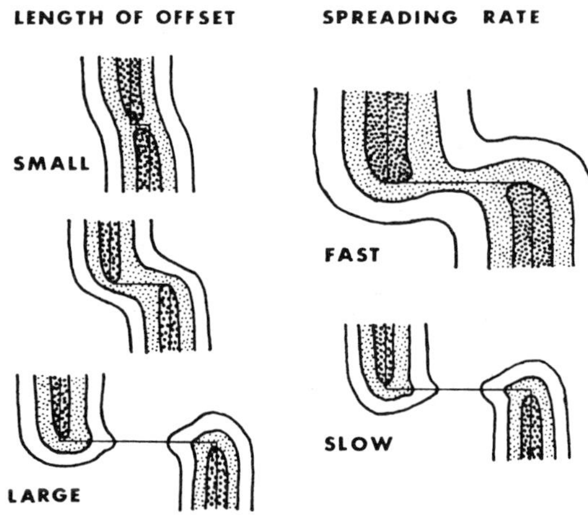

FIG. 6—Stability of fracture zone is influenced by length of offset and spreading rate. Contours are, schematically, thickness of lithosphere or, alternatively, depth to particular isotherm. If offset is very small, obliquely spreading rise may develop and transform fault would no longer exist. Length of offset of minimum stable transform fault would be dependent on spreading rate as with high spreading rates broader region of "thin lithosphere" would enclose rise (from Vogt *et al.*, 1969).

ever, the causative factors behind the maintenance of many short transform faults and the tendency for the rise to form segments perpendicular to the transform faults remain unknown.

Vogt *et al.* (1969) have noted that the minimum observed offset of transform faults is less in the Atlantic than the Pacific. They have devised the model shown in Figure 6 to account for this difference. The contours in this figure schematically represent either isotherms at a given depth (*e.g.*, 10 km), with the hottest temperature at the rise crest, or alternatively they represent the depth to a given isotherm (*e.g.*, 1,000°C). We may say arbitrarily that all material cooler than 1,000°C is lithosphere and any hotter material is asthenosphere; in this case the contours in Figure 6 show the thickness of the lithosphere. These contours are patterned after heat flow models of rises (see McKenzie, 1967) in which it was shown that the distance between the isotherm contours is directly proportional to the spreading rate of the rise. The model in Figure 6 also assumes that heat is generated on the transform faults by frictional heating. Vogt *et al.* (1969) used this diagram to illustrate that, if the offset of a transform fault is "'sufficiently small'" (where "sufficiently small" is directly related to spreading rate),

then the regions of thin lithosphere may merge and an oblique spreading ridge replaces the two short perpendicular segments and the transform fault. Figure 6 also indicates that an offset greater than a certain minimum length will be stable and will not be replaced by an obliquely spreading rise.

The flow pattern chosen by nature must minimize the total dissipation rate. Therefore, the friction per unit length along a transform fault must be markedly less than the tensile resistance of a rise segment, otherwise an oblique rise would be observed more commonly. Further, I conclude that there is more tensile resistance for an oblique rise than a perpendicular rise, or else the dissipation rate of "perpendicular rise and fracture zone" would not be less than "oblique rise." The larger resistance of the oblique rise must be due to increased heat loss through the longer sides resulting in a higher effective viscosity for the oblique rise. Several models to relate these factors were constructed by the writer to attempt to relate the tensile stress to the dissipation rate. However, it was realized that a constant compressive stress produced by buoyant material would have no effect on the model. The buoyant terms are determined by the rate of the upflow, which is the same in the oblique and normal cases; thus the solution of this problem could indicate much about the mechanisms at rise crests but cannot be used to place a limit on the overall tensile or compressive nature of the rises.

I conclude that we do not really know whether rises are characterized by compressive stresses aiding the plate motion or by tensile stresses acting as a brake to plate motion. Inasmuch as tensile stresses at rises would be most favorable to the plume driven model, it seems conservative to expect all rises to generate compressive stresses of about 100–300 bars.

On the question of the tensile (actively pulling down) or compressive (resisting being pulled down) character of trenches, the earthquake mechanism solutions of Isacks *et al.*, (1969) show that earthquakes at 100–200 km depth indicate tensile stress along the sinking slab; the slab above is being pulled down by a density excess at or below this depth. Deeper earthquakes, from 300–700 km, show compressive stresses along the slab; the deeper part is resisting being driven deeper into the mantle by stresses generated above. From earthquake seismic moments (Wyss, 1970a), from surface gravity anomalies (Morgan, 1965), and from calculations based on assumed temperature profiles in the sinking slab (McKenzie, 1969), the stress produced by the sinking slab has been estimated at from a few hundred bars to a thousand bars. The motion of the sinking slab is resisted by the material on the slab boundaries, particularly on the boundary of the highest seismic activity, where the two lithospheric plates rub together. Wyss (1970a) found that 200-bar stresses are indicated by the shallow earthquakes at trenches, the same magnitude of stress as found at rises. There are many questions about the magnitude of the stresses at trenches. How much of the pull of the sinking slab is cancelled by the shallow friction of the two lithospheric plates? Do deep trenches pull with more tension than intermediate depth trenches, or does the deeper slab push into a very resistant media which tend to reduce the pull of the slab? Do areas of continental underthrusting, as in the Zagros and Himalayas, exert any pull at all, or are all stresses in these regions compressive, resisting the closing motions?

I add one new point to the discussions of trenches. Figure 3 (and more precisely if not more accurately, Table 1 of Morgan, in press) shows the absolute rate of each plate over the mantle, as determined by relative spreading rates plus the trajectories of hot-spots. The Nazca plate is moving eastward toward the Peru-Chile trench at about 7 cm/year, whereas the South American plate is moving westward toward this trench at about 1.5 cm/year. When we examine the velocities of plates at other trench systems, we find the following to be a general rule: both plates move toward the trench with the underthrusting plate moving about four times faster than the overriding plate. (This claim is made with great reservations, because, to establish it a much more accurate determination of absolute plate motions is needed.) However, note the special situation of the Philippine Sea. It is moving westward toward Asia and thus is not moving toward the Marianas trench. Also, the Indian plate does not have a component of motion toward the Tonga trench. It thus appears that the areas of anomalous spreading behind trenches (Karig, 1970) can be identified if the absolute motion of the plates is known. Moberly (in press) and Elsasser (1971) have given theoretical explanations of this phenomenon; they suppose the lithospheric slabs are not exclusively sliding obliquely into the mantle but also have a vertical sinking component. The sliding and sinking cause the trench axis to migrate "seaward," and the overthrusting plate must migrate to-ward the trench or a gap will open behind the trench.

This motion of both plates toward a trench appears to clinch the argument in favor of tensile stresses at trenches. In the following discussion I shall assume this horizontal tensile stress has a magnitude of a few hundred bars. I shall neglect the stresses produced by plates rubbing together at great faults, partly because such stresses must surely be reactive and not drive plates, and partly because the model of Vogt *et al.* (1969) on transform faults shows the dissipative stresses on faults must be much less than on rises.

ESTIMATES OF PLUME MAGNITUDES

Estimates of stresses, heat flow, and lead isotope data are now used to estimate the sizes of plumes. A self-consistent set of relations is found if each plume is 150 km in diameter, with an average upward velocity of 2 m/year; 20 such plumes would bring up a total volume of 500 cu km/year. Other values used in this model are a lithosphere 70 km thick and an asthenosphere 200 km thick, with an average viscosity of 3×10^{21} poise. No exactitude should be placed on any of these numbers, as it is the overall effects of the parameters and not the precise value of any one which is significant in the ensuing discussion. My purpose in the use of this model is to show that plumes can provide the stresses needed to move the surface plates and have important implications in the interpretation of heat flow and age of the mantle, and at the same time do not violate any of the "known" values of the earth.

Using equation (3) I found that the shear stress 500 km away from a plume was about 100 bars. I obtain this same magnitude of stress in the plume and thin asthenosphere model specified above. The average velocity $\bar{v}_a$ of the asthenosphere at a distance R from the plume is related to the upward velocity of the plume v_p by $2\pi R D \bar{v}_a = \pi d^2 v_p/4$, where D and d are the thickness of the asthenosphere and diameter of the plume respectively. Using the model values given above, I calculate that the average asthenosphere velocity 500 km from the plume is 5 cm/year. The asthenosphere flow is channeled between the rigid upper plate and the lower mantle (assumed to be slightly more viscous), and the velocity profile in this channel takes the well-known parabolic shape. (If the upper plate is moving, a linear velocity profile will be superposed on this pattern.) The stress at the top boundary (σ) is related to the average velocity in the asthenosphere ($\bar{v}_a$) by $\sigma = 6\eta\bar{v}_a/D$, giving $\sigma = 150$ bars at 500 km radius.

Integration of the effect of this stress on the bottom of a plate determines the total force one plume can exert on a plate. I assume that a rise axis passes right over a plume and integrate the component of force directed away from the rise. I choose as limits of my integration 75 km (the radial velocity is zero directly above a plume and increases to a maximum value a plume radius away) and 1,500 km (roughly half the distance between plumes). This integration yields $F = 1 \times 10^{24}$ dynes for the total force exerted on a plate.

I then compare this magnitude with the stress created by a plate moving over the asthenosphere. With the properties of the asthenosphere given above, a plate moving 3 cm/year over the mantle creates a shear stress of 15 bars. If we assume the plate is a square about 5,000 km on a side, then the total drag on a plate is $F = 4 \times 10^{24}$ dynes. Another comparison is with the stress generated by plate-to-plate interactions. I assume a stress of 100 bars on a plate 70 km thick along a boundary 10,000 km long; the total force acting on the plate is $F = 7 \times 10^{24}$ dynes. The plume generated stress is smaller than the other two, but it could be increased by changing the values of the viscosity and thickness of the asthenosphere, or by increasing the flow up a pipe. My conclusion is that, given the uncertainty in these factors, all three mechanisms (plume, drag, and plate-to-plate) should be considered in a model of plate motion.

The key estimate is the volume of flow up the pipes. There is a clear-cut way to obtain a lower limit on the rate if the plumes are driving the plates. From roughly 40,000 km of rise axis with an average (half) spreading rate of 3 cm/year, I can determine how much new crust is generated each year. An accurate summing of the spreading rates along each rise yields 2.5 sq km/year for this rate (Deffeyes, 1970, p. 214). If this is multiplied by the thickness of the lithosphere, it is apparent that lithosphere is being generated at a rate of 170 cu km/year, and of course destroyed at an equal rate at the trenches. Suppose I had concluded that the total volume brought up by the plumes was only 10 cu km/year, or some other equally small number. Then asthenosphere currents of total flux 10 cu km/year would be spreading horizontally away from several points on or near rise crests, whereas 170 cu km/year would be flowing toward the rises as a counterflow to the lithosphere motion. The net stresses on plate

bottoms would be such as to close up the plates. Thus the total volume emanating from the plumes must be several times larger than 170 cu km/year. The value of 500 cu km/year specified in the model is three times this; one may like a larger multiple, but this value does satisfy the lower bound and fits the lead isotope criteria to be discussed subsequently. In this light, even small plume flow aids the shallow convection in a way not noted in the preceding paragraph, where I found that a plate spreading at 3 cm/year created a viscous drag of 15 bars. If I include in this calculation the assumption that the 200-km-thick asthenosphere must have a net flow to counter the mass transport of the lithosphere, then the stress on the bottom of the plate is 50 bars instead of 15 bars.

These numbers have interesting consequences for the interpretation of heat flow data. Suppose 500 cu km/year (ϕ) brought up by the plumes is on the average 300°C (ΔT) hotter than the nonplume mantle at the same depth. Then using $C_p = 0.25$ cal/g°C, I find that the total upward heat available from the plumes is $Q = \rho \, C_p \, \Delta T \phi = 1.5 \times 10^{20}$ ergs/sec. This number is half the total heat flow of the earth. Here is a mechanism for concentrating all the deep mantle's heat production into predominantly oceanic regions. The correct interpretation of this may show why the oceanic and continental heat flow averages are so nearly equal. There is another surprise in these numbers: the return flow of the plumes, involving the slow sinking of the entire mantle, is at a rate of 0.1 cm/year and downward convection at 0.1 cm/year can dominate over conduction or radiation mechanisms of heat transport. For example, if I assume a temperature (T) of 1,500°K at the base of the asthenosphere, then the downward flux of heat by convection is $q = \rho \, C_p \, T \, v = 3.7 \, \mu\text{cal}/$ sq cm/sec. Thus, heat could leave the lower mantle only at the plumes. How much heat recycles compared with the amount of heat lost at the upper surface is a measure of the efficiency of the heat engine—thus the earth could be regarded as being a moderately efficient heat engine.

The isotopic composition of lead from Tristan da Cunha and St. Helena has been discussed by Oversby and Gast (1970) together with earlier results from Ascension and Gough. They showed that the lead data cannot be interpreted with a simple one-stage growth model: i.e., a mixing of lead and uranium isotopes 4.5 b.y. ago, when the mantle was formed, with no separation or mixing since (except perhaps in the last few million years as the rocks were brought to the surface). Instead, they found a two-stage growth history with lead events at 4.5 b.y. and 1.8 b.y. (plus possible changes in the last few million years). That is, they interpreted their data to show that there was an homogenization and then separation of the lead and uranium isotopes 1.8 b.y. ago in the material that now makes up these islands, and that no further mixing or separation occurred until very recently.

I incorporate this observation into the plume model as follows—I assume that the rocks which now make up these islands were last near the earth's surface 1.8 b.y. ago; *i.e.*, that 1.8 b.y. is the cycle time required for a particle to sink slowly in the mantle and then to rise in a plume back to the asthenosphere (or in this case, for part of the plume to reach the surface). The rate of upwelling of 500 cu km/year fits these data; the total volume of the mantle, 1×10^{12} cu km, divided by this rate gives 2 b.y. If the lead isotope data are interpreted in this manner, it makes two restrictions on the plume model. First, this is evidence that the entire mantle is involved in the overturn—that the plumes extend all the way down to the core-mantle boundary. Second, the rate of upwelling may be estimated most accurately by knowing the period of the mantle overturn—a very straightforward estimate if lead isotope data from other plumes far from the South Atlantic also show the 1.8 b.y. lead event.

In conclusion, the mid-ocean position of most of the plumes and the land evidence of plume activity prior to continental breakup suggest that the plumes produce the stresses which drive the plates apart. An order of magnitude estimate shows that stresses produced by plume currents are comparable to other stresses. The model implies that the entire mantle overturns once each 2 b.y., a conclusion which would require a new interpretation of heat flow and chemical evolution problems.

REFERENCES CITED

Deffeyes, K. S., 1970, The axial valley: a steady-state feature of the terrain, *in* H. Johnson and B. L. Smith, eds., Megatectonics of continents and oceans: New Brunswick, New Jersey, Rutgers Univ. Press, p. 194–222.

Elsasser, W. M., 1969, Convection and stress propagation in the upper mantle, *in* W. K. Runcorn, ed., The application of modern physics to the earth and planetary interiors: London-New York, Wiley-Interscience, p. 223–246.

———— 1971, Sea-floor spreading as thermal convection: Jour. Geophys. Research, v. 76, p. 1101–1112.

Francheteau, J., C. G. A. Harrison, J. G. Sclater, and M. L. Richards, 1970, Magnetization of Pacific seamounts, a preliminary polar curve for the northeastern Pacific: Jour. Geophys. Research, v. 75, p. 2035–2061.

Isacks, B., L. R. Sykes, and J. Oliver, 1969, Focal mechanisms of deep and shallow earthquakes in the Tonga-Kermadec region and the tectonics of island arcs: Geol. Soc. America Bull., v. 80, p. 1443–1470.

Johnson, R. J., 1970, Active submarine volcanism in the Austral Islands: Science, v. 167, p. 977–979.

Karig, D. E., 1970, Ridges and basins of the Tonga-Kermadec island-arc system: Jour. Geophys. Research, v. 75, p. 239–254.

Kaula, W. M., 1970, Earth's gravity field: relation to global tectonics: Science, v. 169, p. 982–985.

McKenzie, D. P., 1967, Some remarks on heat flow and gravity anomalies: Jour. Geophys. Research, v. 72, p. 6261–6273.

———— 1969, Speculations on the consequences and causes of plate motions: Royal Astron. Soc. Geophys. Jour., v. 18, p. 1–32.

Moberly, R., in press, Origin of lithosphere behind island arcs, with reference to the western Pacific, *in* A volume to honor H. H. Hess: Geol. Soc. America Mem. 132.

Morgan, W. J., 1965, Gravity anomalies and convection currents: Jour. Geophys. Research, v. 70, p. 6175–6204.

———— 1971, Convection plumes in the lower mantle: Nature, v. 230, no. 5288, p. 42–43.

———— in press, Plate motions and deep mantle convection, *in* A volume to honor H. H. Hess: Geol. Soc. America Mem. 132.

Oversby, V. M., and P. W. Gast, 1970, Isotopic composition of lead from oceanic islands: Jour. Geophys. Research, v. 75 p. 2097–2114.

Vogt, P. R., O. E. Avery, E. D. Schneider, C. N. Anderson, and D. R. Bracy, 1969, Discontinuities in sea-floor spreading, *in* The world rift system: Tectonophysics, v. 8, p. 285–317.

Wilson, J. T., 1963a, Continental drift: Sci. American, v. 208, p. 86–100.

———— 1963b, Hypothesis of earth's behavior: Nature, v. 198, p. 925–929.

———— 1965a, Evidence from ocean islands suggesting movement in the earth, *in* Symposium on continental drift: Royal Soc. London Philos. Trans., ser. A, v. 258, p. 145–167.

———— 1965b, Submarine fracture zones, aseismic ridges and the International Council of Scientific Unions line; proposed western margin of the East Pacific ridge: Nature, v. 207, p. 907–911.

Wyss, M., 1970a, Apparent stresses of earthquakes on ridges compared to apparent stresses of earthquakes in trenches: Royal Astron. Soc. Geophys. Jour., v. 19, p. 479–484.

———— 1970b, Stress estimates for South American shallow and deep earthquakes: Jour. Geophys. Research, v. 75, p. 1529–1544.

11

MANTLE PLUMES AND PLATE MOTIONS

J. TUZO WILSON
Erindale College, University of Toronto, Toronto, Ont. (Canada)

(Accepted for publication March 27, 1973)

ABSTRACT

Wilson, J.T., 1973. Mantle plumes and plate motions. In: E. Irving (Editor), *Mechanisms of Plate Tectonics. Tectonophysics*, 19(2): 149–164.

This paper elaborates the hypothesis that convection plumes may be rising from the lower mantle to spread out in the asthenosphere and drive lithospheric plates about and thus possibly provide the primary mechanism which governs the behaviour of the earth's surface. The paper notes some characteristics of plumes and identifies more than thirty by the hot spots which overlie them. Most lie close to mid-ocean ridges and have produced aseismic ridges trending away from them on either plate. A few have been overridden by plates to produce single, isolated chains of seamounts and islands. One plume may have uplifted the Colorado Plateau. Such distinctions serve to identify five types of hot spots.

Most plates are in motion over the lower mantle. They are considered to be driven by the plumes, but their paths are influenced by interactions with other plates. Some temporarily become more or less stationary relative to the lower mantle. It is held that stationary plates, of which Africa and Southeast Asia may be present examples, develop special characteristics among which much volcanism, epeirogenic uplift, rifting and the development of basins and swells are diagnostic.

It is well-known that if two plates approach one another at a subduction zone that a continental plate generally overrides an oceanic one. It is here suggested that the question of which plate is more nearly stationary over the mantle is important and determines the character of the continental margin. It is held that, if a continental plate advances over an oceanic one which is fixed over the mantle, a migrating marginal trench and mountains of Andean type with huge batholiths will form on the leading edge of the continent. On the other hand, if a continental plate is fixed and one or more oceanic plates are advancing and sliding under it, island arcs (and, when a collision with another continent occurs, mountains of Appalachian type) will form along each coast towards which a plate is advancing.

INTRODUCTION

The hypotheses that contraction, expansion or convection currents are the cause of mountain building were all stated long ago. In spite of much effort and many claims no clear statement of any of them is yet widely agreed upon; all are open to objection on physical grounds and none has yet been developed to explain the whole sweep of geological phenomena.

This paper, therefore, studies the consequences of accepting a new hypothesis to see whether it can be stated unambiguously, is acceptable physically and can provide an explanation for broad geological observations.

Fig.1. Sketch map of the world showing the location of supposed plumes rising from deep in the mantle. Notice that those reaching the surface are concentrated either beneath plates believed to be fixed (Africa, Nasca and Southeast Asia) or near mid-ocean ridges which are plate boundaries.

The hypothesis is in two parts of which the first is that of relative plate tectonics. For the later part of geological time, at least, this hypothesis defines lithospheric plates precisely and enables their motions relative to one another to be accurately calculated in terms of poles of rotation, sea-floor spreading, magnetic imprinting and the directions of fracture zones (Vine, 1966; Morgan, 1968; Le Pichon, 1968). It is compatible with the findings of paleomagnetism (Phillips and Forsyth, 1972).

The second part is an extension which endeavours to establish the absolute motion of plates relative to the deeper mantle. If the lower mantle is convecting this makes little sense, but there is evidence that it is rigid and Wilson (1963; in Blackett et al., 1965) has suggested that convection plumes rise at fixed locations from the deep mantle and Morgan (1971, in press) has suggested that they spread out radially in the asthenosphere to drive the lithospheric plates (Fig.1).

This paper assumes that hot spots on the earth's surface, marked by volcanism, high heat flow and uplift, are the surface expressions of deep seated plumes rising in the mantle

CHARACTERISTICS, THEORY AND IDENTIFICATION OF HOT SPOTS

The concept of hot spots can be stated in quite precise terms. Several dozen existing hot spots have been identified. Each has had a life which may be as great as some tens of millions or even a hundred million years. Hot spots have diagnostic characteristics which are clearest where they are located in ocean basins.

(1) Each is an uplift, marked by elevated basement rocks on land and shoal water at sea, including two cases in which sea-floor or mantle have been raised above sea-level. (Macquarie Island and St. Paul's Rocks.)

(2) The uplifts are capped by active volcanoes characteristically producing alkaline basalts and rhyolite as well as the tholeiite basalts common on the sea-floor. These lavas have distinctive isotopic ratios and geochemical patterns (Oversby, 1972).

(3) Gravity highs accompany at least some of them, although the data are imperfect (Kaula, 1972).

(4) In the oceans and in some cases on continents one or sometimes two lateral aseismic ridges stretch away from them.

(5) They are areas of high heat flow.

No physical theory has yet been given, but Morgan believes that one may be possible. Similar diapiric uplifts are common in nature (e.g., salt domes, igneous plutons, shale diapirs, volcanic pipes and thunderheads).

Davies and Sheppard (1972) have shown from patterns of seismic delays at the LASA array that there are numerous anisotropic zones deep in the mantle which may be occupied by or which may in future give rise to ascending plumes. Kanasewich et al. (1972) have produced evidence that directly below the Hawaiian hot spot the core—mantle interface is anomalous. The rising-plume theory meets Lliboutry's (1972) condition derived from consideration of energy requirements that more of the mantle than the asthenosphere be involved, but avoids the difficulty imposed on the usual convection-current hypotheses by the very high viscosity of the greater part of the lower mantle.

For purposes of discussion it is useful to divide those existing hot spots which have been identified into five groups.

Hot spots on or close to the Mid-South Atlantic Ridge and East Pacific rise

It was about these two sections of the mid-ocean ridge system which sea-floor spreading proceeded most steadily throughout later Mesozoic and Paleogene time. For this reason Heirtzler et al. (1968) chose the South Atlantic as their standard for magnetic imprinting.

Along these lengths of mid-ocean ridge the following hot spots have been recognized (peculiarities in their locations are given in brackets):

St. Paul's Rocks (large fracture zone)
St. Helena
Tristan da Cunha—Gough Islands
Bouvet Island (triple point)
Juan de Fuca ridge
Revilla Gigedo Islands (fracture zone)
Galapagos Islands (triple point)
Easter Island (large fracture zone)
Balleny Islands (fracture zone)

It is held that the regularity and steady expansion of these two sections of mid-ocean ridge indicate that from Mid-Cretaceous to the end of Oligocene time the plumes rising beneath them were dominant and drove the adjacent plates away from these sections of ridge without interference by other plates. These two sections of ridge could expand without interfering with one another, because subduction zones nearly completely ring the Pacific basin and serve to separate them. Until Miocene time these two sections of mid-ocean ridge remained fixed over the plumes listed above.

Hot spots on or close to other sections of mid-ocean ridges

    Jan Mayen Island (major fracture zone)
    Iceland (major fracture zones)
    Azores (triple point)
    Prince Edward and Marion Islands (fracture zone)
    Réunion Island
    St. Pauls–Amsterdam Islands (triple point)
    Macquarie Island (triple point)

It has long been recognized that while some mid-ocean ridges can remain fixed over the deeper mantle, others must move over the mantle. For example, during the break-up of Pangea, the size of the two rings of ridges which almost encircle Africa and Antarctica must have increased in diameter and moved, or the present continents could not have separated.

From this it follows that there are two types of ocean basins: those like the South Atlantic which from Mid-Cretaceous to Oligocene time grew by spreading from a fixed mid-ocean ridge and those like the Indian Ocean basin and the North Atlantic which have grown about mid-ocean ridges which have moved relative to their plumes.

The distinction may be clearer if we consider poles of rotation. Instant poles may move about, but Bullard et al. (in Blackett et al., 1965) showed that the Sout Atlantic can be described as having opened by the rotation of the American and African plates equally in opposite directions away from the Mid-South Atlantic Ridge about a pole at 44°N and 30°W (Fig.2). They also showed that the North Atlantic opened by the rotation of the same American plate and the Eurasian plate away from the Mid-North Atlantic Ridge about a pole at 88°N and 28°W. Now the American plate could not have rotated about two different poles both fixed relative to the mantle. Therefore one or both the poles of Bullard et al. must have been moving over the mantle and if so the position of the associated mid-ocean ridge has been moving also, in which case it has been forced to migrate away from the plumes that presumably initiated it.

The Sout Atlantic basin is very simple and symmetrical, as Heirztler et al. (1968) recognized when they chose it as their standard for the time-scale of magnetic reversals. We will therefore assume that the Mid-South Atlantic Ridge remained fixed through later Cretaceous and Paleogene time. It follows that both the North Atlantic pole and the

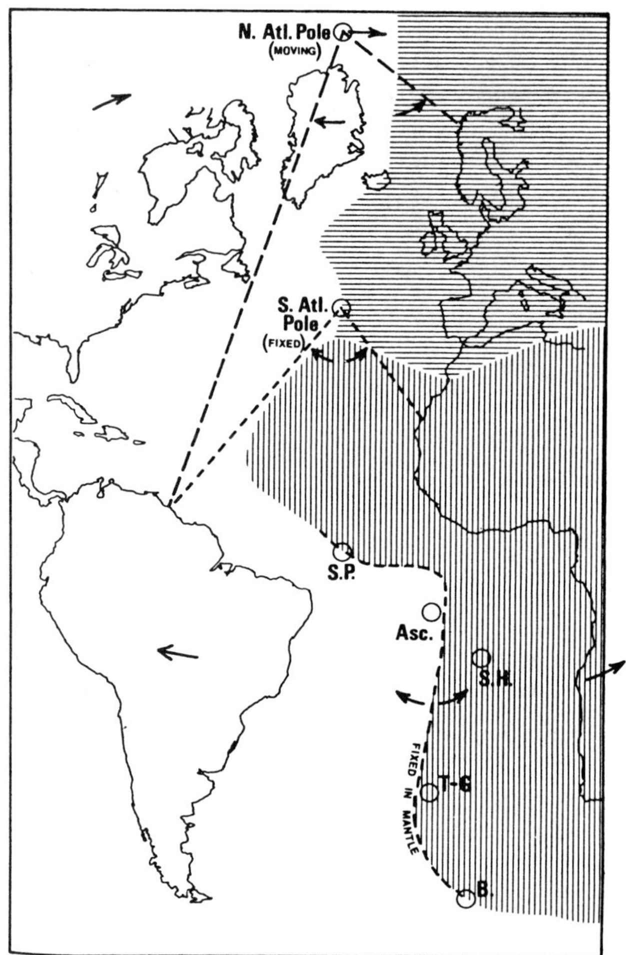

Fig.2. Sketch map of opening of the Atlantic Ocean by rotation of plates. If the Mid-South Atlantic Ridge was fixed relative to the mantle from 110 to 25 m.y. ago, so was the corresponding pole of rotation. Since the Americas plate cannot rotate about two poles fixed relative to the mantle the North Atlantic pole and ridge must then have been moving over the mantle. Islands indicated are Saint Paul's Rocks (S.P.), Ascension (Asc.), Saint Helena (S.H.), Tristan–Gough (T–G) and Bouvet (B).

North Atlantic Ridge have been moving over the mantle and hence this ridge has been forced to migrate away from the plumes that initially generated them (Fig.3 and 4; Vogt et al., 1970; Burke et al., 1972).

The effects of this can be best understood by realizing that mid-ocean ridges appear to form where several strong plumes rise near enough to one another so that the lithosphere cracks between them. Two types of rising matter lift the ridge: at the plumes material rises from the deep mantle and the asthenosphere and in the sections between plumes material rises from the asthenosphere only. Perhaps because of the latter, ridges acquire a life of their own which persists even after they have been pushed off the plumes that

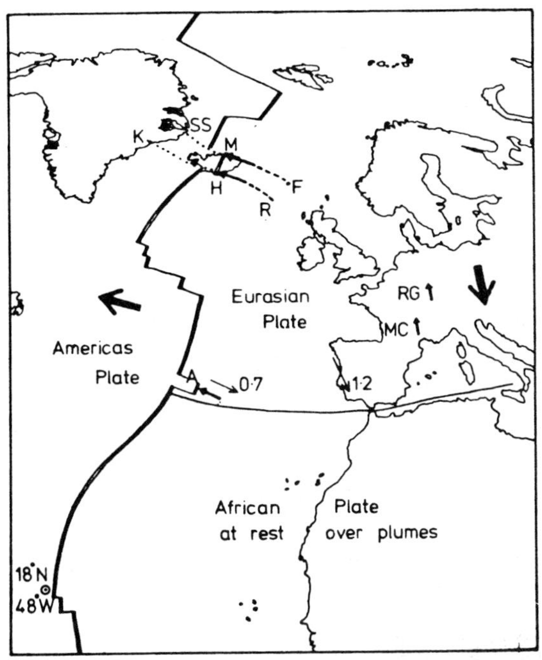

Fig.3. Map showing by broad arrows the direction of motion of the Americas and Eurasian plates since Africa became stationary 25 m.y. ago and by small arrows some resulting volcanic plume traces due to rotation of the Eurasian plate about a pole at 18°N and 48°W. Note how the mid-ocean ridge has moved west off the plumes beneath Azores (A), Hecla (H) and Myvaten (M).

generated them. Presently, however, the plumes start a new section of mid-ocean ridge sub-parallel with the old. This is considered to be the origin of the several inactive and active sections of mid-ocean ridge in the Arctic Sea, Baffin Bay and the Norwegian Sea which have jumped clockwise at 20–40 m.y. intervals throughout the Cenozoic Era.

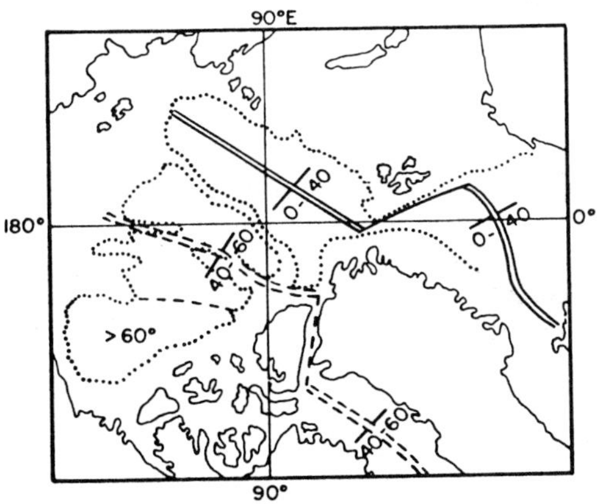

Fig.4. Map of the Arctic regions showing by double lines the active mid-ocean ridge which has been spreading for the past 40 m.y., by dotted lines the position of a ridge spreading between 60 and 40 m.y. ago and a third part older than 60 m.y. (after Vogt et al., 1970).

Similar arguments are thought to apply to the Indian Ocean and to the Tasman Sea and may explain why they have many irregular ridges and complex histories.

Young hot spots and their associated rift valleys

Ethiopia and other East African hot spots (chiefly at triple junctions)
Other African hot spots
Ordos Plateau, China
Lake Baikal, U.S.S.R.

It will be noted that this paper has so far referred to motions from Triassic to Oligocene time and not included Neogene time. The reason is that Burke and Wilson (1972) noted an apparent change in behaviour at about the beginning of Miocene time 25 m.y. ago. They observed that the islands of St. Helena, Tristan da Cunha and Gough lie on magnetic anomaly six, 400 — 500 km east of the Mid-Atlantic Ridge and that the Walvis and Discovery seamount chains and the Cape rise also terminate east of the rise on that anomaly and that in the Indian Ocean another hot spot, Réunion, also lies at the east end of a ridge terminating at anomaly six (Fig.5). In the Pacific the Galapagos and Easter Islands are also east of the East Pacific rise.

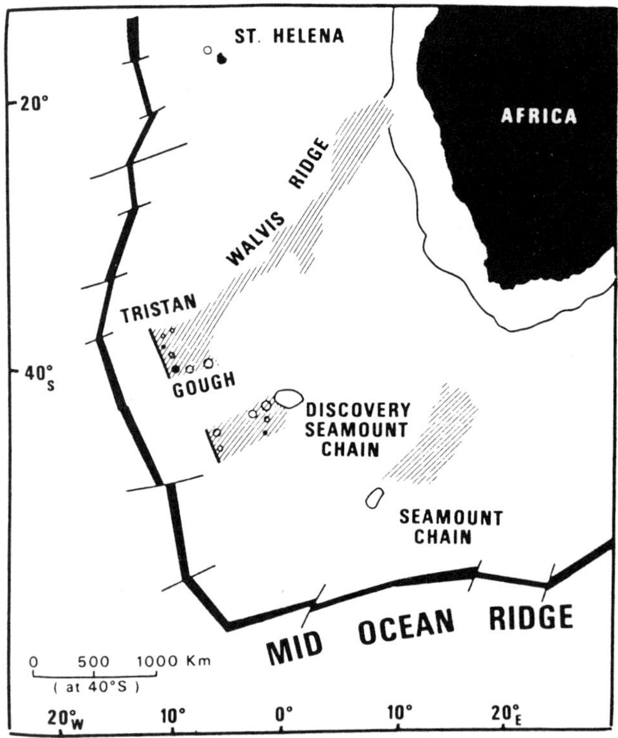

Fig.5. Map of part of the African plate showing volcanic centres over supposed plumes. Until it became stationary 25 m.y. ago the plumes near the Mid-Atlantic Ridge formed trails leading to the continent, but since then the ridge has moved west from the plumes.

They suggested that a change in plate motions marked the beginning of Neogene time about 25 m.y. ago. At that time the hot spots of the East African rift developed, uplifted East Africa and halted the eastward motion of the African plate. The spreading from the Mid-South Atlantic Ridge continued which forced that ridge to begin to migrate away from the hot spots marked by St. Helena, Tristan, Gough and Bouvet Islands. The Indian Ocean and East Pacific patterns were also reorganized and extensive changes occurred in the geology of the Pacific Rim (Dott, 1969).

The Neogene geology of Africa has a number of unusual features not common elsewhere and certainly not present during the Cenozoic Era in Europe, North America or South America. These include regional epeirogenic uplift, hot spots with associated Neogene volcanism, basin-and-swell topography and rift valleys. It is considered that these may be marks of a stationary plate (Burke and Wilson, 1972).

Looking elsewhere we detect in what little is visible of Antarctica a great rift along the Transantarctic Mountains accompanied by Neogene volcanism (Hamilton, 1972) and epeirogenic uplift. Since there is little activity along the section of mid-ocean ridge between Antarctica and Africa, Antarctica also has been stationary or nearly so during the Neogene.

If Africa has been stationary then Eurasia has been rotating slowly southwards about poles which McKenzie (1970) placed near Barbados at 9°N and 46°W and near the Banda Sea. Any part of Eurasia near one of these poles would have been nearly stationary also and Southeast Asia forms such a part. It seems particularly likely to have been stationary since a system of faults crosses Southeast Asia isolating it form the rest of the Eurasian plate.

Along this sytem the Ordos Plateau is incised by graben and is rising rapidly with many earthquakes (Fig.1). There has been much volcanism in China during the later Tertiary. It is suggested that a hot spot may underlie the Ordos.

A similar argument may apply to the Lake Baikal rift zone which again shows uplift, rifting, Tertiary volcanism, high heat flow and recent earthquakes.

Young hot spots on ocean floors that may be stationary

    Canary Islands
    Cape Verde Islands
    Comores Islands
    St. Felix—Ambrosia Islands

Of these the active Canary, Cape Verde and perhaps the inactive Madeira Islands lie on the stationary African plate. The nature of the Comores Islands and the plate behaviour in the Mozambique Channel is still obscure. The St. Felix—Ambrosia Islands lie on the Nasca plate which may also be nearly stationary. Neither these islands nor the inactive Juan Fernandez group have extensive trails.

Older hot spots which have been overridden

> Hawaii
> Austral Islands (MacDonald submarine volcano)
> Society Islands
> Tuamotu Islands
> Marquesas Islands
> Kerguelen Island
> Heard Island
> Colorado Plateau

If plumes drive the lithospheric plates a majority of hot spots should lie along mid-ocean ridges or on stationary plates and few should have been overridden. Among these few the several parallel chains of islands in the Pacific were the first hot spots recognized. (Wilson, 1963; Jackson et al., 1972). The forced migration of the Indian Ocean ridges away from such hot spots as Kerguelen and Heard has already been discussed.

The peculiarities of the Colorado Plateau and of the Basin and Range province have long been recognized and may well be due to the interaction of several causes. Among those most often considered are:

(1) Residual activity in the Antler—Sonoma—Sevrier orogenic belt indicated by continuing seismic activity (Sbar et al., 1972).

(2) Marginal effects from the Nevadan orogeny.

(3) Volcanic activity arising from the Benioff zone caused by overriding the Farallon plate (McKee, 1971; Thompson, 1972).

(4) Rise of a fresh young hot spot.

(5) Overriding of a hot spot formerly on the East Pacific rise.

It is considered that the first two factors probably play a part, that (3) is unlikely because the volcanic activity above other Benioff zones for example in East Asia and the Andes does not extend so far from the associated trench and that (5) is to be preferred over (4) for reasons given below.

Atwater's (1970) proposal that if a continent overrides a mid-ocean ridge, the ridge should die away, may well be true for the sections of ridge that link plumes or hot spots and which never extended below the asthenosphere. On the other hand it seems hardly likely to apply to the overriding of a plume which has risen through the whole mantle. Overriding would seem hardly likely even to alter the location of a plume. If so the map of magnetic anomalies in the northeast Pacific (Atwater and Menard, 1970) can suggest where the "virtual image" of the overridden ridge should lie. Fig.6 shows the locations of anomalies 23 and 8 and of the East Pacific rise and its extensions. A plume is postulated at the junction with the Mendocino fracture zone.

These anomalies were chosen because at latitude 20°N (through the entrance to the Gulf of California) the distance between anomalies 23 and 8 is equal to that between anomalies 8 and 0 (the crest). Fig.7 shows the present situation and where the extension

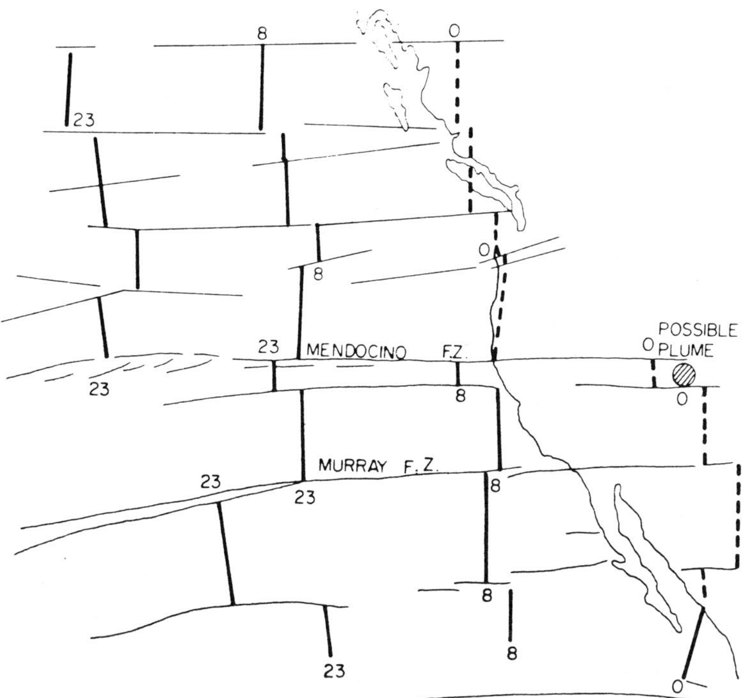

Fig.6. Sketch map of west coast of North America showing location of Mendocino fracture zone and magnetic anomalies 23 and 8 and the crest (0) south of the Gulf of California. Since the distance from 23 to 8 is equal to the distance from 8 to the crest in the south, the location of the rest of the crest (if it had continued to exist) can be projected and also the location of a possible plume at the junction with the Mendocino fracture zone.

of the crest should lie under the continent if overriding has not disturbed it. However, we will assume that only the plumes remain and that any shallow-seated parts of the crest linking one plume to the next have vanished. Recalling that the point of intersection of major fracture zones with crests is a common location for plumes, we suggest that the one part of the East Pacific rise which remains in existence under North America is a plume at the intersection of a line extending the Mendocino fracture zone eastward and the projected crest. This plume underlies the Colorado Plateau and has caused its uplift. En route to its present location that plume would have caused a wave of uplift to travel inland northeasterly from the Pacific coast during the past 30 m.y. That indeed appears to have happened. Prior to about 18 m.y. ago the Basin and Range province was uplifted and the drainage flowed northward from it on to the area of the Colorado Plateau (Lucchitta, 1972). Between 5 and 10 m.y. ago the Colorado Plateau was uplifted with a reversal of the direction of drainage (McKee and McKee, 1972).

Hot spots on continents have not left such well-marked trails of extinct volcanoes aging away from the present location as do hot spots on oceanic plates. Concentric plume traces on the continental part of the Eurasian plate appear restricted to areas where velocity over the plumes is less than about 2 cm/year (Burke et al., 1972). There has been much

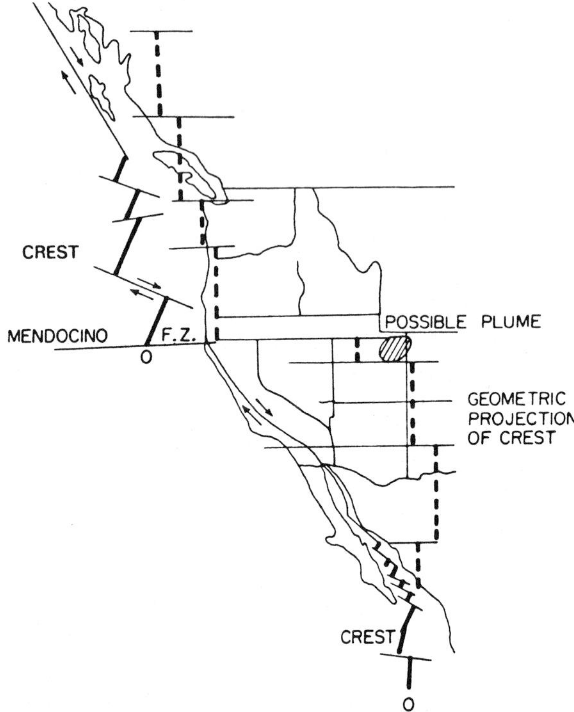

Fig.7. Sketch map of west coast of North America showing location of sections of active mid-ocean ridges and connecting faults as well as geometric projection of crest of ridge from Fig.6. The only part of the projection believed to be active is the plume which is possibly still uplifting the Colorado Plateau.

volcanism in the southwestern United States, but its pattern is not clear and much of it is more than 25 m.y. old. Presumably it arises from a combination of several causes.

PLATE MOTION AND MOUNTAIN BUILDING

It has been observed that two converging plates frequently form a subduction zone along the margin of a continent and that the continental plate commonly overrides the oceanic so that the Benioff zone dips beneath the continent. Two principle causes are possible (Wilson and Burke, 1972):

(1) The oceanic plate may be stationary with respect to the mantle and the continental plate may be advancing and overriding it (Fig. 8 and 9).

(2) The continental plate may be stationary and the oceanic plate may be actively sliding beneath it (Fig.8 and 9).

There will also be cases in which both plates are moving over the mantle and in which there is a lateral as well as a normal direction of motion between the plates, but many of these will approximate to (1) or (2) above.

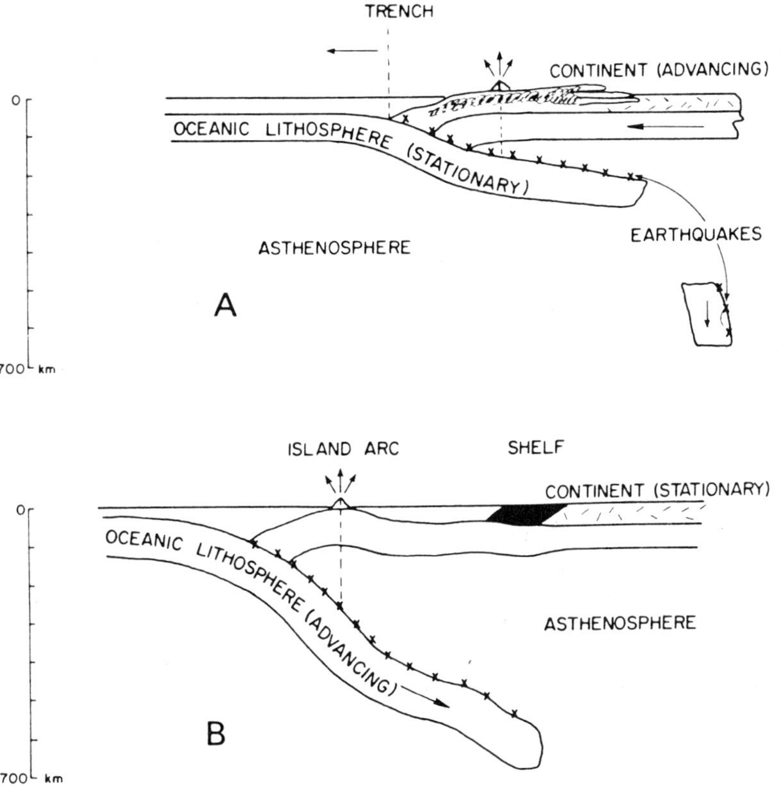

Fig.8. A. Section showing a continent advancing relative to the mantle and to a stationary plate of oceanic lithosphere. In this case the trench and volcanoes form along the coast. B. Section showing an oceanic plate advancing relative to a continent and the mantle. In this case island arcs have moved the trench and volcanoes out from the coast line.

Mountain building where a continent advances over the mantle

If, during the Tertiary, first the Mid-Atlantic Ridge and latterly the African plate have been stationary over the mantle then South America has been advancing over the mantle. The subduction zone at the leading edge in Chile has taken the form of a marginal trench and mountain system along the coast. There are three facies of deposits: filling of the Chile—Peru trench at its extreme north and south; the largely metamorphosed material of the Andes which combines shelf deposits and the products of volcanism with large batholiths and the comparatively narrow folded belts of older sedimentary rocks in the eastern Andes.

Since no volcanoes rise within the trench or east of the Andes all the volcanic products from the Benioff zone have risen near the coast into the Andes becoming mixed with the shelf deposits and metamorphosing them. This is believed to have been a major factor in

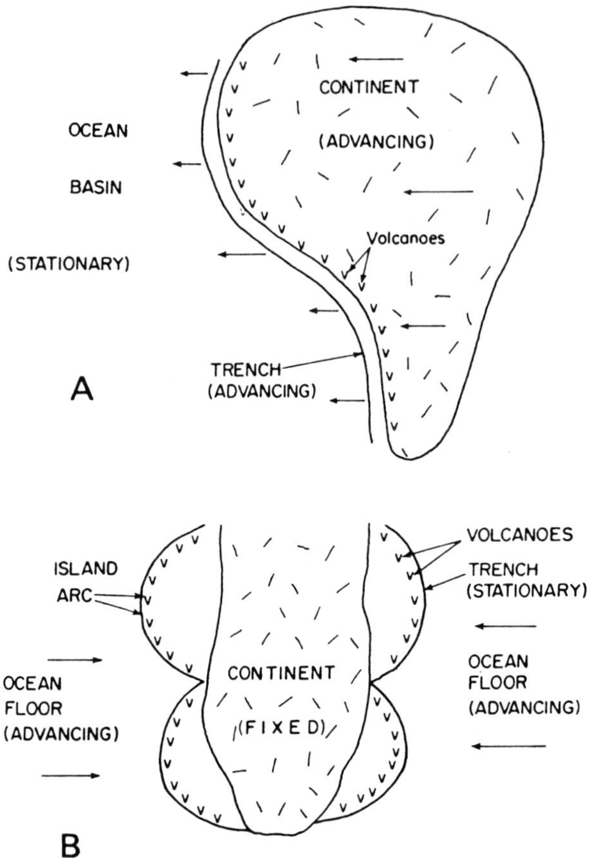

Fig.9. Maps corresponding to the sections in Fig.8 showing that volcanoes and a trench only form on the leading edge of an advancing continent, but can be formed all around a stationary one.

producing the batholiths (Bateman and Eaton, 1967). There has therefore never been a major clear cut distinction between shelf (miogeosynclinal) and island arc (eugeosynclinal) deposits after subduction was initiated and the continual forward movement of the trench and the mountains has resulted in frequent disturbance of the Andean system since it started (James, 1971).

That view may have to be somewhat modified if it can be shown that the Andes underwent a change in type of orogeny. It is possible that after the East Pacific rise started to expand in Late Triassic or Early Jurassic time, and before the South Atlantic began to open in Mid-Cretaceous time that island arcs could have formed in the Andean region.

An Andean orogen can only form along the leading edge of an advancing continent and hence only one such system can form at any one time on a continent. This explains why the Andes have no counterpart on the Atlantic coast.

The same arguments apply to the Cordillera of North America where Gilluly (1969), Hamilton and Myers (1967) and Bateman and Eaton (1967) have described the change of the shelf deposits and Kistler et al. (1971) have emphasized the importance of volcanic additions.

Mountain building where the oceanic plate is advancing over the mantle towards a stationary continent

From Jurassic to at least Oligocene time the East Pacific rise was fixed over the mantle. If it has been moving since, it has migrated westwards. Hence the Pacific plate has been moving rapidly northwest to west-northwest as the trail of the Emperor and Hawaiian Islands indicates (Jackson et al., 1972). Eurasia on the other hand has been rotating only slowly southwards over the mantle, hence the island arcs of East Asia have formed where an oceanic plate has been actively advancing and passing beneath a nearly stationary continent. A similar argument applies to the Burmese—Indonesian arc because the India—Australia plate has been moving rapidly northeastwards over the mantle.

Notice in this case how different the conditions of mountain building have been. The lavas rising from the Benioff zones have formed island arcs which became metamorphosed in place without disturbing the shelf deposits along the continental coast (Emery et al., 1969), so that the miogeosynclinal shelves, the volcanic eugeosynclinal deposits and the trench fillings (as in Timor) have formed distinct facies. Because the leading edges on which mountains form are on the oceanic plates, mountains have been able to form simultaneously on different sides on a nearly stationary continent-laden plate by the advance of more than one oceanic plate towards it. Continental collisions may follow, on several sides.

It is suggested that the Taconic—Acadian—Appalachian—Caledonian, the Ouachita—Marathon, the Antler—Sonoma—Sevrier and the Innuitian orogenies which all formed during the Paleozoic Era on different sides of North America are other examples of this type of orogeny. Cady (1972) thus appears to be correct in pointing out that the Appalachian and Cordilleran orogenies are not entirely homologous, although one need not agree with all of his arguments. It is maintained that the location of former island arcs can still be discerned (Wilson, in Kuiper, 1954), and that the basins and swells of North America during the Paleozoic time may resemble those of the present-day in Africa. Examples are the Michigan and Williston basins.

The shelves off East Asia are not being folded today. If the island arcs there have moved, it has been away from the coast, not towards it (Karig, 1971). Thus it is not at once apparent in the Paleozoic examples what mechanism eventually folded the thick sedimentary shelf deposits and pushed the island arcs up against them. It has been suggested that it was the arrival of other continents or continental fragments borne into the subduction zone by the advancing oceanic plates (Wilson, 1966, 1968). Some such cause as this is still held to be necessary. While the Paleozoic orogenies on different sides of North America are in a general way contemporaneous, it is believed that the arrival of various continental blocks and the pulses of activity in different belts are not so precisely synchronous as has sometimes been argued (Johnson, 1971).

165

It is therefore held that present-day plates and the plumes which drive them can be precisely defined, that former plumes and plates can be identified with increasing difficulty backwards into time, that mantle plumes may be able to draw energy from the whole interior of the earth and thus provide a physically reasonable source of energy and that these processes may provide a good framework upon which to interpret the details of geological and geophysical observations.

ACKNOWLEDGEMENT

I wish to thank Kevin Burke and W.F.S. Kidd for reading this paper and for valuable suggestions.

REFERENCES

Atwater, T., 1970. Implications of plate tectonics for the Cenozoic tectonic evolution of western North America. *Geol. Soc. Am. Bull.*, 81: 3516–3536.

Atwater, T. and Menard, H.W., 1970. Magnetic lineations in the Northeast Pacific. *Earth Planet. Sci. Lett.*, 7: 445–450.

Bateman, P.C. and Eaton, J.P., 1967. Sierra Nevada batholith. *Science*, 158: 1407–1417.

Blackett, P.M.S., Bullard, E.C. and Runcorn, S.K., 1965. A symposium on continental drift. *R. Soc. Lond., Philos. Trans., Ser. A*, 258: 1–323.

Burke, K.C. and Wilson, J.T., 1972. Is the African plate stationary? *Nature*, 239: 387–390.

Burke, K.C., Kidd, W.F.S. and Wilson, J.T., 1972. Plumes and concentric plume traces of the Eurasian plate. *Nature* (in press).

Cady, W.M., 1972. Are the Ordovician Northern Appalachians and the Mesozoic Cordilleran system homologous? *J. Geophys. Res.*, 77: 3806–3815.

Davies, D. and Sheppard, R.M., 1972. Lateral heterogeneity in the earth's mantle. *Nature*, 239: 318–323.

Dott, Jr., R.H., 1969. Circum-Pacific Late Cenozoic structural rejuvenation: Implications for sea-floor spreading. *Science*, 166: 874–876.

Emery, K.O., Hagashi, Y., Hilde, T.W.C., Kobayashi, K., Koo, J.H., Menz, C.Y., Niino, H., Osterhagen, J.H., Reynolds, L.M., Wageman, J.M., Wang, C.S. and Yang, S.J., 1969. Geological structure and some water characteristics of the East China Sea and the Yellow Sea. *Tech. Bull. ECAFE*, 2: 3–43.

Gilluly, J., 1969. Oceanic sediment volumes and continental drift. *Science*, 166: 992–994.

Hamilton, W., 1972. The Hallett volcanic province, Antarctica. *U. S. Geol. Surv. Prof. Pap.*, 456-C: 1–62.

Hamilton, W. and Myers, W.B., 1967. The nature of batholiths. *U. S. Geol. Surv. Prof. Pap.*, 554-C: 1–30.

Heirtzler, J.R., Dickson, J.O., Herron, E.M., Pitman, W.C. and Le Pichon, X., 1968. Marine magnetic anomalies, geomagnetic field reversals, and motions of the ocean floor and continents. *J. Geophys. Res.*, 73: 2119–2136.

Jackson, E.D., Silver, E.A. and Dalrymple, G.B., 1972. Hawaiian–Emperor Chain and its relation to Cenozoic circum-Pacific tectonics. *Geol. Soc. Am. Bull.*, 83: 601–618.

James, D.E., 1971. Plate-tectonic model for the evolution of the Central Andes. *Geol. Soc. Am. Bull.*, 82: 3325–3346.

Johnson, J.G., 1971. Timing and coordination of orogenic, epeirogenic and eustatic events. *Geol. Soc. Am. Bull.*, 82: 3263–3298.

Kanasewich, E.R., Ellis, R.M., Chapman, C.M. and Gutowski, P.R., 1972. Teleseismic array evidence for inhomogeneities in the lower mantle and the origin of the Hawaiian Islands. *Nature, Phys. Sci.*, 239: 99–100.

Karig, D.E., 1971. Origin and development of marginal basins in the Western Pacific. *J. Geophys. Res.*, 76: 2542–2561.

Kaula, W.M., 1972. Global gravity and tectonics. In: E.C. Robertson (Editor), *The Nature of the Solid Earth*. McGraw-Hill, New York, N.Y., pp. 385–405.

Kistler, R.W., Evernden, J.F. and Shaw, H.R., 1971. Sierra Nevada plutonic cycle, 1. Origin of composite granitic batholiths. *Geol. Soc. Am. Bull.*, 82: 853–868.

Kuiper, G.P., 1954. *The Earth as a Planet*. Univ. Chicago Press, Chicago.

Le Pichon, X., 1968. Sea-floor spreading and continental drift. *J. Geophys. Res.*, 73: 3661–3697.

Lliboutry, L., 1972. The driving mechanism, its source of energy, and its evolution studied with a three-layer model. *J. Geophys. Res.*, 77: 3759–3770.

Lucchitta, I., 1972. Early history of the Colorado River in the Basin and Range province. *Geol. Soc. Am. Bull.*, 83: 1933–1948.

McKee, E.H., 1971. Tertiary igneous chronology of the Great Basin of Western United States – Implications for tectonic models. *Geol. Soc. Am. Bull.*, 82: 3497–3502.

McKee, E.D. and McKee, E.H., 1972. Pliocene uplift of the Grand Canyon region – Time of drainage adjustment. *Geol. Soc. Am. Bull.*, 83: 1923–1932.

McKenzie, D.P., 1970. Plate tectonics of the Mediterranean region. *Nature*, 226: 239–243.

Morgan, W.J., 1968. Rises, trenches, great faults and crustal blocks. *J. Geophys. Res.*, 73: 1959–1982.

Morgan, W.J., 1971. Convective plumes in the lower mantle. *Nature*, 230: 42–43.

Morgan, W.J., 1972. Deep mantle convection plumes and plate motions. *Bull. Am. Assoc. Pet. Geol.*, 56: 203–213.

Morgan, W.J., in press. Plate motions and deep mantle convection. *Geol. Soc. Am. Mem. (H.H. Hess volume)*.

Oversby, V.M., 1972. Genetic relations among the volcanic rocks of Réunion: Chemical and lead isotopic evidence. *Geochim. Cosmochim. Acta*, 36: 1167–1179.

Phillips, J.D. and Forsyth, D., 1972. Plate tectonics, paleomagnetism and the opening of the Atlantic. *Geol. Soc. Am. Bull.*, 83: 1579–1600.

Sbar, M.L., Barazangi, M., Dorman, J., Scholz, C.H. and Smith, R.B., 1972. Tectonics of the intermountain seismic belt, Western United States: Microearthquake seismicity and composite fault plane solutions. *Geol. Soc. Am. Bull.*, 84: 13–28.

Thompson, G.A., 1972. Cenozoic Basin Range tectonism in relation to deep structure. *Int. Geol. Congr., 24 Sess.*, Sect.3: 84–90.

Vine, F.J., 1966. Spreading of the ocean floor: New evidence. *Science*, 154: 1405–1415.

Vogt, P.R., Ostenso, N.A. and Johnson, G.L., 1970. Magnetic and bathymetric data bearing on sea-floor spreading north of Iceland. *J. Geophys. Res.*, 75: 903–920.

Wilson, J.T., 1963. A possible origin of the Hawaiian Islands. *Can. J. Phys.*, 41: 863–870.

Wilson, J.T., 1966. Did the Atlantic close and then re-open? *Nature*, 211: 676–681.

Wilson, J.T., 1968. Static or mobile earth: The current scientific revolution. *Am. Philos. Soc. Proc.*, 112: 309–320.

Wilson, J.T. and Burke, K.C., 1972. Two types of mountain building. *Nature*, 239: 448–449.

12

Reprinted from *Jour. Geophys. Research* **79**:3943-3949 (1974)

Tectonism and Volcanism of the Tharsis Region of Mars

Michael H. Carr

The Tharsis region of Mars, where the planet's largest shield volcanoes are located, is at the center of a fracture system that extends over almost half the surface of the planet. The fractures are tensional and aligned roughly radial to Tharsis and form two opposing fanlike arrangements that converge on the line of the Tharsis volcanoes. The fractures are attributed to a broad asymmetric updoming of the Martian crust centered to the southeast of the volcano line. Formation of the dome took place before the formation of the shield volcanoes. The location of the shields was controlled by the fractures, and volcano formation may have reactivated some of the fractures. The updoming, thought to result from mantle convection, separates eras of different volcanism and appears to have been a discrete and unique event in the history of Mars.

The Tharsis region of Mars, in addition to being the planet's most prominent volcanic region, is at the center of an extensive set of fractures. These fractures tend to be concentrated around the periphery of the region and generally preserve an orientation roughly radial to the volcanic center (Figure 1). The set covers a remarkably large part of the planet. Some fractures around Elysium, more than a quarter of the planet's circumference away from Tharsis, still appear to be part of the same set. So widespread is the pattern that it dominates an entire hemisphere, despite local differences in orientation and the presence of other seemingly unrelated fractures. Formation of the fractures was clearly a major event in Martian history, the establishment of an extensive stress system thus being reflected. This paper describes the fracture system and its relation to other geologic features and explores possible mechanisms for its formation. The general conclusions are that the fractures were caused largely by asymmetric epeirogenic upwarping of the Martian crust, centered in the Labyrinthus Noctis region, and probably as a result of mantle convection. These tentative conclusions have some terrestrial implications. It is widely believed that the plates of the earth's crust are driven by mantle convection and that plates move away from points of mantle upwelling. The earth's crust may initially be warped upward over a rise, develop tension cracks, and ultimately break into separate plates to allow independent motion [*Wilson and Burke,* 1973]. On Mars, although the crust appears to have been similarly strained, crustal breakup and plate motion did not follow, so that a record of the initial stress system has been well preserved in the fractures.

Description of the Fracture System

Fractures are evident on the Mariner photographs either as linear escarpments or linear clefts commonly transecting other features such as craters and channels. No example of strike slip movement has been found. Displacements appear everywhere to be normal, and although the dips of the fractures cannot be positively determined from the photographs, the map patterns suggest steep dips. Thus normal faulting appears to be the dominant, if not the only, mode of fracture, tension and extension of the crust thus being implied. This is particularly clear around the circular structure Alba Patera, north of Tharsis, where a series of horsts and grabens cuts across fine channels (Figure 2).

Fractures are most common to the northeast and southwest of the line of three Tharsis shield volcanoes. The fractures diverge in both directions away from Tharsis to form opposing fanlike arrangements. The northeast fan includes the fractures (fossae) around Alba Patera, the Mareotis and Tempe fossae, some fractures north of Olympus Mons, and possibly some structures parallel to Kasei Vallis. Near Alba Patera the fractures diverge around it, and there are at least three sets. An early closely spaced set trends NNE–SSW and is cut by a later closely spaced north-south set. The third set of fractures also trends NNE–SSW; its individual fractures are spaced widely apart and commonly have numerous depressions along their lengths. All these fractures curve around Alba Patera to form a fracture ring. Farther south, at approximately 25°N, in areas of fractured terrain almost completely surrounded by nonfractured terrain the fractures are so closely spaced that no level terrain occurs between them. Fractures are well developed east of Alba Patera, forming the Mareotis and Tempe fossae, which line up with the Tharsis shields. Here the fractures cut old volcanic plains and densely cratered terrain. West of Alba Patera are mostly young volcanic plains that are largely unfractured. However, where older terrain is exposed, as it is in a vague circular structure north of Olympus Mons at 38°N, fractures are present and lined up in the direction expected from the fanlike arrangement. It is clear from the geologic map pattern [*Carr et al.,* 1973] that in these northern latitudes we are seeing only part of the fracture system. Over much of the area it has been covered by younger deposits.

The opposing fan of fractures south of the Tharsis region is much less well developed than that to the north. To the southwest, along the line of the three Tharsis shield volcanoes, are the Sirenum Fossae and the Memnonia Fossae, which are visible for a length of over 3000 km. At high resolution these grabens lack the crisp definition of those around Alba Patera and appear to be partially filled. Directly south of Tharsis are the Claritas Fossae, in what is probably the most complexly fractured region on the planet. The most intense fracturing is confined to an elongated zone trending NNW–SSE, within which is mostly primitive cratered terrain. This zone is crossed by numerous closely spaced fractures mostly trending NNE–SSW, radial to Labyrinthus Noctis. However, in the northern part of the fracture zone closer to Labyrinthus Noctis this trend becomes less prevalent, and some of the fractures appear to be radial to local centers. At the extreme northern end of the zone the fractures curve northeast into Labyrinthus Noctis. South of the Claritas Fossae a more regular pattern is preserved as the fractures fan out over a wide area. Numerous

168

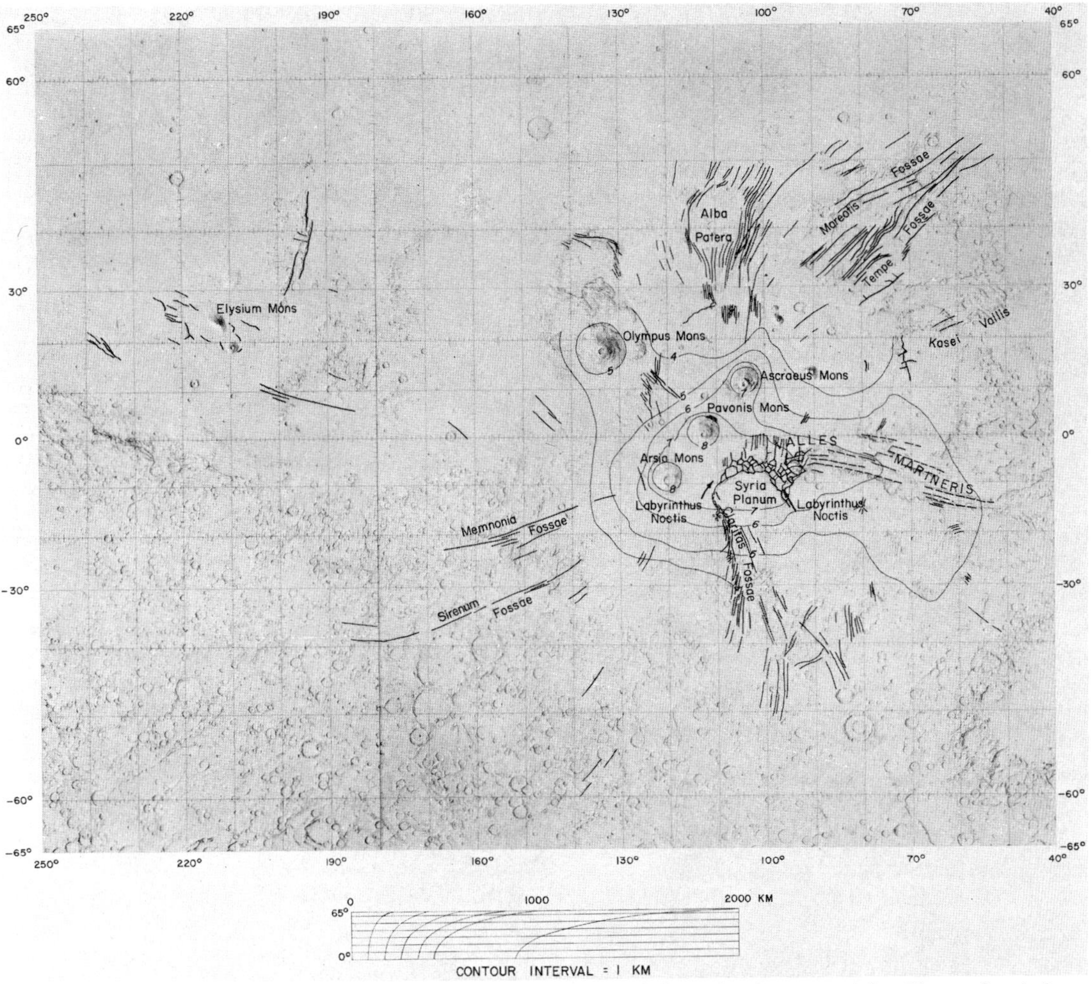

Fig. 1. Generalized fracture map of Mars between 40°W and 250°W. Dashed lines indicate where fractures are inferred from preferred orientation of scarps. Contours show the elevation of the surface in kilometers above the 6.1-mbar level [*Hord et al.*, 1974].

Tharsis radials also parallel the Claritas fault zone in the plains east of it.

Toward the center of the main fracture system, in the general region of Tharsis and Labyrinthus, the fractures form a pattern much less coherent than the one in the more peripheral areas just described. Much of the terrain in this region is sparsely cratered, a relatively young surface thus being suggested; fractures are visible only where older terrain is exposed. Just north of Labyrinthus Noctis an intensely fractured surface is truncated abruptly by seemingly younger, sparsely cratered plains roughly along the equator. The dominant trend of the fractures is north-south. A similar trend occurs in a patch of fractured terrain just north of the equator at 90°W. At this same latitude but 10° further east, yet another small area of fractured terrain is exposed, but here the fractures have a northeast trend. These three areas are of note in that the fractures are not radial to Tharsis; they are more closely radial to Labyrinthus Noctis. Another large area of fractured terrain occurs between Olympus Mons and Pavonis Mons. Two main directions of faulting can be discerned, a

northwest-southeast set that could be interpreted as radial to Tharsis or Labyrinthus Noctis and a roughly north-south set that certainly is not.

One final area of fracturing worth noting is Elysium. The main volcanic shield, Elysium Mons, is surrounded by a vague circular structure approximately 1000 km across. Within this structure are several grabens concentric with the main volcano, but outside the ring the trend is quite different. Several prominent grabens trend northwest-southeast in line with some grabens to the southeast that are roughly radial to Tharsis and follow a curving line similar to the previously mentioned Memnonia and Sirenum fossae. The Elysium fractures thus appear to have been affected by the same stress system and appear to be the most distant from Tharsis, being over 6000 km away.

In addition to the obvious lineaments so far described, there are other features whose orientations appear to have been controlled by the fractures. In Labyrinthus Noctis numerous grabenlike features can be traced into the main part of Labyrinthus, where they become much deeper and wider or

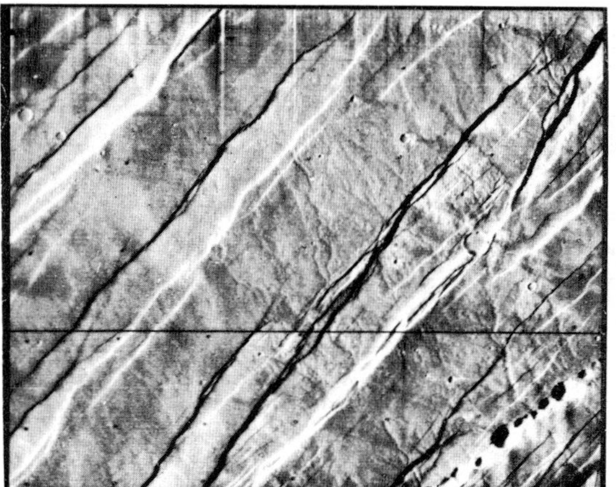

Fig. 2. Fractures cutting fine channels on the flanks of Alba Patera (DAS 08443059).

end up as lines of circular depressions. Irrespective of the origin of the intersecting crevasselike features, their orientation is clearly fracture controlled. Similarly, the orientation of the main canyon, Valles Marineris, and some of its parallel equivalents appears to be fracture controlled. Lines of depressions, the long central spine, and a general elongation radial to Tharsis all suggest structural control. The control is particularly striking at the eastern end of Ophir Chasma, where two narrow elongate strings of depressions parallel the

main valley. Structural control of Kasei Vallis north of Lunae Planum is much less convincing, but the rectilinear pattern of intersecting crevasselike features at the western edge of Lunae Planum is very suggestive, and these features are indicated in Figure 1. One last feature that may be related to the fracture system is a broad indistinct flat-floored depression that trends north-south just west of Lunae Planum. The trend of this feature is parallel to fractures on either side, and so it may also be tectonically controlled.

The extent to which the relief of the canyon and crevasselike features just described is due to tectonic movement is unknown. The fractures may have resulted in very little differential movement, the present relief being caused by unrelated processes such as excavation by wind or water erosion. Nevertheless, strong structural control is evident in the orientation of the features, so that they can be used in evaluating the origin of the fracture system.

AGE OF THE FRACTURE SYSTEM

The remarkable consistency of the fracture pattern over such a large area suggests that the fractures did not result from different local events but formed as a result of a stress system established by a discrete event centered in the general region of Tharsis. In places there may have been several episodes of fracturing (e.g., around Alba Patera), but the different episodes appear to be part of the same event. The time relations are well displayed in the Mars quadrangle MC-9 (Figure 3). Within the quadrangle are Olympus Mons, the two northernmost of the large Tharsis shields (Ascreus Mons and Pavonis Mons), and several smaller shield volcanoes. Surrounding these features are plains that are fractured and cratered to different degrees.

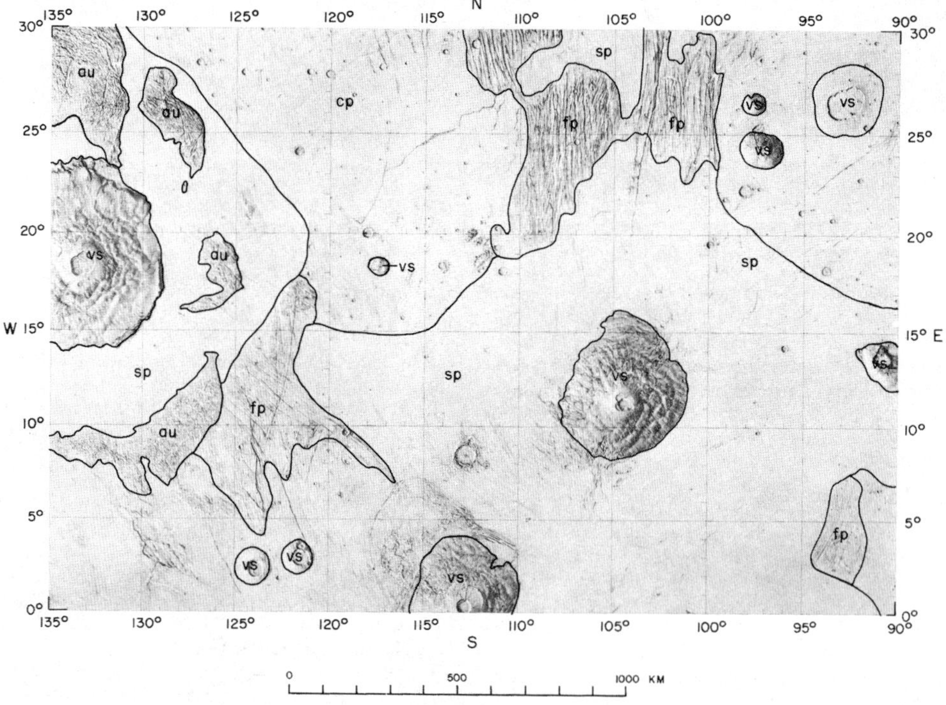

Fig. 3. Simplified geologic map of MC-9 showing volcanic shield (vs), Olympus Mons aureole (au), fractured plains (fp), cratered plains (cp), and sparsely cratered plains (sp).

They have been divided into three units: fractured plains, cratered plains, and sparsely cratered plains. The fractured plains and cratered plains have similar numbers of superimposed craters, whereas the sparsely cratered plains have substantially fewer (Figure 4). The implication is that the fractured plains and cratered plains are similar in age and somewhat older than the sparsely cratered unit. Where the fractured plains and sparsely cratered plains come in contact the fractures terminate abruptly against the contact, a relation that is consistent with the older fractured unit's being partially buried by a younger nonfractured unit. The relation between the fractured plains and the cratered plains is less clear. The crater ages are very similar, so that they could be the same unit, a part being fractured. There is some evidence against this interpretation. First, west of the center of the quadrangle the fractured and cratered plains are in contact, and the fractures are abruptly truncated. This truncation strongly suggests that the cratered unit is younger than the fractured unit. Second, the fractured unit has a much lower albedo than the cratered unit, suggesting different materials (see Mariner 9 frame DAS 12994387). This argument is not strong because the lower albedo could result directly from the presence of fractures. Third, it is difficult to visualize a stress pattern as wide in areal extent as the one being discussed causing the same unit to respond so differently over such a short distance, as is implied by the map pattern in the northern part of MC-9. Burial of the fractured unit by a slightly younger unit is a much simpler explanation of the observed relations. If this explanation is true, fracturing over a relatively short period of time is implied because the two units have such similar crater densities.

An estimate of the absolute age of the fracture system can be made from the crater counts. If one assumes the model of *Soderblom et al.* [1974] for the flux of bodies within the solar system, the crater densities of both the fractured and the

cratered plains give crater ages close to 1.2 b.y., which suggest that fracturing terminated rapidly at this time. The sparsely cratered plains have ten times fewer craters, a correspondingly younger age of approximately 10^8 years thus being indicated. This latter age is somewhat deceptive. Of the 17 B frames in the sparsely cratered plains in MC-9, two include 60% of all craters counted on the unit at B frame scale. Thus unless some other factor, such as destruction of small craters by wind, has affected the crater frequency curves, the sparsely cratered plains formed over a period of time longer than that implied by their average age of 10^8 years. From the spread of crater densities an age range of 10^7–3×10^8 is likely. For comparison, ages of 800 m.y. for Arsia Mons, 400 m.y. for Pavonis Mons, and 200 m.y. for Olympus Mons are derived by means of the flux model of Soderblom et al. *Hartmann* [1973] independently finds ages of 250–300 m.y. for the sparsely cratered plains and 200, 80, 100 and 130 m.y. for the four volcanoes. These ages confirm that the volcanoes postdate the fracturing.

Thus the fracturing event was largely over 1.2 b.y. ago, although a small amount of fracturing continued well after that date, as is indicated by isolated fractures that cut the sparsely cratered plains unit. The start of the fracturing cannot be satisfactorily dated from the crater counts. However, fracturing appears to have peaked around 1.2 b.y. ago because, at least around Alba Patera and in MC-9, the pre-1.2-b.y. surface was completely destroyed by the fractures.

ORIGIN OF THE FRACTURE SYSTEM

Any suggested origin for the fracture system must explain the opposing fanlike arrangement of tension fractures that converge upon and line up with the Tharsis volcanoes and the several areas where the fractures are radial not to Tharsis but to the west end of Labyrinthus Noctis. These characteristics can be explained reasonably by the following sequence of events. Mantle convection causes upwarping of the crust centered in the Labyrinthus Noctis region. The upwarp is asymmetric. To the northwest the slope is steeper, so that a monoclinelike feature forms. Because flexure is the greatest here, fracturing is concentrated along the monocline, and the large Tharsis shield volcanoes are thereby caused to be preferentially located there. Where the monocline dies out to the northeast and southwest, a fanlike pattern of fractures develops that may later be accentuated by the formation of the shield volcanoes. Each aspect of this proposed origin will be examined in detail in the rest of the paper.

At the center of the fracture system are two major features of the planet, the volcanic region of Tharsis and a broad region of high elevation centered southeast of the main volcanic area. The term Tharsis ridge has been widely used informally to refer to both the line of Tharsis volcanoes and the broad rise. Since the two are not coincident, use of the same term is confusing. The ultraviolet spectrometer (UVS) data [*Hord et al.*, 1974] show that if the volcanoes are ignored, the surface elevations are highest in the region of Labyrinthus Noctis and the northern part of Syria Planum. Elevations at the center of the upwarp are 10–12 km above the 6.1-mbar level, and the plains slope gently away in all directions. To distinguish the broad upwarp from the line of volcanoes, it will here be referred to as the Syria rise. Slopes appear steepest on the northwest flanks of the rise, along the line of the volcanoes, although even here the regional slopes are all less than 0.5°. Northeast of Pavonis Mons an anticlinal bulge extends from

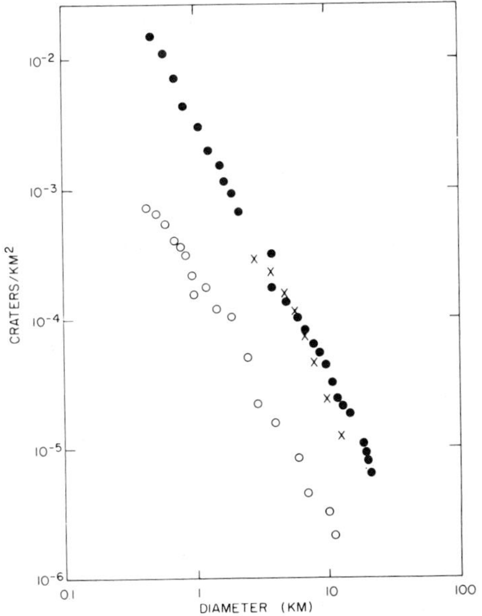

Fig. 4. Crater frequency distributions on different plains units in MC-9 showing the close similarity in crater frequencies for the fractured and cratered plains. Solid circles indicate cratered plains; open circles, sparsely cratered plains; crosses, fractured plains.

the dome proper and plunges to the northwest. This general picture of the configuration of the Syria rise is tentative since the UVS data are widely spaced, considerable latitude being allowed in how the contours are drawn.

Possibly the earth's closest equivalents to the Syria rise are the broad epeirogenic upwarps 1–10 km high and up to a few thousand kilometers across that occur widely in both continental and oceanic areas [Menard, 1974; Hartmann, 1973]. Wilson [1972] has proposed that mantle convection causes the initial updoming. As the doming continues, radial rifts develop and extend outward. Further development may be arrested at this point; however, in some cases the outwardly extending rifts intersect others to divide the lithosphere into new plates. The Red Sea, the Gulf of Aden, and the Ethiopian rifts are examples of rifts roughly radial to a dome in the Arabian Nubian shield. A similar but much smaller example is provided by the Rhine graben. A slightly different analogy is provided by May [1970], who postulated that Triassic and Jurassic basaltic dikes in the Americas and Africa belong to a set that was formerly radial to a point between the continents and that they formed during the initial breakup of the lithosphere to form the Atlantic Ocean. Unfortunately, epeirogenic upwarps rarely have such a clear fracture pattern associated with them. More commonly, the upwarp is long lived in comparison with plate motion, and complicated patterns develop as the lithosphere rides over the swell.

A better indication of the type of fracture patterns that are likely to develop from domical uplift is given by smaller domes. Small domes are relatively rare on earth because horizontal forces tend to dominate and produce elongate structures. Where domes do occur, however, such as over diapiric structures like salt plugs and igneous stocks, radial fracture patterns are common. For example, the Hawkins oil field [Wendtland, 1951] and the Van oil field [Nevin, 1942], both over salt diapirs, have well-developed radial fractures. In both cases the fracture pattern tends to become less regular toward the center of the dome, as the Tharsis radials do as they approach Labyrinthus Noctis. Similar patterns can develop around an igneous stock, such as that at Mount Monadnock, Vermont [Chapman, 1954]. Radial patterns can also be produced experimentally by doming [Cloos, 1939; Bucher, 1933]. Thus ascribing the Tharsis radials to doming seems reasonable.

The Syria rise is approximately 2000 km in radius, and at the center the surface is approximately 7 km above the level of the surroundings. The radial and circumferential extension of the surface caused by a spherical dome of this dimension was estimated by assuming radial movement out from the center of the sphere. The radial extension of the surface to the edge of the dome is 2.2 km, and the circumferential extension reaches a maximum of 9.6 km 1250 km from the center (approximately 0.1% extension). This value can be compared with the extension as estimated from photographs of the grabens. Graben walls such as those near Alba Patera (Figure 2) are probably talus slopes that bury a more steeply dipping fault plane. When reasonable values for the talus slope (30°) and the dip of the postulated fault plane (60°) are assumed, the displacements along the faults can be estimated from the width of the talus slope on the frame. In several B frames of the Alba Patera region that were examined, the faults appeared to average about 250 m of vertical displacement. Since there was no UVS coverage of this area, this value cannot be confirmed. The total horizontal extension, obtained by counting the number of faults, is generally about 2%. Thus the crustal extension north and northeast of the Tharsis volcanoes appears to be larger than that expected from the simple doming mechanism. To the southwest, where faults are far fewer, doming appears adequate.

If the fracturing resulted simply from doming, a radially symmetric pattern would develop. The actual pattern is clearly not symmetrical; the fractures tend to converge along the volcano line and to be far more common to the north. The asymmetry probably resulted from the asymmetrical shape of the dome, although a preexisting planetwide northeast-southwest stress system suggested by Hartmann [1973] could have affected the fractures. The slope of the dome flanks appears to be steepest along the volcano line. Here crustal flexure is greatest, so that fracturing should be most intense. Unfortunately the presence of the volcanoes and other young volcanics prevents surfaces older than approximately 1 b.y. from being observed along the line of volcanoes. Indirect evidence suggests, however, that the volcanoes are lined up along a northeast-southwest fracture zone. Alcoves and fractures on the northwest and southwest parts of each shield volcano [Carr, 1973] and a rough northeast-southwest alignment of slope discontinuities of the volcano flanks all suggest deep northeast-southwest-trending fractures. Alignment of volcanic features along an epeirogenic crustal flexure of this type is not unknown on earth. Wager and Deer [1938] describe the alignment of dikes along a monoclinal flexure that closely follows the east coast of Greenland, and Du Toit [1929] describes a similar dike swarm along the Lebombo flexure in southeast Africa.

The reason that the updoming of the crust was asymmetric is not known, but a reasonable possibility is that it is related to the hemispheric asymmetry of the planet. Mars can be divided into two strikingly different hemispheres [Carr, 1973]. One is mostly primitive cratered terrain; the other contains most of the young volcanic plains. The nature of the boundary between the two hemispheres is not known. It could be merely erosional, but, more likely, it could represent some primitive fractionation of the crust. The boundary cuts across the Syria rise (Figure 1) and may have resulted in its asymmetric development. Another possibility is that the asymmetry developed as a result of the aforementioned postulated planetwide stress system.

The volcanoes in their very early stages of development may have contributed to the formation of the fracture system. Radial patterns around intrusions are not uncommon. Because dikes tend to follow extension fractures, the orientations of dikes and fractures can generally be equated. Probably the best-documented examples of radial dike patterns around volcanic intrusives are the Spanish Peaks in Colorado [Knopf, 1936; Odé, 1957] and the island of Rum in northwest Scotland [Richey, 1948]. These dikes are generally only a few tens of kilometers long at the most; however, the volcanic center of northwest Scotland as a whole is associated with a dike swarm that is considerably more extensive. Dikes several hundred kilometers long attest to a stress system of wide areal extent and one that was in some way related to the volcanic activity. Although the pattern is not radial, the dikes converge on the volcanic center in much the same way that the Martian fractures converge on Tharsis.

Odé, in his analysis of the Spanish Peaks, was able to reproduce theoretically the observed pattern of dikes around the peaks. He computed the stresses that would result if a radial stress pattern around the west Spanish Peak intrusion were superimposed on a regional stress pattern. The direction of the principal stresses so computed agreed remarkably well with the observed dike pattern. Odé's analysis was adapted to determine the stress field that would result from the intrusion of the three Tharsis shield volcanoes. Following Odé, it was assumed that associated with each volcanic center is a stress function $\phi = A \ln r$, where r is the radial distance from the center and A is a constant. This assumption is equivalent to treating each volcano as a hole in a semi-infinite plate, a hydrostatic pressure being maintained within the hole. The constant A was assumed to be the same for each volcano, this identity implying that each center was active at the same time and affected the stress field equally. These assumptions are almost certainly not valid, but they suffice for this preliminary analysis. For this simple model the orientations of the maximum and minimum stresses in the plane of the surface were computed and are shown in Figure 5. Details of the analytical procedure are given by *Odé* [1957] and *Johnson* [1970]. The solid lines at right angles to the least principal stress are the directions along which tension fractures should occur. In three dimensions the greatest principal stress can be assumed to be vertical, so that normal faults would result.

The validity of applying the type of fracture analysis used by Odé to areas as wide in extent as the area encompassed by the Tharsis fractures is questionable. Transmittal of stress over several thousand kilometers implies an enormous pressure within the volcanic pipe. Odé's computed values for the Spanish Peaks are very high. The pressures required here would be orders of magnitude higher. Although there are insufficient data to calculate the actual pressure, it could be well above the crushing strength of rocks. The only fracture patterns of comparable dimensions known to the writer are the radial fractures around some of the large basins on the moon. These are widely believed to have formed as a result of impact, for which the central pressures were probably in the region of several hundred kilobars. Such pressures could not be generated close to the surface by volcanic processes. Nevertheless, because of the uncertainties, it is worth examining how well the predicted pattern mimics the observed pattern.

The predicted fracture pattern only superficially resembles the pattern observed around Tharsis. The roughly concentric pattern is reproduced, as are the opposing fanlike arrangements to the northeast and southwest. Details of the pattern are, however, difficult to reconcile with the model. A northwest-southeast trend should dominate in the Labyrinthus area, yet at the north end of the Claritas Fossae, the fractures are at right angles to this direction. Similar discrepancies occur immediately north of Labyrinthus and in some isolated patches of fractured terrain south of Tharsis Tholus. Significant also is the lack of fractures in the sector between the Claritas Fossae and Vallis Marineri. Here northwest-southeast fractures radial to Pavonis Mons should be well developed. Different patterns would result if different values for the constant A were chosen for each volcano. Decreasing the central pressure for either Arsia Mons or Ascreus Mons would have the effect of shifting the isotropic points toward that volcano and increasing the curvature of the maximum principal stress trajectories around it. Conversely, if the effect of Pavonis Mons were decreased, the isotropic points would move toward it, and the curvature of stress trajectories around the outside centers would be decreased. Neither of these modifications improves the agreement with the observed pattern significantly. Furthermore, Olympus Mons has no obvious radial fracture pattern associated with it, so that the efficacy of the volcanic mechanism is in doubt.

We can conclude that the volcanoes did not cause the fractures. On the contrary, the volcanoes probably formed as a result of the fractures. This conclusion is in agreement with the relative ages of the fractures and the volcanoes. The three Tharsis volcanoes all appear to be younger than the fracture system. The old volcanic feature, Alba Patera, may have been active when the fractures formed. It would then have had a stress system associated with it that caused the stress trajectories to be deflected around it to form the fracture ring that we now observe. The effect of the Tharsis volcanoes, however, probably was only to cause reactivation of appropriately aligned fractures at some time after the main fracturing event.

SUMMARY

It appears that roughly 1 b.y. ago a major tectonic event terminated on Mars. A broad upwarping of the crust had taken place in the Labyrinthus-Tharsis region. The upwarp, 3000-4000 km across and about 7 km high in the center, was asymmetric and resulted in extensive fracturing of the crust. At the time of the upwarp some volcanic features, such as Alba Patera, were in the process of formation, but the main shield-building period in Tharsis occurred later, the shields being preferentially aligned along the zone of most intense fracturing. Mantle convection appears to be the most likely cause of the initial updoming and the ensuing volcanism in the region.

The Syria rise appears to be unique on Mars. Although there are other high-standing areas, none have radial fractures, together with a well-developed volcanic center, associated with them. The Elysium volcanic region in particular lacks the extensive fracture system that occurs around the Syria rise (Figure 1). Development of the Syria rise and its fractures also separates eras of quite dissimilar volcanism. The postrise

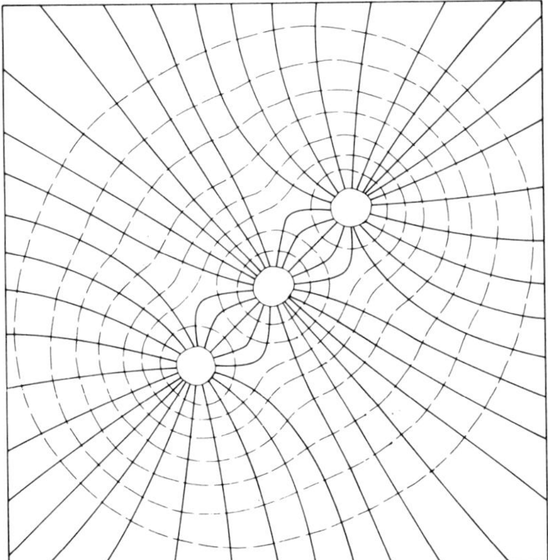

Fig. 5. Stress trajectories for the three Tharsis shield volcanoes on the assumption that they formed simultaneously with the same stress field. Solid lines are at right angles to the least principal stress and are the direction along which fracturing is expected.

volcanic centers are large high-standing shields, deep magma sources and a thick (approximately 200 km) rigid lithosphere thus being implied. The prerise volcanic features, although not yet well studied, appear not to grow as high, implying shallower depths of origin of the magma or more fluid magma. Several questions concerning the rise remain to be addressed. Whether the rise is still being actively supported or whether it is in isostatic equilibrium or sustained by a thick rigid lithosphere is of particular interest, as is the question of what implications the rise has for the interior of Mars. These topics will be discussed in a later paper.

Acknowledgments. I wish to acknowledge the contributions of K. Howard and D. Milton of the U.S. Geological Survey, both of whom made numerous helpful suggestions. The work was supported by the Planetology Programs Office, Office of Space Science, NASA Headquarters, contract W 13204, and by the Jet Propulsion Laboratory, contract WO 8122.

REFERENCES

Bucher, W. H., *The Deformation of the Earth's Crust: An Inductive Approach to the Problems of Diastrophism,* p. 121, Princeton University Press, Princeton, N. J., 1933.

Carr, M. H., Volcanism on Mars, *J. Geophys. Res., 78,* 4049–4062, 1973.

Carr, M. H., H. Masursky, and R. S. Saunders, A generalized geologic map of Mars, *J. Geophys. Res., 78,* 4031–4036, 1973.

Chapman, R. W., Criteria for the mode of emplacement of the alkaline stock at Mt. Monadnock, Vermont, *Geol. Soc. Amer. Bull., 65,* 97–114, 1954.

Cloos, H., Hebung-Spaltung-Vulkanismus, *Geolog. Rundsch., 30,* 401–525, 1939.

Du Toit, A. L., The volcanic belt of the Lebombo—A region of tension, *Trans. Roy. Soc. S. Afr., 18,* 189–217, 1929.

Hartmann, W. K., Martian surface and crust: Review and synthesis, *Icarus, 19,* 550–575, 1973.

Hord, C. W., K. E., Simons, and L. K. McLaughlin, Mariner 9 ultraviolet spectrometer experiment: Pressure-altitude measurements on Mars, *Icarus,* in press, 1974.

Johnson, A. M., *Physical Processes in Geology,* pp. 401–428, Freeman, Cooper, San Francisco, Calif., 1970.

Knopf, A., Igneous geology of the Spanish Peaks region, Colorado, *Geol. Soc. Amer. Bull., 47,* 1727–1784, 1936.

May, P. R., Pattern of Triassic-Jurassic diabase dikes around the North Atlantic in the context of predrift position of the continents, *Geol. Soc. Amer. Bull., 82,* 1285–1292, 1970.

Menard, H. W., Epeirogeny and plate tectonics, *Eos Trans. AGU, 54,* 1244–1255, 1974.

Nevin, C. M., *Principles of Structural Geology,* p. 96, John Wiley, New York, 1942.

Odé, H., Mechanical analysis of the dike pattern of the Spanish Peaks area, Colorado, *Geol. Soc. Amer. Bull., 68,* 567–576, 1957.

Richey, J. E., *Scotland: The Tertiary Volcanic Districts,* p. 86, Her Majesty's Stationery Office, London, 1948.

Soderblom, L. A., R. A. West, B. M. Herman, T. J. Kreidler, and C. D. Condit, Martian planet-wide crater distribution: Implications for geologic history and surface processes, *Icarus,* in press, 1974.

Wager, L. R., and W. H. Deer, A dyke swarm and crustal flexure in east Greenland, *Geol. Mag., 75,* 39–46, 1938.

Wendtland, E. A., Hawkins field, in *Occurrence of Oil and Gas in Northeast Texas, Publ. 5116,* pp. 153–157, Univ. of Texas, Austin, 1951.

Wilson, J. T., New insights into old shields, *Tectonophysics, 13,* 73–94, 1972.

Wilson, J. T., and K. Burke, Plate tectonics and plume mechanics (abstract), *Eos Trans AGU, 54,* 238, 1973.

(Received April 2, 1974;
revised April 24, 1974;
accepted April 25, 1974.)

Part IV

ROCKS OF THE MANTLE
AND THEIR RHEOLOGY

Editor's Comments
on Papers 13, 14, and 15

13 CARTER, BAKER, and GEORGE
Seismic Anisotropy, Flow, and Constitution of the Upper Mantle

14 BOULLIER and NICOLAS
Classification of Textures and Fabrics of Peridotite Xenoliths from South African Kimberlites

15 POST and GRIGGS
The Earth's Mantle: Evidence of Non-Newtonian Flow

Knowledge of the rheology of the mantle is essential for understanding its flow, but is difficult to study. Rough estimates can be based on study of isostatic readjustments, but a more straightforward approach is experimental study of likely constituents of the mantle. However, the extreme conditions, and especially the natural low strain rates, cannot be reproduced in the laboratory. Therefore, experimental results must be extrapolated to the appropriate conditions, or the traces of strain left in rocks have to be interpreted. Fortunately, xenoliths that occur in alkaline basalts and in kimberlites provide a sampling of deep-seated materials, and they bear traces of deformation. Interpretation of their petrographic textures, based on experimental results, can be used to obtain valuable insight into the mechanics of their source regions. Additional insight can be gained from exposed Alpine-type peridotite bodies, which are probably uplifted slices of the mantle and resemble the xenoliths in many ways.

To extrapolate laboratory studies to mantle conditions, the deformation mechanisms must be known on an atomic scale. Two main mechanisms have been proposed for the steady-state creep of solids at the high temperatures and pressures that occur in the mantle (e.g., Weertman, 1970; Weertman and Weertman, 1975; Kirby and Raleigh, 1973; Stocker and Ashby, 1973; Carter, 1976). At very high temperatures (close to the melting point) and slow strain rates, creep occurs predominantly by diffusion of ions and shift of lattice vacancies through crystals (Nabarro–Herring creep). In this process the relation between strain rate and stress is linear, so that the material behaves as a Newtonian fluid. This relation depends also on grain size which is, however, stress dependent, so the strain rate may still depend on a power of the stress. Another mechanism is the migration of crystal defects or dislocations. In this case the strain rate is proportional to approximately the third power of the stress (power-law rheology). The process is accompanied by orientation of crystals, so that strong fabrics are produced. Recrystallization also occurs, but its role as a deformation mechanism is not clear; it may represent a recovery process, with the new crystals inheriting the orientation of their strained precursors. These processes may operate simultaneously. The problem is to determine which of them leads to a higher strain rate at the P-T and stress conditions in the mantle. In both cases the effective viscosity depends strongly on the temperature and pressure.

Carter, Baker, and George (Paper 13) reviewed the problem using experimental data on likely mantle minerals as well as data on ultramafic rocks. They concluded that over most of the upper mantle conditions migration of dislocations is the dominant creep mechanism, so there is a power-law relation between strain rate and stress, and flow in the mantle is, therefore, non-Newtonian. Deformation of grains and recrystallization in the laboratory produce typical textures and, especially, fabrics that are identical to those observed in mantle-derived peridotite nodules in basalts and in peridotite massifs, implying identical deformation mechanisms. The deformation fabrics in peridotites (but not in eclogites) can also account for the observed anisotropy of seismic velocities in the upper mantle. Thus, extrapolation of laboratory measurements to upper mantle conditions seems to be justified. These results were confirmed by additional experimental work (Carter, 1976; Kohlstedt, Goetze, and Durham, 1976; Post, 1977). The properties of olivine were studied most extensively, because it is the major, and very probably the most deformable, constituent of the outer part of the mantle. Therefore, it may well control the overall deformation. An important conclusion from the laboratory measurements is that because of its P-T dependence the effective viscosity, for likely strain rates, has a minimum of about 10^{20}–10^{21} poise in the low velocity zone (about 100–200 km depth), which thus coincides with the asthenosphere. This minimum is less pronounced than that predicted by Nabarro–Herring creep.

Ultramafic xenoliths in basalts are predominantly spinel lherzolites, whose stability range corresponds to conditions within the lithosphere (Paper 9). On the other hand, garnet-bearing peridotites are common as xenoliths in kimberlites. Such mineral assemblages form at deeper levels, some at depths of 150 km or more (Boyd, 1973; Mercier, 1980). Therefore these xenoliths more likely represent the lowest parts of the lithosphere, and probably also the underlying asthenosphere.

Boullier and Nicolas (Paper 14) describe textures of peridotite xenoliths in kimberlites. Many types, though recrystallized to some degree, have prominent mineral fabrics, similar to fabrics produced in creep experiments by migration of dislocations, implying a power-law rheology for these xenoliths. Other xenoliths are very strongly sheared. Except for a few types, the textures of the xenoliths from kimberlites resemble in a general way those found in peridotite xenoliths in basalts (Mercier and Nicolas, 1975; Pike and Schwartzman, 1976). Studying the structures of peridotite xenoliths, Nicolas (1978) estimated that the stresses under which they were deformed were in the range of 10–1000 bars. On the other hand, Goetze (1975) concluded from the density of dislocations in grains of highly sheared xenoliths that they were deformed under stresses of a few kilobars. This implies, given any reasonable effective viscosity, high strain rates which could have occurred only during limited periods and/or in restricted volumes, for example, at the margins of mantle diapirs. It was also suggested that the highly sheared xenoliths have undergone superplastic flow, which occurs at relatively slow strain rate and low stresses as a result of slippage between grains (Boullier and Gueguen, 1975; Twiss, 1976). The deformation apparent in the mantle xenoliths may have occurred while the lithosphere was generated, or shortly before the xenoliths were extracted (Papers 13 and 14). Such interpretations, however, are not easily reconciled with isotopic evidence that some peridotite xenoliths were recrystallized long ago (Allsop, Nicolaysen, and Hahn-Weinheimer, 1969; Steuber and Ikramudin, 1974; Morioka and Kigoshi, 1975). Is one dealing with different populations, or with ancient deformation events that were locally preserved? These problems show that some aspects of xenolith deformation are not clear. In particular, the significance of stresses inferred from the xenoliths (if reliably estimated), and whether the xenoliths record the "average" mantle deformation, is not known.

Strictly speaking, the study of xenoliths applies directly only to the outer 350–400 km of the mantle. At greater depths olivine and the other mantle minerals break down to denser phases in which the relative importance of different deformation mechanisms may be different from the low pressure minerals. There are no direct observations on the deformation of the high pressure phases that constitute the deeper mantle. The effective viscosity may increase, but probably not very much (O'Connell, 1977; Sammis et al., 1977). Still, it appears likely that flow of much of the mantle is governed by a non-Newtonian (power law) rheology. Therefore, fluid dynamic models assuming a Newtonian behavior should be used with care. However, some studies of large-scale convection and of movement of plumes (Parmentier, Turcotte, and Torrance, 1975, 1976) suggest that this is not of fundamental importance. The distribution of internal heat sources and temperature- and pressure-dependent variations of viscosity appear to be more important.

The rheology of the mantle, especially of its deep parts, can also be inferred from its response to surface loading and unloading, mainly following deglaciation. Early studies of the rebound of Fennoscandia (Haskell, 1936; Van Bemmelen and Berlage, 1935) showed that the mantle under the lithosphere, i.e., the asthenosphere, had a viscosity of 10^{20}–10^{21} poise. Later studies, which also treated the much larger Laurentian uplift (McConnell, 1968; Lliboutry, 1971; Walcott, 1973; Cathles, 1975; Peltier, 1976) showed that the asthenosphere may have an effective viscosity as low as 4×10^{20} poise, or even less, and that it may be a few hundred km thick, though its depth is not well constrained. Crittenden (1963) found an even lower viscosity under former Lake Bonneville, but this may reflect the abnormally high temperature under the Basin and Range province. Regarding the viscosity of the lower mantle, O'Connell (1971), Cathles (1975) and Peltier (1976) favor a uniform viscosity of about 10^{22} poise on the basis of post-glacial rebound data, but a greater increase of the viscosity was also advocated (Walcott, 1973). The possibility that the viscosity is rather uniform in most of the mantle seems surprising, since viscosity is a very sensitive function of temperature and pressure. However, convection in the lower mantle can provide a mechanism which regulates the viscosity and does not allow it to vary very much (Tozer, 1965, and Paper 27). An upper limit of 10^{24} poise for the viscosity of the lower mantle was inferred by Goldreich and Toomre (1969) on the basis of the rate of polar wandering.

In all these studies, the mantle underneath the lithosphere is considered to behave as a Newtonian fluid, and it is considered that the data on the postglacial rebound favor such a behavior. On the other hand, the rebound is strongly influenced by flow in the upper mantle where a non-Newtonian rheology is expected. Post and Griggs (Paper 15) addressed this problem. They argued that the available data could be interpreted as favoring a non-Newtonian mantle, behaving in accordance with experimental data on deformation of olivine. Brennen (1974) also favored a non-Newtonian mantle. This problem was not much discussed in later papers.

Thus, the rheology of the mantle is still imperfectly known. It is encouraging that the experimental data and the studies of isostatic rebound yield broadly similar values for the effective viscosity in the upper mantle; both types of data favor a low-viscosity zone, an asthenosphere, underneath the lithosphere, but its possible lateral variations require more study. There remain also uncertainties regard-

179

ing the effective viscosity and the dominant mechanism of deformation in the lower mantle. On the other hand, much evidence favors a non-Newtonian rheology in the upper mantle. Finally, it should be stressed that all mechanisms of high-temperature creep are such that there is no finite strength, that is, the material of the mantle will creep at any stress difference, however small. Neither is there strain hardening. These properties differ from low-temperature deformation.

REFERENCES

Allsop, H. L., Nicolaysen, L. O., and P. Hahn-Weinheimer, 1969, Rb/K Ratios and Sr Isotopic Compositions of Minerals in Ectogitic and Peridotitic Rocks, *Earth and Planetary Sci. Letters* **5:**231–244.

Boullier, A. M. and Gueguen, Y., 1975, SP-Mylonites: Origin of Some Mylonites by Superplastic Flow, *Contr. Mineralogy and Petrology* **50:**93–104.

Boyd, F. R., 1973, A Pyroxene Geotherm, *Geochim. et Cosmochim. Acta* **37:**2533–2546.

Brennen, C., 1974, Isostatic Recovery and the Strain Rate Dependent Viscosity of the Earth's Mantle, *Jour. Geophys. Research* **79:**3993–4001.

Carter, N. L., 1976, Steady State Flow of Rocks, *Rev. Geophys. Space Phys.* **14:**301–360.

Cathles, L. M., 1975, *The Viscosity of the Earth's Mantle,* Princeton University Press, Princeton, N. J., 386p.

Crittenden, M.D., 1963, Effective Viscosity of the Earth Derived from Isostatic Loading of Pleistocene Lake Bonneville, *Jour. Geophys. Research* **68:**1865–1880.

Goetze, C., 1975, Sheared Iherzolites: From the Point of View of Rock Mechanics, *Geology* **3:**172–173.

Goldreich, P., and A. Toomre, 1969, Some Remarks on Polar Wandering, *Jour. Geophys. Research* **74:**2555–2567.

Haskell, N. A., 1936, The Motion of a Viscous Fluid under a Surface Load, *Physics* **7:**56–61.

Kirby, S. H. and C. B. Raleigh, 1973, Mechanisms of High-Temperature Solid-State Flow in Minerals and Ceramics and Their Bearing on the Creep Behavior of the Mantle, *Tectonophysics* **19:**165–194.

Kohlstedt, D. L., Goetze, C., and W. B. Durham, 1976, Experimental Deformation of Single Crystal Olivine with Application to Flow in the Mantle, in *The Physics and Chemistry of Minerals and Rocks,* R. G. J. Sterns, ed. Wiley, New York, pp. 35–49.

Lliboutry, L. A., 1971, Rheological Properties of the Asthenosphere from Fennoscandian Data, *Jour. Geophys. Research* **76:**1433–1446.

McConnell, R. K., 1968, Viscosity of the Mantle from Relaxation Time Spectra of Isostatic Adjustment, *Jour. Geophys. Research* **73:**7089–7105.

Mercier, J. C. C., 1980, Single-Pyroxene Thermobarometry, *Tectonophysics* **70:**1–37.

Mercier, J. C., and A. Nicolas, 1975, Textures and Fabrics of Upper Mantle Peridotites as Illustrated by Basalt Xenoliths, *Jour. Petrology* **16:**454–487.

Morioka, M., and K. Kigoshi, 1975, Lead Isotopes and Age in Hawaiian Iherzolite Nodules, *Earth and Planetary Sci. Letters* **25:**116–120.

Nicolas, A., 1978, Stress Estimates from Structural Studies of Some Mantle Peridotites, *Royal Soc. London Philos. Trans.*, ser. A, **288**:49–57.

O'Connell, R. J., 1971, Pleistocene Glaciation and the Viscosity of the Lower Mantle, *Royal Astron. Soc. Geophys. Jour.* **23**:299–327.

O'Connell, R. J., 1977, On the Scale of Mantle Convection, *Tectonophysics* **38**:119–136.

Parmentier, E. M., D. L. Turcotte, and K. E. Torrance, 1975, Numerical Experiments on the Structure of Mantle Plumes, *Jour. Geophys. Research* **80**:4417–4124.

Parmentier, E. M., D. L. Turcotte, and K. E. Torrance, 1976, Studies of Finite Amplitude Non-Newtonian Thermal Convection with Application to Convection in the Earth's Mantle, *Jour. Geophys. Research* **81**:1839–1846.

Peltier, W. R., 1976, Glacial-Isostatic Adjustment II, The Inverse Problem, *Royal Astron Soc. Geophys. Jour.* **46**:669–706.

Pike, J. E. and F. C. Schwartzman, 1976, Classification of Textures in Ultramafic Xenoliths, *Jour. Geology* **85**:49–61.

Post, R. L., 1977, High Temperature Creep of Mt. Burnet Dunite, *Tectonophysics* **42**:75–110.

Sammis, C. G., J. C. Smith, G. Schubert, and D. A. Yuen, 1977, Viscosity-Depth Profile of the Earth's Mantle: Effects of Polymorphic Phase Transitions, *Jour. Geophys. Research* **82**:-3747–3761.

Stocker, R. L. and M. F. Ashby, 1973, On the Rheology of the Upper Mantle, *Rev. Geophys. Space Phys.* **8**:145–168.

Stueber, A. M. and M. Ikramudin, 1974, Rubidium, Strontium and the Isotopic Composition of Strontium in Ultramafic Nodule Minerals and Host Basalts, *Geochim. et Cosmochim. Acta* **38**:207–216.

Tozer, D. C., 1965, Heat Transfer and Convection Currents, *Royal Soc. London. Philos. Trans.*, ser. A, **258**:252–271.

Twiss, R. J., 1976, Structural Superplastic Creep and Linear Viscosity in the Earth's Mantle, *Earth and Planetary Sci. Letters* **33**:86–110.

Van Bemmelen, R. W., and H. P. Berlage, 1935, Versuch einer mathematischen Behandlung geotektonischer Bewegung unter besonderer Beruchsichtung der Undationstheorie, *Gerlands Beitr. Geophysik* **43**:19–55.

Walcott, R. I., 1973, Structure of the Earth from Glacio-Isostatic Rebound, *Ann. Rev. Earth and Planetary Sci.* **1**:15–37.

Weertman, J., 1970, The Creep Strength of the Earth's Mantle, *Rev. Geophys. Space Phys.* **8**:145–168.

Weertman, J., and J. R. Weertman, 1975, High Temperature Creep of Rocks and Mantle Viscosity, *Ann. Rev. Earth and Planetary Sci.* **3**:293–315.

13

Reprinted from pages 167–190, 337, and 339–351 of *Flow and Fracture of Rocks,*
Geophysical Monograph Series, vol. 16, H. C. Heard, I. Y. Borg, N. L. Carter and
C. B. Raleigh, eds., American Geophysical Union, Washington, D. C., 1972, 356p.

Seismic Anisotropy, Flow, and Constitution of the Upper Mantle

NEVILLE L. CARTER

*Department of Earth and Space Sciences, State University of New York
Stony Brook, New York   11790*

DAVID W. BAKER

*Department of Geological Sciences, University of Illinois at Chicago Circle
Chicago, Illinois   60680*

RICHARD P. GEORGE, JR.

*Department of Earth and Space Sciences, State University of New York
Stony Brook, New York   11790*

Experiments on syntectonic recrystallization of compacted powders of enstatite and diopside were conducted in axial compression at a confining pressure of 15 kb, temperatures of 900°–1300°C, and strain rates from 7.8×10^{-4} to 7.8×10^{-7} sec^{-1}. Both recrystallized enstatite and diopside aggregates have preferred orientations with [010] oriented parallel to σ_1, and the remaining principal axes and crystallographic axes measured are arranged in girdles in the $\sigma_2 \simeq \sigma_3$ plane. A similar fabric was obtained previously for experimental syntectonic recrystallization of olivine, and such fabrics have common counterparts in naturally deformed peridotites, dunites, and some garnet clinopyroxenites. Variations in seismic body wave velocities have been calculated for the experimentally recrystallized aggregates and eight naturally deformed peridotites, dunites, and garnet clinopyroxenites by using the method described in the previous paper (Baker and Carter, this volume). The calculated compressional wave velocities and anisotropies compare well with measured values when allowance is made primarily for alteration products. The average mean P-wave velocity for these and certain other peridotites and dunites is 8.3 km/sec, and the average maximum anisotropy is 0.76 km/sec, as compared with average mean velocities and maximum anisotropies for garnet clinopyroxenites of 7.8 and 0.22 km/sec, respectively. The average mean P-wave velocity in the upper mantle of the northeast Pacific is 8.1 km/sec, and the anisotropy ranges from 0.3 to 0.7 km/sec. Thus the observed anisotropies can not be produced by a dominantly garnet pyroxenite upper mantle but are easily accounted for by peridotite and/or dunite with preferred crystal orientations produced primarily by syntectonic recrystallization. If syntectonic recrystallization, accompanied by plastic deformation and recovery, is the dominant mechanism governing power law creep in the upper mantle, our results support the hypothesis that thermal convection is the driving force for the sea floor spreading process.

The constitution of the earth's upper mantle and its flow properties and processes are important questions that have been debated extensively, especially since the discovery of the sea floor spreading process. Indirect information concerning possible compositions comes from experimental and natural propagation of seismic waves and free oscillation data combined with the known mass and moment of inertia and inferences drawn from high pressure–temperature experimental petrology and also from meteoritic and cosmic elemental abundances. This information severely limits permissible upper-mantle compositions, but to date it has not been sufficiently sensitive to allow a definitive decision between a dominantly eclogitic [e.g., *Ito and Kennedy,* 1970] or a dominantly peridotitic [e.g., *Ringwood,* 1969] upper mantle. More direct methods include systematic petrological and chemical investigations of inclusions in basalts known to have emanated from the upper mantle [e.g., *Jackson and Wright*, 1970] or in diamond-bearing kimberlite pipes [e.g., *MacGregor and Carter*, 1970]. The generally great preponderance of inclusions of peridotite over those of eclogite, combined with the indirect information mentioned above, has convinced most workers that the upper mantle is dominantly peridotitic.

Equally important to determinations of the constitution of the upper mantle are determinations of the mechanical equation of state governing the high-temperature creep and of the atomic processes responsible for it. The stress-strain rate relationship generally has been assumed to be linear (for mathematical simplicity), but recent theoretical [*Weertman*, 1970] and empirical [*Carter and Avé Lallemant*, 1970; *Raleigh and Kirby*, 1970; *Post*, 1970] studies have indicated that flow over most of the upper mantle is probably governed by power law creep. This conclusion, which is of great importance to estimates of the depth dependence of viscosity, is supported by similarities in the deformational processes that have operated in materials derived from the upper mantle and those observed in the high-temperature steady-state experiments. These similarities led to the suggestion [*Avé Lallemant and Carter*, 1970] that the dominant mode of flow during steady-state creep in the upper mantle is probably syntectonic recrystallization (hot working).

An independent test of this hypothesis is now possible because of the careful seismic refraction studies of the upper mantle in the northeast Pacific by *Raitt et al.* [1969], *Morris et al.* [1969], and *Keen and Barrett* [1971]. Their results have revealed a systematic variation of compressional wave velocity with azimuth, the greatest velocity being nearly normal to the ridge axis and the velocity anisotropy being, in the range 0.3–0.7 km/sec. Any hypothesis concerning the constitution of the upper mantle and/or its flow properties must also account for these most important results. Using the methods discussed in the previous paper [*Baker and Carter*, this volume] and other arguments, we show that a dominantly peridotitic upper mantle alone, having crystals with preferred orientations induced primarily by syntectonic recrystallization, is most consistent with these as well as other observations.

FLOW PROPERTIES AND PROCESSES IN THE UPPER MANTLE

There can no longer be any serious question concerning the importance of high-temperature creep in at least the upper mantle in being primarily responsible for the distributions of continents with time, the origin of their first-order structures, and the evolution and destruction of island arcs and ocean basins. Some important remaining questions pertain to the driving mechanisms and their vertical extent and to the flow laws governing the creep and the atomic processes responsible for it. Most theoretical attempts to model motions in the mantle, when various initial boundary conditions are assumed (for a review, see *Knopoff* [1969]), have also assumed a Newtonian viscous behavior of the material, mainly for mathematical convenience. Such an approach has been justified frequently by citing certain data from creep studies on some very fine-grained metals and ceramics near melting [*Garafalo*, 1965; *Weertman*, 1968; *Sherby and Burke*, 1967; *McKenzie*, 1968] for which the atomic deformational process is supposed to be stress-induced vacancy migration (diffusion or Nabarro-Herring creep). Assuming that Nabarro-Herring creep dominates in the mantle, *Gordon* [1965] found the increase in viscosity with depth below the low velocity zone to be so rapid as to preclude the possibility of mantle-wide convection. Other arguments against mantle-wide convection are based mainly on possible chemical stratification [e.g., *Anderson et al.*, 1971] and phase changes [*Knopoff*, 1967], although these arguments are by no means incontrovertible [*Wang*, 1970; *Ringwood*, 1969; *Knopoff*, 1969].

More recently, *Weertman* [1970], *Raleigh and Kirby* [1970], *Carter and Avé Lallemant* [1970], *Green* [1970], and *Green and Radcliffe* [1972b] have pointed out that Nabarro-Herring creep must be of limited extent and that the law governing flow over most of the upper mantle is probably nonlinear. The steady-state creep data available for dunite and lherzolite [*Carter and Avé Lallemant*, 1970; *Raleigh and Kirby*, 1970; *Post*, 1970] are best fit by a power creep equation of the form

$$\dot{\epsilon} = A \exp\left(-Q/RT\right)\sigma^n \qquad (1)$$

where, in the absence of externally released H_2O, a material constant A is near 10^{10} when σ

is expressed in kilobars, the creep activation energy Q is near 110 kcal/mole, and the stress exponent n is near 5. The values of these constants are generally somewhat lower for dunite and lherzolite deformed in the presence of H_2O, which is released during the alteration of the talc confining medium [*Carter and Avé Lallemant*, 1970; *Post*, 1970].

There is a considerable theoretical [e.g., *Weertman*, 1968] and empirical [e.g., *Sherby and Burke*, 1967] basis for a flow law of the form of (1) in the high-temperature creep regime, and the constants determined for the dry dunite and lherzolite are in reasonable accord with those found in extensive studies on high-temperature creep of metals and metal compounds. In such experiments the activation energy for creep is near that for the self-diffusion of the least mobile atomic species, and generally it can be shown that diffusion is rate controlling, although a similar result can be obtained if the creep rate is controlled by the glide of jog-dragging screw dislocations [*Weertman*, 1968, 1970]. An equation of the form of (1) is also best fit by high-temperature steady-state creep data for ice [*Kamb*, 1964], halite single crystals [*Carter and Heard*, 1970], halite aggregates [*Heard*, this volume], and marble [*Heard and Raleigh*, 1972]. (*Goetze* [1971] obtained a different result in his wet creep experiments on Westerly granite, but, because of the low strain amplitudes used (10^{-3}–10^{-5}), it is unlikely that steady-state flow was achieved.)

In Figure 1a we have estimated the variation with depth of stress and viscosity (solid lines) for dry dunite deformed at a constant rate of 10^{-14} sec^{-1}, using (1) and the constants determined by *Carter and Avé Lallemant* [1970]. We have also used the empirical relation employed by *Weertman* [1970; see also *Sherby and Simnad*, 1961].

$$\dot{\epsilon} \simeq f(\sigma) D = f(\sigma) D_0 \exp(-aTm/T) \quad (2)$$

where D is the diffusion coefficient and $f(\sigma)$ is a constant that depends primarily on stress. This equation assumes, with some experimental justification [*Weertman*, 1970], that the pressure effect on D is given approximately by its effect on the melting temperature Tm or on the ratio of Tm/T in (2). The constant $a = Q/RT$ at melting is 28.7, as was determined by using our activation energy (120 kcal/mole) and the

melting temperature of forsterite at 15 kb (about 2100°K [*Davis and England*, 1964]). The coefficient of oxygen diffusion in olivine (expected to be rate limiting) is not known, but, for most materials near melting, $D \simeq 10^{-8}$ cm^2 sec^{-1} [*Shewmon*, 1963]. With this value and our constant a, $D_0 = 2.9 \times 10^4$ cm^2/sec.

The stress estimate ($\sigma_1 - \sigma_3$) in Figure 1a, calculated at a representative geological strain rate of 10^{-14} sec^{-1}, is based on the equation

$$\sigma^{4.8} = \frac{\dot{\epsilon}}{A \exp(-28.7Tm/T)}$$

$$= \frac{8.38 \times 10^{-25}}{\exp(-28.7Tm/T)} \quad (3)$$

where Tm/T (Figure 1a) is taken as the ratio of the melting temperature of forsterite with depth [*Davis and England*, 1964] to the ambient temperature as given by the oceanic geotherm of *Ringwood* [1969]. Over the depth interval of 100–400 km the shearing stresses range from 7 to 16 bars. The effective viscosity ($\eta = \sigma/3\dot{\epsilon}$; [*Griggs*, 1939]) in the same interval ranges from 5×10^{20} to 5×10^{21} poises and increases only slowly with depth below 200 km. Both the stress and the viscosity estimate are in excellent accord with estimates based on other geophysical considerations [e.g., *Crittenden*, 1967; *McConnell*, 1968].

For comparison we have calculated the effective viscosity from a Nabarro-Herring relationship of the form

$$\eta = \frac{\sigma}{\dot{\epsilon}} = \frac{L^2 kT}{5 V_a D_0 \exp(-28.7Tm/T)}$$

$$= \frac{95 L^2 T}{\exp(-28.7Tm/T)} \quad (4)$$

where L is the grain diameter, V_a is the atomic volume (about 10^{-23} cm^3), $D_0 = 2.9 \times 10^4$ cm^2/sec, and k and T have their usual meaning. The viscosities shown (dashed lines in Figure 1a) were calculated for average grain diameters of 0.5 and 5 cm, which encompass the range in grain size observed in peridotites and dunites derived from the upper mantle [*Raleigh and Kirby*, 1970; *Avé Lallemant and Carter*, 1970]. In the depth interval 100–400 km the viscosities range from 7×10^{19} to 3×10^{23} poises, depending critically on grain size and increasing rapidly at depths below 200 km for both grain sizes.

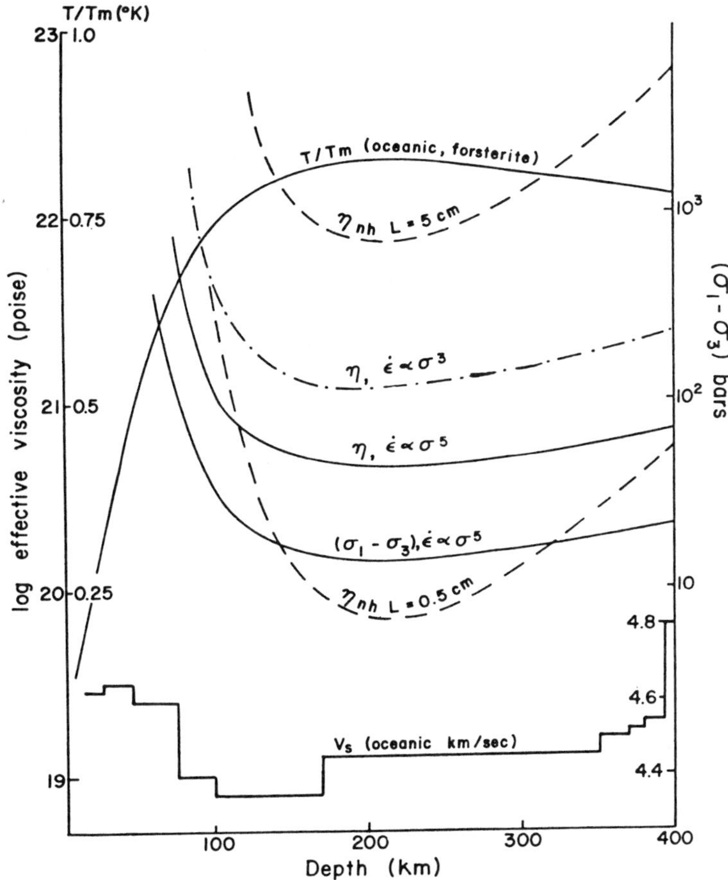

Fig. 1a. Stresses and effective viscosities of dunite in the upper mantle under the oceans, calculated at a strain rate of 10^{-14} sec^{-1}. The ratio of T/Tm in degrees Kelvin is calculated from the oceanic geotherm [*Ringwood*, 1969] and the melting of forsterite [*Davis and England*, 1964]. Shear wave velocity data showing the low-velocity channel are from *Anderson* [1967]. The solid lines are calculated from (3), the dashed-dotted line from (6), and the dashed lines from (4).

Steady-state creep cannot be achieved by the Nabarro-Herring mechanism, nor can large creep strains, because, as the diffusion path becomes longer, the strain rate decreases and viscosity increases [*Weertman*, 1970; *Green*, 1970]. The elongate grains must ultimately break down to subgrains, which may continue to creep by the vacancy diffusion mechanism provided that the subgrain boundaries are good sources and sinks for vacancies and that the subgrain size is independent of stress. However, for metals and alloys [*Sherby and Burke*, 1967; *Weertman*, 1968] the subgrain size is a function of stress, the dependence being given by

$$L = L_0(\mu/\sigma) \qquad (5)$$

where L_0 is a constant and μ is the shear modulus. *Raleigh and Kirby* [1970] also found that the subgrain size developed during the experimentally produced polygonization of olivine in lherzolite was dependent on stress, the constant being $L_0\mu = 2.4 \times 10^7$ dynes cm^{-1} (incorrectly printed as 2.4×10^5 in their paper). Substituting (5) into (4) and using $L_0\mu$ gives the creep equation

$$\sigma^3 = \frac{\dot{\epsilon}kT(L_0\mu)^2}{5V_aD_0 \exp{(-28.7Tm/T)}}$$

$$= \frac{5.47 \times 10^2 T}{\exp{(-28.7Tm/T)}} \qquad (6)$$

where now the creep rate is proportional to σ^3

and independent of grain size. The effective viscosity calculated for subgrain creep ranges from 1.2×10^{21} to 4.3×10^{21} poises in the depth interval 100–400 km.

Thus all three creep equations give reasonable effective viscosities near 10^{21} poises in the low-velocity zone, if we assume an intermediate grain size in the range 1–3 cm for the Nabarro-Herring mechanism. The values of the viscosity calculated depend rather critically, of course, on the temperature gradient chosen (through the Tm/T terms in (3), (4), and (6)). For such independent processes the process that gives the fastest creep rate at constant stress and T/Tm should be the dominant one [*Sherby and Burke*, 1967; *Weertman*, 1970]. In Figure 1b we plot the creep rate versus stress at $T/Tm = 0.75$; for intermediate grain size (≥ 1 cm), σ^5 creep should dominate at shearing stresses of ≥ 10 bars and at strain rates between 10^{-15} and $\geq 10^{-14}$ sec^{-1}.

The relationships at higher T/Tm values can be visualized by shifting the abscissa in Figure 1b to the right; thus for $T/Tm = 0.825$ (maximum in Figure 1a) the position of $\dot{\epsilon} = 10^{-14}$ sec^{-1} would be approximately that shown by the vertical bar. Under these conditions it is not certain which of the processes would dominate. At still higher T/Tm values, Nabarro-Herring creep should dominate at all reasonable strain rates, stresses, and grain sizes, but, as was mentioned above, the crystals must ultimately break down to give subgrain or power law creep. Thus it would appear that power law creep should govern flow over most upper-mantle conditions, but for confirmation we must compare the atomic processes in the steady-state experiments and naturally deformed upper-mantle materials available.

Steady-state creep in the experiments is accomplished by plastic deformation (dislocation

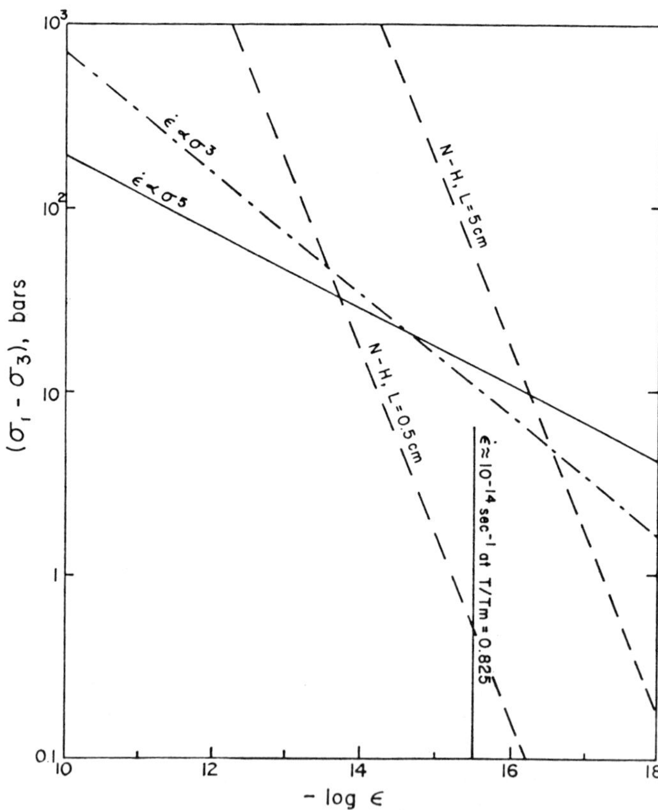

Fig. 1b. Calculations of creep rate versus stress at $T/Tm = 0.75$ for dunite, based on (3), (4), and (6). The process giving the fastest creep rate at constant stress and T/Tm should dominate. The vertical bar indicates the shift of the abscissa to the right at the higher T/Tm of 0.825.

glide), polygonization (dislocation climb), and recrystallization, any process of which can dominate, depending on the physical conditions. In general, all three processes are observed, and in most instances there is evidence that diffusion is important and probably rate controlling. The plastic-deformation mechanisms in olivine change with increasing temperature and pressure and decreasing strain rate (Figure 2) from T (slip plane) = {110} and t (slip direction) = [001] through {0kl}[100] to (010)[100] [*Raleigh*, 1968; *Carter and Avé Lallemant*, 1970]. Those systems were determined by optical microscopy but have since been confirmed by transmission electron microscopy [*Phakey et al.*, this volume; *Green and Radcliffe*, 1972b]. The pencil glide system {0kl}[100] is most commonly observed in naturally deformed olivine [*Raleigh and Kirby*, 1970] and apparently occurs by the conservative motion of rectangular loops of unit dislocations [*Green and Radcliffe*, 1972b]. There is some evidence of slip on the high-temperature system (010) [100], but diffusion-controlled processes probably dominate at moderate depths in the upper

mantle because of the rapid increase in temperature along the geotherm (Figure 2).

At the higher temperatures and lower strain rates and in natural deformations, plastic flow of orthopyroxenes occurs by kinking and slip on the system {100}[001] [*Raleigh et al.*, 1971]. Under similar conditions, clinopyroxenes deform primarily by twin gliding and translation gliding on {100}[001] [*Raleigh and Talbot*, 1967], although slip may also occur on other systems (S. H. Kirby, personal communication, 1972).

At temperatures above about 1000°C at a strain rate of 10^{-3} sec^{-1} in the experiments the diffusion-controlled process of polygonization becomes increasingly important (Figure 2) [*Raleigh and Kirby*, 1970; *Carter and Avé Lallemant*, 1970]. Excellent examples of the polygonization process, whereby subgrains form by dislocation climb and by cross slip during recovery, have been described for both experimentally and naturally deformed olivine by *Raleigh and Kirby* [1970]. The subgrains are rectangular with their sides parallel to (100) and (001). According to *Green and Radcliffe*

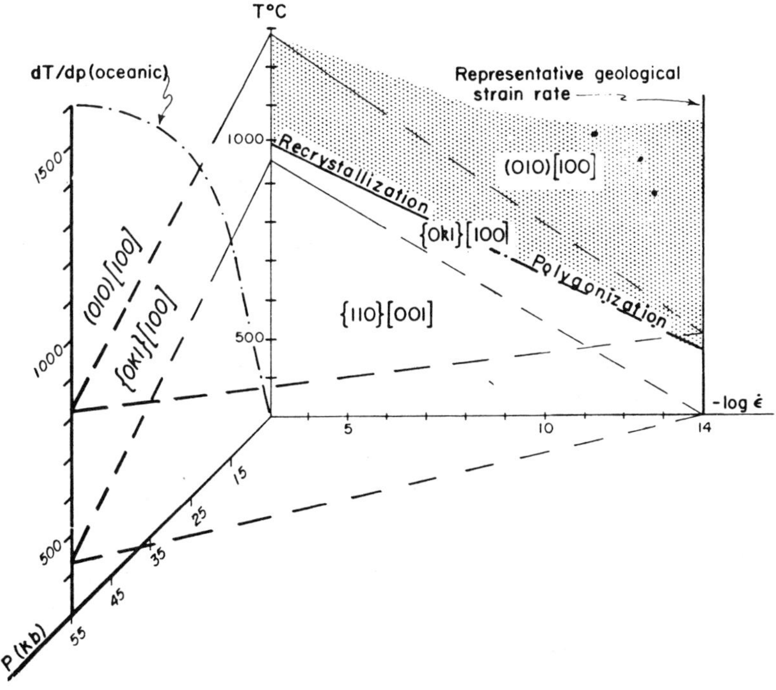

Fig. 2. Deformational processes in olivine as functions of temperature, pressure, and strain rate. The data are from *Carter and Avé Lallemant* [1970] and *Avé Lallemant and Carter* [1970]. The geotherm is from *Clark and Ringwood* [1964].

[1972*b*], who studied the same specimens by electron petrography, edge components of the {0kl}[100] dislocation loops climb into walls parallel to (100), and the screw components cross-slip into walls parallel to (001). Polygonization is also observed in the pyroxenes but is much less common and requires higher temperatures and/or lower strain rates.

At temperatures and strain rates near those required for polygonization, syntectonic recrystallization (hot working) also becomes important and in many instances clearly dominates. This diffusion-controlled process, whereby new, relatively strain-free grains grow at the expense of old, strained ones, leads ultimately to strong preferred crystal orientations that are controlled by the orientations of the principal stress axes. For coarse-grained starting material the new grains are first observed at grain boundaries, and at higher temperatures (or lower strain rates) new grains appear within the host crystal with orientations related to the host [*Avé Lallemant and Carter*, 1970]. Finally, grain boundary recrystallization advances until the host and the new grains oriented unfavorably with respect to the stress are consumed. For fine-grained starting material only the grains oriented favorably (statistically) with respect to the stress grow at the expense of the others. The final preferred orientations produced by using both fine-grained and coarse-grained aggregates are similar to each other and to fabrics commonly observed in olivine of peridotites from the upper mantle (with allowance for the fact that σ_2 need not equal σ_3 in natural deformations, as is required in the axially symmetric tests), a topic discussed more fully later.

Similarities of orientations of olivine crystals induced experimentally by syntectonic recrystallization and those observed in upper-mantle peridotites led *Avé Lallemant and Carter* [1970] to suggest that this process might dominate in the upper mantle, a hypothesis that we still favor and one that is tested in this paper. We emphasize, however, that syntectonic recrystallization is generally accompanied by plastic deformation and polygonization [*Raleigh and Kirby*, 1970; *Green and Radcliffe*, 1972*b*; *Nicolas et al.*, 1972], any process of which might dominate under certain physical conditions. The similarity in these flow processes observed in the steady-state experiments and those observed in peridotites presumably derived from the upper mantle, as well as other considerations outlined above, lead us to suspect that power law creep prevails over most of the upper mantle (or at least over those portions available to us). It should be pointed out, however, that processes such as grain boundary sliding (ultimately controlled by internal deformation) and Nabarro-Herring creep, if they have operated, would be very difficult to detect in naturally deformed peridotites and dunites.

EXPERIMENTAL SYNTECTONIC RECRYSTALLIZATION

Syntectonic-recrystallization experiments were conducted on compacted pellets of powdered ($<37\ \mu$) Mount Addie dunite and ground enstatite and diopside single crystals in the manner described by *Avé Lallemant and Carter* [1970]. All experiments were done in Griggs's solid-pressure-medium apparatus [*Griggs*, 1967; *Green et al.*, 1970] at constant strain rate, and the physical conditions and extent of grain growth are indicated in Figure 3. The specimens were deformed in compression and shortened by 10–20% after compaction. The results for olivine (Figure 3*a*) are from *Avé Lallemant and Carter* [1970], and those for diopside and enstatite (Figure 3*b*) are from the present study. Under comparable conditions, olivine recrystallizes and grows more readily than the pyroxenes. This difference is more pronounced for the deformation of coarse-grained aggregates of dunite and lherzolite in which the olivine can recrystallize totally, whereas the pyroxenes can only slip and polygonize. In all experimentally recrystallized specimens a foliation develops normal to σ_1 and is defined by the shapes of the new grains that are shortest parallel to σ_1.

In Figure 4*a* we present an inverse pole figure for syntectonically recrystallized olivine powder, a pattern that is typical of that observed for olivine over the entire range of pressure-temperature-strain rate examined [*Avé Lallemant and Carter*, 1970]. Pole figures for crystallographic axes of grains in this specimen (N-154) are given in Figure 5*a*. The dominant features are statistical alignment of [010] parallel to σ_1 (the axis of maximum principal com-

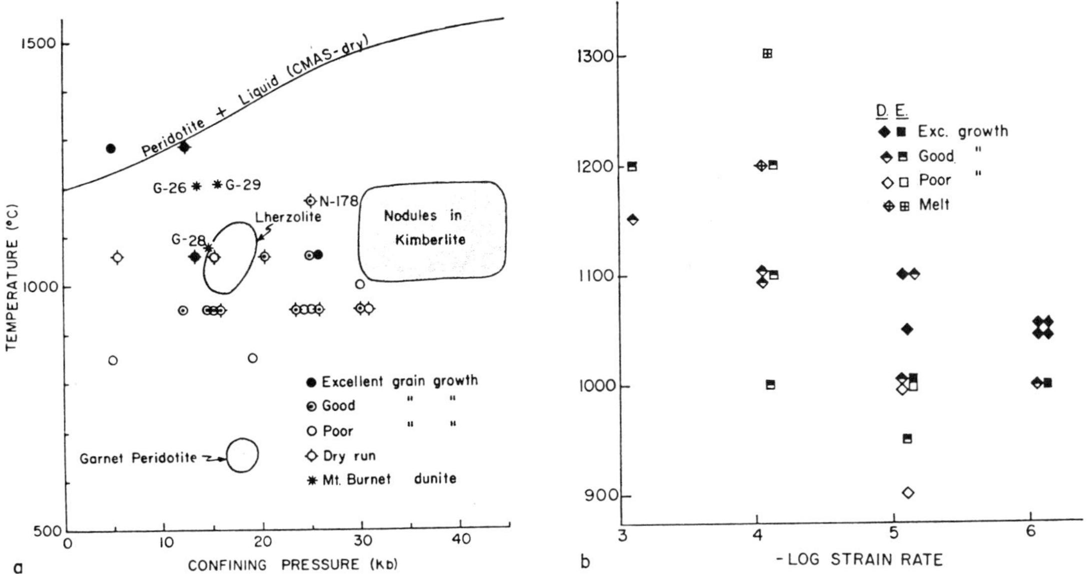

Fig. 3. Physical conditions of experiments on syntectonic recrystallization and the extent of grain growth for compacted powders of (a) olivine, deformed at a strain rate of about 8×10^{-6} sec^{-1} [Avé Lallemant and Carter, 1970], and (b) pyroxenes, enstatite E and diopside D, deformed at 15-kb pressure. Fields of lherzolite, garnet peridotite, and nodules in kimberlite and peridotite solidus in (a) are from O'Hara [1957].

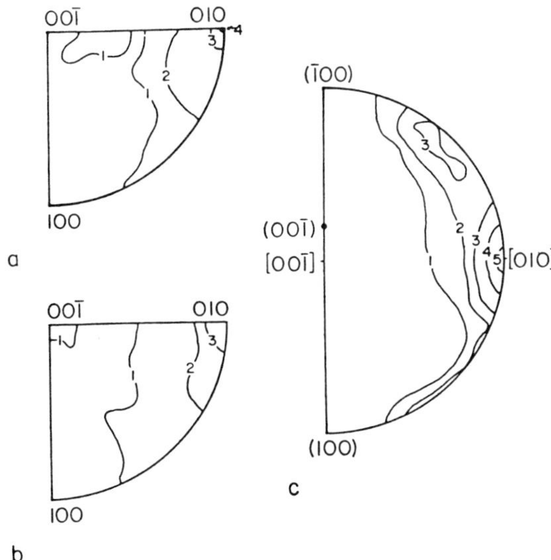

Fig. 4. Inverse pole figures for experimentally recrystallized powders: (a) olivine, N-154, (b) enstatite, N-352, and (c) diopside, N-367, showing variation in concentration of σ_1 axes in relation to crystal axes. The contours are in multiples of the expected number for a random distribution. The aggregates were deformed at 1000°C, 15 kb, and 7.8×10^{-7} sec^{-1}. Equal-area projections are shown in the lower hemisphere.

pressive stress; north-south in Figures 5 and 6) with [001] and [100] girdles in the $\sigma_2 = \sigma_3$ plane (Figure 5a). Underpopulated areas in the girdles of Figures 5 and 6, at about 45° from the periphery, are ascribed to sampling errors inherent in universal-stage measurements of principal axes of optical indicatrices of crystals in fine-grained aggregates. The fabrics shown in Figures 4a and 5a were observed both for powders and for recrystallized coarse-grained starting material [Avé Lallemant and Carter, 1970].

Fewer experiments have been done on pyroxenes, and even fewer have been analyzed for fabric; the results should therefore be regarded as being preliminary. The results for a specimen of powdered enstatite compressed at 1000°C at a strain rate of 8×10^{-7} sec^{-1} are presented in Figures 4b and 5c; another specimen deformed under similar conditions showed the same fabric. As was observed for olivine, [010] axes of recrystallized enstatite grains are oriented statistically parallel to σ_1 with poorly defined partial girdles of [100] and [001] axes in the $\sigma_2 = \sigma_3$ plane. The [010] axes of both olivine and bronzite are most compliant elastically [Baker and Carter, this volume, Figure 1],

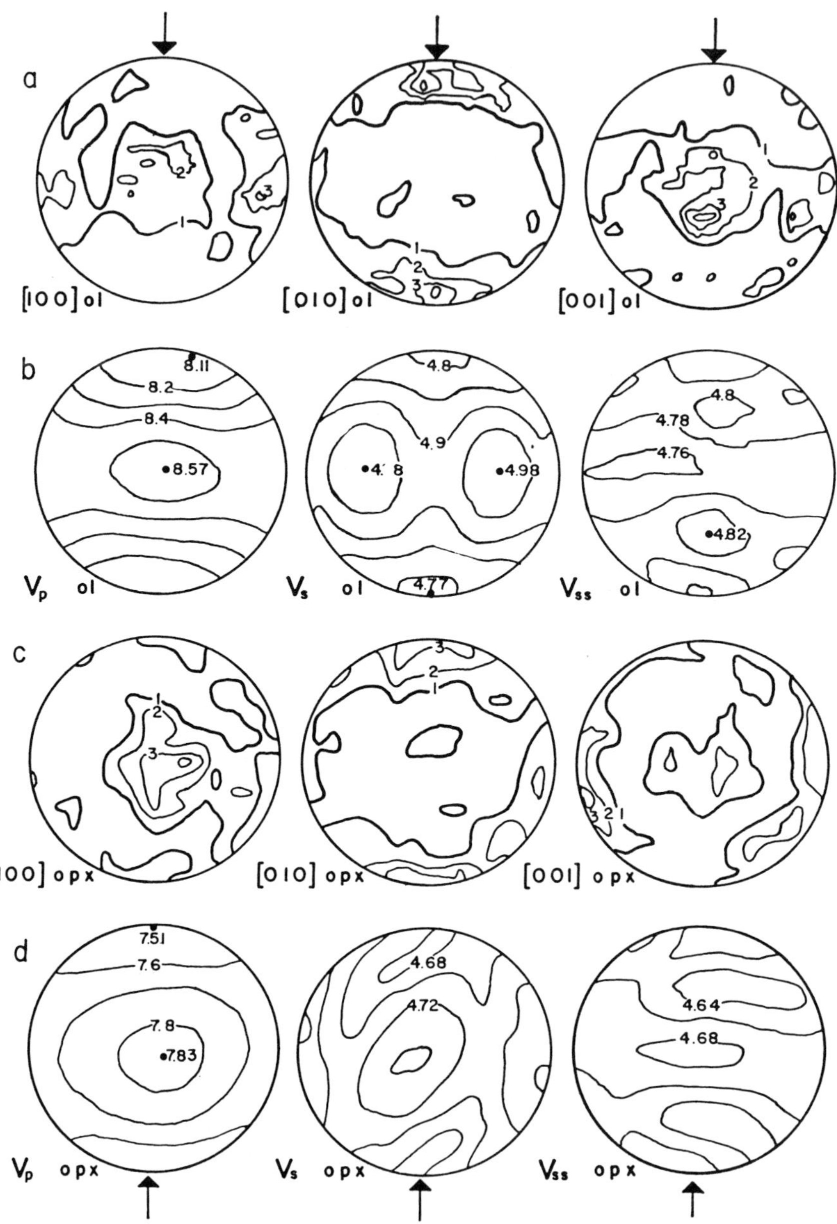

Fig. 5. Pole figures and calculated seismic velocities for experimentally recrystallized olivine and enstatite; equal-area projections are shown in the lower hemisphere. Axis σ_1 is oriented north-south. (a) Pole figures for 338 olivine grains in specimen N-154 deformed at 1000°C, 13.4 kb, and 7.8×10^{-7} sec^{-1} [*Avé Lallemant and Carter*, 1970]. Contours are in multiples of the expected number for a random distribution; the counting area is 1.2%. (b) Seismic velocities calculated for specimen N-154. (c) Pole figures for 200 enstatite grains in specimen N-352 deformed at 1000°C, 15 kb, and 7.8×10^{-7} sec^{-1}. The contours are in multiples of the expected number for a random distribution; the counting area is 2%. (d) Seismic velocities calculated for specimen N-352.

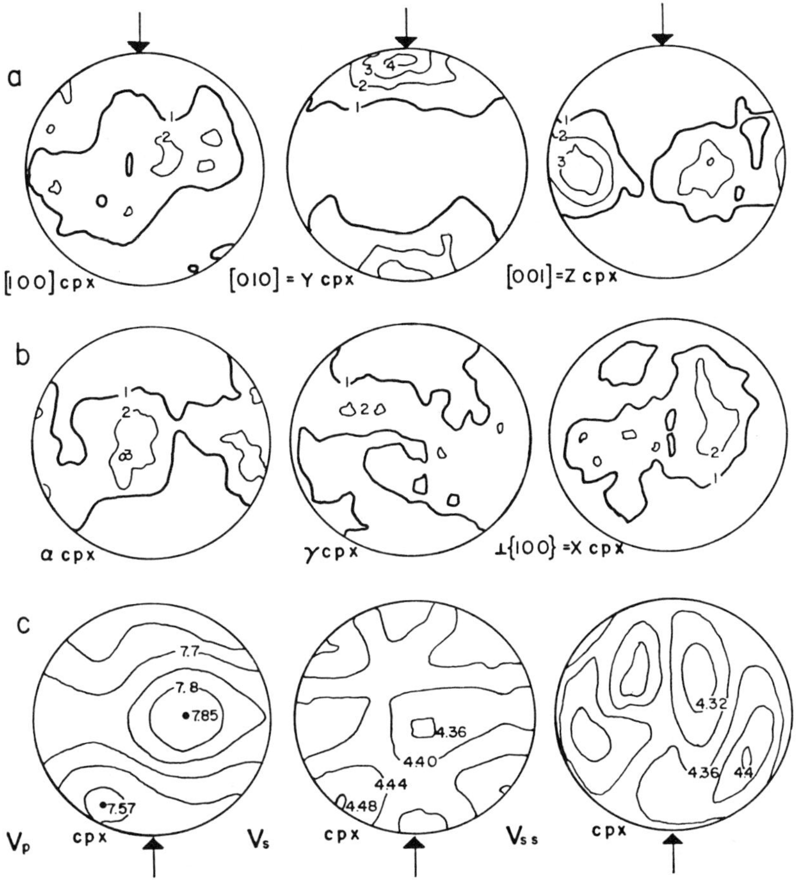

Fig. 6. (a and b) Pole figures and (c) calculated seismic velocities for 76 diopside grains in specimen N-367 deformed at 1000°C, 15 kb, and 7.8×10^{-7} sec^{-1}. Equal-area projections are shown in the lower hemisphere. Axis σ_1 is oriented north-south. The contours in (a) and (b) are in multiples of the expected number for a random distribution; the counting area is 4.9%.

and in the presence of an interstitial fluid they should orient parallel to σ_1 in accord with *Kamb's* [1959a] thermodynamic theory as it is applied to ultramafic minerals by *Hartman and den Tex* [1964].

Only one recrystallized diopside specimen has been fully analyzed to date, and the results are given in Figures 4c, 6a, and 6b. Again, [010] axes of the new grains align parallel to σ_1, although a distinct concentration is observed at about 45° to σ_1 (Figure 4c). The remaining subfabric elements determined (Figure 6a, b) form girdles with maximums in the $\sigma_2 = \sigma_3$ plane. For diopside, however, [010] is not the most compliant direction [*Aleksandrov et al.*, 1964; *Kumazawa*, 1969; *Baker and Carter*, this volume, Figure 2] at STP, but its magnitude is near that of the most compliant direction

(nearly parallel to the indicatrix axis α). Temperature and pressure derivatives of elastic constants of clinopyroxenes have not yet been determined, and it is possible that [010] is the most compliant direction at the elevated temperatures of our experiments.

Variations in compressional and shear wave velocities for these experimentally recrystallized specimens (Figures 5b, d and 6c) have been calculated in the manner described in the previous paper [*Baker and Carter*, this volume]. For the olivine (Figure 5b) the compressional wave velocity anisotropy ΔV_p is 0.46 km/sec, the minimum velocity (8.11 km/sec) being near σ_1 and the maximum (8.57 km/sec) being in the $\sigma_2 = \sigma_3$ plane. The shear wave velocity anisotropies ΔV_s and ΔV_{ss} are 0.21 and 0.06 km/sec, respectively. The enstatite fabric produces a maximum ΔV_p of 0.32 km/sec (Figure 5d), maximum ΔV_s and ΔV_{ss} both being 0.04 km/sec. Experimentally recrystallized diopside yields a maximum ΔV_p of 0.28 km/sec (Figure 6c), the maximum ΔV_s and ΔV_{ss} being 0.12 and 0.08 km/sec, respectively.

Thus syntectonic recrystallization of olivines and pyroxenes in these axially symmetric tests leads to strong preferred crystal orientations that are related to the orientations of principal stress axes. These preferred orientations give rise to appreciable anisotropies in velocitites of seismic compressional waves. In the next section we compare these fabrics and velocity anisotropies with those observed for naturally deformed ultramafic aggregates, most of which may have been derived from the upper mantle.

FABRICS AND SEISMIC ANISOTROPY OF NATURAL PERIDOTITES AND GARNET CLINOPYROXENITES

Peridotite

A great deal of fabric work has been done on olivine in peridotites and dunites from many sources and localities since the early work of *Andreatta* [1934] (for recent reviews, see *den Tex* [1969], *Avé Lallemant and Carter* [1970], and *Lappin* [1971]). With the exception of some recent studies, pyroxenes in these rocks have received considerably less attention, presumably because they form relatively minor volume percentages. In particular, clinopyroxenes form only a few per cent of the most common peridotites, and hence they will be ignored in the discussions and the analyses that follow.

We have selected two lherzolites, a garnet peridotite, a chlorite peridotite, and a dunite (discussed in the previous paper) to represent the range of olivine and orthopyroxene fabrics and seismic anisotropies expected in upper-mantle aggregates. Of these fabrics, only those of the dunite and the two lherzolites are typical of fabrics observed in olivine-rich tectonites. Modal analyses of these specimens are presented in Table 1 with the recalculated modes used in the Voigt-Reuss-Hill (VRH) averages of the elastic constants [*Baker and Carter*, this volume]. All the specimens are tectonites having fabrics, on the basis of textural and structural evidence and comparisons with the experiments, that are believed to have originated by syntectonic recrystallization. There is evidence of

TABLE 1. Modal Analyses of the Specimens

| Mineral | Peridotite | | | | | | | | Garnet Pyroxenite | | | |
| | L-62 | | John Day | | Ar-26 | | Ar-18(17) | | OFS-7 | | OFS-8 | |
	Mode	Hill	Mode*	Hill	Mode	Hill	Mode	Hill	Mode	Hill	Mode	Hill
Olivine	60.0	73.2	35.6	93.9	61.0	81.2	60.0	82.2				
Orthopyroxene	22.0	26.8	6.1	6.1	14.0	18.8	13.0	17.8				
Clinopyroxene	10.3				5.0				49.1	64.4	47.5	54.8
Garnet					11.0				27.1	35.6	39.2	45.2
Amphibole					9.0		13.0					
Spinel	4.8		0.3									
Serpentine			57.8				8.0					
Chlorite							6.0					
Micas and alteration									23.3		13.2	
Miscellaneous	2.8		0.2						0.5			

* A more recent modal analysis of the specimen by H. G. Avé Lallemant, who has distinguished between host crystals of the serpentine, gives 76.8% olivine, 14.6% orthopyroxene, 8.2% clinopyroxene, and 0.4% spinel.

mild to moderate plastic deformation according to the system {0kl} [100] in the specimens, but this deformation has been superimposed on a pre-existing fabric.

Alpine lherzolite L-62 from the French Pyrenees. An olivine fabric commonly observed in peridotites is that shown by specimen L-62, a type spinel lherzolite from Etang de Lers in the French Pyrenees [*Avé Lallemant*, 1967]. The fabric (Figure 7a) has a [010] maximum, spreading into a girdle, normal to a foliation *s* that contains more or less well-developed [100] and [001] partial girdles. The foliation is defined by the shape of olivine grains whose shortest semi-axis is normal to the foliation plane, as is typical of many olivine tectonites. This fabric is similar to that obtained for 200 crystals from the same specimen by *Avé Lallemant* [1967, Figure 27], although the girdles are developed somewhat better in his diagrams. The fabric for the enstatite crystals (Figure 7b) is stronger than but similar to that of the olivine. The orientations of both olivine and enstatite are remarkably similar to those produced experimentally (Figure 5a, c) and are consistent with σ_1 having been oriented normal to *s* with $\sigma_2 \simeq \sigma_3$.

In Figure 7c we show the longitudinal and transverse wave velocities for this specimen by using the recalculated mode (Table 1) and the averaging scheme and the procedure discussed in the previous paper [*Baker and Carter*, this

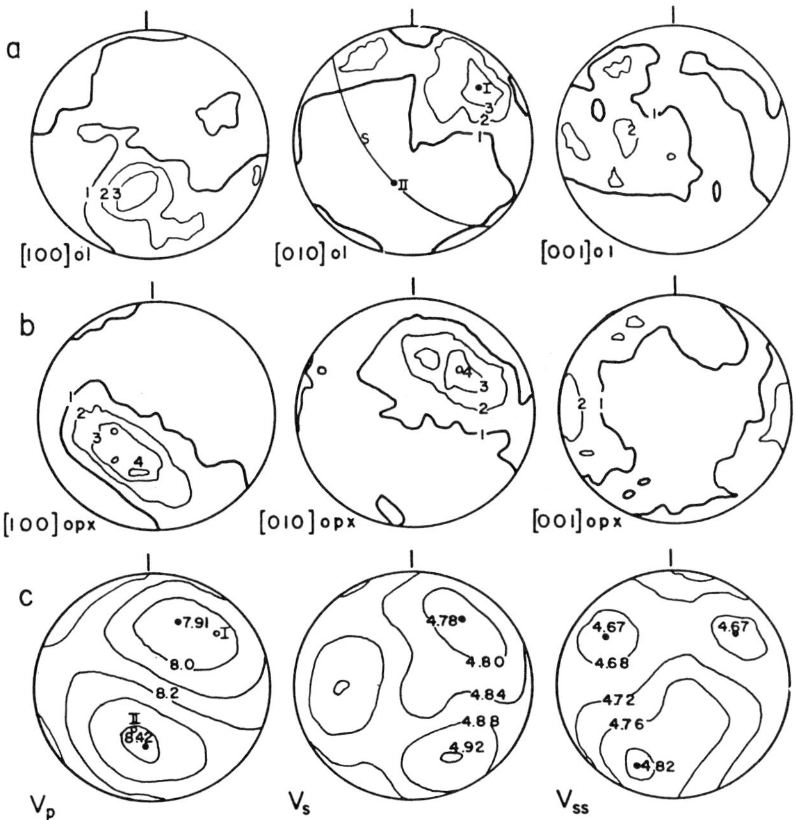

Fig. 7. Pole figures and seismic velocities for lherzolite specimen L-62 from Etang de Lers, French Pyrenees (specimen supplied by E. den Tex). Equal-area projections are shown in the lower hemisphere. (a) Pole figures for 98 olivine grains with contours in multiples of the expected number for a random distribution; the counting area is 3.9%. The great circle *s* shows the orientation of foliation, and points I and II show the orientations of specimens cored for seismic measurements (Table 3). (b) Pole figures for 49 enstatite crystals with contours in multiples of the expected number for a random distribution; the counting area is 7.5%. (c) Seismic velocities calculated from the data in (a) and (b) and the VRH mode (Table 1).

volume]. The minimum *P*-wave velocity, 7.91 km/sec, is subparallel to the [010] maximum, and the maximum velocity, 8.42 km/sec, is subparallel to the [100] concentration, as was expected. The maximum ΔV_p is 0.51 km/sec, and maximums ΔV_s and ΔV_{ss} are both 0.14 km/sec. The fabric and body wave velocity anisotropy are therefore somewhat stronger than those for the experimentally deformed specimens.

John Day lherzolite from the Canyon Mountain complex, Oregon. A somewhat different fabric has been determined by H. G. Avé Lallemant (personal communication, 1971) for the John Day lherzolite from the Canyon Mountain complex in Oregon. The specimen is a highly serpentinized lherzolite in contact with dunite having the same olivine fabric (H. G. Avé Lallemant, personal communication, 1971). The olivine fabric (Figure 8a) has a strong [100] maximum parallel to the long dimensions of the olivine grains normal to which are well-developed [010] and [001] girdles with maximums; no foliation is evident in this specimen. Such mineral elongation lineations in peridotites and dunites are commonly parallel to [100] concentrations that we believe form parallel to σ_3 [*Avé Lallemant and Carter*, 1970; H. G. Avé Lallemant, personal communication, 1972]. This

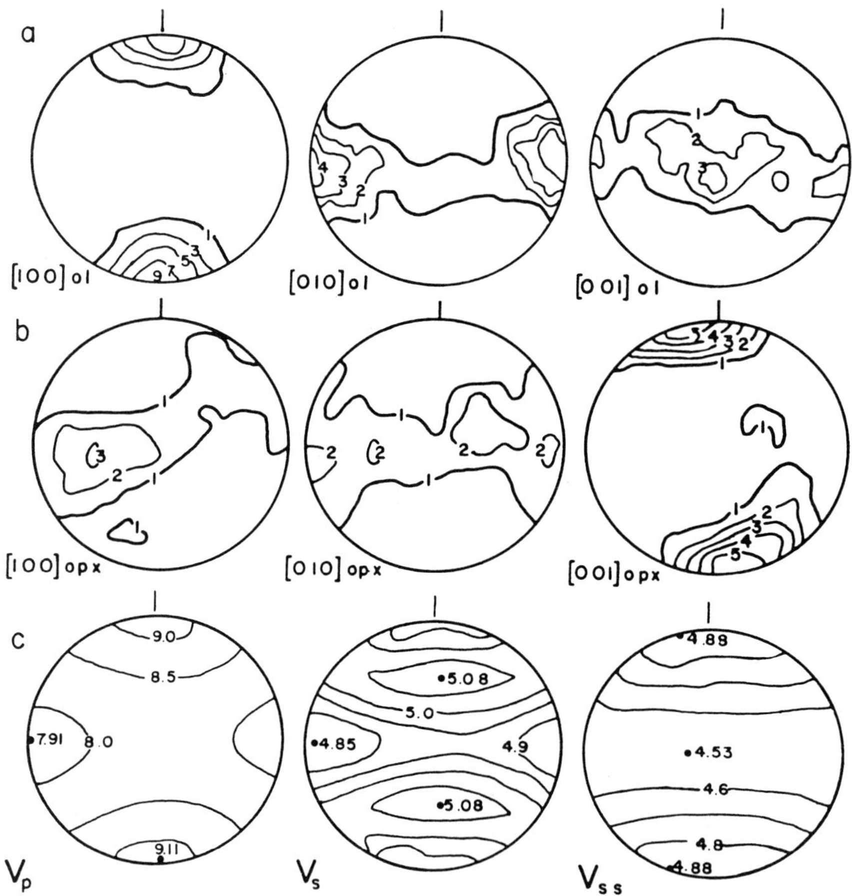

Fig. 8. Pole figures and calculated seismic velocities for the John Day lherzolite from the Canyon Mountain complex in Oregon (data provided by H. G. Avé Lallemant, private communication, 1971). Equal-area projections are shown in the lower hemisphere. (*a*) Pole figures for 100 olivine grains with contours in multiples of the expected number for a random distribution; the counting area is 3.8%. The specimen has a mineral elongation lineation parallel to the [100] concentration. (*b*) Pole figures for 41 enstatite grains with contours in multiples of the expected number for a random distribution; the counting area is 8.9%. (*c*) Seismic velocities calculated from the data in (*a*) and (*b*) and the VRH mode (Table 1).

fabric is, therefore, interpreted to be a variant of the previous one (Figure 7a), σ_3 parallel to [100] being the unique stress axis and σ_1 being approximately equal to σ_2.

The strong enstatite fabric (Figure 8b), although apparently related symmetrically to the olivine fabric, is not parallel to it, in contrast to the previous specimen. This inconsistency has also been observed for other peridotites [e.g., *Raleigh*, 1965b; *Avé Lallemant*, 1967; *Nicolas et al.*, 1971] and is not fully understood to date. The relative ease with which olivine recrystallizes could account for this discrepancy [*Helmstaedt and Anderson*, 1969; *Avé Lallemant and Carter*, 1970; *Carter*, 1971]. The rather common parallelism of $[100]_{ol}$ and

$[001]_{opx}$ could be produced, during recrystallization of the olivine, by slip parallel to [001] in the orthopyroxenes and/or bodily rotation of the orthopyroxene crystals that are inherently elongate parallel to [001].

In Figure 8c we plot the variations of compressional and shear wave velocities calculated for this specimen. Because of the strong fabric the maximum ΔV_p (1.2 km/sec) is very high, and ΔV_s (0.23 km/sec) and ΔV_{ss} (0.35 km/sec) are also fairly high. The actual velocities and anisotropy for this specimen would, of course, be much lower because of the extensive serpentinization.

Garnet peridotite Ar-26 from the Alpe Arami, Switzerland. The diagrams in Figure 9 show

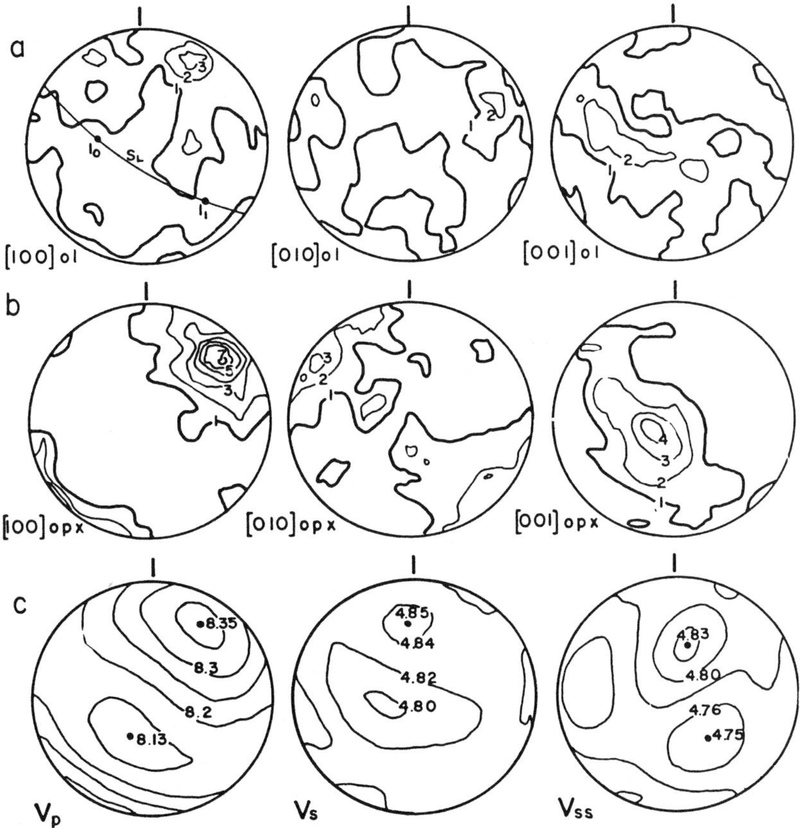

Fig. 9. Pole figures and calculated seismic velocities for garnet peridotite Ar-26 from the Alpe Arami, Switzerland (data provided by J. R. Möckel, private communication, 1971). Equal-area projections are shown in the lower hemisphere. (a) Pole figures for 200 olivine grains with contours in multiples of expected number for a random distribution; the counting area is 2%. The great circle s_L is the layering plane, l_o is a mineral elongation lineation, and l_1 is a striation lineation. (b) Pole figures for 200 enstatite grains with contours in multiples of the expected number for a random distribution; the counting area is 2%. (c) Seismic velocities calculated from the data in (a) and (b) and the VRH mode (Table 1).

olivine (Figure 9a) and orthopyroxene (Figure 9b) fabrics in a partly altered garnet peridotite specimen, Ar-26, from the Alpe Arami, Switzerland [*Möckel*, 1969; data provided by J. R. Möckel, private communication, 1971]. The layering plane s_L is defined by pyroxene streaks and mineral elongations, and l_0 is a mineral elongation lineation. The striation lineation l_1 may be due to the low-angle intersection of s_L and a later alpine cleavage, but the preferred crystal orientations are related to the earlier layering s_L and lineation l_0 [*Möckel*, 1969]. The olivine fabric is rather weak, but it is nevertheless distinctly different from that found for most lherzolites and for dunites associated with garnet peridotites from Norway [*Lappin*, 1967]. The main features are a [100] maximum spreading into two partial girdles normal to s_L and a [001] girdle with a maximum near l_0 parallel to the layering. As for lherzolite L-62 (Figure 7), the enstatite (Figure 9b) shows a somewhat similar but much stronger fabric. All experimental attempts to produce the [100]-maximum olivine fabric shown in Figures 9a and 10a have been unsuccessful to date [*Avé Lallemant and Carter*, 1970]. This fabric is predicted by the theory of *Kamb* [1959a] as it is applied to olivine by *Hartman and den Tex* [1964] for the case in which interstitial fluids are absent.

Body wave velocities calculated for this specimen are presented in Figure 9c. Because of the weak olivine fabric and the presence of garnet the maximum P-wave velocity anisotropy is only 0.22 km/sec, a maximum velocity of 8.35 km/sec occurring subparallel to [100] concentrations in olivines and orthopyroxenes. Maximum anisotropies for the shear waves are, of course, correspondingly low, i.e., 0.05 km/sec for ΔV_s and 0.08 km/sec for ΔV_{ss}.

Chlorite peridotite Ar-18 from the Alpe Arami, Switzerland. The results for a specimen of chlorite peridotite from the Alpe Arami, Switzerland [*Möckel*, 1969; data provided by J. R. Möckel, private communication, 1971] are shown in Figure 10. The mode for this specimen is not available, and we have assumed that the mode is similar to that for chlorite peridotite Ar-17 (Table 1), although Ar-17 may be more highly serpentinized than Ar-18. The presence of spinel clouds within amphibole in Ar-18 indicates the earlier presence of clinopyroxene [*Möckel*, 1969]. The plane s is a schistosity

probably parallel to an alpine cleavage and layering plane in which orthopyroxenes and amphiboles are elongated, and s_{01} is a plane containing the long semiaxes of olivine crystals. The mineral elongation lineation l_0 is due to the alignment of amphibole and orthopyroxene, which, according to Möckel, is not parallel to olivine elongations.

The olivine fabric for Ar-18 (Figure 10a) is similar to but stronger than that for Ar-26 and has a [100] maximum normal to s_{01}, which contains [001] and [010] partial girdles. As was observed before, the enstatite fabric is much stronger and clearly related to s and l_0, but it does not appear to be related to the olivine fabric. Thus it appears that the olivine and orthopyroxene fabrics originated during two different deformational episodes, but there is no unambiguous means for determining the relative chronology. *Möckel* [1969] believes that, because of the preservation of s and l_0, the olivine fabric is the earlier one, but, on the basis of the experimental observations, we prefer the opposite interpretation, i.e., that the orthopyroxene fabric is the earlier one.

Body wave velocity patterns for Ar-18 are dominated by the large percentage and fairly strong fabric of the olivine, the maximum P-wave velocity being 8.49 km/sec, parallel to the [100] maximum. The maximum ΔV_p is 0.48 km/sec, whereas the maximum ΔV_s and ΔV_{ss} are both 0.11 km/sec.

Garnet clinopyroxenites and eclogites

In contrast to work on peridotites and dunites, very little fabric work has been done on crustal and mantle garnet clinopyroxenites and eclogites. To our knowledge, the only published reports of fabrics of clinopyroxenes in such rocks are those for xenoliths in kimberlite-bearing breccia pipes in Arizona and Utah [*Kumazawa et al.*, 1971]. Partial fabrics of four specimens were presented in that study, and, although the strength of the fabric appears to be moderate, the calculated and measured anisotropies are low (discussed more fully later). According to *Watson and Morton* [1969], the eclogite inclusions in the kimberlites in Arizona may have come from an in situ layer in the deep crust or the upper mantle.

For our studies we have selected one crustal eclogite (GAL-143, supplied by D. E. Vogel)

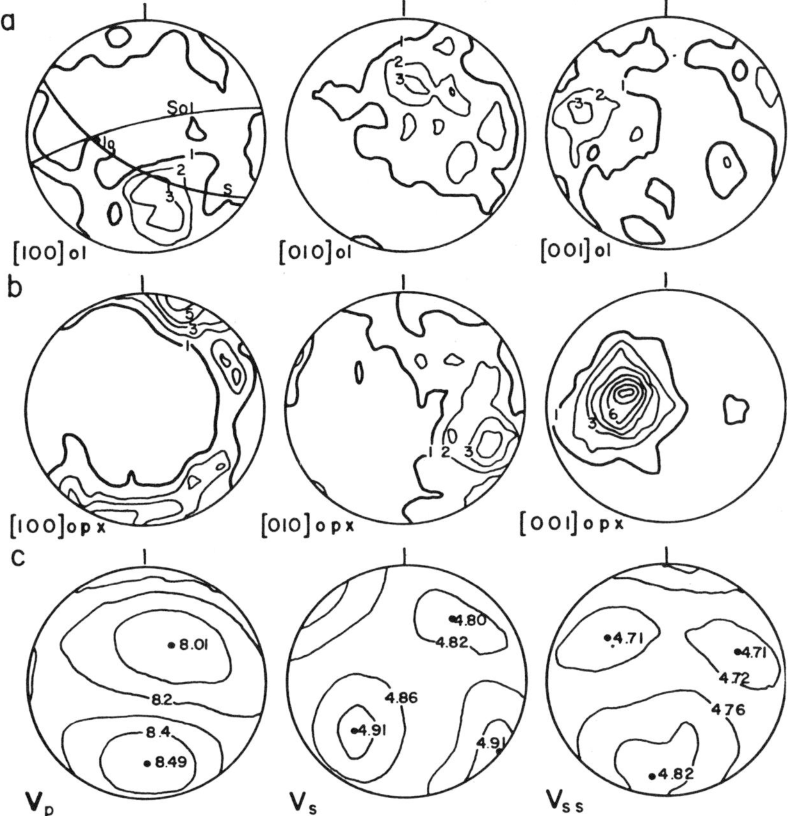

Fig. 10. Pole figures and calculated seismic velocities for chlorite peridotite specimen Ar-18 from the Alpe Arami, Switzerland (data provided by J. R. Möckel, private communication, 1971). Equal-area projections are shown in the lower hemisphere. (*a*) Pole figures for 200 olivine grains with contours in multiples of the expected number for a random distribution; the counting circle is 2%. The plane marked s_{01} is an olivine shape foliation, and l_0 is a mineral elongation lineation. The plane *s* is an alpine schistosity. (*b*) Pole figures for 200 enstatite grains with contours in multiples of the expected number for a random distribution; the counting area is 2%. (*c*) Seismic velocities calculated from the data in (*a*) and (*b*) and the VRH mode (Table 1).

and two mantle garnet clinopyroxenite inclusions (OFS-7 and OFS-8) from the Roberts Victor mine, Orange Free State, South Africa (specimens supplied by E. Maske). In the analyses that follow we have treated the clinopyroxene elastically as augite with the constants at STP of *Aleksandrov et al.* [1964]. The crustal eclogite was discussed in the previous paper [*Baker and Carter,* this voume]; this specimen was analyzed mainly for comparison with the mantle specimens. The clinopyroxene fabric in GAL-143 is very strong, having a [010] maximum normal to a strong foliation defined by layering and grain shape and containing well-developed girdles or partial girdles of the remaining subfabric elements measured [*Baker and Carter,* this volume, Figure 4]. This fabric is similar to the experimentally recrystallized diopside specimen (Figure 6) and thus consistent with the orientation of σ_1 normal to the foliation and with $\sigma_2 \simeq \sigma_3$. Voigt-Reuss-Hill averages for the pyroxene and amphibole fabrics combined with the elastic constants for garnet, weighted proportionately, lead to a calculated maximum ΔV_p of 0.47 for this specimen, ΔV_s and ΔV_{ss} both being 0.12 km/sec.

Mantle garnet pyroxenites OFS-7 and OFS-8 are similar mineralogically and texturally, although proportions of the minerals are different and the mineral layering observed in OFS-7 is absent in OFS-8. The pyroxene and garnet crystals are both large (averaging about 3 mm), and therefore two parallel sections were cut from each rock for the required number of measurements. The anhedral to subhedral clinopyroxenes are sodic-aluminous augites (Table 2) with an average $c \wedge \gamma = 37°$ and $2V_\gamma = 59°$. They show profuse lamellar mechanical twinning according to the system {100} [001], which is the high-temperature and low strain rate twinning mechanism in clinopyroxenes [*Raleigh and Talbot*, 1967]. The garnets are dominantly pyrope (Table 2), have kelyphitic rims, and are partially altered internally. Chlorite and phlogopite, which are generally interstitial to the garnet and pyroxene and internal alteration products in both garnet and pyroxene, form a substantial part of both specimens (Table 2) and should be the major factors contributing to errors in the calculated velocities. The over-all texture, petrology, and chemistry of these specimens is similar to that described by *MacGregor and Carter* [1970] as 'type 1 eclogites,' although these rocks have been deformed and metamorphosed.

Clinopyroxene fabrics of OFS-7 and OFS-8 are given in Figures 11 and 12, respectively. The fabric for OFS-7 is rather weak and seems to be dominated by a c-axis maximum inclined at 15° to the mineral layering s_L. The remaining subfabric elements form girdles or partial girdles around the c-axis maximum. An analysis of this fabric combined with garnet leads to a maximum ΔV_p of 0.17 km/sec, the maximums ΔV_s and ΔV_{ss} being 0.1 and 0.15 km/sec, respectively.

The clinopyroxene fabric for OFS-8 (Figure 12) is also fairly weak, having a maximum of c axes spreading into an east-west partial girdle. Other subfabric elements are weakly arranged in girdles and partial girdles. The maximum compressional wave velocity anisotropy calculated for this specimen is also 0.17 km/sec, ΔV_s being 0.09 km/sec and ΔV_{ss} being 0.04 km/sec. Calculated P-wave velocity anisotropies of these mantle garnet pyroxenites are, therefore, substantially lower than the anisotropy of the crustal eclogite and all except one of the peridotites analyzed in the last section.

RESULTS

As was pointed out by *Avé Lallemant and Carter* [1970], the olivine fabrics produced by syntectonic recrystallization in our axially sym-

TABLE 2. Electron Probe Analyses of Pyroxenes and Garnets in Garnet Clinopyroxenites*

	OFS-7		OFS-8	
	Pyroxene	Garnet	Pyroxene	Garnet
SiO_2	56.3	43.1	53.6	40.3
Al_2O_3	2.62	21.9	1.81	22.0
TiO_2	0.14	0.18	0.07	0.12
FeO	3.06	9.2	3.31	11.1
MnO	<.1	0.20	0.05	0.41
MgO	17.9	21.5	16.5	18.3
CaO	19.6	3.8	21.4	4.9
Na_2O	1.48		1.29	
Cr_2O_3	<0.1	0.03		0.21
Total	101.1	99.91	98.03	97.34
Cation Proportions Based on 6 Oxygens				
Si	1.99	1.53	1.99	1.52
Al	0.01		0.01	
Al	0.10	0.92	0.069	0.92
Fe	0.09	0.27	0.101	0.33
Mg	0.95	1.14	0.91	1.03
Ca	0.75	0.15	0.85	0.20
Na	0.10		0.093	
Ti	0.003	0.005	0.002	0.004
Cr		0.001		
Mn		0.005	0.003	0.013

* Averages of four analyses by Louis A. Fernandez.

metric tests have common counterparts in naturally deformed peridotites (e.g., Figure 7). For natural deformations, however, two of the principal stresses need not be equal, and, if they are not equal, maximums should develop in the girdles or a three point maximum fabric should result, as is observed for the Twin Sisters dunite [*Baker and Carter*, this volume, Figure 3]. To date, we have definite experimental evidence only that [010] in olivine orients parallel to σ_1, but other observations on the experimentally and naturally deformed specimens [*Avé Lallemant and Carter*, 1970] and preliminary extrusion experiments (H. G. Avé Lallemant, personal communication, 1972) suggest that [100] orients preferably parallel to σ_3. We therefore interpret the John Day lherzolite olivine fabric (Figure 8) as having originated by recrystallization during nearly uniform extension ($\sigma_1 \simeq \sigma_2 > \sigma_3$) parallel to the [100] concentration. *Lappin* [1971] has emphasized the few fabrics (also pointed out by *Avé Lallemant and Carter* [1970]) in which [001] axes are parallel to mineral elongation lineations and fold axes. Lappin equates lineations parallel to both [100]

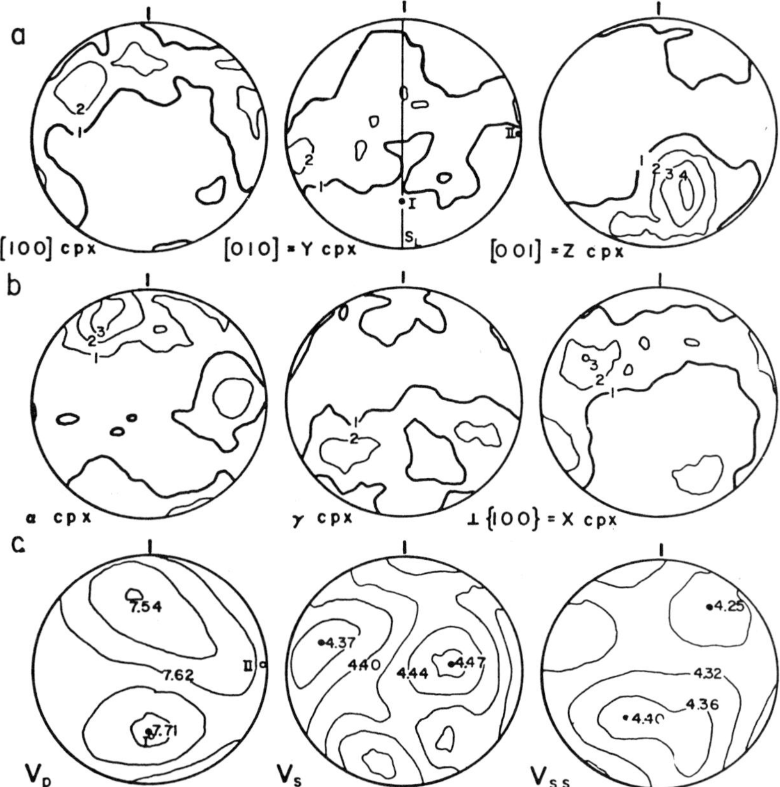

Fig. 11. Pole figures and calculated seismic velocities for garnet clinopyroxenite specimen OFS-7 from the Roberts Victor Mine, South Africa (specimen supplied by E. Maske). Equal-area projections are shown in the lower hemisphere. (*a* and *b*) Pole figures for 86 clinopyroxene grains with contours in multiples of the expected number for a random distribution; the counting area is 4.4%. The north-south plane in the center sphere of (*a*) shows the orientation of mineral layering, and points I and II show the orientations of specimens cored for seismic measurements (Table 3). (*c*) Seismic velocities calculated from the data in (*a*) and (*b*) and the VRH mode (Table 1).

and [001] to fabric *b* axes, which he assumes are parallel to σ_2. His interpretation lacks verification except for possible analogies with some open crustal folds in quartzitic and calcitic sequences [*Dieterich and Carter*, 1969]. As was mentioned above, the origin of the [100]-maximum (normal to foliation) fabrics is not yet understood, but such fabrics are not common and hence should not affect our major conclusions.

Data for fabrics produced experimentally by syntectonic recrystallization of the pyroxenes are fewer, but again similar fabrics have been observed in naturally deformed peridotites and garnet clinopyroxenites. The orthopyroxene fabrics produced are similar to the olivine fabrics (Figures 4 and 5), and this correlation has also

been observed in peridotites L-62 (Figure 7) and Ar-26 (Figure 9) as well as in other peridotites. However, there are also several peridotites for which the olivine and orthopyroxene fabrics do not coincide (e.g., Figures 8 and 10). This discrepancy probably can be accounted for generally by recrystallization of the olivine during a later (lower temperature) deformational episode, as was suggested by the experimental evidence.

The clinopyroxene fabric of crustal eclogite GAL-143 [*Baker and Carter*, this volume, Figure 4] is similar to that produced experimentally (Figure 6), and this fabric probably originated by recrystallization during compression normal to the foliation. The amphibole fabric of GAL-143 is similar and probably also originated by

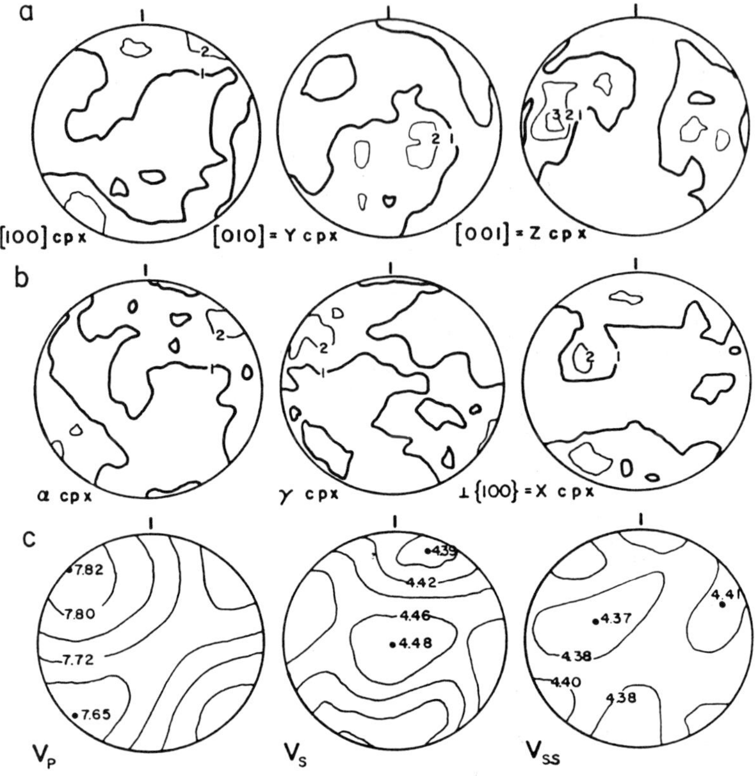

Fig. 12. Pole figures and calculated seismic velocities for garnet clinopyroxenite specimen OFS-8 from Roberts Victor Mine, South Africa (specimen supplied by E. Maske). Equal-area projections are shown in the lower hemisphere. (*a* and *b*) Pole figures for 75 clinopyroxene grains with contours in multiples of the expected number for a random distribution; the counting area is 4.8%. (*c*) Seismic velocities calculated from the data in (*a*) and (*b*) and the VRH mode (Table 1).

this mechanism during the retrograde metamorphism of the rock [*Vogel*, 1967]. In contrast, the weaker clinopyroxene fabrics in garnet clinopyroxenites OFS-7 and OFS-8 (Figures 11 and 12) may be primary fabrics that originated during the accumulation process, in agreement with the interpretation of *MacGregor and Carter* [1970].

In Table 3*a* we give a summary of the *P*-wave velocities and maximum calculated anisotropies in the eight specimens analyzed. We no longer consider the shear wave velocity anisotropies in this paper, since they are low and have not been reported in the seismic refraction studies at sea. In Table 3*b* we compare the calculated velocities and anisotropies with those measured at 25°C and 4 kb by N. I. Christensen. The values are, of course, not strictly comparable, because alteration products in the specimens will

lower measured values, and the values calculated at 225°C and 3 kb should be lower than those measured at 25°C and 4 kb by a few per cent, provided that all cracks are closed. In addition, eclogite and garnet clinopyroxenite inclusions are typically porous, and all pores may not be closed at 4 kb or even at 10 kb (N. I. Christensen, personal communication, 1971), so that the values calculated at STP may be higher than those measured. Nevertheless, the correlation is excellent for lherzolite L-62, which shows very little alteration. The measured values and anisotropy for the Twin Sisters dunite are substantially lower than those calculated on the basis of 100% olivine, as is expected because of the 12% serpentine and the 5% other phases in the specimen [*Christensen and Ramananantoandro*, 1971]. Similarly, the measured values for garnet clinopyroxenite specimen OFS-7 are lower than

TABLE 3a. Compressional Wave Velocities, km/sec

Specimen	ρ (Calculated)	Maximum	Minimum	Mean	ΔVp (Maximum)
Peridotite Velocities Calculated at 225°C and 3 kb					
Lherzolite, L-62	3.315	8.42	7.91	8.17	0.51
Lherzolite, John Day	3.305	9.11	7.91	8.51	1.20
Garnet peridotite, Ar-26	3.311	8.35	8.13	8.24	0.22
Chlorite peridotite, Ar-18	3.311	8.49	8.01	8.25	0.48
Dunite, Twin Sisters	3.300	9.27	7.88	8.58	1.39
Garnet Clinopyroxenite Velocities Calculated at STP					
Mantle, OFS-7	3.627	7.71	7.54	7.63	0.17
Mantle, OFS-8	3.710	7.82	7.65	7.74	0.17
Crust, GAL-143	3.652	7.86	7.39	7.63	0.47

those calculated, probably because of the presence of micas and garnet and pyroxene alteration products and possible voids, but, surprisingly, the measured anisotropy is appreciably higher. The reason for this result is not yet known, but we note that a similar one was obtained at 4 kb for a specimen of garnet clinopyroxenite by *Kumazawa et al.* [1971, specimen GR-3]. The single measurement from crustal eclogite GAL-143, a very compact rock, is in excellent accord with the calculated value. Therefore, when allowances are made mainly for alteration products and also for other phases and possible voids, the correspondence between the measured and calculated values is rather good.

The data of Table 3 and those from other sources are presented graphically in Figure 13, in which we plot mean *P*-wave velocity against maximum (or nearly so) velocity anisotropy. We have selected specimens from the literature only for which fabric data are available and for which the measured anisotropies should be near maximum. Two generalizations can be drawn immediately from Figure 13. The first is that the mean velocity for the garnet clinopyroxenites (7.8 km/sec) is, on the average, lower by 0.5 km/sec than that for the peridotites and dunites (8.3 km/sec) even though their average measured density is higher (3.40 versus 3.29 g/cm³), a result anticipated by *Jackson and Wright* [1970]. The average mean velocity of eclogite and garnet clinopyroxenite specimens (GR-3, GR-3A, GR-23, and ME 1-17) measured by *Kumazawa et al.* [1971] is higher by 0.1 km/sec at 10 kb, and the average ΔV_p is lower by 0.07 km/sec. Mean velocities measured in the uppermost mantle in the northeast Pacific in kilometers per second are: (1) Flora, 8.14; Quartet, 7.98 [*Raitt et al.*, 1969]; (2) Hawaiian arch, 8.16 [*Morris et al.*, 1969]; and (3) British Columbia, 8.07 [*Keen and Barrett*, 1971]. These values and their average, 8.09 km/sec, are between the mean velocities observed for the peridotites and the dunites and for the garnet clinopyroxenites and hence do not provide a

TABLE 3b. Comparison between Calculated and Measured *P*-Wave Velocities, km /sec

Specimen	ρ (Measured)	Core 1		Core 2		Core 3		ΔVp	
		Calculated	Measured	Calculated	Measured	Calculated	Measured	Calculated	Measured
L-62	3.36	7.95	7.97	8.41	8.45			0.46	0.48
Twin Sisters	3.24	8.0	7.96	8.1	7.91	9.2	8.60	1.2	0.69
OFS-7	3.31	7.71	7.54	7.63	7.26			0.08	0.28
GAL-143	3.45	7.85	7.88						

The measurements were made by N. I. Christensen at 25°C and 4 kb. For core orientations see Figures 7 and 11 and *Baker and Carter* [this volume, Figures 3, 4a, and 5c.]

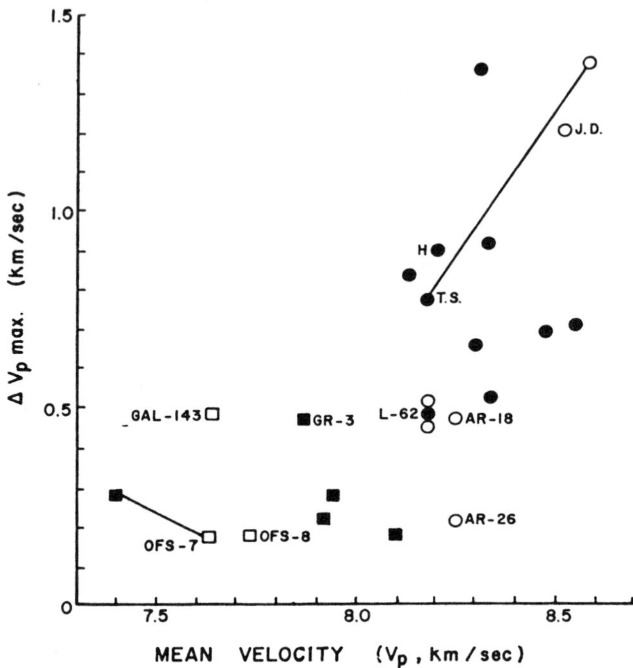

Fig. 13. Mean compressional wave velocity versus maximum *P*-wave velocity anisotropy for measured (solid symbols) and calculated (open symbols) velocities for peridotites and dunites (circles) and garnet clinopyroxenites and eclogites (squares). The unmarked solid circles represent measurements by *Christensen* [1966, 1971] and *Christensen and Ramananantoandro* [1971] on specimens of Twin Sisters dunite, except for the circle in the lower left, which is from serpentinized Mount Addie dunite. The solid circle H is from measurements at STP by *Kasahara et al.* [1968] on the Hidaka Horoman dunite from Japan; all other measurements were made at 4 kb and 25°C. With the exception of OSF-7 (measurements by N. I. Christensen), the solid squares are measurements on eclogites and garnet clinopyroxenites by *Kumazawa et al.* [1971], who also calculated anisotropies with the method of *Kumazawa* [1964] but did not take the garnet content into account.

means for deciding whether peridotite or garnet clinopyroxenite dominates.

The second generalization is that maximum *P*-wave velocity anisotropies for the peridotites and the dunites (averaging 0.76 km/sec) are, on the average, higher by about 0.5 km/sec than those for the garnet clinopyroxenites (averaging 0.29 km/sec). If we disregard crustal eclogite GAL-143 and specimen GR-3, which has a fabric [*Kumazawa et al.*, 1971, Figure 7] similar to that of GAL-143 and may also be a crustal rock, the average anisotropy for the garnet clinopyroxenites drops to 0.22 km/sec. Thus even the maximum *P*-wave velocity anisotropies of the mantle garnet clinopyroxenites are incapable of producing the 0.3- to 0.7-km/sec anisotropies observed in the northeast Pacific. The peridotites and dunites (with the exception of Ar-26), however, have maximum anisotropies in the range 0.48–1.37 km/sec and are therefore easily capable of producing the anisotropies observed. The maximum calculated anisotropies (and mean velocities) will, of course, be lowered by serpentinization and the presence of other phases. Both measured and calculated maximum values will be reduced by an amount depending on the in situ orientation of the fabric in the upper mantle, a topic discussed next.

Avé Lallemant and Carter [1970] discussed in a preliminary way the orientations of olivine fabrics and anisotropies produced with reference to the two most seriously regarded hypotheses advanced to explain plate motions. These hypotheses are (1) that the displacements result from drag by creep of convecting material below (model 1) and (2) that the displacements are due to rigid translation by the push from injecting hot material in the vicinity of ridges, by the pull from the relatively cold and dense sinking slab near trenches, gravity sliding, or by a combination of these forces (model 2). For model 2, rigid translation is opposed by drag due to creep in the mantle below. In Figure 14a we have reproduced the schematic flow pattern, orientations of planes of maximum shearing stress, and fabrics expected from syntectonic recrystallization for the convection model [*Avé Lallemant and Carter*, 1970, Figure 7a]. (The idealized fabrics shown in Figure 14a are based on the assumption that the crystals are oriented with respect to the principal stress axes rather than the principal strain axes. Inasmuch as the stress and strain tensors coincide in the experiments to date, it is not possible to distinguish between these two possibilities; extrusion experiments are currently underway to determine whether stress or strain controls the orientations. If the crystals are oriented with respect to the principal strain axes during simple shear, the foliation plane would be inclined to the $\sigma_2 - \sigma_3$ (zero shearing stress) plane, and thus slip could take place parallel to the [100] maximum accompanying recrystallization, as is commonly observed. This possibility does not, however, alter appreciably the analyses in this section or the conclusions that follow.) The shearing sense for the rigid translation model (model 2) is just reversed, so that the orientations of $\alpha = [010] = \sigma_1$ and $\gamma = [100] = \sigma_3$ would be opposite those shown in Figure 14a [*Avé Lallemant and Carter*, 1970, Figure 7b]. We also show schematically on the figure the approximate location of the 500°C isotherm (assuming that the plate is about 50 km thick and the velocity is about 4 cm/yr), as suggested by studies of *Langseth et al.* [1966] and *Oxburgh and Turcotte* [1968] for the area near the ridge, *Clark and Ringwood* [1964] for the ocean basin, and *McKenzie* [1969b], *Griggs* [1972], and *Toksöz et al.* [1971] for the descending slab near the trench.

Syntectonic recrystallization is expected to be the dominant flow mechanism below the 500°C isotherm, as was indicated by the experiments (Figure 2). The fabric shown in the upper right of Figure 14a originates during the ascent of mantle peridotite or dunite and could be 'frozen in' and carried along passively thereafter. The fabric shown in the center would be expected for steady-state flow above 500°C between the ridge and the trench. The fabric in the upper left could arise from bodily rotation of the frozen-in fabric shown in the center or by continued recrystallization above 500°C in the downgoing slab. Inasmuch as the seismic refraction observations have been made between the ridge and the trench in the uppermost mantle of the Pacific plate at about 225°C and 3 kb, we might expect to sample fabrics having orientations near that shown in the upper right of Figure 14a, although the orientations could range to that shown in the center of the figure.

The maximum *P*-wave velocity anisotropies, calculated in the horizontal plane, are shown in Figure 14b as a function of tilt about concentrations of [001] axes in the dunite and the two lherzolites regarded as being typical of mantle tectonites. The anisotropies given are the differences in velocities parallel to [001] concentrations (normal to the plane of the diagram in Figure 14a) and the horizontal velocity in the east-west direction. The values, with increasing inclination to the horizontal of $\gamma = $ [100] concentrations, from the ridge to the trench (right to left), are taken as positive if the velocities normal to the ridge are greater than those parallel to it, as was observed in the refraction studies. For comparison, the inset in Figure 14b shows the ΔV_p between [001] and the east-west direction ($\perp$ ridge) as a function of inclination of [100] for single-crystal forsterite at 225°C and 3 kb. It is evident that the fabrics for both the Twin Sisters dunite and the John Day lherzolite could easily account for the observed 0.3- to 0.7-km/sec anisotropy for the range of orientations considered ([100] $\wedge$ horizontal = 0°–45°). For lherzolite L-62 it appears that the correct sign and magnitude of the anisotropy would be given only if the frozen-in fabric (Figure 14a, right) were sampled.

Inasmuch as the shearing sense and hence the fabrics are reversed for the rigid translation model (model 2), the curves in Figure 14b corresponding to fabric orientations for this model, from the ridge to the trench, should be read from left to right. The observed anisotropies could be produced only near the limiting orientation shown in the center of Figure 14a (but reversed) or under other very special circumstances. Therefore, if syntectonic recrystallization is the dominant mineral orienting process, our results clearly favor the convection model. **A similar conclusion, based on geological con-**

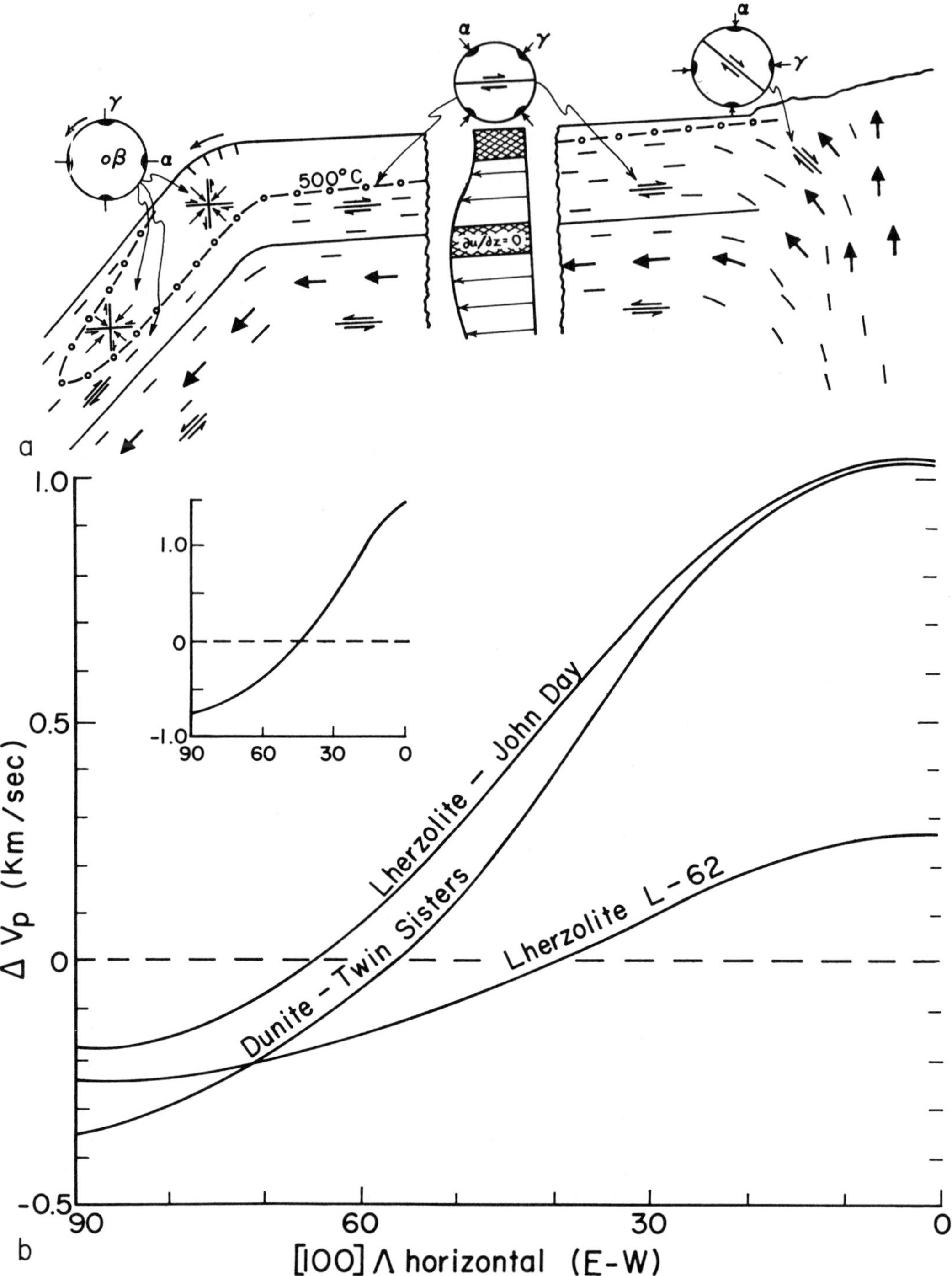

Fig. 14.   (a) Schematic flow model showing orientations of planes of maximum shearing stress and fabrics expected for syntectonic recrystallization during thermal convection [*Avé Lallemant and Carter*, 1970] (see the text). (b) Seismic *P*-wave velocity anisotropy as a function of inclination of [100] concentrations to the horizontal for three natural specimens and (inset) olivine single crystal (see the text).

siderations and the results of *Avé Lallemant and Carter* [1970], was reached by *Christensen* [1971] in his study of the Twin Sisters dunite.

Finally, *Francis* [1969] proposed that plastic deformation in olivine according to the system {0kl} [100], during the ascent of convection cells near ridges, might be the dominant mineral orienting process. In this instance, for very large homogeneous strains the [100] axes would tend to align parallel to the least principal stress axis, σ_3. Fabrics produced in this manner and sampled by the refraction studies would also be frozen in, since the system {0kl} [100] is expected to operate only above about 300°C at geological strain rates (Figure 2). Therefore, if large plastic deformation is the dominant mineral orienting process, the fabrics expected would be the same as those for syntectonic recrystallization during convection. (The grain shape fabric should, in general, be quite different and would have to be considered in the seismic velocity calculations.) However, there is no evidence for the strong plastic deformation required to produce the preferred orientations in any of the specimens that we have studied. *Nicolas et al.* [1971] have found such evidence for specimens near the periphery of the Lanzo massif and a few other peridotite bodies in Europe, but the strong plastic deformation may have post-dated steady-state flow in the upper mantle.

On the basis of similarities in fabrics and deformational processes in mantle peridotites and dunites and those observed in the steady-state experiments, we conclude that power law creep governs flow over most of the upper mantle. The dominant deformational process is believed to be syntectonic recrystallization, and the preferred crystal orientations produced by this process are regarded as the primary cause for the seismic P-wave velocity anisotropy observed in the upper mantle of the northeast Pacific. Maximum anisotropies found for mantle garnet clinopyroxenites are insufficient to produce the observed magnitudes (0.3–0.7 km/sec), whereas, in general, those found for peridotites and dunites can easily account for the observations. Hence the upper mantle, at least in the northeast Pacific, is probably composed dominantly of peridotite and/or dunite. If syntectonic recrystallization, accompanied by plastic deformation and recovery, is indeed the dominant creep mechanism in the upper mantle, our results support the hypothesis that thermal convection is the primary driving force for the sea floor spreading process.

Acknowledgments. We are indebted to N. I. Christensen for measurements of compressional wave velocities in cores from some of our specimens. J. R. Möckel kindly supplied us with his data for Ar-18 and Ar-26, and H. G. Avé Lallemant supplied the data for the John Day lherzolite. E. den Tex provided lherzolite L-62, and E. Maske supplied the mantle garnet clinopyroxenites. L. A. Fernandez permitted us to use his unpublished microprobe data on the minerals of the garnet clinopyroxenites. We benefitted greatly by discussions, suggestions, and criticisms from H. G. Avé Lallemant, I. Y. Borg, N. I. Christensen, H. C. Heard, A. Nicolas, and C. B. Raleigh.

This research was supported by National Science Foundation grant GA-15412.

References

Alexsandrov, K. S., and T. V. Ryzhova, The elastic properties of rock-forming minerals, 1, Pyroxenes and amphiboles, *Bull. Acad. Sci. USSR, Geophys. Ser.,* Engl. Transl., no. 9, 871–875, 1961.

Anderson, D. L., Latest information from seismic observations, in *The Earth's Mantle,* edited by T. F. Gaskell, pp. 354–420, Academic, New York, 1967.

Anderson, D. L., C. Sammis, and T. Jordan, Composition and evolution of the mantle and core, *Science, 171,* 1103–1112, 1971.

Andreatta, C., Analisi strutturali di rocce metamorfiche, 5, Olivinite, *Period. Mineral., 5,* 237–253, 1934.

Avé Lallemant, H. G., Structural and petrofabric analysis of an 'Alpine-type' peridotite, The lherzolite of the French Pyrenees, *Leidse Geol. Meded., 42,* 1–57, 1967.

Avé Lallemant, H. G., and N. L. Carter, Syntectonic recrystallization of olivine and modes of flow in the upper mantle, *Geol. Soc. Amer. Bull., 81,* 2203–2220, 1970.

Baker, D. W., and N. L. Carter, Seismic velocity anisotropy calculated for ultramafic minerals and aggregates, this volume.

Carter, N. L., Static deformation of silica and silicates, *J. Geophys. Res., 76,* 5514–5540, 1971.

Carter, N. L., and H. G. Avé Lallemant, High temperature flow of dunite and peridotite, *Geol. Soc. Amer. Bull., 81,* 2181–2202, 1970.

Carter, N. L., and H. C. Heard, Temperature and rate dependent deformation of halite, *Amer. J. Sci., 269,* 193–249, 1970.

Christensen, N. I., Elasticity of ultrabasic rocks, *J. Geophys. Res., 71,* 5921–5931, 1966.

Christensen, N. I., Fabric, seismic anisotropy, and tectonic history of the Twin Sisters dunite, *Geol. Soc. Amer. Bull., 82,* 1681–1694, 1971.

Christensen, N. I., and R. Ramananantoandro, Elastic moduli and anisotropy of dunite to 10 kilobars, *J. Geophys. Res., 76,* 4003–4010, 1971.

Clark, S. P., Jr., and A. E. Ringwood, Density distribution and constitution of the mantle, *Rev. Geophys. Space Phys., 2,* 35–88, 1964.

Crittenden, M. D., Jr., Viscosity and finite strength of the mantle as determined from water and ice loads, *Geophys. J. Roy. Astron. Soc., 14,* 261–279, 1967.

Davis, B. T. C., and J. L. England, The melting of forsterite up to 50 kilobars, *J. Geophys. Res., 69,* 1113–1116, 1964.

den Tex, E., Origin of ultramafic rocks, their tectonic setting and history: A contribution to the discussion of the paper 'The origin of ultramafic and ultrabasic rocks' by P. J. Wyllie, *Tectonophysics, 7,* 457–488, 1969.

Dieterich, J. H., and N. L. Carter, Stress-history of folding, *Amer. J. Sci., 267,* 129–154, 1969.

Francis, T. J. G., Generation of seismic anisotropy in the upper mantle along the mid-oceanic ridges, *Nature, 221,* 161–165, 1969.

Garofalo, F., *Fundamentals of Creep and Creep Rupture in Metals,* MacMillan, New York, 258 pp., 1965.

Goetze, C., High-temperature rheology of Westerly granite, *J. Geophys. Res., 76,* 1223–1230, 1971.

Gordon, R. B., Diffusion creep in the earth's mantle, *J. Geophys. Res., 70,* 2413, 1965.

Green, H. W., Diffusional flow in polycrystalline materials, *J. Appl. Phys., 41,* 3899–3902, 1971.

Green, H. W., and S. V. Radcliffe, Dislocation mechanisms in olivine and flow in the upper mantle, *Earth Planet. Sci. Lett., 15,* 239–247, 1972b.

Green, H. W., D. T. Griggs, and J. M. Christie, Syntectonic and annealing recrystallization of fine-grained quartz aggregates, in *Experimental and Natural Rock Deformation,* edited by P. Paulitsch, pp. 272–335, Springer-Verlag, Berlin, 1970.

Griggs, D. T., Creep of rocks, *J. Geol., 47,* 225–251, 1939.

Griggs, D. T., Hydrolytic weakening of quartz and other silicates, *Geophys. J. Roy. Astron. Soc., 14,* 19–31, 1967.

Griggs, D. T., The sinking lithosphere and the focal mechanism of deep earthquakes, in *Nature of Solid Earth,* edited by E. C. Robertson, pp. 361–384, McGraw-Hill, New York, 1972.

Hartman, P., and E. den Tex, Piezocrystalline fabrics of olivine in theory and nature (abstract), *Int. Geol. Congr. Rep. Sess. India 22nd,* 42–43, 1964.

Heard, H. C., Steady-state flow in polycrystalline halite at pressure of 2 kilobars, this volume.

Heard, H. C., and C. B. Raleigh, Steady-state flow in marble at 500° to 800°C, *Geol. Soc. Amer. Bull., 83,* 935–956, 1972.

Helmstaedt, H., and O. L. Anderson, Petrofabrics of mafic and ultramafic inclusions from kimberlite pipes in southeastern Utah and northeastern Arizona (abstract), *Eos Trans. AGU, 50,* 345, 1969.

Ito, K., and G. C. Kennedy, The fine structure of the basalt-eclogite transition, *Mineral. Soc. Amer. Spec. Pap. 3,* 77–84, 1970.

Jackson, E. D., and T. L. Wright, Xenoliths in the Honolulu volcanic series, Hawaii, *J. Petrol., 11,* 405–430, 1970.

Kamb, B., Theory of preferred crystal orientation developed by crystallization under stress, *J. Geol., 67,* 153–170, 1959a.

Kamb, B., Glacier geophysics, *Science, 146,* 353–365, 1964.

Kasahara, J., I. Suzuki, M. Kumazawa, Y. Kobayashi, and K. Tida, Anisotropism of P-wave in dunite, *J. Seismol. Soc. Jap., 21,* 222–228, 1968.

Keen, C. E., and D. L. Barrett, A measurement of seismic anisotropy in the northeast Pacific, *Can. J. Earth. Sci., 8,* 1056–1064, 1971.

Knopoff, L., Thermal convection in the earth's mantle, in *The Earth's Mantle,* edited by T. F. Gaskell, pp. 171–196, Academic Press, New York, 1967.

Knopoff, L., Continental drift and convection, in *The Earth's Crust and Upper Mantle, Geophys. Monogr. Ser.,* vol. 13, edited by P. J. Hart, pp. 683–689. AGU, Washington, D.C., 1969.

Kumazawa, M., H. Helmstaedt and K. Masaki, Elastic properties of eclogite xenoliths from diatremes of East Colorado Plateau and their implication to the upper mantle structure, *J. Geophys. Res., 76,* 1231–1247, 1971.

Langseth, M. G., Jr., X. LePichon, and M. Ewing, Crustal structure of the mid-ocean ridges, 5, Heat flow through the Atlantic Ocean floor and convection currents, *J. Geophys. Res., 73,* 5321, 1966.

Lappin, M. A., The petrofabric orientation of olivine and seismic anisotropy of the mantle, *J. Geol., 79,* 730–740, 1971.

MacGregor, I. D., and J. L. Carter, The chemistry of clinopyroxenes and garnets of eclogite and peridotite xenoliths from the Roberts Victor Mine, South Africa, *Phys. Earth Planet. Interiors, 3,* 391–397, 1970.

McConnell, R. K., Viscosity of the earth's mantle, in *The History of the Earth's Crust,* edited by R. A. Phinney, pp. 45–57, Princeton Univ. Press, Princeton, N.J., 1968.

McKenzie, D. P., The geophysical importance of high temperature creep, in *The History of the Earth's Crust,* edited by R. A. Phinney, p. 28, Princeton Univ. Press, N. J., 1968.

McKenzie, D. P., Speculations on the consequences and causes of plate motions, *Geophys. J. Roy. Astron. Soc., 18,* I, 1969*b*.

Möckel, J. R., The structural petrology of the garnet peridotite of Alpe Arami (Ticino, Switzerland), *Leidse Geol. Meded., 42,* 61–130, 1969.

Morris, G. B., R. W. Raitt, and G. G. Shor, Velocity anisotropy and delay-time maps of the mantle near Hawaii, *J. Geophys. Res., 74,* 4300–4316, 1969.

Nicolas, A., J. L. Bouchez, F. Boudier, and J. C. Mercier, Textures, structures and fabrics due to solid state flow in some European lherzolites, *Tectonophysics,* in press, 1972.

O'Hara, M. J., Mineral parageneses in ultrabasic rocks, in *Ultramafic and Related Rocks,* edited by P. J. Wyllie, p. 393, John Wiley, New York. 1967.

Phakey, P., G. Dollinger, and J. Christie, Transmission electron microscopy of experimentally deformed olivine crystals, this volume.

Post, R. L., Jr., The flow laws of Mt. Burnette dunite at 750°–1150°C (abstract), *Eos Trans. AGU, 51,* 424, 1970.

Raitt, R. W., G. G. Shor, T. J. G. Francis, and G. B. Morris, Anisotropy of the Pacific upper mantle, *J. Geophys. Res., 74,* 3095–3109, 1969.

Raleigh, C. B., Mechanisms of plastic deformation in olivine, *J. Geophys. Res., 73,* 5391–5406, 1968.

Raleigh, C. B., and S. H. Kirby, Creep in the upper mantle, *Mineral. Soc. Amer. Spec. Pap. 3,* 113–121, 1970.

Raleigh, C. B., and J. L. Talbot, Mechanical twinning in naturally and experimentally deformed diopside, *Amer. J. Sci., 265,* 151–165, 1967.

Raleigh, C. B., S. H. Kirby, N. L. Carter, and H. G. Avé Lallemant, Slip and the clinoenstatite transformation as competing rate processes in enstatite, *J. Geophys. Res., 76,* 4011–4022, 1971.

Ringwood, A. E., Composition and evolution of the upper mantle, in *The Earth's Crust and Upper Mantle, Geophys. Monogr. Ser.,* vol. 13, edited by P. J. Hart, pp. 1–17, AGU, Washington, D.C., 1969.

Sherby, O. D., and P. M. Burke, Mechanical behavior of crystalline solids at elevated temperature, *Progr. Mater. Sci., 13,* 324–390, 1967.

Sherby, O. D., and M. T. Simnad, Prediction of atomic mobility in metallic systems, *Trans. Amer. Soc. Met., 54,* 227–240, 1961.

Shewmon, P. G., *Diffusion in Solids,* 272 pp., McGraw-Hill, New York, 1963.

Toksöz, M. N., J. W. Minear, and B. R. Julian, Temperature field and geophysical effects of a downgoing slab, *J. Geophys. Res., 76,* 1113–1138, 1971.

Vogel, D. E., Petrology of an eclogite- and pyrigarnite-bearing polymetamorphic rock complex at Cabo Ortegal, NW Spain, *Leidse Geol. Meded., 40,* 121–213, 1967.

Wang, C., Density and constitution of the mantle, *J. Geophys. Res., 75,* 3264–3284, 1970.

Watson, K. D., and D. M. Morton, Eclogite inclusions in kimberlite pipes at Garnet Ridge, northeastern Arizona, *Amer. Mineral., 54,* 267–285, 1969.

Weertman, J., Dislocation climb theory of steady-state creep, *Trans. ASME, 61,* 681–694, 1968.

Weertman, J., The creep strength of the earth's mantle, *Rev. Geophys. Space Phys., 8,* 145–168, 1970.

14

CLASSIFICATION OF TEXTURES AND FABRICS OF PERIDOTITE XENOLITHS FROM SOUTH AFRICAN KIMBERLITES

ANNE MARIE BOULLIER and ADOLPHE NICOLAS
Laboratoire de Géologie Structurale, B.P. 1044, Nantes Cedex 44037, France

ABSTRACT

A previous classification of textures and fabrics in peridotite xenoliths from kimberlites is summarized and extended by addition of two new types: the secondary tabular type which results from recrystallization following an intense flowage; the fluidal mosaic type, with distinctive tectonic stripes which could involve superplasticity. The coarse granular and tabular types, previously defined, are strongly recrystallized and recovered; they could represent a mantle that is rigid or deforming at slow strain rates. The porphyroclastic and mosaic types would correspond to peridotites flowing with a faster strain rate at the time they were extracted by the kimberlite magma. The textures and fabrics of these two highly deformed types have been generated by plastic flow; olivine recrystallization has been minor in the former type and complete in the latter. These data and interpretations argue for a non-steady-state origin for the kink in the pyroxene geotherm proposed by other authors.

INTRODUCTION

Assuming thermodynamic equilibrium, it is now possible to assign a position in a P, T diagram to the peridotite xenoliths from kimberlites by a study of the chemistry of their pyroxenes. This has led BOYD and NIXON (1973) and MACGREGOR (in press) to discover that, when plotted in a P, T diagram, these peridotite xenoliths lie on a curved line that has a pronounced kink. They also observed that texturally the xenoliths were different on each side of the kink. The "granular" xenoliths plot along a curve interpreted as the continental geotherm in the lithosphere below South Africa in the Cretaceous when the kimberlite brought up the xenoliths. The "sheared" xenoliths plot below the kink in this curve with a steeper slope interpreted as the geotherm in the upper asthenosphere modified by shear heating. When our classification of textures and fabrics in peridotite xenoliths from Lesotho became available (BOULLIER and NICOLAS, 1973a), representative samples of each group were also plotted on the P, T diagram. The points aligned according to increasing deformation and helped to define what has been considered as a transition between rigid lithosphere and flowing asthenosphere (BOYD and NIXON, 1973).

Though our classification is still provisional, this remarkable coincidence between conditions of equilibration and textures has created some need for a paper with the following objectives: (1) to recall the definitions of the different textural types and to illustrate them; the fabrics typical of these textural types will not be described here (see BOULLIER and NICOLAS, 1973a and b) except for two new types which were not presented previously; (2) to compare briefly the textural types in peridotites from kimberlites with their counter-

208

parts in basalts and in large Alpine-type massifs; (3) to derive conclusions about the environmental conditions prevailing in the mantle when the textures and fabrics developed; and from that (4) to discuss the general model of BOYD and NIXON (1973) and MACGREGOR (in press).

MAIN TEXTURES

Table 1 summarizes the principal data on the different structural types met in peridotite xenoliths from South African kimberlites.

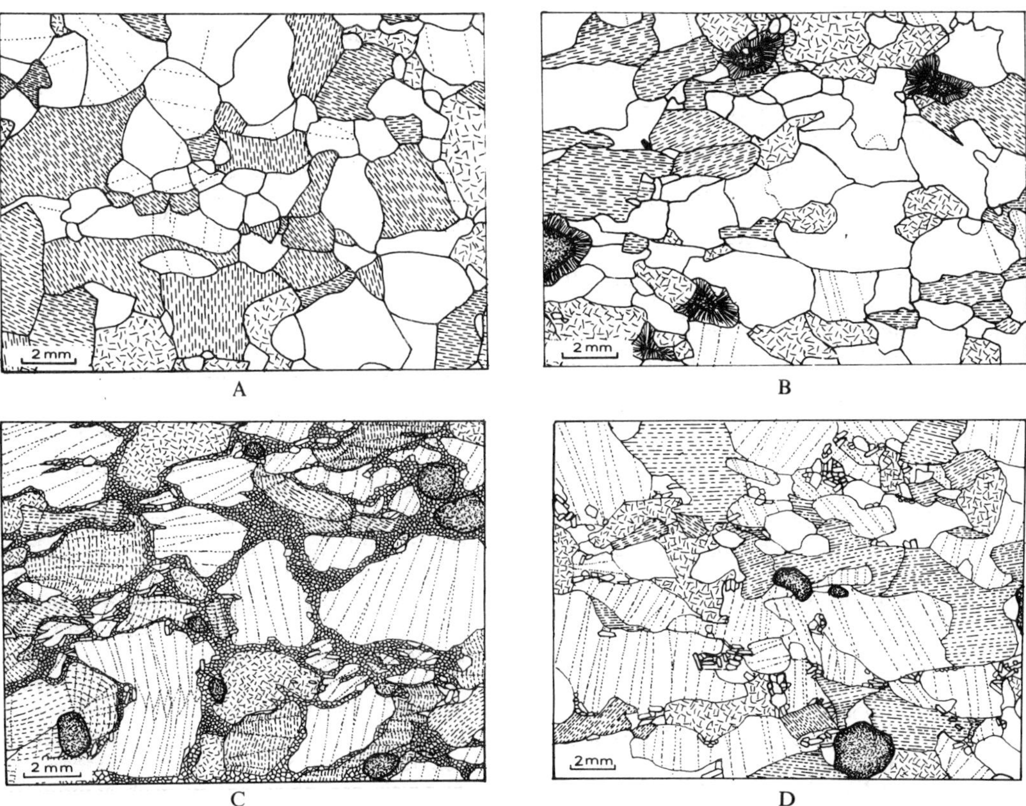

FIGS. 1 and 2. Drawings after micrographs of the typical textures and of some intermediate ones; the foliation, if any, is E.W. The micrographs and the fabrics of the typical textures have already been published (BOULLIER and NICOLAS, 1973a). Olivine: blank, except for the KBB represented by lines of small dots; orthopyroxene: aligned dashes parallel to the (100) plane when visible, otherwise random dashes; clinopyroxene: largely spaced dots; garnet: closely spaced dots, thick contours with occasional kelyphite-rims.

FIG. 1. Scale bars = 2 mm. A. Coarse granular texture (harzburgite 69 Ki 26, Kimberley). B. Coarse tabular texture (garnet and spinel lherzolite PHN 1595, Thaba Putsoa). See KBB details on Fig. 3B. C. Porphyroclastic texture (garnet harzburgite, DB2, De Beers Mine). The recrystallization of the olivine porphyroclasts in small neoblasts is important. Only a few euhedral tablets are present. D. Texture probably transitional between the coarse tabular and the porphyroclastic ones (garnet harzburgite, K7, Kamfersdam). The porphyroclasts are less deformed and recrystallized than in a typical porphyroclastic texture. Affinities in textures and fabrics with the coarse tabular texture argue for such an origin. Starting with a nearly random fabric as in the coarse granular texture, it seems improbable that the moderate deformation evidenced in the rock could be responsible for its strong preferred orientation.

TABLE 1

Textures	Coarse granular (Fig. 1A)	Coarse tabular (Fig. 1B)	Porphyroclastic (Figs. 1C,D; Fig. 2A)	Mosaic		Secondary tabular (Fig. 2D)
				Normal (Fig. 2B)	Fluidal (Fig. 2C) descriptions restricted to non-penetrative stripes	
Former and other terminologies	Coarse grained (BOULLIER and NICOLAS, 1973)	Tabular olivine and enstatite (BOULLIER and NICOLAS, 1973)				
	GRANULAR			SHEARED (BOYD and NIXON, 1972)		
Number of different generations of olivine (ol.) and enstatite (en.)	1 ol. 1 en.	1 ol. 1 en.	2 or 3 ol. 1 or 2 en.	1 ol. 2 en.	1 en.	1 or 2 ol. 1 en.
Grain shape and size	isometric 6 mm	tabular 5 × 2 mm	ol. and en. porphyroclasts: elongated, ol.: 5 × 2 mm; en.: 3 × 1 mm neoblasts: isometric ol.: 0.2 mm en.: 0.02 mm ol. tablets: tabular 1 × 0.3 mm	en. porphyroclast: elongated, 4 × 2 mm neoblasts: isometric ol.: 0.3 mm en.: 0.02 mm	en. grains, isometric 0.01 mm concentrated in stripes 0.03 to 0.01 mm thick	ol. tabular: 2 × 1 mm and irregular: 2 mm en. isometric: 0.3 mm
Grain boundaries	ol./ol. straight or curved 120° triple points ol./en. curved	ol./ol. straight or curved 120° triple points ol./en. curved or straight when parallel to foliation	ol. and en. porphyroclasts: irregular, invaded by neoblasts ol. and en. neoblasts: straight or rounded ol. tablets: straight rational faces	en. porphyroclasts: irregular, invaded by neoblasts ol. and en. neoblasts: straight, 120° triple points	en., straight or rounded	straight and embayed
Deformation in crystals (KB = kink band; KBB = kink band boundary)	almost none; largely spaced and clear KB, sharp KBB converging in triple points with grain boundaries	almost none; largely spaced and clear KB, sharp KBB converging in triple points with grain boundaries	porphyroclasts: intense; closely spaced and wavy KB, blurred KBB neoblasts: invisible tablets: absolutely none	en. porphyroclasts: intense distortion in lattices neoblasts: ol., almost none en., invisible	en., invisible	almost none, wavy KB
Structures in the nodule (foliation S = plane of mineral flattening; lineation L = mineral elongation in S)	no foliation no lineation	good foliation weak lineation	good foliation good lineation	good foliation excellent lineation	excellent non-penetrative "foliation" excellent lineation	excellent foliation excellent lineation
Fabrics olivine and enstatite preferred orientations (⊥ = perpendicular; // = parallel)	ol.: weakly orthorhombic X ol. ⊥ S if any en.: almost random	ol.: good X ol. ⊥ S; Y ol. and Z ol., girdles in S weak maxima en.: X en. ⊥ S or girdle ⊥ S	porphyroclasts: strong X ol. ⊥ S, Z ol. // L; X en. girdle ⊥ S Z en. // L neoblasts: X ol. ⊥ S weak Z ol. // L tablets: strong X ol. ⊥ S, Z ol. // L	ol. neoblasts: almost random, weak Z ol. // L strong local ol. subfabrics	unknown	tabular olivine: X ol. ⊥ S, Z ol. // L en: random
Dominant petrographic type. See restrictions p. 100. * Jagersfontein field count	garnet- and spinel-poor harzburgite 72%*	garnet- and spinel-poor harzburgite	garnet lherzolite	garnet lherzolite 83%.*		garnet lherzolite and harzburgite
Numerical importance. See restrictions p. 100. * Jagersfontein field count	50% 66%*	29% coarse tabular	13% and porphyroclastic 11%*	8% 22%*		

210

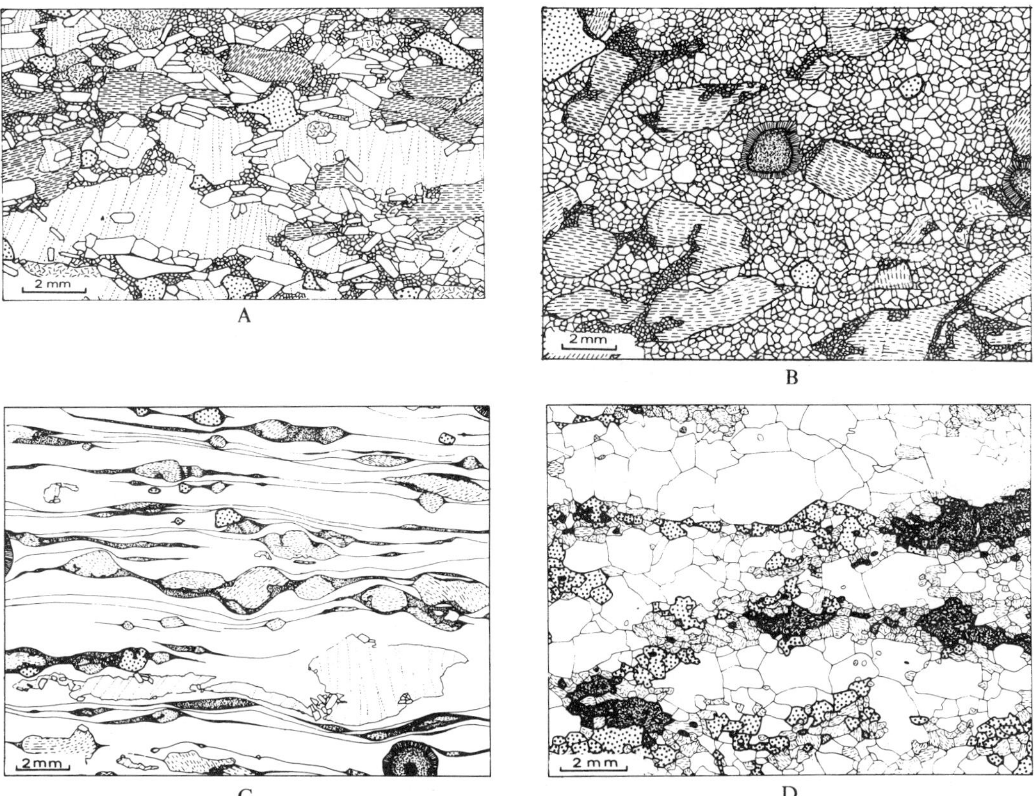

FIG. 2. Scale bars = 2 mm. A. Porphyroclastic texture (garnet lherzolite 69 Ki 14, Kimberley). Mineral lineation E.W. Note the three kinds of olivine grains: the large deformed porphyroclasts, the small neoblasts and the euhedral tablets postdating the deformation. B. Mosaic texture (garnet lherzolite PHN 1925, Mothae). Mineral lineation E.W. The olivine is entirely recrystallized while the enstatite displays porphyroclasts partly recrystallized in tiny neoblasts. C. Fluidal mosaic texture (garnet lherzolite, KAO2, Kao). Mineral lineation E.W. A large and unique olivine porphyroclast proves that the mosaic texture derived from a porphyroclastic one. The fluidal character is superimposed on the mosaic texture. See Fig. 3A. D. Secondary tabular texture (garnet lherzolite, PHN 1654, Matsoku). Mineral lineation E.W. The olivine tends to be segregated in bands where it has a tabular shape with some inclusions; these bands contrast with those where the other minerals dominate in small and scattered grains. For the fabric, see Fig. 4.

The main textures and fabrics have been described in detail in previous papers (BOULLIER and NICOLAS, 1973a and b). They were based at that time on the study of only twenty-five xenoliths. We have now examined 120 xenoliths. This larger basis verifies that all transitions exist between the types, as had been predicted from a comparable study of peridotite xenoliths from basalts (MERCIER, 1972; MERCIER and NICOLAS, in press). As in the case of basaltic xenoliths, the provisional names were assigned considering only olivine, although the investigation on textures and fabrics has also included enstatite which is the other major mineral of peridotites. For example, a texture with olivine porphyroclasts (Fig. 2A) is called "porphyroclastic"; another with the olivine entirely as small polygonal grains (Fig. 2B) is called "mosaic", whether or not the enstatite constitutes porphyroclasts. We propose here to modify (1) the former provisional name of "coarse grained" into "coarse granular" which seems more appropriate and suggests a comparison with the protogranular texture described in basaltic xenoliths (see below); (2) the former provisional name of "tabular olivine and enstatite" into "coarse tabular" because it is as coarse (sometimes coarser) than the coarse granular type and suggests between the two types an association which is verified in the position in the P, T diagram and in the petrology (see Table 1).

Estimations of the relative abundance of the different types would require a systematic study. Our only attempt in this direction is a field count of 133 xenoliths made with M. G. WILSHIRE and M. PRINZ on the Jagersfontein dumps. It is reported in Table 1 together with the proportions measured in our collection which does not claim to be representative.

Two other modifications of our earlier, tentative classification appear in Table 1. The non-penetrative deformation already described (BOULLIER and NICOLAS, 1973a, p. 65) is commonly superimposed on the mosaic texture and is therefore worthy of being distinguished as a subtype—"the fluidal mosaic texture" of the mosaic texture. It is characterized by narrow stripes (0.01–0.03 mm) that are intensely deformed and parallel to the foliation in the mosaic groundmass (Fig. 2C and Fig. 3A). They originate in the recrystallized boundaries of enstatite porphyroclasts where the neoblasts are evenly grained and tiny (0.01 mm). From there the neoblasts of enstatite, and not of olivine as previously thought, are scattered along the stripes which connect several enstatite porphyroclasts or end in the olivine mosaic

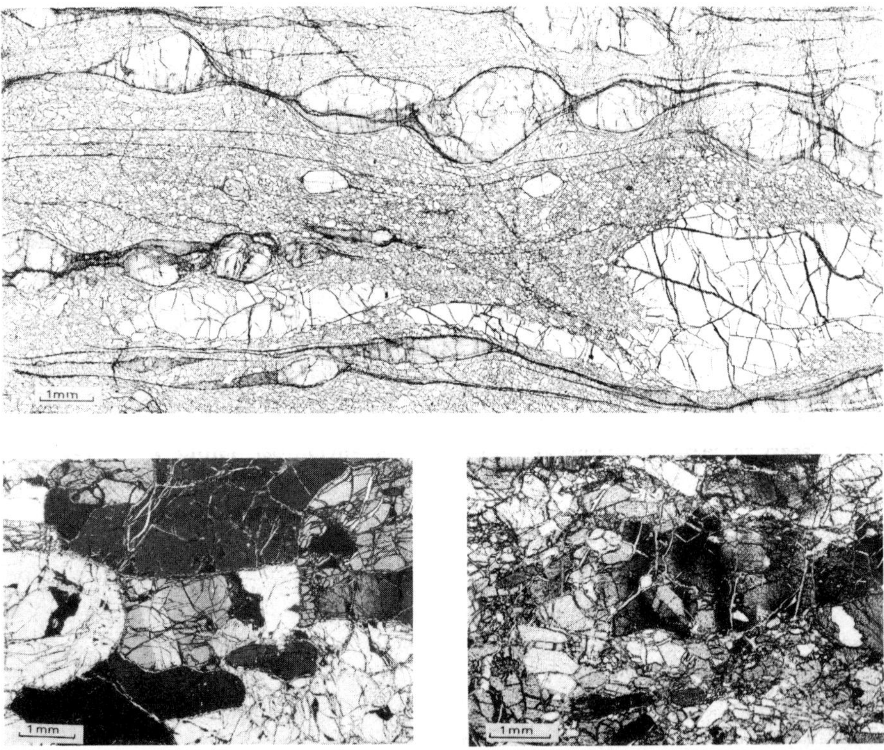

FIG. 3. Scale bars = 1 mm. Photomicrograph of the fluidal mosaic texture (detail of the drawing Fig. 2C). The narrow stripes are superimposed on a mosaic texture. They connect beads of enstatite porphyroclasts, originating in their recrystallized rims and carrying away the 0.01 mm neoblasts. They end up as microfaults in the olivine mosaic matrix. B. Photomicrograph (crossed polarizers) of a detailed area in the coarse tabular texture of Fig. 1B. The kink bands in olivine are highly recovered with sharply cut kink band boundaries C. Photomicrograph (crossed polarizers) of a detailed area in the porphyroclastic texture of Fig. 2A. The wavy kink bands evidence a weak recovery.

where they sharply cut through the crystals as microfaults (Fig. 3A). The extremely fine grain size has made it impossible to measure the fabric in the stripes so far.

The other modification is the introduction of a new textural type, the "secondary tabular" texture to account for the special textures described by Cox *et al.* (1973) in the Matsoku pipe (Table 1, Fig. 2D). Observation on these samples shows that the peridotite has been intensely deformed with development of strong foliation and lineation, underlined by the scattering of the garnet in trails of small grains. One sample 20 cm thick shows a beautiful deformation gradient from flattened garnets to garnets scattered as described a few millimeters across, some richer in olivine, some in the other phases where enstatite, diopside and garnet are in somewhat comparable proportions. In these bands, the minerals are recrystallized but remain in small grains (0.3 mm). By contrast, in the olivine-rich bands, the recrystallization has resulted in larger tabular crystals (2 × 1 mm) which commonly include small crystals of the other phases*. The number of inclusions in tabular olivine grains is greater than in the coarse tabular texture. This is a first criterion to distinguish these two textures (the coarse tabular and the secondary tabular). Better ones are provided

(1) by the shape of the orthopyroxene which is small and irregular in the secondary tabular texture, and large and tabular in the coarse tabular texture;

(2) by the comparison of structures in the sample: strongly foliated and lineated in the secondary texture, and mildly foliated and non-lineated in the primary texture.

The fabric is orthorhombic for the tabular olivine grains with diffuse point maxima: Xol normal to the foliation, Yol and Zol in the foliation, the latter being parallel to the mineral lineation. The enstatite orientations are inconsistent (Fig. 4).

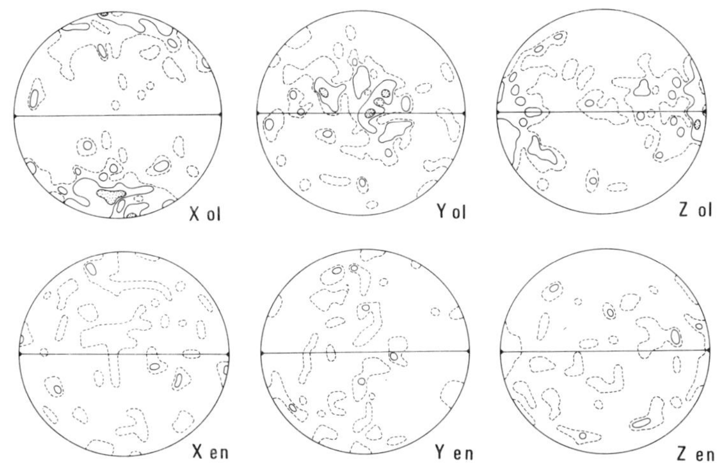

FIG. 4. Olivine (ol) and enstatite (en) fabrics in the secondary tabular texture of Fig. 2D. Equal area projections on the lower hemisphere. Contours: 1, 2, 4%; 100 olivine grains, 200 enstatite grains. Foliation horizontal, mineral lineation E.W.

* It is commonly observed in peridotites which are recrystallized that the grain size is smaller when the number of phases present in the rock is larger. In similar environments dunites can be formed by olivine grains of a few centimeters across, harzburgites by grains of 5 to 10 mm, and lherzolites by grains of 3 to 5 mm. Assuming that recrystallization takes place by grain boundary migration, with some highly mobile boundaries favoring the growth of given grains, these observations can be explained by comparing the possibilities of boundary migration in the cases of an olivine rich and of a mixed band. In the first case the growth is impeded only when highly mobile olivine boundaries meet together and compete; in the other case it is soon impeded when each boundary of a given phase comes in contact with grains of different stable phases.

COMPARISON WITH PERIDOTITES FROM BASALTIC
XENOLITHS AND MASSIFS

An extensive study of textures and fabrics in peridotite xenoliths in basalts from different environments has resulted in a new classification (MERCIER, 1972; MERCIER and NICOLAS, in press), which has a wider basis and correlated the fabric more closely with the textures than the earlier classification of BROTHERS and RODGERS (1969). The different types are also compared with textures and fabrics observed in peridotites from massifs. Their origin is investigated in the light of recent work on the problem of flow and orienting mechanisms in peridotites (AVE'LALLEMANT and CARTER, 1970; NICOLAS et al., 1971, 1973).

The protogranular type is the oldest one recorded in peridotite xenoliths from basalts; it can be identified in some peridotite massifs despite the modification of textures and fabrics imposed by the plastic flow occurring during their intrusion. In many aspects it can be compared with the coarse granular type from kimberlite described here. In the basaltic xenoliths, every transitional stage is observed between the protogranular and the porphyro-clastic type; this second type differs from the one in kimberlite xenoliths mainly on its larger amount of recovery (see below). Again, through transitions, one grades into the equigranular types; the equant subtype compares with the mosaic type described here except for the enstatite habit. In the basaltic xenoliths, the enstatite is recrystallized in the same fashion as the olivine while in the kimberlite xenoliths, it is not. In the kimberlite xenoliths, no equivalent of the equigranular tabular subtype has yet been observed. Conversely, the coarse tabular type and the fluidal mosaic subtype of kimberlite peridotite are unknown in basaltic xenoliths. Finally the secondary textures, exemplified in kimberlites by the Matsoku series, are well represented in basaltic xenoliths where a secondary cycle is initiated by formation through an annealing process of secondary protogranular types. The inter-pretation is the same as the one proposed by HARTE et al. (1973), i.e. a secondary recrystal-lization, post-dating a deformation cycle, which builds up a new texture and fabric in many ways comparable to the primary one. This has been documented in basaltic xenoliths using as a strain marker the progressive scattering of spinel–clinopyroxene–orthopyroxene clusters characterizing the protogranular texture. Thus it has been demonstrated that the sequence of our description of basaltic types corresponds to increasing deformation. Though the evidence is more tenuous in xenoliths from kimberlites it is believed to be also true for them (Figs. 1 and 2).

The peridotites from alpine-type massifs commonly show the porphyroclastic textures with different intensity of recovery and recrystallization (NICOLAS et al., 1972); this can ultimately lead to the equigranular equant type comparable to our mosaic one. It is mainly restricted to peridotite massifs incorporated in granulitic terranes like in the Ivrea Zone of northern Italy (LENSCH, 1971).

INTERPRETATION OF THE TEXTURES AND FABRICS

Many more accurate studies need to be carried on, using, for instance, the electron-microscope techniques (GREEN and BOULLIER, in preparation) before those textures and fabrics are clearly understood. We will restrict ourselves to a few general points.

The coarse granular and coarse tabular types (like the protogranular one in basaltic xenoliths) have been recrystallized with a strong tendency toward the minimization of grain-boundary energy (KRETZ, 1966). The only evidence of plastic deformation is given by the rare kink bands (Fig. 3B). They indicate that a strong recovery taking place at the same time as boundary migration minimizes the grain-boundary energy, as shown by convergence of grain boundaries with triple points at 120°.

It is not known whether these textures were developed during flow or in a static state. The lack of any orientation in the coarse granular type would indicate a static environment. The foliation and definite fabrics of the coarse tabular type could be due to some deformation but the process would have to be distinct from plastic flow because neither the shape of the grains nor the enstatite fabric are compatible with it. A stress-controlled recrystallization might explain it (CARTER *et al.*, 1972), but we do not wish to exclude the possibility of an origin by a cumulate or other magmatic process, followed by some recrystallization. HARTE *et al.* (1973) consider the possibility that the coarse-grained textures can be regenerated from fine-grained and deformed ones by recrystallization. This is suggested by the recrystallization, subsequent to a deformation, in xenoliths from the Matsoku pipe. The differences between such a secondary type and the coarse granular or tabular types has been stressed (p. 3); they would certainly be attenuated by prolonged annealing which cannot be ruled out as an alternative hypothesis. However it seems improbable because, in basaltic xenoliths where this recrystallization is better documented (MERCIER and NICOLAS, in press), the minor phases, spinel, diopside and enstatite, which have been scattered in the olivine matrix during flow, do not concentrate again significantly and are often present as inclusions in olivine. Considering the relatively low temperature of equilibration of these types (1000°, BOYD and NIXON, 1972; MACGREGOR, in press) and the textural features described above, we assume that the corresponding peridotites were probably static in the mantle or deforming at a slow strain rate compared to the ones displaying the other types.

The interpretation of the porphyroclastic and mosaic types has already been presented (BOULLIER and NICOLAS, 1973a). The former type was attributed to plastic flow with minor recrystallization and the second with complete recrystallization in olivine, whereas the less ductile enstatite was still in the porphyroclastic stage. This recrystallization is not considered as stress guided (AVE'LALLEMANT and CARTER, 1970) but as "strain guided", that is belonging to the general recovery processes due to the generation and movement of dislocations in crystals (plastic flow). There is a striking contrast between the amount of strain and paucity of recovery in the porphyroclasts on the one hand, and the total absence of any optically visible strain in the euhedral tablets on the other hand (Fig. 3C). Small tablets are also observed in the similar type in basaltic xenoliths but they are not euhedral and are often slightly strained. This special shape can tentatively be explained by a growth in the presence of a liquid film between the growing tablet and the consumed porphyroclast or, as suggested by H. W. GREEN (oral communication), by growth in a porphyroclast with such a high density of dislocations that it behaves as an isotropic matrix with regard to the growing tablet. It is also remarkable that those xenoliths which were deformed at higher temperatures than the coarse granular and tabular ones (up to 1400° for the mosaic ones; BOYD and NIXON, 1972; MACGREGOR, in press) show far less recovery in their porphyroclasts (Fig. 3B and C). Basalt xenoliths showing primary textural types, were deformed at comparable temperatures and they show the same recovery as the coarse granular and tabular xenoliths. More instructive is the case of the Lanzo massif in the Alps in which also the peridotites show a strong recovery and recrystallization. It was intruded in connection with the plate movements responsible for the alpine orogeny (NICOLAS *et al.*, 1972). Therefore we assume that the mean strain rate is in the range of 10^{-14}/sec which is compatible with plate movements. The plastic flow occurred at 1200° when the peridotites were in the 7kbar range (BOUDIER and NICOLAS, 1972).

All this strongly suggests that the porphyroclastic and mosaic textured peridotites were deformed at *strain rates larger than the plate movements ones* and that *they were flowing at the time they were extracted* and brought up in the kimberlites. Otherwise, taking into account their high temperature, they would rapidly recover and probably recrystallize. GREEN and RADCLIFFE (1972) have argued similarly for a basaltic xenolith from Lunar Crater, Nevada. The absence of visible strain in the tablets cutting through the porphyroclasts suggests that they grew after the deformation, that is after the rock was incorporated into the kimberlite. The available fluids would favor their rapid growth and possibly explain their euhedral habit. On the other hand, the coarse granular and tabular textured rocks may come from areas in the mantle which are not flowing or are flowing (by a process distinct of plastic flow) at much smaller strain rates, and possibly due to plate movements.

A minor but interesting problem is raised by the fluidal mosaic texture. The scattering of tiny grains of orthopyroxene along non-penetrative planes suggests that they may have been deforming in the superplasticity field (HAYDEN *et al.*, 1972). The flow which can be considerable takes place mainly by intergranular slip. The peripheral recrystallization of the porphyroclasts of enstatite into the tiny grains would necessarily precede and prepare this new stage by creating grains small enough to be able to deform by this process. This hypothesis is under investigation.

DISCUSSION OF THE RECENT MODELS OF THE UPPER MANTLE BELOW SOUTH AFRICA

We have already discussed (BOULLIER and NICOLAS, 1973a) the point regarding the composition of the minerals used by BOYD and NIXON (1972) and MACGREGOR (in press) to construct their geotherms. The strain in their sheared nodules is probably sufficient to mechanically mix layers of different compositions to form a homogeneous lherzolite as has been observed in comparable rocks from peridotite massifs. Therefore, it is not certain that the minerals which have been analyzed originally belonged to these sheared peridotites. This is not critical if it can be proved than they are now chemically equilibrated with the deforming peridotite.

The interpretation recently proposed by BOYD and NIXON (1973) and MACGREGOR (in press) mentioned above must be modified in the light of the conclusions reached in the preceding section. Two possibilities are envisaged. Either

(1) The general model is correct; the porphyroclastic and mosaic textured xenoliths represent the top of the moving asthenosphere. In this case, we think that the plate movement responsible for the development of porphyroclastic and mosaic textures is not continuous; it is a "staccato" movement as SHAW (1973) has implied for the plate movement below Hawaii with consequent basalt generation by shear heating. A comparable relationship should be expected with kimberlite generation, since we have argued that the xenoliths had to be extracted from the mantle during their deformation; or

(2) A diapiric intrusion is responsible for the distribution of xenoliths in the P, T diagram as proposed by GREEN and GUEGUEN (in press). The xenoliths above the kink of the normal geotherm belong to the diapir which, originating at a greater depth,

is hotter than the surrounding rocks at a given level in the mantle. The xenoliths coming from its margins can be deforming at any strain rate depending on the total velocity and the velocity gradient in the diapir. The generation of kimberlite is connected with the intrusion and the xenoliths can be incorporated in this fluid at any time during their deformation. This model fits better with our general constraints as well as with details like the observation of a sharp deformational gradient in a xenolith.

ACKNOWLEDGEMENTS

We thank I. D. MacGREGOR and P. H. NIXON who provided most of our samples and H. W. GREEN who critically reviewed the manuscript.

REFERENCES

AVE'LALLEMANT, H. G. and CARTER, N. L. (1970) Syntectonic recrystallization of olivine and modes of flow in the Upper-mantle. *Geol. Soc. Amer. Bull.* **81**, 2203–20.

BOUDIER, F. and NICOLAS, A. (1972) Fusion partielle gabbroïque dans la lherzolite de Lanzo (Alpes Piémontaises). *Bull. Suisse Miner. Petrogr.* **52/1**, 39–56.

BOULLIER, A. M. and NICOLAS, A. (1973a) Texture and fabric of peridotite nodules from kimberlite at Mothae, Thaba Putsoa and Kimberley. In *Lesotho Kimberlites* (editor, P. H. NIXON), pp. 56–66.

BOULLIER, A. M. and NICOLAS, A. (1973b) Texture and fabric of peridotite nodules from kimberlite at Mothae, Thaba Putsoa and Kimberley. *1st Int. Conf. Kimberlite, Extended Abstracts*, pp. 43–46.

BOYD, F. R. and NIXON, P. H. (1972) Ultramafic nodules from the Thaba Putsoa kimberlite pipe. *Carnegie Inst. Wash. Yearb.* **71**, 362–73.

BOYD, F. R. and NIXON, P. H. (1973) Origin of the lherzolite nodules in the kimberlites of Northern Lesotho. *1st Int. Conf. Kimberlite, Extended Abstracts*, pp. 47–50.

BROTHERS, R. H. and RODGERS, K. A. (1969) Petrofabric studies of ultramafic nodules from Auckland, New Zealand. *J. Geol.* **77**, 452–65.

CARTER, N. L., BAKER, D. W. and GEORGE, R. P. (1972) Seismic anisotropy, flow and constitution of the upper-mantle. In: *Flow and Fracture of Rocks* (the Griggs Volume, A.G.U.) **16**, 167–90.

COX, K. G., GURNEY, J. J. and HARTE, B. (1973) Xenoliths from the Matsoku pipe. In: *Lesotho Kimberlites* (editor P. H. NIXON), pp 76–100.

GREEN, H. W. II and RADCLIFFE, S. V. (1972) Deformation process in the upper-mantle. In: *Flow and Fracture of Rocks* (the Griggs Volume, A.G.U.) **16**, 139–56.

GREEN, H. W. II and GUEGUEN, Y. Kimberlite pipes: origin by diapiric upwelling in the upper mantle. Nature (in press).

HAYDEN, H. W., FLOREEN, S. and GOODELL, P. D. (1972) The deformation mechanisms of superplasticity. *Metall. Trans.* **3**, 833–42.

HARTE, B., COX, K. G. and GURNEY, J. J. (1973) Petrography and geological history of upper-mantle xenoliths from the Matsoku Kimberlite. *1st Int. Conf. Kimberlite, Extended Abstracts*, pp. 155–8.

KRETZ, R. (1966) Interpretation of the shape of mineral grains in metamorphic rocks. *J. Petrol.* **7**, 68–94.

LENSCH, G. (1971) Die Ultramafitite der Zone von Ivrea. *Annales Universitatis Saraviensis* **9**, 5–146.

MACGREGOR, I. D. The system $MgO–Al_2O_3–SiO_2$—solubility of Al_2O_3 in enstatite for spinel and garnet peridotite compositions. *Amer. Mineral.* (in press).

MERCIER, J. C. (1972) Structures des péridotites en enclaves dans quelques basaltes d'Europe et d'Hawaï. Regards sur la constitution du manteau supérieur. Thèse Nantes, 229 pp.

MERCIER, J. C. and NICOLAS, A. Textures and fabrics of the upper-mantle peridotites as illustrated by basalt xenoliths. *J. Petrol.* (in press).

NICOLAS, A., BOUCHEZ, J. L., BOUDIER, F. and MERCIER, J. C. (1971) Textures, structures and fabrics due to solid state flow in some European lherzolites. *Tectonophysics* **12**, 55–86.

NICOLAS, A., BOUCHEZ, J. L. and BOUDIER, F. (1972) Interprétation cinématique des déformations plastiques dans le massif de Lanzo (Alpes Piémontaises). Comparaison avec d'autres massifs. *Tectonophysics* **14**, 143–71.

NIXON, P. H., BOYD, F. R. and BOULLIER, A. M. (1973) The evidence of kimberlite and its inclusions on the constitution of the outerpart of the earth. In: *Lesotho Kimberlites* (editor P. H. NIXON), pp. 312–18.

SHAW, M. R. (1973) Mantle convection and volcanic periodicity in the Pacific; evidence from Hawaii. *Geol. Soc. Amer. Bull.* **84**, 1505–26.

15

THE EARTH'S MANTLE: EVIDENCE OF NON-NEWTONIAN FLOW

Robert L. Post, Jr. and David T. Griggs

During the last decade, it has been conclusively shown that flow on a truly massive scale is occurring in at least the earth's upper mantle. Evidence for this flow is the global pattern of continental drift and sea floor spreading. If a quantitative understanding of this flow is to be achieved, the rheological behavior of mantle rock must be known. The rheological behavior is characterized by a flow law relating the strain-rate field to the stress field and to environmental variables, notably temperature, hydrostatic pressure, and water content.

A key assumption almost universally made by investigators seeking quantitative models of flow in the mantle is that the earth's mantle behaves like a Newtonian fluid (*1–4*). This report challenges the assumption that the mantle is Newtonian. The data on postglacial rebound and experiments on hot creep of dunite both suggest non-Newtonian flow in the mantle.

The melting of the ice cap in Fennoscandia (Finland, Sweden, Norway, and Denmark) about 10,000 years ago left a surface depression that could be only partially removed by immediate elastic rebound. The history of the return toward isostatic equilibrium is shown in Fig. 1a, and the present rate of return in Fig. 1b. The depression has kept a nearly constant shape and the present rate of uplift is everywhere proportional to the amount of remaining depression. This leads us to believe that postglacial rebound belongs to a special class of behavior: "proportional relaxation" of stressed bodies. The ratio of the effective shear stress at any point to that at any other point remains constant with time, while the amplitudes of both stresses diminish asymptotically to zero. The validity of applying this concept to Fennoscandia is discussed below.

We assume that the mantle obeys a general flow law of the form

$$\dot{\varepsilon}_{\aleph} = B(z)\tau^{n} \qquad (1)$$

where $\dot{\varepsilon}_{\aleph}$ is the steady-state effective strain rate, τ is the effective shear stress, B depends only on depth (z), and n is a constant exponent which is to be determined from postglacial rebound (*5, 6*). The effective viscosity is

$$\eta_{eff} = 1/[2B(z)\tau^{n-1}]$$

which reduces to: $\eta_{eff} = \eta = 1/[2B(z)]$ for a Newtonian fluid ($n = 1$). It is assumed that the shear stress resulting from the depression is greater than shear stresses in the mantle due to other causes (for example, plate tectonics). If the latter stresses were much greater the effective n would be 1, no matter what the intrinsic value (*7*). Modern plate (and plume) theory yields small relative motion of Fennoscandia and the deep mantle in recent times, so in this locale it is believed that the assumption is justified.

In a material undergoing proportional relaxation and obeying Eq. 1, where B is a function of position but not of time, it can be shown that

$$\langle \dot{\varepsilon}_{\aleph} \rangle = A \langle \tau \rangle^{n} \qquad (2)$$

where $\langle \dot{\varepsilon}_{\aleph} \rangle$ and $\langle \tau \rangle$ are the averages over any finite space of the effective shear strain rate and stress, respectively, and A is a constant (*8*). It follows immediately for the case of postglacial uplift that $\dot{\zeta} \propto \langle \dot{\varepsilon} \rangle$ and $\zeta \propto \langle \tau \rangle$, so that

$$\dot{\zeta} = C\zeta^{n} \qquad (3)$$

where ζ is the amount of remaining uplift at the center of the uplifting area, $\dot{\zeta}$ is the rate of uplift at that point, and C is a constant which depends on $B(z)$ and the areal extent of the mass deficiency. It will be noted that Eqs. 1, 2, and 3 are equally valid for Newtonian flow ($n = 1$) and for flow of the type characteristically observed in crystalline solids at high temperatures where Eq. 1 is obeyed and $n \geqslant 3$.

We now consider the validity of proportional relaxation in postglacial uplift. If the mantle is assumed to be incompressible (with no elastic component), the stress distribution in the mantle must be consistent with the surface depression and also subject to the constraint that the total viscous dissipation be a minimum at all times. Provided that the flow law in the mantle has the form of Eq. 1, the shape of the stress distribution in the mantle will depend on the shape of the surface depression (including its horizontal extent) but not on the magnitude of the actual surface elevations. (This would not be

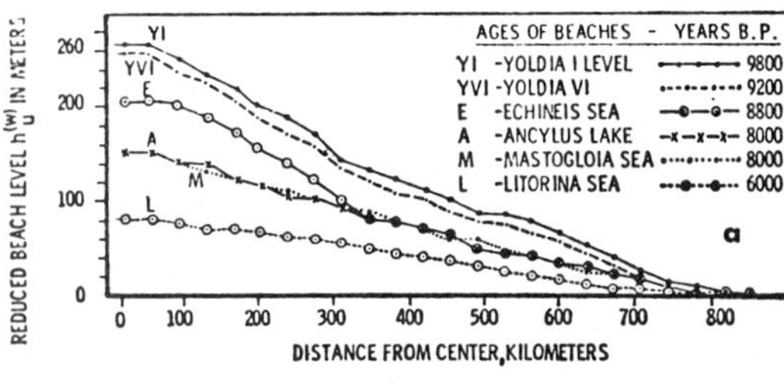

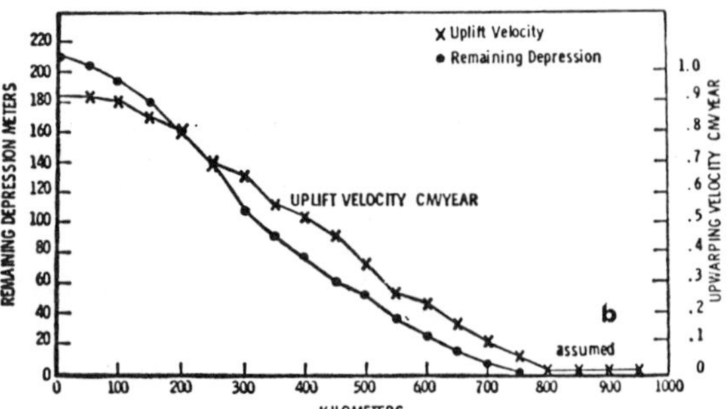

Fig. 1. (a) Beach levels and (b) remaining depression and present rate of uplift in Fennoscandia. Reproduced with permission of McConnell (*2*).

218

true if n were dependent on stress.) If a particular stress distribution in the mantle results in minimizing the total viscous dissipation for a particular shape of the surface elevation the stress distribution in the mantle resulting from a multiplication of the surface elevations by a constant factor F will result in a minimum of viscous dissipation if the original stresses are also multiplied by the same factor F. The viscous dissipations ($\tau \dot{\epsilon} dV$, where dV is a volume element) will all be multiplied by F^{n+1}. In other words, if the surface depression keeps a constant shape with time, as observed in Fennoscandia, the stress distribution in the mantle will also keep a constant shape, and proportional relaxation is achieved. Consequently, Eq. 3 should be valid for the Fennoscandian uplift, if an incompressible mantle is assumed (9).

To apply Eq. 3 to Fennoscandia, ζ is obtained by adding the remaining uplift to the elevation of past beach levels from Fig. 1a. The remaining central uplift has been estimated as 180 m (3), 210 m (10), and 150 m (11). For 180 m, the values of ζ are given in Table 1. In order to check these against Eq. 3, we integrate the latter

$$\zeta^{1-n} = \zeta_{\max}^{1-n} + C(1-n)t, \quad n \neq 1 \quad (4)$$
$$\ln\zeta = \ln\zeta_{\max} + Ct, \quad n = 1 \quad (5)$$

where $\zeta_{\max}$ is the central depression at $t = 0$, the time of glacier retreat (446 m).

Equation 5 applies for Newtonian viscous flow, in which case a plot of the logarithm of ζ against time should yield a straight line. Figure 2 shows that this is not true. Fitting Eq. 4 to these data to minimize the root-mean-square deviations yields

$$\zeta^{-2.21} = \zeta_{\max}^{-2.21} + 9.629 \times 10^{-10}t \quad (6)$$

which on differentiation yields

$$\dot{\zeta} = -4.36 \times 10^{-10}\zeta^{3.21} \quad (7)$$

A comparison of Eq. 6 with the observations is shown in Table 1 and Fig. 2. The present central uplift rate from Eq. 7 is 7.6 mm/year, compared to Kääriäinen's (12) estimate of 9 mm/year from geodetic leveling measurements over the last few decades.

Altering the estimate of the amount of uplift remaining from 180 m to 210 or 150 m does not affect the quality of the fit, but the preferred exponent n is changed from 3.21 to 3.58 and 2.85, respectively. The calculated rate of present uplift remains the same within 2 percent because the change in ζ is

Table 1. Observed and calculated amounts of central uplift (ζ) remaining at various times.

Time (years ago)	ζ (m)	
	Observed (2)	Calculated
9320*	446	446
9200	437	430
8800	385	388
8000	331	333
6000	261	260
0	180	180

* Corrected from McConnell's (2) value of 9800 years. Subsequent uplift indicates that removal of the ice cap was not complete 9800 years ago.

offset by the change in n. It thus appears that this proportional relaxation model of the Fennoscandian postglacial uplift requires non-Newtonian flow with a stress exponent of 3.2 ± 0.3 (13). Newtonian flow seems to be excluded by the curvature of the data in Fig. 2.

Postglacial rebound data also exists for a number of areas in Canada (14). An interpretation by the above method is, however, severely hampered by the lack of knowledge of the remaining uplift and lack of data at the centers of the uplifting regions. A preliminary analysis indicated non-Newtonian flow with the stress exponent $n \cong 3.5 \pm 1.0$ (15).

We now examine the consistency of the above finding for mantle flow with experimental evidence on the flow of presumed mantle rock. In recent years, the "plastic" flow behavior of a number of such rocks has been investigated at high temperatures and pressures

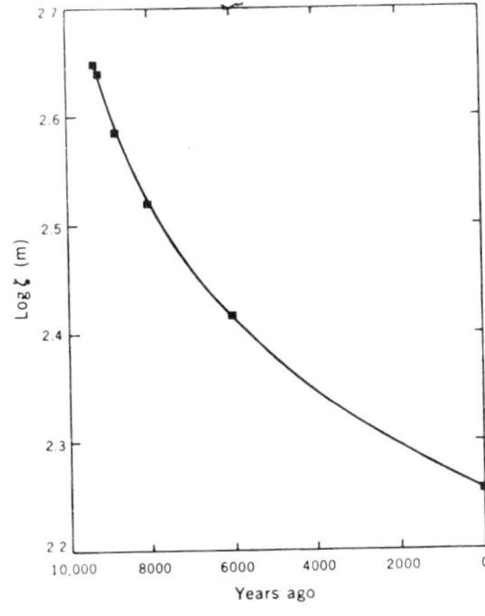

Fig. 2. Fennoscandia postglacial rebound: remaining uplift (ζ) versus time. (■) Observed values from Table 1 (7); (solid line) Eq. 6, where $n = 3.21$. Note that Newtonian flow would give a straight line on this plot for proportional relaxation.

(16, 17). Most experiments have involved compression, employing three types of deformation techniques: (i) constant strain rate; (ii) creep, or constant compressive stress; and (iii) relaxation (the advance of a deformation piston is stopped, and the nonhydrostatic stress in the sample is relieved through conversion of the elastic strain, in both the apparatus and the sample, into plastic strain in the sample). The deformational behavior of dunite (composed of olivine plus minor accessory minerals) is of particular interest, as olivine is likely to be the primary constituent of the upper mantle. Dunite, experimentally deformed at high temperatures and relatively low stresses, develops textures similar to those found in olivine-rich xenoliths believed to have come from the mantle (18).

Creep and relaxation experiments on Mt. Burnett dunite in both hydrous and anhydrous environments were performed by Post (15). Creep experiments in which steady-state flow was attained at a particular stress and relaxation tests at temperatures (T) of 1000° to 1600°K were fitted best by the following equation, when the nonhydrostatic compressive stress was less than 3 to 4 kbar

$$\dot{\epsilon}_s = 1.7 \times 10^9 \exp(-31.8T_m/T) \times \tau^{3.18 \pm 0.18} \quad (8)$$

where $T(°K)$ is the sample temperature; $T_m(°K)$ is the estimated melting temperature, as influenced by varying water content and hydrostatic pressure; and τ(kbar) and ϵ_s (sec^{-1}) are defined in Eq. 1. The activation energy of undried samples of Mt. Burnett dunite deformed in jackets of dehydrated talc is 93.1 ± 2.5 kcal/mole. In a completely anhydrous environment, the activation energy is 130 kcal/mole (15, 17).

The stress exponent of 3.18 ± 0.18 in Eq. 8 is in good agreement with the 3.2 ± 0.3 determined from the analysis of the uplift data. Thus, the available experimental and geophysical data agree in predicting a non-Newtonian mantle where $\dot{\epsilon}_s$ is proportional to τ^3. With an assumed profile of T/T_m in the mantle, and the assumption that the elastic stress field is similar to the actual field, Eq. 8 can be used to estimate the rheological behavior. The elastic stress distribution beneath the Fennoscandian region can be calculated from the observed depression. With that stress field the resulting central uplift rates predicted from Eq. 8 are within a factor of about 2 of the observed central uplift rates obtained from the raised beachlines and survey measurements (15). While agreement as close as this must be fortuitous, considering the number of assumptions involved in predicting the uplift rates, the experimental and geophysical data appear consistent with each other, and support the contention that the mantle, at least beneath shield areas, is non-Newtonian (19).

In conclusion, recent evidence indicates a non-Newtonian mantle with $n \simeq 3$, at least under Fennoscandia. This suggests that all studies of mantle motion, notably the search for the driving force for plate tectonics, and all inferences about the viscosity of the earth should endeavor to incorporate non-Newtonian flow of this kind.

References and Notes

1. N. A. Haskell, *Physics* 6, 265 (1935).
2. R. K. McConnell, Jr., *J. Geophys. Res.* 73, 7089 (1968).
3. F. A. Vening Meinesz, *Proc. Kon. Ned. Akad. Wetensch.* 40, 654 (1937).
4. L. A. Lliboutry, *J. Geophys. Res.* 76, 1433 (1971).
5. Equation 1 describes the creep behavior of most metals and ceramics at high temperatures and low to intermediate stresses (6). The value B varies directly with the diffusion coefficient for the rate-controlling atomic (or molecular) species. The stress exponent n ranges from 3 to 5 for most coarse-grained polycrystalline materials. Equation 1 is, however, equally valid for Newtonian flow.
6 O. D. Sherby and P. M. Burke, *Progr. Mater. Sci.* 13, 325 (1968); F. Garofalo, *Fundamentals of Creep and Creep Rupture in Metals* (Macmillan, New York, 1965).
7. J. F. Nye, *Proc. Roy. Soc. London Ser. A* 219, 477 (1953); *ibid.* 239, 113 (1957).
8. The proof of Eq. 2 reduces to proving that $\langle \tau^n \rangle = k \langle \tau \rangle^n$, where k is a constant. The definition of proportional stress relaxation is: $r = \tau(r,t) = \tau(r,0) f(t)$, where r is a vector defining position and $f(t)$ is the appropriate function of time, invariant with position. Consequently

$$\frac{\langle \tau^n \rangle}{\langle \tau \rangle^n} = \frac{\langle [\tau(r,t)]^n \rangle}{\langle \tau(r,t) \rangle^n} = \frac{\langle [\tau(r,0)]^n \rangle [f(t)]^n}{\langle \tau(r,0) \rangle^n [f(t)]^n} = k$$

9. The inclusion of an elastic component into a model of a non-Newtonian mantle is definitely a complicating factor. When the elastic component is taken into account, the viscosity values calculated from the 6000- to 9000-year-old raised beachlines (for an assumed Newtonian mantle) are about half the values that result when the elastic component is ignored [D. P. McKenzie, *Geophys. J. Roy. Astron. Soc.* 14, 297 (1967)]. The main reasons for concluding that the elastic component should not seriously affect the validity of Eq. 3 are that (i) the stress distribution resulting from a surface load on an elastic half-space is generally quite similar to that resulting from the same load on a nonelastic half-space [H. Jeffreys, *The Earth* (Cambridge Univ. Press, Cambridge, England, ed. 2, 1952), chapter 6]; (ii) the stress distribution in the elastic case keeps a constant shape if the surface load keeps a constant shape, as in the purely viscous case; and (iii) the stress distribution in a viscoelastic mantle should be intermediate between the purely elastic and purely viscous (both Newtonian and non-Newtonian) cases.
10. E. Niskannen, *Ann. Acad. Sci. Fenn. Ser. A* 53, 1 (1939).
11. W. A. Heiskanen and F. A. Vening Meinesz, *The Earth and Its Gravity Field* (McGraw-Hill, New York, 1958).
12. E. Kääriäinen, *Publ. Finn. Geod. Inst.* 42 (1953).
13. J. Weertman [*Rev. Geophys.* 8, 145 (1970)] concluded (p. 163) that for stresses greater than 0.01 bar, $n = 3$ in the mantle. He also showed that the effect of pressure can be taken into account by the expression $\exp(-GT_m/T)$ as in Eq. 8 (G is a constant). Lliboutry (4) realized that the uplift data required $n = 3$. He showed that van Bemmelen and Berlage's diffusion model gives $n = 3$ if it is assumed that Newtonian flow is confined to the asthenosphere. His treatment fails to account for the observed facts for two reasons. (i) The shape of the ice cap was more nearly circular than linear, as he assumed. The diffusion model gives $n = 2$ for a circular load. (ii) Ancient beach levels constructed from Lliboutry's specific model exhibit downwarping beyond 500 km, in conflict with the observations.
14. R. I. Walcott, *Rev. Geophys.* 10, 849 (1972); J. T. Andrews, *A Geomorphological Study of Post-Glacial Uplift with Particular Reference to Arctic Canada* (Institute of British Geographers, London, 1970), p. 117.
15. R. L. Post, Jr., thesis, University of California at Los Angeles (1973).
16. C. B. Raleigh and S. H. Kirby, *Mineral. Soc. Amer. Spec. Pap.* 3 (1970), p. 113; S. H. Kirby and C. B. Raleigh, *Tectonophysics*, in press. Recent relaxation tests on dunite give $n \simeq 3$.
17. N. L. Carter and H. G. Avé Lallemant [*Geol. Soc. Amer. Bull.* 81, 2181 (1970)] found $n = 2.4$ and 4.8 for wet and relatively dry dunite, respectively. The corresponding activation energies were 80 and 120 kcal/mole. It is believed that the current results are more reliable because the effect of transient (primary) creep was completely evaluated and the transition from low stress ($n = 3$) to high stress ($\dot{e} \propto \sinh \tau$) was determined by Post (15), but not by Carter and Avé Lallemant. Moreover, their experiments were done at constant strain rate rather than in creep or relaxation. When these factors are taken into account, the two sets of experimental data are consistent within the experimental error.
18. C. L. Ross, M. D. Foster, A. T. Myers, *J. Mineral. Soc. Amer.* 39, 693 (1954); E. D. Jackson and T. L. Wright, *J. Petrol.* 11, 405 (1970); H. G. Avé Lallemant and N. L. Carter, *Geol. Soc. Amer. Bull.* 81, 2203 (1970).
19. B. Gjevik [*Phys. Earth Planet. Interiors* 5, 403 (1972); *ibid.*, in press] has considered the possibility that mantle transitions might govern the rate of return to isostatic equilibrium. To first order, a mantle transition does not affect the shape of the flow field, but can exert a "back pressure" if the transition is significantly displaced from its equilibrium level by the flow. We believe that in this circumstance the back pressure would be proportional to the surface displacement, so that the conditions of proportional relaxation would be maintained. We estimate that the maximum back pressure would reduce the flow velocity by about 30 percent. As Gjevik points out, however, the dynamics of such transitions are difficult to analyze except in certain simple cases, so this effect will bear more examination.
20. The simple proof of Eq. 2 given in (8) was suggested by W. G. McMillan. N. L. Carter, B. Gjevik, and R. I. Walcott critically read the manuscript. We thank R. K. McConnell for permitting reproduction of his figures. Discussions with R. L. Shreve were stimulating and helpful. Publication No. 1191 of the Institute of Geophysics and Planetary Physics, University of California at Los Angeles. Supported by NSF grants GA-1394, GA-26027, and GA-36077x.

14 May 1973; revised 20 July 1973

Part V

MECHANICS OF THE LITHOSPHERE AND THE FORCES THAT DRIVE THE PLATES

Editor's Comments
on Papers 16 Through 20

The early theories on mantle convection (e.g. Papers 1 and 2), as well as many later studies, assumed or implied that the surficial plates were propelled by the flow of the underlying material, as if on top of conveyor belts. However, it is difficult to apply this approach because it is not very meaningful when the oceanic lithosphere is a part of the convective flow, being its cold boundary layer. An alternative approach is to treat the lithospheric plates as a separate subsystem and to study the forces arising in this system, and the forces acting on it as a result of mechanical interaction with the rest of the mantle. Such an approach allows us to discuss the forces that drive the plates independently of the flow in the mantle. One important force—the "slab pull"—results from the negative bouyancy of the descending lithospheric slabs, which arises because the slabs are colder than the surrounding mantle, and because the olivine-spinel phase transition within them occurs at a shallower depth than in the surrounding hotter mantle (Schubert and Turcotte, 1971; Toksoz, Minear, and Julian 1971; Toksoz, Sleep, and Smith, 1975; Ringwood, 1972; Schubert, Yuen, and Turcotte, 1975). Other forces arising in the plates can also influence their motions.

Elsasser (1969) suggested that because the lithosphere was much more rigid than the underlying asthenosphere it could act as a stress guide, so that negative buoyancy of the cold descending slabs could drag the plates and control their motions. An updated account of these ideas (1971) is given in Paper 16. In this paper Elsasser also introduced the concept of retrograde subduction, which describes the migration of the subduction zones, where the descending plate bends in relation to the underlying mantle. He does not dwell, however, on the actual extent of the tensile stresses in the lithosphere, which is required by his model.

Isacks and Molnar (Paper 17) tested these ideas by using the focal mechanisms of deep earthquakes as indicators of the state of stress in the descending slabs. They showed that when slabs do not reach great depths they are in a general state of tension; this is compatible with gravitational sinking of the slabs. When the slabs reach a depth of about 700 km, compression along their dips takes over, as if the slabs encounter some resistance. They also pointed out the importance of the olivine–spinel transition at 350–400 km. The apparent resistance to motion at depth may be related to the spinel–post-spinel phase transition at about 650 km, which can act in several ways: the denser material at depth may be more viscous and resist motion, or it may signify a compositional boundary (see below) or the thermodynamics of the phase change (negative dT/dp) may delay it in the slab and make its descent difficult. The state of stress in the descending slabs is a basic observation which must be accounted for by any acceptable model of plate dynamics.

In addition to the negative bouyancy of the descending slabs, the high-standing mid-oceanic ridges produce a force that pushes the plates (Hales, 1969; Lliboutry, 1969). Jacoby (Paper 18) estimated the magnitudes of these forces—the "slab pull" and the "ridge-push"— and showed that they are indeed large enough to overcome the viscous resistance of the underlying astenosphere. Hence these forces can actually drive the lithospheric plates. The pushing effect of the ridges was further discussed by Lister (1975) and Hager (1978).

Intraplate stresses have important implications for models of forces acting on the plates. Sykes and Sbar (Paper 19) used focal mechanisms of intraplate earthquakes to infer the stress fields within the plates. They found that compressive stresses are common, which shows that the pull of the descending slabs cannot be the only force that controls the motions of plates. Additional forces may originate from the push of ridges or from drag at the base of plates, but no one of these forces is dominant.

An important advance is the application of the laws of rigid-body mechanics to the lithosphere. As changes in plate motions are negligibly slow, the net torque acting on any single plate, and hence also on the entire lithosphere, is zero. This leads to a relation between the forces acting on the plates. Some forces (e.g., viscous drag at their base) are velocity dependent, so the known plate motions can be used to test and to constrain models of force systems acting on the plates.

Solomon and Sleep (Paper 20) used this approach to calculate the motion of the plates relative to the mantle, that is, the "absolute" plate motions. They considered the torque acting on the entire lithosphere, so the symmetric forces at ridges and transform faults cancel out; the remaining forces are viscous drag at the base of plates, which are assumed to be moving over a static mantle, and the pull at subduction zones. Given the relative plate motions, it is possible to deduce their motions relative to the mantle. These are found to be similar to the absolute motions of the plate relative to the hot spots (Paper 10), the results being quite insensitive to the driving mechanism assumed. This conclusion gives a physical interpretation to the "absolute" motions, but removes the rationale from the assumptions that plumes underneath hot spots drive the plates. A similar, but less detailed, calculation was done by Lliboutry (1974).

The papers included in this section introduce an approach that was developed in many later works. A more elaborate treatment is to consider the balance of torques acting on each plate individually. This procedure was followed by Harper (1975). Assuming a simple set of forces, he deduced plate motions, that compare favorably with the known relative motions. A more extensive system of forces was considered by Forsyth and Uyeda (1975). Using the "absolute" plate motions, they obtained the relative magnitudes of the various forces: the forces acting on the descending slabs, that is, their negative buoyancy and the viscous resistance to their motion, are much stronger than the others, but these two forces nearly cancel out, so the net pull on the lithosphere is similar to the other forces (e.g. push of ridges, resistance along transform faults, interaction between colliding plates, and "suction" of overthrusting plates toward ridges). Second in importance is the drag at the base of the lithosphere, especially under continents. Chapple and Tullis (1977) explored a somewhat different model. They estimated directly the slab-pull, and then determined the sign and magnitude of the other forces necessary to account for the plate motions. In this model there is a small resistive force at the base of continents, but elsewhere the drag at the base of the lithosphere is negligible; there is also a force that pulls overriding plates toward subduction zones. All other forces are insignificant in this model.

An important step of testing the models of plate driving forces was taken by Solomon, Sleep and Richardson (1975) and especially by Richardson, Solomon, and Sleep (1976, 1979) who related these models to the observable intraplate stresses, extending the work of Paper 13. Treating the plates as elastic shells they modeled the intraplate stresses on the assumption that they result from the action of the various driving forces on plate margins. In the best models, which are, however only in moderate agreement with the observations, the push at ridges is of comparable magnitude to the net pull at subduction zones. In addition, it must be assumed that the overriding plates are pulled toward subduction zones as strongly as the consumed plates. Models in which the asthenosphere passively resists plate motions are in better agreement with observed stresses than models in which the plates are driven by drag at their base.

In all these works the asthenosphere is assumed to oppose the plate motions, or it is practically decoupled from the lithosphere. This is compatible with the common estimates that the resisting forces that act on transform and convergent plate boundaries are relatively small and are overcome or nearly balanced by the ridge-push and slab-pull forces that drive plate motion. However, Davies (1977, 1978) pointed out that if the stresses acting on such boundaries have magnitudes on the order of a kilobar, which is possible, then the asthenosphere must provide the force to propel the plates. In such a case, different stresses act at the base of different plates. That plate motion is resisted by asthenospheric drag (i.e., it drives the flow at depth) is also compatible with the properties of thermal convection. The negative buoyancy of the material cooled at the surface and its descent are as important for driving the flow as the buoyancy of the ascending material that is heated at depth. This is clearly seen in simple models such as those of Paper 5 in which the upper boundary layer moves faster than the underlying material and thus, in a sense, it drives the motion of the interior of the cell. This is especially pronounced when heating from within is important and the cold boundary layer is accentuated (Paper 6). Thus, the negative buoyancy arising in the cooling plates not only propels the plate themselves, but it also overcomes the viscous drag at their base and helps to drive the sub-lithospheric mantle. It is desirable, therefore, to relate the dynamics of plate motions to the underlying flow as was done by Richter (1973, 1977) and Richter and McKenzie (1977). Their results are similar to those of works that treated the plates alone. However, some complications can be seen. Richter (1973) and Richter and McKenzie (1977) showed that the sinking slabs can also excite a flow under the overriding plates, and the shear in this flow is such that it contributes to the motion of these plates toward the subduction zones. In a study of the possible coupling between mantle convection and overlying, rigid, moving plates, Lux et al. (1979) found that the flow may move the plates in some circumstances, especially if whole-mantle convection takes place (this is a controversial topic: see below). These studies show that lithospheric plates may interact with the asthenosphere in different ways, so this problem clearly needs more study.

It appears, therefore, that despite much work various aspects of the mechanism that drives plate are not yet fully understood. The importance of forces arising in the lithosphere itself, mainly the pull of the descending slabs and the push of the mid-oceanic ridges, is well established. Quite good models can be derived, but important problems remain. Thus, most models include an important trenchward pull of the overriding plates, but the origin of this force is not clear; in particular it is not clear whether it can be related to small convection cells excited by the descending slabs. The nature of the lithosphere-asthenosphere interaction is still incompletely known. Another problem is that there is no detailed model of the stresses in the subducted plates where they bend and descend into the mantle. Knowledge of the stress distribution in these areas is needed in order to understand how the slab pull is actually transformed into a horizontal force. Finally, there is not yet a satisfactory explanation of the intraplate stresses, though they provide an important observational constraint on any model of plate dynamics. Nevertheless, in spite of these difficulties, it seems that now the problem is well posed and the dynamics of plate motions is understood in principle.

In the discussions of forces driving the plates, their "absolute" motion relative to the underlying mantle has a fundamental significance. The near coincidence of such motions as deduced from mechanical models with the motions relative to the hot spots is a strong argument for the validity of the approach. These motions have several noteworthy features (Solomon, Sleep, and Richardson, 1975; Forsyth and Uyeda, 1975). Plates with large descending slabs attached to them (which now are mostly oceanic) move faster than other plates, the motion being toward the subduction zones. The overriding plates also move toward suduction zones, but much more slowly. Plate boundaries generally move relative to the mantle, but the absolute velocity is such as to minimize the motions of the plate boundaries (Kaula, 1975). At present, the largely continental plates move much slower than the oceanic plates. However, modelling (Jurdy, 1978) and paleomagnetic data (Gordon, McWilliams and Cox, 1979) show that in the geologic past this was not so, and continents moved as fast as the oceanic plates move at present. Because of these properties of absolute plate motions, their study is of great interest and can yield much insight into the behavior of the plates.

REFERENCES

Chapple, W. M., and Tullis, T. E., 1977, Evaluation of the Forces that Drive the Plates, *Jour. Geophys. Research* **82:**1967–1984.

Davies, G. F., 1978, The Roles of Boundary Friction, Basal Shear Stress and Deep Mantle Convection in Plate Tectonics, *Geophys. Res. Lett.* **9:**161–164.

Davies, G. F., 1977, Viscous Mantle Flow under Moving Lithospheric Plates and under Subduction Zones, *Royal Astron. Soc. Geophys. Jour.* **49:**557–563.

Elsasser, W. M., 1969, Convection and Stress Propagation in the Upper Mantle, in *The Application of Modern Physics to the Earth and Planetary Interiors,* S. K. Runcorn, ed. Wiley-Interscience, New York, pp. 223–246.

Forsyth, D. and S. Uyeda, 1975, On the Relative Importance of the Driving Forces of Plate Motion, *Royal Astron. Soc. Geophys. Jour.* **43:**163–200.

Gordon, R. G., M. O. McWilliams and A. Cox, 1979, Pre-Tertiary Velocities: A Lower Bound From Paleomagnetic Data, *Jour. Geophys. Research* **84:**5480–5486.

Hager, B. H., 1978, Oceanic Plate Motions Driven by Lithospheric Thickening and Subducted Slabs, *Nature* **276:**156–159.

Hales, A. L., 1969, Gravitational Sliding and Continental Drift, *Earth and Planetary Sci. Letters* **6:**31–34.

Harper, J. F., 1975, On the Driving Forces of Plate Tectonics, *Astron. Soc. Geophys. Jour. Royal* **40:**465–474.

Jurdy, D. M., 1978, An Alternative Model for Early Tertiary Absolute Plate Motion, *Geology* **6:**469–472.

Kaula, W. M., 1975, Absolute Plate Motions by Boundary Velocity Minimizations, *Jour. Geophys. Research* **80:**244–248.

Lister, C. R. B., 1975, Gravitational Drive on Oceanic Plates caused by Thermal Contraction, *Nature* **257:**663–665.

Lliboutry, L., 1969, Sea-Floor Spreading, Continental Drift, and Lithosphere Sinking with an Asthenosphere at Melting Point, *Jour Geophys. Research* **74:**6525–6540.

Lliboutry, L., 1974, Plate Movement Relative to Rigid Lower Mantle, *Nature* **250:**298–300.

Lux, R. A., G. F. Davies, and J. H. Thomas, 1979, Moving Lithospheric Plates and Mantle Convection, *Royal Astron. Soc., Geophys. Jour.* **58:**209–228.

Richardson, R. M., S. C. Solomon, and N. H. Sleep, 1976, Intraplate Stress as an Indicator of Plate Tectonic Driving Forces, *Jour. Geophys. Research* **81:**1847–1856.

Richardson, R. M., S. C. Solomon, and N. H. Sleep, 1979, Tectonic Stress in Plates, *Rev. Geophys. Space Phys.* **17:**981–1019.

Richter, F. M., 1973, Dynamical Models for Sea-Floor Spreading, *Rev. Geophys. Space Phys.* **11:**223–287.

Richter, F. M., 1977, On the Driving Mechanism of Plate Tectonics, *Tectonophysics* **38:**61–88.

Richter, F. M., and D. McKenzie, 1978, Simple Plate Models of Mantle Convection, *Jour. Geophys.* **44:**441–471.

Ringwood, A. E., 1972, Phase Transformations and Mantle Dynamics, *Earth and Planetary Sci. Letters* **14:**233–241.

Schubert, G., and D. L. Turcotte, 1971, Phase Changes and Mantle Convection, *Jour. Geophys. Research* **76:**1424–1432.

Schubert, G., Yuen, D. A., and D. L. Turcotte, 1975, Role of Phase Transitions in a Dynamic Mantle, *Royal Astron. Soc. Geophys. Jour.* **42:**705–735.

Solomon, S. C., N. H. Sleep, and R. M. Richardson, 1975, On the Forces Driving Plate Tectonics: Inferences from Absolute Plate Velocities and Intraplate Stress, *Royal Astron. Soc. Geophys. Jour.* **42:**769–801.

Solomon, S. C., N. H. Sleep, and D. M. Jurdy, 1977, Mechanical models for absolute plate motions in the early Tertiary, *Jour. Geophys. Research* **82:**203–212.

Toksoz, M. N., J. W. Minear, and B. R. Julian, 1971, Temperature Field and Geophysical Effects of Downgoing Slab, *Jour. Geophys. Research* **76:**1113–1138.

Toksoz, M. N., N. H. Sleep, and A. T. Smith, 1975, Evolution of the Downgoing Lithosphere and the Mechanisms of Deep Focus Earthquakes, *Royal Astron. Soc. Geophys. Jour.* **35:**285–310.

16

Reprinted from *Jour. Geophys. Research* **76**:1101–1112 (1971)

Sea-Floor Spreading as Thermal Convection

WALTER M. ELSASSER

MODEL ASSUMPTIONS

Past analysis of sea-floor spreading and continental drift has been mainly kinematical. Here we outline a somewhat specialized model that we hope will assist us in advancing toward a more dynamical theory.

The driving forces of any circulation in the mantle are both thermal and chemical, but the proportion in which these contribute is still largely unknown; very likely the thermal effects dominate. The thermal processes can be considered reversible (in a mechanical but not in a rigorous thermodynamical sense), whereas effects due to chemical differences (rising of lighter material into the crust) are radically irreversible. Some authors have presumed that convective motions in the upper mantle are caused primarily by heating from below, but there is no direct evidence for this. A process that we know must operate is the cooling of the topmost layers of the oceanic mantle by simple upward conduction of heat. Since the ocean bottom is practically at a constant temperature, it furnishes a boundary condition for the temperature distribution in the oceanic upper mantle. Because of their radioactivity, the continents, on the other hand, act as heating pads. This heating effect is probably confined to the topmost few hundred kilometers.

The resulting temperature gradients must generate convective motions in the upper mantle. The simplest assumption would be that the material under the continents rises toward the oceans under a slant, but observation shows that exactly the reverse motion occurs. The top stratum of the mantle under the oceans, the oceanic lithosphere, which is cold, goes down under a slant toward a region that is usually below a continent. What this amounts to is a sort of upside-down convection in which motion is engendered by a cold top layer sinking rather than by a hot bottom layer rising. The thermodynamical situation is schematically represented in Figure 1. The assumption made here is that a certain instability (a steeper-than-adiabatic temperature gradient) is first produced by thermal conduction before large-scale convective motions begin. The more usual case, in which heating occurs from below, is shown by curve I; the alternate case here proposed, in which there is cooling from above, is illustrated by curve II.

Quantitative information about mechanical properties of the upper mantle was first obtained from studies of the post-glacial uplift, for which the upper mantle was treated as a homogeneous, extremely viscous fluid. Recently, *Walcott* [1970] has used a more refined model: in it the lithosphere is a plate assumed to respond to forces by elastic flexure; the underly-

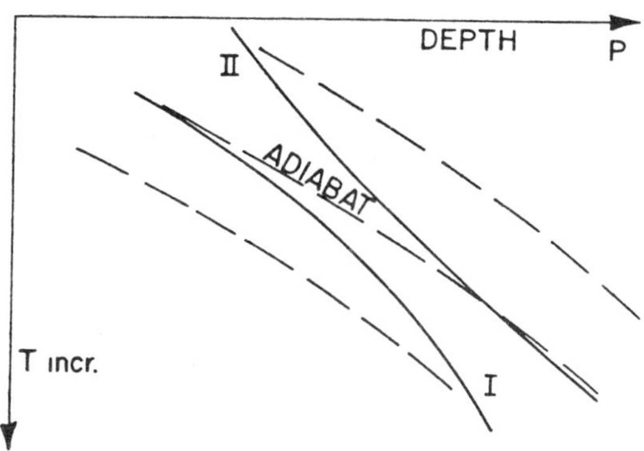

Fig. 1. Comparative temperature-depth distributions for heating from below and for cooling from above.

ing asthenosphere yields much more readily, and it enters into Walcott's calculations only by way of the buoyancy forces it exerts on the lithosphere, not through its 'viscosity.' From Walcott's data, the most likely thickness of the lithosphere is 110 km under continents and >75 km under oceans. Long ago the geologist *Barrell* [1914–1915] first inferred the existence of a soft layer at some depth and created the terms lithosphere and asthenosphere.

Since the deformation of the upper mantle is by creep, it is hard to treat mathematically except in the special case where the relationship between stress and strain rate is linear, for which the equations of motion reduce to the Navier-Stokes equations of hydrodynamics. This form of relationship has not infrequently been assumed in the past, but *Orowan* [1965] and then later again *Weertman* [1970] have indicated in detail that with the material of the mantle the predominant creep mechanism is likely to be a nonlinear one. Figure 2 has been drawn from Weertman's data, the ordinate being the equivalent Newtonian viscosity, $\eta_N = \sigma/\dot{\epsilon}$, in the nonlinear case not a constant but a function of ϵ even for steady-state flow. (A change has been made from Weertman's graphs in that we assume a possible strain rate of 10^{-14} sec^{-1} in the mantle, as against Weertman's 10^{-16}, which is likely to be a good average over the whole earth. Linear creep of the Nabarro-Herring type is shown by the dashed curves for two different grain sizes. Weertman's type of creep, on the other hand, is independent of grain size.)

The very steep decline of the equivalent viscosity with depth, as shown in Figure 2, results

simply from the rapid increase of the temperature with depth that is generally admitted to occur in the lithosphere. The creep viscosity, in turn, varies exponentially with temperature. We may therefore expect that the low viscosity of the asthenosphere (assumed to extend from 100 to 300-km depth) is a general property of most if not all creep models of the upper mantle. Roughly, the viscosity of the asthenosphere is here of order 10^{19} to 10^{20}. If this is so, the lithosphere is only very loosely coupled to the middle and lower parts of the mantle so far as horizontal shear forces are involved.

Undoubtedly, however, the assumptions that lead to the curve of Figure 2 represent an oversimplification. To obtain this curve, Weertman first assumes that the atomistic basis of creep lies in diffusion-type processes. One can hardly quarrel with this assumption. Next, however, he assumes that the diffusion coefficient varies according to the simple formula

$$D = D_0 \exp\left(-\gamma T_m/T\right)$$

where T_m is the melting point and D_0 and γ are constants for any one given substance. This empirical formula has been known to metallurgists for some time and is very well satisfied by a dozen or so metals for which it has been experimentally checked. On the other hand, metals form monatomic crystals; one may well ask whether, for compounds of some complexity, the extreme simplification of a thermodynamical relationship of two independent variables (p, T) to one variable (T_m/T) is at all legitimate. *Bridgman* [1949] long ago made experiments that throw some light on the pressure dependence of the viscosity of liquids: He showed that for a very simple, monatomic liquid such as mercury the viscosity changes very little with pressure; on the other hand, for organic liquids whose molecules are of even moderate structural complexity, the effect of pressure on viscosity is tremendous, the viscosity may change by many powers of ten over a pressure range of the order of some kilobars. From the viewpoint of our physical intuition, this behavior is plausible enough.

One can expect that the diffusive behavior of crystalline rocks is intermediate between that of metals and that of complex organic compounds. The use of the preceding formula for estimating the diffusion coefficient and hence the creep rate in the asthenosphere will appear doubtful. It seems likely that the familiar breakdown under pressure of silicates, from complicated structures to simple, cubic ones, tends to bring the creep-viscosity of crystals closer to that of metals for which the above formula holds. But it would be most difficult to make any quantitative estimates. There are, however, other purely chemical phenomena that should occur at the depth of the asthenosphere and that should lead to a very pronounced lowering of the creep viscosity. We have in mind primarily the action of water which, in certain circumstances, can have a powerful effect as a 'plasticizer' or 'fluxing agent' (according to a number of petrologists, carbon dioxide has similar effects, but we shall not explicitly consider the latter here). Under various thermodynamical conditions, water can simply be incorporated into a lattice as crystal water, leading sometimes (e.g., in serpentine) to a very much softened rock. More important now is the fact that water frequently depresses the melting point of magmas. If even a small fraction of magmatic melt exists in the asthenosphere, the average viscosity may well be lowered by several powers of ten.

Though the existence of an asthenospheric layer with low creep viscosity was, to begin with, deduced from purely empirical, geological observations, there are now a number of general, geophysical observations that make the assumption of such a nearly singular, 'soft' layer almost inevitable. The leading argument derives from the conclusion that the near-surface, directly measured temperature gradient, which is steeper than the melting-point gradient, must necessarily diminish as one goes farther down, for several reasons too well known to be detailed.

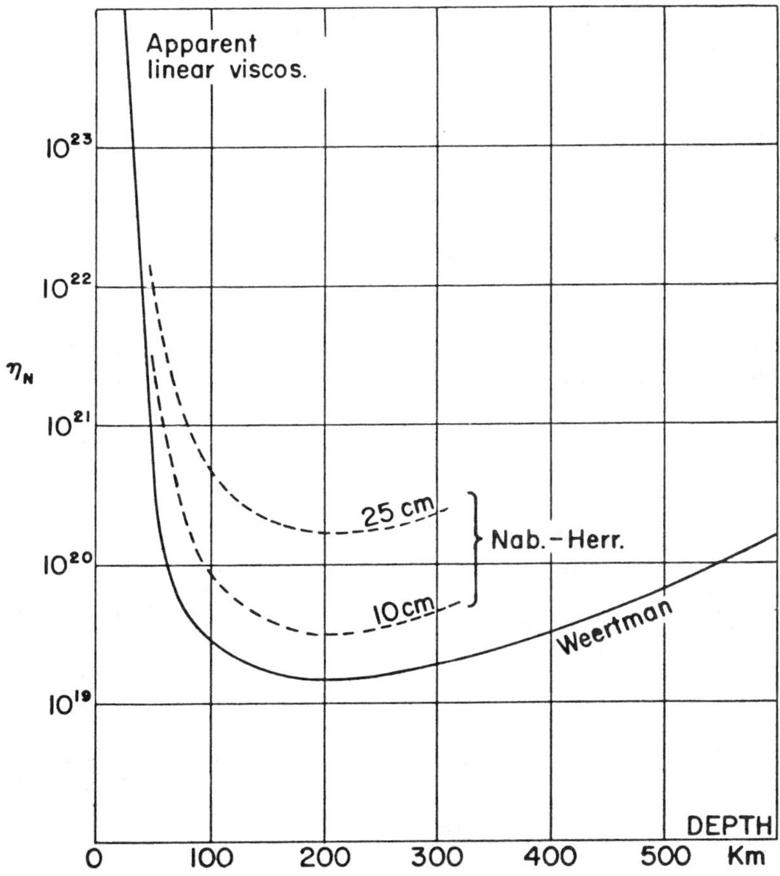

Fig. 2. Equivalent Newtonian viscosity as a function of depth. After *Weertman* [1970], modified.

231

Hence there is a region of closest approach to the melting point of mantle materials; this can be taken as a physical definition of the asthenosphere, as distinct from that drawn from purely geological inferences. Even a small degree of partial melting in such a layer will have a decisive influence upon its effective creep viscosity.

It is tempting to identify the asthenosphere with the seismic low-velocity layer. *Press* [1970] has recently described data which indicate that the drop in the transverse seismic velocity is quite sharp at a depth just below 70 km. He thinks that at this depth one reaches the solidus, that is, the boundary between completely solid material and a phase containing fractions of melt. The complete identification of the asthenosphere, the layer of reduced creep viscosity, with the low-velocity layer, the layer of reduced seismic velocity, seems an unwarranted speculation, since there is a factor of about 10^{12} between seismic velocities and creep velocities. Nevertheless, it is apparent that, though not identical, the two layers are very closely related. *Lambert and Wyllie* [1968], two petrologists who have studied these questions, believe that one can 'provide a model for the low-velocity zone of a layer of anhydrous minerals with interstitial hydrous silicate magma, sandwiched between two crystalline layers in which water is contained in hydrous minerals. Such a zone would have many of the characteristics necessary to current concepts of diapiric movements in magma genesis and in ocean-floor spreading, and of a decoupling zone between the lithosphere and the mantle.'

Though none of these observations has yet reached the stage of a definitive proof of a very low viscosity in the asthenosphere, we have tried to show that the inducement to assume this is very strong indeed. Mechanical softness will not, of course, prevent the asthenosphere from contributing to isostatic readjustments, by vertical motion. But one can presume, then, that no strong horizontal stresses can be transmitted by shear from the asthenosphere to the lithosphere. It follows that if strong horizontal stresses of either sign are observed in the lithosphere, they must have been generated in the lithosphere itself. This requirement acts as a severe constraint on the possible models of thermal convection in the upper mantle.

HORIZONTAL STRESS AND STRAIN

We shall now be concerned with the horizontal stresses and strains in the lithosphere. These can be either compressive or tensile. These two types require rather different treatment for several reasons on which we shall not dwell now. Compressive stresses have often been held responsible for the folding of mountain belts, but it is quite possible, and in fact more likely on mechanical grounds, that such folding is largely produced by accretion of roots from below [*Elsasser*, 1970]. In addition gravity sliding [*Hubbert and Rubey*, 1959] gives rise to compressive features. All such events are observed primarily on the continents, whereas simpler and largely tensile features are most noticeable on the ocean bottoms (though extensive tensile features are also observed on land; these will be mentioned presently). Here we are dealing with the tensile features, leaving the discussions of horizontal stresses that are compressive, or purportedly so, for a subsequent publication. We are trying, so far as we know for the first time, to sketch a mechanism that can generate large-scale horizontal tension on the earth; it seems clear that, if there is some truth to this model, many traditional ways of thinking in terms of compression will have to be abandoned.

That submarine ridges are dominated by tensile features has been noted by *Menard* [1969] and also stressed by *Orowan* [1969]. According to Orowan, it is nearly impossible on mechanical ground to conceive of a curtain of hot material rising and pushing the lithosphere apart. In the first place, such a curtain would not be stable; it would break up into a series of rising columns. The frequent offsets of the center lines of ridges connected by transform faults, sometimes fairly long ones, would require that the rising curtain be similarly offset or at the least have a pronouncedly sinuous shape already below the lithosphere, but no clear physical reason for this has been given. On the other hand, all these features become readily understandable if the formation of a ridge is contingent upon the existence of a horizontal tension in the lithosphere. One feature, in particular, seems to make an interpretation in terms of horizontal tensile stress or strain imperative: This is the symmetry of the magnetic lineations first discovered by *Vine and Matthews* [1963] and since then so amply confirmed by numerous similar observations. It seems hardly conceivable that this symmetry could be understood if the basaltic material that carries the lineations were pushed up and then sideways by forces acting locally or else below the lithosphere. Any such problems disappear on assuming that the lithospheric plate is spread apart by horizontal tension while it rests on an asthenosphere of relatively low viscosity. The lithosphere will be brittle only in its upper part (say 20–30 km). Below this it will react plastically, the more so the higher the temperature. This reaction will express itself by 'necking' in the hottest region. But the hottest part in a horizontal traverse of the ridge is where material has come up most recently; hence the lithosphere will continue to neck and split at the same material location of the plate [see also *F. C. Frank*, 1968]. If the asthenosphere has a low enough viscosity, the process of splitting and consequent outpouring of crustal material will be dominated by the mechanical condition of the lithosphere and will be symmetrical in a coordinate system fixed in the latter; it should be independent of the relative motion of lithosphere and asthenosphere.

Moreover, if the material coming out at or near the ridge crest is hot and ductile only to begin with and is later cooling and becoming hard, the well-known 'median' position of some ridges, such as the Atlantic ridge, which is centered between the continental shelves, becomes intelligible: This median position is then the 'frozen-in' remnant of the historical process of sea-floor spreading in the Atlantic; it is not the product of later plastic adjustments. Similarly, the curved ridge that surrounds South Africa at a roughly constant distance can be understood as representing the contemporary state of this ridge in the process of its gradual recession from the shores of Africa.

Models of this type lead to serious geometrical difficulties only when one assumes that the material of the lithosphere remains constant. However, the lithosphere steadily loses material at the trenches and gains material at the ridges, and over geological periods the aggregated amount is not small compared to the total mass of the lithosphere. Some geometrical rearrangements of the lithosphere will be required, mainly by rotation of 'plates' [*Morgan*, 1968; *Isacks et al.*, 1968], but also by some predominantly horizontal, observed plastic deformations.

We wish to be clearly understood here: We do not claim that a state of horizontal tensile stress and strain of the lithosphere is universal; it is merely a very widespread condition. The stresses and strains are of variable strength over the earth. There appears to be enough freedom so that the rearrangements required of plates of the lithosphere can lead to extensive horizontal, 'transcurrent' shears, as well as to the incidental development of conditions in which the horizontal stresses and strains are compressive rather than tensile. The present model is an elaboration of the notion of the upper mantle as a horizontal 'stress guide' [*Elsasser*, 1969], a concept based on the empirical observation that transcurrent faults at the ocean bottom often run over very long distances, especially in the Pacific (e.g., the Clipperton fault can be followed for more than 90° along nearly a great circle).

Evidence of tensile stress or strain is not confined to the ocean bottom or to the neighborhood of the large ridges. One well-known example of horizontal tension on a continent is offered by western North America. The East Pacific rise can be followed into the Gulf of Mexico, and there is little doubt that the narrow strip of Baja California has been split off the mainland by tensile strains. The region of the continent to the north of the Gulf, the Basin and Range Province of the western United States, is full of geological evidence expressing tensile features. It has often been said that this region is most likely a northerly extension of the East Pacific rise. If this is true, it must seem likely that the tensile strain is propagating to the north by necking of the lower and hotter part of the lithosphere.

The best known and most extensive continental region of horizontal strain is in East Africa and the adjacent territory (East African Rift Valley, Red Sea). In view of this state of affairs, we shall assert without going into geological details that the low-latitude part of the lithosphere from the eastern coast of South America all the way to the eastern side of the Indian Ocean seems almost entirely tensile.

Demonstrably tensile regions on land therefore cover a non-negligible part of the continental surface. The amount of area covered by ridges (if these are considered tensile throughout) can be estimated: The over-all length of ridges is about 7.10⁴ km; if we assume an average width of 10³ km, we obtain an area of 7.10⁷ km², or 14% of the earth's surface. If we add to this the areas of continents or ocean demonstrably tensile or very likely so, we see that a significant fraction of the entire earth's surface is under tensile stress. Now it is very hard to doubt that such horizontal strains as have for instance produced the Red Sea or the Gulf of California, in recent geological times, have also acted in the more remote past to produce the splitting of the original Gondwana continent.

CAUSES OF LITHOSPHERIC TENSION

We next proceed to a discussion of mechanisms that may produce large-scale horizontal tension. On searching for such mechanisms, one finds that the alternatives are severely limited by our requirement stated in the beginning, namely, that the asthenosphere be soft enough so that it cannot transmit significant horizontal stresses. The mechanism to be discussed is based on the existence of a pattern of flow in the mantle, or at least in the upper mantle, which is a form of thermal convection. In the Appendix we report on the hypothesis advanced by some serious investigators that the earth as a whole is expanding (a cosmological effect).

Our present knowledge of the dynamics of the mantle, limited as it may be, allows us to say conclusively that, if there is a convective circulation, lithospheric material goes down along the Gutenberg-Benioff faults and the lithosphere is in turn replenished by material that rises at the center clefts of the submarine ridges. This does not as yet specify a pattern of the flow at depth. Under the conditions schematized in Figure 1 (cooling from the top), most of

the convection-generating forces are exerted by parts of the oceanic lithosphere that have been cooled at the free surface, which is being kept at the constant temperature of about 4°C; this relatively cold material then sinks down and thereby serves as the primary driving agency of the circulation. As the material sinks, it will create a potentially vacant region in the lithospheric plate, but a vacancy will not really arise; material flowing in horizontally will exhibit tensile strains and stresses.

It is well known that all trenches and their associated Gutenburg-Benioff faults are found at or very near the margin of the Pacific Ocean. With the exception of the Java trench and its fault zone, which are practically perpendicular to this margin, all the slabs go down so as to slant away from the Pacific. If the Atlantic and the Indian Ocean have opened up since the Mesozoic, it follows that of necessity the Pacific must have shrunk by a corresponding area since that time.

We are therefore confronted with the curious kinematic situation illustrated in Figure 3. If the lithospheric plate goes down toward P_3, there is a one-parameter manifold of motions by which this can be realized. Two limiting cases of this are as follows: In motion V_1, the horizontal part P_1 of the lithosphere slides to the left and then down; that is, P_1 and P_3 slide together like a conveyor belt. There is no force, then, on P_2. The other limit is given by motion V_2, the chief part of which is described by the slanting arrows in Figure 3. The upper one of the two arrows is such as to carry the rhomboid $ABCD$ into the rhomboid $DEFG$. The lower arrow is parallel to the first. On such a motion, not only the lithosphere at P_2 but also the asthenosphere underneath it must be pulled to the right. This will be discussed in somewhat more detail presently. We can now write for the general motion at or near the fault zone, in a linear approximation,

$$V = \alpha V_1 + \beta V_2 \qquad (1)$$

We must, in particular, realize that the ratio β/α cannot be small, i.e., the motion V_2 must be of appreciable magnitude. This follows quite simply from the inferred shrinkage of the Pacific area during the last 100–200 m.y. The shrinkage can only be understood if the motion has a large V_2, that is retrograde, component.

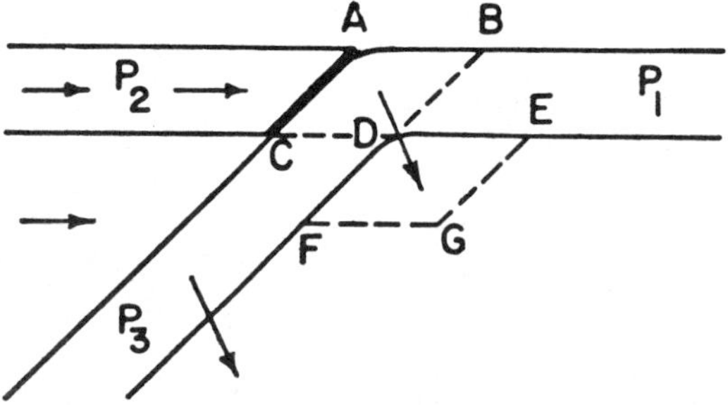

Fig. 3. 'Retrograde' motion of a lithospheric slab at a Gutenberg-Benioff fault zone.

Although we do not yet fully understand the dynamics of this motion, the fact that such a pattern must exist can hardly be questioned.

Now though one can feel sure of the current interpretation so far as the loss of material from the lithosphere near trenches and the gain of material in the middle of ridges is concerned, the detailed mechanisms are still very obscure. There seems to be general agreement on one point: that the descending slab is pulled down by its own gravity, it being colder than its surroundings at all levels. As was just noted, the descent of the material is achieved by combining the forward or conveyor-belt motion V_1 with the retrograde motion V_2; in general, neither of the two will be small. The retrograde motion, in particular, will require more dynamical analysis: The downgoing slab must push the asthenospheric material on the right of it (see Figure 3) farther to the right; this is equivalent to a horizontal compressive stress acting on the oceanic asthenosphere. One may then suspect that in the retrograde motion there could develop a cleavage to the left of the descending slab; mantle material could rise there. But nothing of the sort is observed; instead, the cleavage and ascent of material occur at large distances, under ridges.

While we cannot as yet provide a satisfactory dynamical model of these phenomena, some remarks make them appear less strange: The tensile stresses that are clearly required to pull the material on the left of Figure 3 toward the regressing slab are deviator stresses only. At all but the smallest depths, they are small compared to the locally prevailing hydrostatic pressure; thus the horizontal tensile stress is a small anisotropic deficit of a static compression.

This of course does not give any reason why a tearing apart and filling with material that rises from below does actually occur much farther away. One, and so far the only, argument we have found for such a behavior lies in the fact that, in the region of ridges, usually rather far away from trenches, the tensile stresses generated by several, or all, of the descending slabs are superposed. If a horizontal pull is exerted at the edge of a plate, the stress generated will decrease only slowly with distance from the edge (for an infinitely long edge of a plane plate, the stress will be independent of distance). On the other hand, if the stress-

to-strain-rate relationship is nonlinear (Weertman suggests $\dot{\epsilon} \sim \sigma^3$), it seems possible that strain is larger in the region of mid-ocean ridges than in the very neighborhood of trenches. These arguments are at present highly speculative and require confirmation by more quantitative analysis, but airing them seems of some benefit because of the past intense fixation of geological reasoning on compressive stress patterns.

RETURN FLOW

Figure 4 is a highly schematic diagram of worldwide thermal convection. The margin of the Pacific has been replaced by a cone (of opening half-angle at the earth's center, $\theta = 70°$). The material descending at the conical fault zone can then be swept toward the upper part of Figure 4, located, broadly speaking, near the region beneath Africa. This requires the lower layers of the mantle to have not too large a creep viscosity, so that rearrangement of material is possible over some millions of years. *Goldreich and Toomre* [1969], in their detailed analysis of polar wandering, have discussed this question of the lower-mantle creep viscosity and arrived at an upper limit for the kinematic viscosity, η/ρ, of about $1\cdot5 \times 10^{25}$ cgs units, but they add at once that this is probably an order of magnitude too large. With an effective

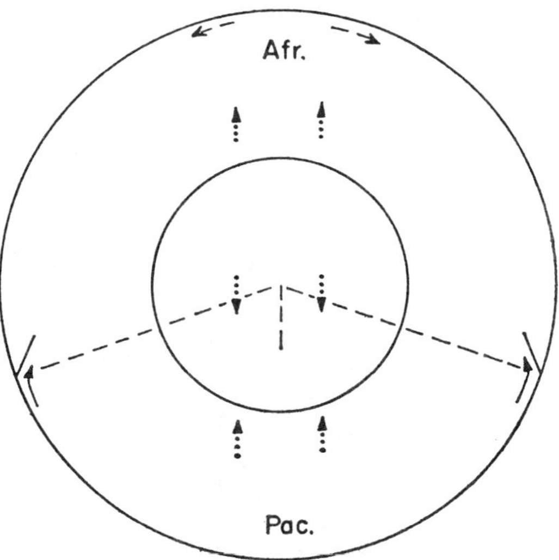

Fig. 4. Circum-pacific fault zones schematized by a cone of 70° opening angle. Dotted arrows are motions in Orowan's 'transvection' model.

ordinary viscosity of 10^{24} or even somewhat more, the completion of a convective circuit through the lower mantle offers no serious difficulty.

In Figure 4, a circle defined by the intersection of a cone of half-angle 70°, whose apex is at the earth's center, divides the spherical shell of the lithosphere into two parts. If there is a large retrograde motion, V_2, the upper partial shell (opening angle 110°) will be subjected to horizontal tension at its boundary. The shell is thin (100-km thick) as compared to the earth's radius; elastic problems of thin shells can readily be treated by rigorous analysis, provided that only the shell is rotationally symmetrical. Details can be found in suitable texts and reference works, but the mathematics is just heavy enough so that in order to avoid a conspicuous display of learning we will do with simple numerical estimates. Let L be the length of the bounding circle, A the area of the shell, and v the horizontal velocity component of lithosphere that disappears by sinking; then we estimate the order of the strain rate as

$$\dot{\epsilon} = vL/A$$

Numerically we set $v = 10^{-7}$ cm/sec (3 cm/year), $L = 2\pi R_0 \sin \theta$, $A = 2\pi R_0^2(1 + \cos \theta)$, where R_0 is the earth's radius and $\theta = 70°$; then $\dot{\epsilon} = 10^{-16}$ sec^{-1}.

There seems no simple way to determine the elastostatic stress from this except on assuming a linear, Maxwell body. From such a model, we can determine a stress relaxation time as

$$\tau = \epsilon/\dot{\epsilon} = \eta/E$$

where E is Young's modulus. From the second equality, we have

$$\sigma = \epsilon E = \dot{\epsilon}\eta$$

so that a knowledge of the horizontal stress requires only the determination of the linearized viscosity. Letting, for example, $\sigma = 10$ bar (10^7 cgs), we find $\eta = 10^{23}$ (for the lithosphere), a quite reasonable value, corresponding to a stress relaxation time of nearly 2000 years. Such a model implies of course, as we indicated in the beginning, that the asthenosphere has a low enough viscosity so that it does not prevent horizontal stresses of the lithosphere from extending over tangential distances of the order of, say, an earth's quadrant.

There is a subsidiary mechanism that will also produce horizontal tension, but to a much smaller degree than the process just discussed. Consider the material that is moving away from the Pacific area in the middle or lower mantle; it must accumulate under the lithosphere, which is antipodal to the Pacific (top region of Figure 4). This must expand the mantle of that hemisphere, resulting again in a horizontal tensile stress of the lithosphere. To get a quantitative estimate, we note that the lithosphere (either of the earth as a whole or of any cone whose apex is in the earth's center) makes up about 5% of the entire volume. The mantle, on the other hand, constitutes about 84% of such a volume, the remaining 16% being core. From these numbers, we can readily compute that tensions related to direct horizontal strain in the lithosphere will be far larger than any tensions that may result from a piling up of material farther down in the mantle.

If neither V_2 nor V_1 is negligible in equation 1, there will be horizontal tensile strains on both sides of a descending lithospheric slab. This situation agrees with the observed fact that the East Pacific rise indicates the presence of horizontal tension in the Pacific area itself.

It is not to be understood, of course, that the present schematic model can be applied so literally that uniform large-scale stresses could be assumed to prevail. The lithosphere, to all appearances, reacts to the stress or strain pattern in an irregular fashion, mainly by breaking up into plates that readjust themselves to the stresses or strains prevailing. A large variability of horizontal tension can be expected, as well as an occasional appearance of relatively smaller regions characterized by horizontal compression.

Let us note in this connection that the relationship of stress to strain is often quite complicated. Consider, for instance, a slab of plastic material, say pitch, spreading horizontally on a flat base, say a plate of glass. There is then no horizontal stress but only a horizontal strain, which represents the Poisson tensile strain generated by the vertical compressive stress that prevails in the slab by its own weight (I am indebted to E. Orowan for pointing this out to me). However, many kinds of inhomogeneity or of boundary conditions can convert a strain into a corresponding stress. The problem is formally somewhat complex and seems to require a more detailed treatment, especially since such a treatment is absent from the standard texts on elasticity. We forego a quantitative analysis here, for now we have circumvented the problem purely verbally on using the term 'tension' without specifying whether we mean stress or strain.

SEISMIC ANISOTROPY

A curious phenomenon that might be quite closely related to the existence of large-scale horizontal stress and strain is the seismic anisotropy observed on the bottom of the Pacific Ocean. The existence of this phenomenon was first indicated by *Hess* [1964] on the basis of measurements made by Raitt between a seismic transmitter and receiver located on two ships some distance apart. According to Raitt, the signals showing anisotropic propagation travel at an average depth below ocean bottom of 6–8 km. Determining the anisotropy effect observationally turns out to be a difficult and painstaking matter, since several trajectories under different angles are required to establish the effect with certainty and quantitatively above noise [*Morris et al.*, 1969]. Such anisotropy has been established definitely for a number of locations in the northeastern quadrant of the Pacific; the anisotropy is of the order of a few percent, ranging from zero at a couple of stations through some intermediate values to 7 and 8%, respectively, for the two sets of measurements giving the largest values. To appreciate the surprising magnitude of this effect, we note that the P velocities along the a, b, c crystallagraphic axes are 9.87, 7.73, and 8.65 km/sec, respectively [*Hess*, 1964]; this gives 1.68 km/sec for the excess of the a axis velocity over the mean of the b and c velocities, that is, a maximum anisotropy for a set of single crystals of 19%, provided that only the a axes are lined up. In the samples of the ocean bottom so far studied by Raitt and his associates [*Raitt*, 1970; see also *Morris*, 1969], the direction of the maximum velocities was west-east; practically it was so within the limits of observational error. This is perpendicular to the magnetic striations in this part of the world, hence parallel to the direction of the indicated motion of the lithosphere.

A mechanism for the formation of seismic anisotropy has been sketched by *Francis* [1969] and others. It is based on the assumption that the anisotropy is the result of plastic shear that occurs when mantle material rises in the middle of the ridges and then curves about with a small radius of curvature to flow off horizontally. Experiments have been made [*Raleigh*, 1968] which show that crystal anisotropy can be induced by plastic flow in the laboratory. Since the detection method was optical, these data cannot be used for quantitative comparisons. One can object to this as being the prevailing mechanism for the lithosphere on two grounds: In the first place, a mechanism that fits much better the general model of this paper is one in which a dike rises in the middle of the ridge, the dike eventually splits in the middle where it is hottest, the two halves move apart while a new dike forms, and so on for geologically long periods, as is evidenced by the magnetic striations that appear in the basalt that seems to segregate upward during this dike formation. Our second objection arises out of the extraordinarily high level of some of the observed anisotropies, namely, 7 to 8% as against an absolute maximum of 19%. It is difficult to see how such a high value could be achieved otherwise than by a very slow, nearly reversible process of recrystallization.

The simplest such process, first proposed by *Kumazawa and Shimazu* [1967] involves recrystallization into lined-up crystals under (tensile) strain. This strain would presumably be east-west on the bottom of much of the northeast Pacific. Slow recrystallization can clearly lead to much higher degrees of anisotropy than any mechanism of plastic deformation. The recrystallization process would largely take place near the ridge centers where the material is hot. This model of the origin of seismic anisotropy is corroborated by certain more indirect observations (communicated to me by Donald Anderson): Olivine bombs that are found ejected from volcanoes, e.g., in Hawaii, appear on inspection to be composed of crystallites that are very strongly lined up parallel to each other. This is what one should expect if these pieces of olivine originally formed part of the Pacific Ocean bottom.

OROWAN'S 'TRANSVECTIVE' MODEL

At one time many geophysicists believed that the lower mantle must be extremely 'hard,' too hard indeed to admit of creep motions. This view had been inferred from the fact, observed seismologically, that the elastic constants of the lower mantle are extremely high, of the order of those of diamond [*Bullen*, 1965]. At laboratory temperatures, this effect, due to extreme compression, would indeed prevent creep, but the temperature in the lower mantle is presumably of order 3000°K. For fixed stress the creep strain rate increases exponentially with rising temperature as the melting point is approached, provided that at least the molecular structure is simple enough [*Weertman*, 1970]. The material may then be relatively soft in spite of high elastic constants. As was mentioned before, *Goldreich and Toomre* [1969] have given semi-empirical evidence in favor of a deformable lower mantle.

Nevertheless, the desire to complete a convective circuit when the lower mantle is too hard to be deformed offers a challenge to ingenuity. *Orowan* [1969] has tried to meet this challenge, and his ideas seem well worth reporting briefly, since they introduce an interesting new element into the discussion.

Orowan describes a process which he calls 'transvection' and which is represented by the dotted arrows in Figure 4. This mode of motion is assumed to be driven by radioactive heating in the continental hemisphere (top of Figure 4) jointly with cooling in the Pacific hemisphere. The lower mantle moves as a solid block (dotted arrows), the circuit being completed by horizontal motions in the top layers of the mantle, convergent under the Pacific, divergent under Africa. Such a motion would clearly displace the core from its equilibrium position at the earth's center, and without a countervailing motion of the core the transvective mechanism would soon be stopped. Orowan assumes that some mantle material enters the fluid of the core at the boundary underlying the Pacific, and that an equivalent amount of material is deposited at the antipodal boundary. The mantle material can either go into true solution in the core or else there could be grains breaking loose at one side to be transported to the other side, where they are precipitated again. However, the size of such grains must be small (a fraction of a millimeter), because, with the appreciable density difference, $\Delta\rho \sim 4$, between core and mantle, the grains will be unable to leave the boundary if they are large enough. By Stokes' formula, the steady-state velocity of rising of a grain under buoyancy is

$$v = (2/9\eta)a^2\Delta\rho$$

where a is the radius of a (spherical) grain. This velocity must be small compared to the average fluid velocity in the core (0.1–1 mm/sec) if the grain is to be carried away from the boundary.

Now the amount of material turned over by upper-mantle convection is rather large; about half the lithosphere is replaced in 150 m.y. An amount of material of the lower mantle of the order of several percent of this would have to be carried across the core in the same time. It does not seem to us that one can assert this possibility without prefacing it by a very detailed investigation of a physico-chemical kind.

On the other hand, this brings to mind a very remarkable phenomenon: the unusually low amplitude of the geomagnetic secular variation in the region of the Pacific [*Doell and Cox* 1965]. Since no immediate cause for such an asymmetry of an otherwise earthwide phenomenon is apparent, its existence has been doubted (for some references see *Stacey* [1969, p. 139]). Now it is quite likely that what Orowan calls a transvective motion does exist as a limited small component of the return flow of upper-mantle convection, in which case an asymmetry of the geomagnetic secular variation, like that between the central Pacific and its antipodal region, would indeed make good sense.

APPENDIX: COSMOLOGICAL EFFECTS

The famous theoretical physicist *Dirac* [1938] suggested that what is called the universal constant of gravitation ($G = 6.7 \times 10^{-8}$ cgs units) is not really a constant but decreases slowly with time on a cosmological scale. This hypothesis was soon applied to the earth, since Dirac's paper appeared a few decades after the confidence of earth scientists in the ancient contraction ('shrinking apple') model of mountain building had begun to be shaken. A number of authors elaborated models of an expanding earth (for ample references, see *Runcorn*, [1969, part A]).

It seems difficult, however, to deal with such a cosmological effect on the same basis as one does with mantle convection. In the cosmological case, we must assume a slow, quasi-static increase of the earth's radius, which leads to corresponding tensile stresses, but there is no way of regenerating such stresses against stress relaxation, as is the case in the mechanism for convective overturn.

In spite of the speculative character of this subject, we prefer not to be swayed here by dogmatism. As a suitable warning, one may remember the historical battle between earth scientists and physicists nearly a century ago: The physicists, led by Lord Kelvin, claimed that the solar system could not possibly be more than a very few million years old, because after that time the heat content of bodies like the sun would be exhausted and 'heat death' would result. On the other hand, Charles Darwin, on purely empirical grounds, arrived at a time span of 300 m.y. for the aggregated length of the geological periods. Since this time is now known to be very close to 600 m.y. (Precambrian geology was, of course, inaccessible to Darwin), the estimate is remarkably, accurate. It is well known how Lord Kelvin's speculative ideas were superseded by the discovery of radioactivity.

Some simple quantitative estimates of the effect of earth expansion are in order: Let us start from the formula

$$\Delta p = -\rho G \int \frac{M(r)}{r^2} \, dr$$

where p is pressure, ρ is density, and $M(r)$ is the total mass contained in a sphere of radius r. We extend the integration over an appreciable fraction of the outer layers of the earth, say over a length $R = \frac{1}{2}R_0$, where R_0 is the earth's radius. We assume that ρ is constant inside the earth, but let the system be compressible with a fixed compressibility; these crude approximations should yet yield the right orders of magnitude. Now with $M(r) = (r\pi/3)\rho r^3$, we have

$$\Delta p \approx -G\rho^2(2\pi/3)R^2 \approx -2G\rho^2 R^2$$

On differentiation, while letting $d(\Delta p) = dp$ as an order-of-magnitude estimate, we have

$$dp = 2R^2(2\rho G d\rho + \rho^2 \, dG)$$

Numerical estimates show that the first term in the parenthesis is smaller than (about 10% of) the second term; it will be neglected, leaving us with

$$dp = -2R^2\rho^2 \, dG$$

We introduce the bulk modulus (incompressibility) by

$$k = \rho \, dp/d\rho$$

and get

$$d\rho/\rho \, dG = -2R\rho^2/k$$

Introducing numerical values, all in cgs units, we have $R^2 = 10^{17}$, $\rho^2 = 30$, $k = 4.10^{12}$ [*Bullen*, 1965], and, remembering that a linear strain is represented by $-\frac{1}{3}d\rho/\rho$, we let

$$\epsilon_G = d\rho/3\rho \, dG = 5.10^5$$

From this we obtain the rate of change of this strain,

$$\dot{\epsilon}_G = 10^{-2} \, dG/G \, dt$$

It is known that the place in the universe where variations of G are most readily observable is in the energy production of stars, which is proportional to G^8 [*Pochoda and Schwarzschild*, 1964]. In order not to raise havoc with the well-established scheme of stellar evolution in terms of the Hertzsprung-Russell diagram, dG/G must be small compared to unity for times of order, say, 10^8 years. This can be expressed by taking, roughly, $\dot{\epsilon}_G < 10^{-16}$ sec^{-1}.

To ascertain the physical effects of the corresponding expansion of the earth, we must again know the stress relaxation time. This time is likely to be much higher in the brittle zone near the surface than it is below, including the lower part of the lithosphere, where the temperature is high enough to lead to plastic ductility. If we take for the relaxation time $\tau = 10^{11}$ sec (3000 years), close to the preceding estimate, and recall that $\tau = \epsilon/\dot{\epsilon}$ and that $\dot{\epsilon} < 10^{-16}$ sec^{-1}, as just previously estimated, we see that $\epsilon < 10^{-5}$ cgs units, or $\sigma < 20$ bar. This effect is not small, but by its nature it cannot give rise to overturn in the way temperature differences do.

It will be interesting to watch the future development of these cosmological investigations, but for a reason just mentioned, namely that the body-force field created by any $dG/dt \neq 0$ is irrotational, one would not expect cosmological effects of this kind to have a major influence on the nature of circulatory motions inside the mantle. This would be so even if the cosmological change in gravity would give rise to observable horizontal tensile strains at the earth's surface. If such an effect can be made plausible by other astrophysical means, the geophysical equivalent would be a most interesting object of observation.

Acknowledgment. The author is indebted for stimulating discussions to several of his colleagues, particularly to H. W. Menard, W. J. Morgan, E. Orowan, R. W. Raitt, and H. Yoder. This work was supported by the National Science Foundation under grant NSF GA 20740.

REFERENCES

Barrell, J., The strength of the earth's crust (in 8 installments), *J. Geol.*, *22*, 1914, and *23*, 1915.

Bridgman, P. W., *The Physics of High Pressure*, G. Bell & Sons, London, 1949.

Bullen, K., *An Introduction to the Theory of Seismology*, 3rd edition, Cambridge University Press, London, 1965.

Dirac, P. A. M., A new basis for cosmology, *Proc. Roy. Soc. London*, *165*, 199–208, 1938.

Doell, R. R., and A. Cox, Paleomagnetism of Hawaiian lava flow, *J. Geophys. Res.*, *70*, 3377–3405, 1965.

Elsasser, W. M., Convection and stress propagation in the upper mantle, in *The Application of Modern Physics to the Earth and Planetary Interiors*, edited by W. K. Runcorn, pp. 223–246, Interscience, New York, 1969.

Elsasser, W. M., The so-called folded mountains, *J. Geophys. Res.*, *75*, 1615–1618, 1970.

Francis, T. J. G., Generation of seismic anisotropy in the upper mantle along the mid-oceanic ridges, *Nature*, *221*, 162–165, 1969.

Frank, F. C., Two-component flow model for convection in the earth's upper mantle, *Nature*, *220*, 350–352, 1968.

Goldreich, P., and A. Toomre, Some remarks on polar wandering, *J. Geophys. Res.*, *74*, 2555–2567, 1969.

Hess, H. H., Seismic anisotropy of the uppermost mantle under oceans, *Nature*, *203*, 629–631, 1964.

Hubbert, M. K., and W. W. Rubey, Role of fluid pressure in mechanics of overthrust faulting, 1, *Bull. Geol. Soc. Amer.*, *70*, 115–166, 1959.

Isacks, B., J. Oliver, and L. R. Sykes, Seismology and the new global tectonics, *J. Geophys. Res.*, *73*, 5855–5900, 1968.

Kumazawa, M., and Y. Shimazu, Preferred mineral orientation by recrystallization under non-hydrostatic stress and geotectonic stress in the upper mantle, *Geophys. J.*, *14*, 57–60, 1967.

Lambert, I. B., and P. J. Wyllie, Stability of hornblende and a model for the low velocity zone, *Nature*, *219*, 1240–1241, 1968.

Menard, H. W., The deep-ocean floor, *Sci. Amer.*, *221*(3), 127–142, 1969.

Morgan, W. J., Rises, trenches, great faults, and crustal blocks, *J. Geophys. Res.*, *73*, 1954–1982, 1968.

Morris, G. B., Velocity anisotropy and crustal structure of the Hawaiian arch, *Marine Phys. Lab. Scripps Inst. Oceanog. Tech. Rep. 69–26*, 1969.

Morris, G. B., R. W. Raitt, and G. G. Shor, Jr., Velocity anisotropy and delay-time maps of the mantle near Hawaii, *J. Geophys. Res.*, *74*, 4300–4316, 1969.

Orowan, E., Convection in a non-Newtonian mantle, continental drift, and mountain building, *Phil. Trans. Roy. Soc. London*, *A258*, 284–314, 1965.

Orowan, E., The origin of the oceanic ridges, *Sci. Amer. 221*(5), 103–119, 1969.

Press, F., How thick is the lithosphere? (abstract), *Trans. AGU*, *51*, 363, 1970.

Pochoda, P., and M. Schwarzschild, Variations of the gravitational constant and the evolution of the sun, *Astrophys. J.*, *139*, 587–593, 1964.

Raitt, R. W., to be published.

Raleigh, C. B., Mechanisms of plastic deformation of olivine, *J. Geophys. Res.*, *73*, 5391–5406, 1968.

Runcorn, W. K., *The Application of Modern Physics to the Earth and Planetary Interiors*, Interscience, London, 1969.

Stacey, R. D., *Physics of the Earth*, Wiley, New York, 1969.

Vine, F. J., and D. H. Matthews, Magnetic anomalies over oceanic ridges, *Nature*, *199*, 947–949, 1963.

Walcott, R. I., Flexural rigidity, thickness, and viscosity of the lithosphere, *J. Geophys. Res.*, *75*, 3941–3954, 1970.

Weertman, J., The creep strength of the earth's mantle, *Rev. Geophys.*, *8*, 145–168, 1970.

(Received August 27, 1970.)

Mantle Earthquake Mechanisms and the Sinking of the Lithosphere

by

BRYAN ISACKS

ESSA, Earth Sciences Laboratories,
Lamont-Doherty Geological Observatory,
Palisades, New York 10964

PETER MOLNAR

Lamont-Doherty Geological Observatory,
Palisades, New York 10964

Downgoing slabs of lithosphere may exert a pull on the portions of plate left at the surface.

THE concept of the lithosphere as a stress guide[1-3] suggests that information about the major driving forces in global tectonics can be obtained from knowledge of the state of stress within the lithosphere. Most shallow earthquakes seem to occur between plates of lithosphere and therefore yield information about the direction of motion of one plate with respect to the other rather than the stress within a plate. But analysis of focal mechanism solutions from the Tonga and Japanese regions[2] suggested that deep and intermediate depth earthquakes occur within downgoing slabs of lithosphere in response to stresses within the plates. A chief purpose of this article is to show that a comprehensive and worldwide survey (our unpublished results) of focal mechanism solutions supports this interpretation of deep and intermediate depth earthquakes. We can therefore determine the orientations of stress within the various downgoing slabs on a worldwide basis.

Gravitational Forces on Downgoing Slabs

The main result of this survey is that the stress in those portions of the mantle seismic zones that have a relatively simple inclined planar configuration, that is, in those portions removed from remarkable contortions or changes in trend, is often oriented such that either the axis of maximum compressive stress or the axis of minimum compressive stress is approximately parallel to the dip of the inclined seismic zone. We find clear evidence for down-dip extensional stress within slabs at intermediate depths in at least six regions. These regions seem to be also characterized by prominent gaps in the seismicity at depths between about 300 and 500 km or by an absence of earthquakes deeper than about 300 km. On the other hand, down-dip compressional stress is predominant at depths greater than 300 km in all regions studied.

These results are summarized in Figs. 1 and 2, and a simple model that attempts to account for them is given in Fig. 3. In this model a heavy slab sinks into the asthenosphere, exerts a pull on the portion of the plate remaining on the surface, and eventually hits bottom in such a way that the support for the load of excess mass within the slab is transferred from above the load to beneath the load. Although the two-dimensional model is certainly too simple to explain all the data, the general predominance of down-dip stress orientations and, especially, the widespread occurrence of down-dip extensional stress, lend encouragement to the ideas that gravitational body forces on the downgoing slabs are important forces in determining the stress within the lithosphere and may be important forces in driving the global system of plate movements.

Axes of Stress

The data include, primarily, focal mechanism solutions of the double-couple type for deep and intermediate depth earthquakes that are large enough $(M \gtrsim 5 \cdot 5)$ to obtain reliable first-motion data from the World-Wide Standardized Seismograph Network (WWSSN). Although these data are restricted to earthquakes that occurred since 1962, reliable solutions for previous earthquakes are also used for certain regions. The data are discussed in more detail elsewhere (our unpublished results). In this article we focus attention primarily on the results for fourteen island arcs and arc-like structures for which the orientations of the inclined seismic zones are relatively simple and well defined. This selection is required because the basic analysis consists of comparing the orientation of the double-couple mechanism solutions with the local orientation of the inclined seismic zone.

Some of the excluded earthquakes occur in zones or portions of zones that are remarkably contorted or complex in structure, such as the northern end of the Tonga arc. The mechanism solutions in these and similarly complex regions reflect the complexities of structure and are reported on in detail elsewhere (ref. 3 and our unpublished results). Others that are excluded occur in regions where the data are insufficient clearly to define a planar zone, such as the Mediterranean and Himalayan regions. Nevertheless, about 60 per cent of the earthquake mechanism solutions are included, and the results apply to most of the major island-arc structures of the world.

The overall results are shown in Fig. 1 in which the stress axes and the poles of nodal planes of the most reliably determined double-couple solutions are plotted relative to the local orientations of the seismic zones. One of the nodal planes of a mechanism solution is the fault plane of the equivalent shear dislocation, and the pole of the other nodal plane gives the direction of slip[4,5]; the seismic data do not, however, distinguish which is which. The compressional (P), tensional (T) and null (B) axes coincide with the axes of maximum, minimum and intermediate compressive stress in the medium only if the nodal planes define planes of maximum shear stress in the medium. This must be taken as an assumption of the analysis. If a Coulomb–Navier type of fracture process is operative for deep and intermediate depth earthquakes and the effective coefficient of internal friction is less than one[2,6], the P, T and B axes will still be good approximations (within the uncertainties of the analyses) to the principal stress axes.

Fig. 1d shows clearly that neither of the possible fault planes is parallel to the seismic zone. This result supports the contention[2,3] that intermediate and deep earthquakes do not indicate slip along a major thrust fault defined by the inclined seismic zone. Figs. 1a, 1b and 1c show that, instead, the inferred stress axes of the double-couple solutions tend to be parallel or perpendicular to the planar geometry of the zones. This result suggests that the downgoing slabs act as stress guides. If the lithosphere can support stresses considerably larger than those in the asthenosphere, so that the shear stresses are relatively small along the upper or lower boundaries of the plate, then the principal axes of stress within a thin plate of lithosphere will be approximately parallel and perpendicular to the plate.

Deviatoric Stresses

This interpretation is strongly supported by recent estimates[7,8] of the magnitude of the deviatoric stresses involved in deep and intermediate depth earthquakes. These stresses appear to be one to two orders of magnitude greater than the stresses estimated[8,9] for zones of shallow earthquakes where movements between plates of lithosphere are apparently accommodated along zones of weakness. Also, Stacey[10] has pointed out that thermodynamic problems arise if stresses much greater than 10 bars are associated with the large strain rates in the mantle that are required by sea-floor spreading and continental drift. These problems may be circumvented if the earthquakes in the mantle indicate large stresses within lithospheric plates rather than the smaller stresses within the weaker asthenospheric material where most of the mechanical work is being done.

The predominance of simple down-dip stress orientations shown in Figs. 1a and 1b suggests that in those portions of the zones characterized by relatively simple planar geometry the stress may be a result chiefly of the forces directly involved in the downward motion of the plate. The presence of down-dip extensional stress suggests that the slabs may be pulled into the mantle as a result of a negative buoyancy of the slab as suggested by Elsasser[1]. At rates of descent of the order of 10 cm/yr, temperatures in the slab may be of the order of 1,000° C cooler than the adjacent mantle, and positive density anomalies of the order of 0·1 g/cc would result.

Hatherton[11] and Morgan and Smith[12] find support for such excess mass in analyses of gravity data. If this load were supported chiefly by tractions along the large areas beneath the surficial plates of lithosphere, then simple calculations show that the deviatoric stress within the downgoing plate would be extensional, of the order of several kilobars, and would decrease with depth (Fig. 3a). As the slab penetrates into stronger material beneath the asthenosphere the support for the load of excess mass is transferred from above to below the load, and the stress in the lithosphere would depend on the variation of excess mass within the slab and the variation of strength with depth. In general, one might imagine a stage in which part of the load is supported from below and part from above (Fig. 3b) such that the stress changes from extension at intermediate depths to compression at great depths,

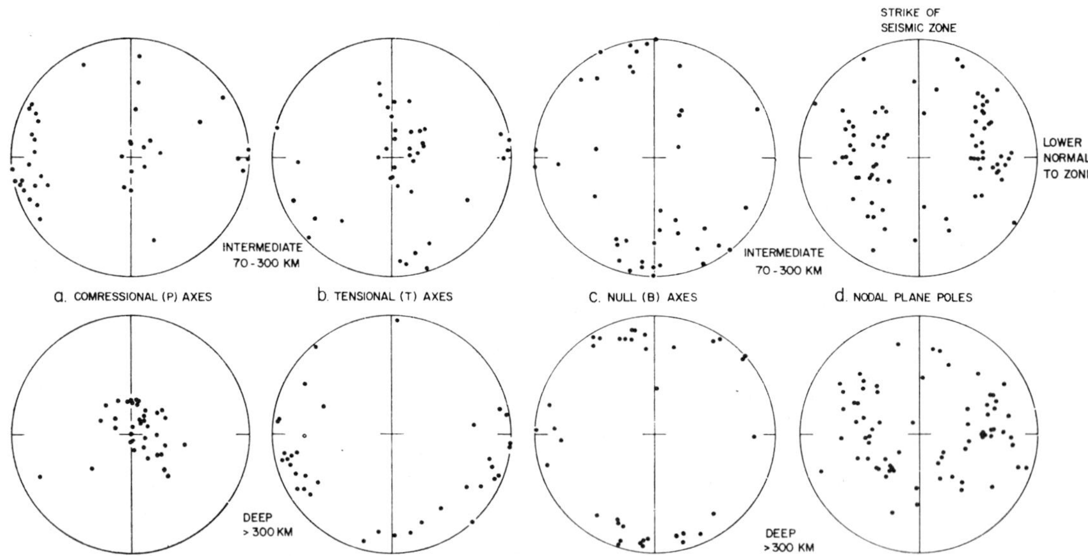

Fig. 1.   Equal-area projections of the axes of compression (P), axes of tension (T), null axes (B) and the poles of the nodal planes of the most reliably determined double-couple mechanism solutions of deep and intermediate depth earthquakes. The centre of each plot is in a direction parallel to the dip of the inclined seismic zone; the strike of the zone and the direction normal to the zone (on the lower side) are as shown in Fig. 1d. The regions and number of earthquake mechanisms included in this plot are as follows: Tonga[3,21,22] (15° S–26° S), 14; Kermadec[3,21] (26° S–36° S), 1; North Island, New Zealand[23,24] (36° S–41° S), 1; New Hebrides[19] (10° S–22° S), 9; Sunda (unpublished results of B. I. and P. M. and of T. Fitch and P. M.) (100° E–127° E), 9; Marianas[25] (12° N–24° N), 2; Izu-Bonin[25–29] (24° N–35° N), 8; Philippines[19] (3° N–12° N), 3; Ryukyu[25] (24° N–33° N), 2; northern Honshu[26–29] (35° N–43° N), 4; Kuriles[21] (43° N–52° N), 8; Aleutians[20] (145° W–170° E), 1; Middle America[30] (14° W–105° W), 5; Chile[19] (15° S–30° S), 8. The data on focal mechanisms and the structures of the seismic zones are described in more detail in the references indicated here.

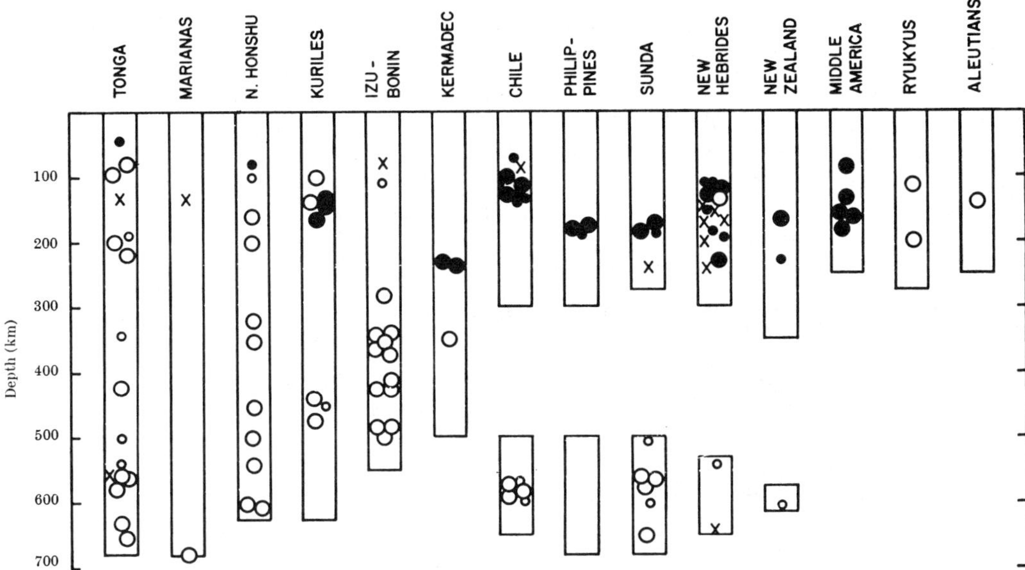

Fig. 2. Down-dip stress type plotted as a function of depth for fourteen regions (see caption for Fig. 1 for locations of regions and sources of data). A filled circle represents an orientation such that the axis of tension T is within 20–30 degrees of the local dip of the zone, that is, down-dip extension; an unfilled circle represents an orientation such that the axis of compression P is within 20–30 degrees of the local dip of the zone, that is, down-dip compression; and the Xs represent orientations that satisfy neither of the preceding. Smaller symbols represent less reliable determinations. The enclosed rectangular areas approximately indicate the distribution of earthquakes as a function of depth by showing the maximum depths and the presence of gaps for the various zones. In addition to the references listed in the caption to Fig. 1, data from Barazangi and Dorman[19] and Gutenberg and Richter[18] were used for each of the fourteen regions. The zones are grouped (from left to right) according to whether the zone is continuous to depths of 500–700 km, discontinuous with a gap between intermediate depth and deep earthquakes, or continuous but reaching depths less than 300–400 km.

with zero deviatoric stress somewhere between. When the load is fully supported from below, the stress becomes compressional throughout. Fig. 3d shows an alternative development where, as a result of the extensional stress and a possible stoppage in the movements, a piece of lithosphere breaks off, sinks independently, hits bottom and thereby comes under compression. In Figs. 3b and 3d gaps in the distribution of earthquakes as a function of depth might occur between extensional type mechanisms at intermediate depths and compressional type mechanisms at great depths. Where the seismic zone is continuous, that is, where earthquakes occur throughout the range of depths to 600–700 km, the slab would be under compression throughout.

Seismicity and Depth

The model shown in Fig. 3 thus predicts a correlation between the respective depth variations of focal mechanism and seismicity. Fig. 2 shows that this correlation is partially supported by the data. The regions where down-dip extensional stress predominates, such as Middle America, the New Hebrides and Chile, are characterized by notable gaps in seismicity as a function of depth or, in the case of Middle America, an absence of deep earthquakes. On the other hand, in regions such as Tonga, Izu-Bonin and northern Honshu the zones are apparently continuous and exhibit down-dip compressional stress at intermediate depths as well as great depths. Thus Tonga, Izu-Bonin and northern Honshu would be represented by Fig. 3c; Middle America would be represented by Fig. 3a; and Kermadec, Chile, North Island of New Zealand, New Hebrides, Sunda and Philippines would be represented by Fig. 3b or 3d.

Anderson[13] and others have proposed a phase change at depths near 350–400 km where the density increases by 9–10 per cent. If dT/dP for the phase boundary is positive (as proposed by Anderson), and the change occurs with

sufficient rapidity, the transition to greater density in the lithospheric plate might occur above the level of the transition in the adjacent mantle. This added load applied near 300–400 km depth might help to explain the prevalence of compression everywhere below 300 km as well as the prevalence of minima or terminations (Fig. 3) in the various distributions of seismicity as a function of depth.

Certain other areas not included in the present analysis because of uncertainties in the configuration of the subcrustal zone may offer further support for the models of Fig. 3. Intermediate depth earthquakes in the Hindu Kush, for example, are characterized by T axes that dip gently northward[14,15]. If this zone of intermediate depth earthquakes is a remnant of a larger slab underthrust

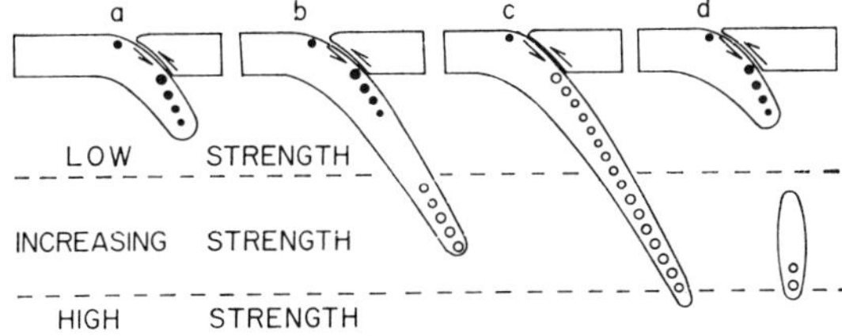

Fig. 3. A cartoon showing possible distributions of stresses in slabs of lithosphere that sink into the asthenosphere (a) and hit bottom (b and c). d represents the case where a piece of lithosphere has broken off. The symbols are the same as in Fig. 2; the filled circles represent down-dip extension and the unfilled circles represent down-dip compression. The size of the circles qualitatively indicates the amount of activity at the respective depths. In b and d gaps in the seismicity would be expected. Also shown is the underthrusting and the extensional stresses near the upper surface of the slabs due to the bending of the slab beneath the trenches. These features are inferred from the mechanisms of shallow earthquakes[2]. The lower boundary where the slabs hit bottom might correspond to the transition region or discontinuity near 650–700 km[18,31].

beneath the Himalayas, then the focal mechanism orientations might be indicative of down-dip extensional stress. The deep earthquakes beneath western Brazil are located nearly vertically beneath intermediate depth earthquakes and are apparently associated with the island arc-like features of the Peru trench. The deep mechanisms are characterized by nearly vertical P axes (refs. 16 and 17 and our unpublished results). Although the structure of the seismic zone is not well defined, the locations of earthquakes[18,19] suggest that the deep zone also dips very steeply beneath western Brazil.

Limitations of the Two-dimensional Model

Further comparisons of Figs. 2 and 3 show that the correlation breaks down in certain regions. In particular, the down-dip compressional stresses in the Aleutians and Ryukyus, the presence of both compressional and extensional stresses at intermediate depths in the Kuriles and New Hebrides, and the solutions denoted by "X" are not explained by the two-dimensional models of Fig. 3. These discrepancies, as well as the complex mechanisms and structures excluded in the analyses of Figs. 1 and 2, indicate the unsurprising result that the two-dimensional model of Fig. 3 does not explain all the data. Certainly, contortions and disruptions of the lithospheric plates would be expected to affect the stresses, although a search for such effects does not yield any simple relationship between the mechanism orientations and the contortions of the zones.

Nevertheless, one interesting association can be pointed out. The down-dip compressional stresses present in the Aleutians, the New Hebrides and the Ryukyu arc (Fig. 2) are located where the curvature of the arc is appreciable. In each case the T axis, instead of the B axis, is nearly parallel to the strike of the seismic zone; this orientation of the extensional stress might result from the type of deformation of the slab suggested by Stauder[20]. In the less arcuate portion of the New Hebrides the T axes are predominantly parallel to the dip of the seismic zone and the B axes are parallel to the strike. Many of the mechanisms represented by Xs in Fig. 2 may reflect other unresolved contortions of the downgoing slab or other sources of stress such as thermal gradients within the plates or local density variations.

Nevertheless, the consistent pattern shown in Fig. 2 for zones that are relatively uncontorted offers support for the interpretation that gravitational body forces are a major source of stress in the lithosphere. If we make this interpretation, then the results require that in many regions the slabs are sinking and are exerting a downwards pull. The results also require the complications of Fig. 3 in which the sinking slabs are supported from above or below, or both. Thus if we use the results to extract information on the forces that drive the global plate movements, we can infer two important effects: (1) the downgoing slabs can exert a pull on the surface portions of the plates, and (2) as the downgoing slabs "hit bottom" beneath the lithosphere the downward pull on a surface plate is significantly decreased. These interpretations suggest that the pull of the descending slabs may be an important contribution to the driving forces of global tectonics. Moreover, hiatuses or changes in the rates and direction of sea-floor spreading and continental drift might result when the descending slabs reach depths of 500–700 km.

We thank T. Fitch, D. T. Griggs, K. Jacob, M. Katsumata, D. P. McKenzie, J. Oliver, F. Stacey and L. R. Sykes for discussions. J. Oliver and L. R. Sykes critically read the manuscript. The work was supported by grants from ESSA (E-22-53-69G), the US National Science Foundation (GA11152) and the Advanced Research Projects Agency of the Department of Defense, and was monitored by the Air Force Cambridge Research Laboratories under Contract No. F19628-C-0341.

Received July 9, 1969.

[1] Elsasser, W. M., *The Application of Modern Physics to the Earth and Planetary Interiors*, NATO Institute, Newcastle (in the press, 1969).
[2] Isacks, B., Oliver, J., and Sykes, L. R., *J. Geophys. Res.*, **73**, 5855 (1968).
[3] Isacks, B., Sykes, L. R., and Oliver, J., *Bull. Geol. Soc. Amer.*, **80** (in the press).
[4] Burridge, R., and Knopoff, L., *Bull. Seis. Soc. Amer.*, **54**, 1875 (1964).
[5] Savage, J. C., *Bull. Seis. Soc. Amer.*, **55**, 263 (1965).
[6] Raleigh, C. B., and Paterson, M. S., *J. Geophys. Res.*, **70**, 3965 (1965).
[7] Berckhemer, H., and Jacob, K. H., *Investigation of the Dynamical Process in Earthquake Foci by Analyzing the Pulse Shape of Body Waves* (Inst. of Meteorology and Geophysics, Univ. of Frankfurt, Germany, 1968).
[8] Wyss, M., and Brune, J. N., *Trans. Amer. Geophys. Union*, **50**, 237 (1969).
[9] Brune, J. N., and Allen, C. R., *Bull. Seis. Soc. Amer.*, **57**, 501 (1967).
[10] Stacey, F. D., *Nature*, **214**, 476 (1967).
[11] Hatherton, T., *Nature*, **221**, 353 (1969).
[12] Morgan, W. J., and Smith, M. L., *Trans. Amer. Geophys. Union*, **50**, 316 (1969).
[13] Anderson, D. L., *Science*, **157**, 1165 (1967).
[14] Ritsema, A. R., *Tectonophysics*, **3**, 147 (1966).
[15] Chander, R., and Brune, J. N., *Bull. Seis. Soc. Amer.*, **55**, 805 (1965).
[16] Ben-Menahem, A., Jarosch, H., and Rosenman, M., *Bull. Seis. Soc. Amer.*, **58**, 1899 (1968).
[17] Khattri, K. N., *Bull. Seis. Soc. Amer.*, **59**, 691 (1969).
[18] Gutenberg, B., and Richter, C. F., *Seismicity of the Earth* (Princeton University Press, Princeton, 1954).
[19] Barazangi, M., and Dorman, J., *Bull. Seis. Soc. Amer.*, **59**, 369 (1969).
[20] Stauder, W. S. J., *J. Geophys. Res.*, **73**, 7393 (1968).
[21] Sykes, L. R., *J. Geophys. Res.*, **71**, 2981 (1966).
[22] Sykes, L. R., Isacks, B., and Oliver, J., *Bull. Seis. Soc. Amer.*, **59** (in the press).
[23] Adams, R. D., *New Zealand J. Geol. Geophys.*, **6**, 209 (1963).
[24] Hamilton, R. M., and Gale, A. W., *J. Geophys. Res.*, **73**, 3859 (1968).
[25] Katsumata, M., and Sykes, L. R., *J. Geophys. Res.*, **73** (in the press).
[26] Honda, H., Masatsuka, A., and Emura, K., *Sci. Rep. Tohoku Univ. Ser. 5, Geophys.*, **8**, 186 (1956).
[27] Hirasawa, T., *Bull. Earthquake Res. Inst. Tokyo Univ.*, **44**, 919 (1966).
[28] Ritsema, A. R., *Bull. Earthquake Res. Inst. Tokyo Univ.*, **43**, 39 (1965).
[29] Katsumata, M., *J. Seismol. Soc. Japan (Zisin)*, **20**, 75 (1967).
[30] Molnar, P., and Sykes, L. R., *Bull. Geol. Soc. Amer.*, **80** (in the press).
[31] Engdahl, E. R., and Flinn, E. A., *Science*, **163**, 177 (1969).

18

Reprinted from *Jour. Geophys. Research* **75**:5671–5680 (1970)

Instability in the Upper Mantle and Global Plate Movements

Wolfgang R. Jacoby

It is shown that the upper mantle is probably unstable. The instability expresses itself in diapirism under the ridges and sinking of the lithosphere under the island arcs. Both processes exert a force on the plate that can move it against the viscous drag from the flow of the asthenosphere. Order of magnitude estimates of the forces and of the required energy demonstrate that the process is plausible and that the drift rate can be determined from the resistance the plate encounters at its edges, particularly below the island arcs where it plunges into the mesosphere. The proposed mechanism can be tested by studying, e.g., the deep structure of the ridges, the stress field in the plates, and the gravity field over ridges and trenches.

1. Introduction

In the last few years a new concept of global plate movements has developed. The earth's surface is considered to consist of rigid plates, bounded by mid-ocean ridges (and continental rifts), transform faults, and deep-sea trenches associated with island arcs and active orogenic belts. The plates are generated as oceanic lithosphere at the ridges and move toward the trenches, where they disappear into the mantle [*Isacks et al.*, 1968; *Le Pichon*, 1968; *Morgan*, 1968]. The concept is illustrated in Figure 1.

The driving force for the plate movements presents a major problem. Mantle-wide convection cells have been proposed as a mechanism, but this assumption leads to difficulties and is physically unconvincing [*Knopoff*, 1969; *Orowan*, 1969]. The plates are passively dragged along on top of the convection cells in this type of model, independent of the assumption of constant or variable viscosity in the mantle.

In this paper an alternative mechanism is suggested. It is proposed that the region of greatest gravitational instability may be the upper mantle, and that the lithosphere and the asthenosphere participate actively in the movements. One aspect of this mechanism is the active diapirism of the asthenosphere under the mid-ocean ridges exerting a force on the plates away from the ridges. Another aspect is the sinking of the lithosphere under the island arcs,

as proposed by *Elsasser* [1967] and by *Isacks and Molnar* [1969]. Both processes must be considered together. Furthermore, matter has to move back to the ridges to maintain continuity of mass. It is suggested that the most likely channel of transport is the asthenosphere because of its mobility.

In this model the plates are not dragged by deeper convection, but are driven by the temperature-induced density difference between lithosphere and asthenosphere. In this sense the mechanism is still a form of thermal convection, but it greatly differs from Rayleigh-Benard convection [*McKenzie*, 1969].

To test whether this model is plausible, we have to make quantitative estimates of the inherent driving forces, the resistances, and the required energy. Admittedly, we have little quantitative knowledge about the upper mantle, but it appears reasonable to make the following assumptions:

The lithosphere is approximately 100 km thick (it may be thicker under shield areas, thinner under the ocean basins, and much thinner under tectonically active regions; *Walcott*, 1970*b*; *Kanamori and Press*, 1970]. It overlies the asthenosphere, which is a few hundred kilometers thick and which in turn overlies the mesosphere (or barysphere). The lithosphere is assumed to be a Maxwell solid [*Walcott*, 1970*b*] with a high viscosity, and the asthenosphere is assumed to have a viscosity several

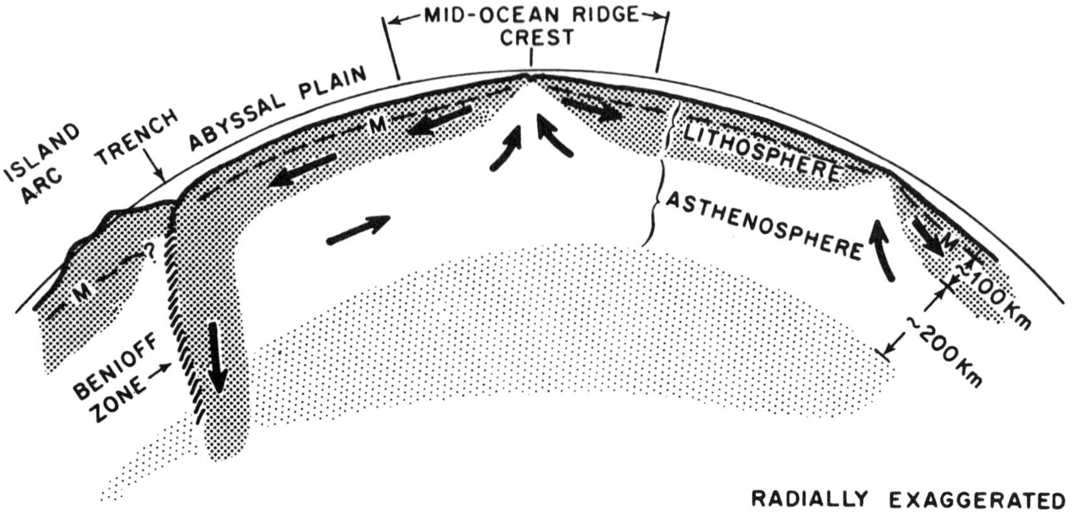

Fig. 1. The assumed structure of the upper mantle.

orders of magnitude lower ($10^{19\pm}$ N sec/m² = 10^{20} poises), whereas the viscosity is assumed to rise again in the mesosphere. Viscosity may not be a completely adequate description of the behavior of the upper mantle; however, it makes the calculations simple and we hope not grossly wrong [*McKenzie*, 1968].

It may be argued that we have a rather weak case because the assumptions are still uncertain. The author admits this weakness frankly and only wants to show that the proposed model will probably work if the assumptions, which are not entirely unreasonable, should prove to be correct. If we accept the present model for the observed plate movements, it can provide some constraints on the acceptable density and viscosity structure of the upper mantle.

2. Active Diapirism under the Mid-Ocean Ridges

The active diapirism of the asthenosphere under the ridges is a process that does not yet appear to have been considered in the literature. The term 'diapirism' has been applied in a passive sense to the filling of the space between the spreading ridges by mantle material [e.g., *Vogt et al.*, 1969, p. 599]. However salt or granite diapirs rise actively into the denser overburden in response to the gravitational instability of the mass distribution. This is the essential analogy between the asthenospheric diapir and salt or granite domes, though the shape is quite different. As the concept of sea-floor spreading implies that the lithosphere is generated at the ridges by cooling and solidification of the asthenosphere, the plates should thicken when they move away from the ridges and the asthenosphere should form an upward-pointing wedge with its apex under the crest [*Jacoby*, 1970]. The rise of the seismic discontinuities [*Ewing*, 1969; *Talwani et al.*, 1965] and the crestal rift or graben of mid-ocean ridges are features that support this concept of diapirism.

Diapirism under the ridges requires an unstable mass distribution of the crust and upper mantle under the ridges. The mass distribution is unstable (1) if there is a lateral density variation or a density reversal with depth, and (2) if movements to allow the reduction of the potential energy of the masses are possible [*Jeffreys*, 1959, p. 339].

It appears that both conditions are satisfied:

1. The free-air gravity anomalies over the ridges are nearly zero [*Kaula*, 1969] and indicate that the topographic mass surplus of the ridge is essentially compensated by a mass deficiency at depth (gravity over the ridges is further discussed in section 7). If we assume that this deficiency is localized in the upward-pointing wedge of the asthenosphere, and that it and the lithosphere are each homogeneous, we can estimate the lateral density contrast $\Delta\rho = \rho_a - \rho$ from the relation

$$(\rho - \rho_w)e_c + \Delta\rho h_c \approx 0 \qquad (1)$$

where ρ, ρ_a, and ρ_w are the densities of the lithosphere, asthenosphere, and water, respectively, and e_c and h_c are the uplift of the ridge from the abyssal plains and the uplift of the top of the asthenosphere from its 'normal' level (Figure 2). With $\rho \approx 3$ g/cm³, $e_c \approx 1$ to 3 km, and $h_c \approx 100$ km, we obtain

$$\Delta\rho \approx -0.02 \quad \text{to} \quad -0.06 \text{ g/cm}^3$$

This does not necessarily imply a true density reversal with depth, since our $\Delta\rho$ is only a difference from any assumed normal density-depth curve. There may, however, be a true density reversal with depth everywhere [*Press*, 1969, 1970; *Anderson*, 1969].

2. The possibility of movements is basic to the new concept. The lithospheric plates are assumed to move relative to each other, seismic zones marking their boundaries. Asthenospheric flow is assumed in the present model. Flow below the lithosphere is also demonstrated by the absence of large free-air gravity anomalies in the presence of large mass transport on the earth's surface, e.g. sediment transport or accumulation and removal of ice sheets.

Hence, diapiric or buoyant rising of the mobile asthenosphere appears likely. If it occurs, it actively lifts up the lithospheric plates on either side of the ridge. Under the force of gravity the plates tend to slide down over an inclined surface, as has already been suggested by *Hales* [1969].

We estimate the force from the inclined part of the plate on the remaining part for a static situation. This assumption is justified because the term of inertia is negligible for the plate motions (see section 6). Let us consider the two-dimensional block of Figure 2. The force (H in Figure 2) is a function of the weight W of the block, of the dip α of sliding, and of the coupling between the two parts of the plate. The plate is bent from the dipping to the horizontal direction by a vertical guiding force G (arising from the plunge of the plate below its local equilibrium level). The horizontal force H is the resultant of the force $F = W \sin\alpha$ in the instantaneous dipping direction and of the guiding force G as shown in Figure 2(1).

$$H = F \cos\alpha = W \sin\alpha \cos\alpha \qquad (2)$$

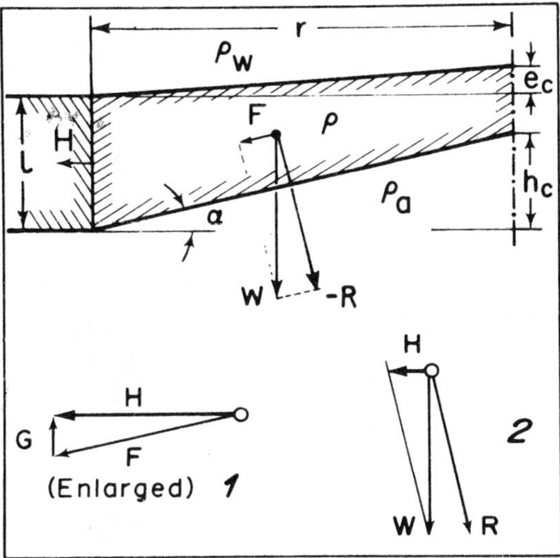

Fig. 2. The block sliding down the ridge.

As α is small, $\sin\alpha \cos\alpha \simeq h_c/r$. (The force $H = Wh_c/r$ corresponds exactly to a coupling of the two parts of the plate over a vertical section without shear strength; see Figure 2(2)). We further write $W = V\rho g = r(l + \frac{1}{2}(e_c - h_c))\,\rho g$, where V is the volume, ρ is the block density, and g is gravity. Both W and V refer to a slab of unit thickness along the strike. Because $e_c \ll h_c$, we can simplify:

$$W \simeq r(l - h_c/2)\rho g$$

Hence, we obtain

$$H \simeq (h_c l - h_c{}^2/2)\rho g \qquad (3)$$

For $h_c \approx l$, the expression is even simpler:

$$H \simeq (h_c{}^2/2)\rho g \qquad (4)$$

The numerical value of the force for a ridge with the dimensions shown in Figure 3 (top diagram) is approximately 1.5×10^{14} N/m (newton per meter of length normal to the section). This estimate is a maximum; for steady movement, the average angle of sliding would be smaller than that assumed in Figure 3. Also the variation of the force H with a variation of the dimensions of the ridge is shown. The width of the ridge and relative thinning of the lithosphere at the crest do not as effectively influence the force as a variation of the over-all thickness of the lithosphere.

The proposed force away from the ridges is

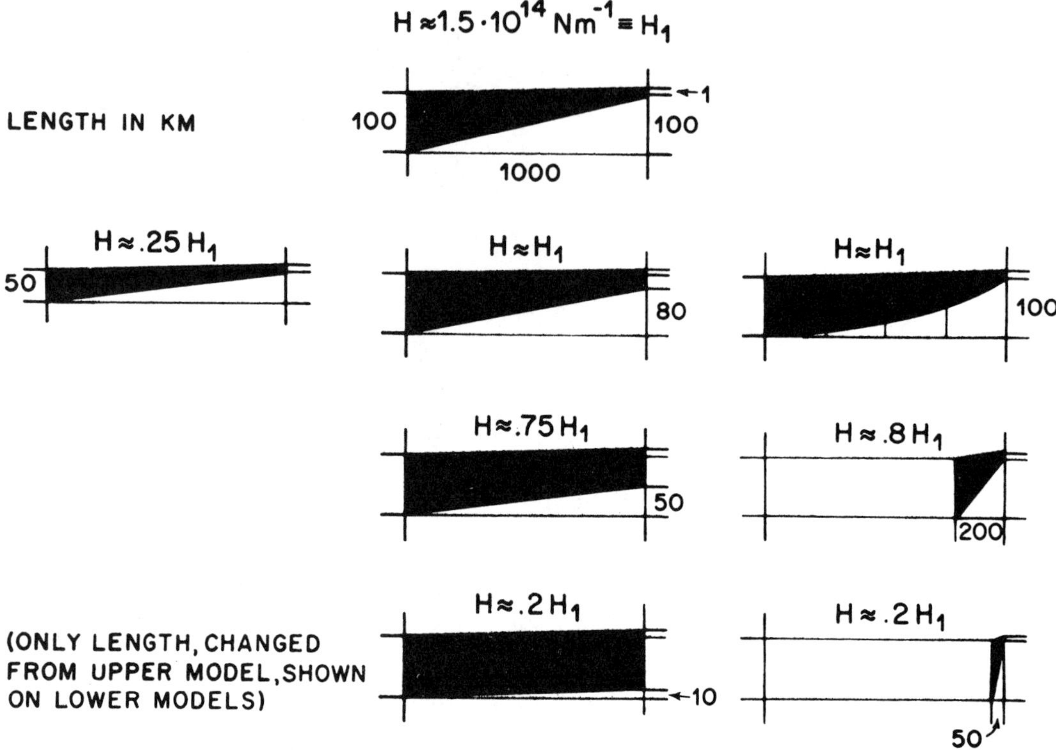

Fig. 3. The numerical value of the horizontal force H, calculated with (2).

important for the understanding of the plate motions. Some plates appear to be 'pushed' by ridges without being 'pulled' by sinking slabs (see section 3). The mid-Atlantic ridge is actively spreading and the plates on either side do not appear to sink anywhere in a major fashion but appear to override the East Pacific plate or to be partly bounded by another ridge or rifting system, respectively. Furthermore, a diapir rises buoyantly only into 'denser' material. It will not rise efficiently into continental lithosphere. This may explain why the East African rift system has not had much movement.

3. Sinking of the Lithosphere under the Island Arcs

Elsasser [1967] and *Isacks and Molnar* [1969] have suggested that the lithosphere is sinking under the island arcs and exerting a downward pull. This requires a positive density contrast $\Delta\rho$ of the sinking slab; such an assumption is supported by thermal arguments and by the observed gravity anomalies [*Hatherton*, 1969a, b].

To estimate the buoyant force on the slab, we consider again a static case. The vertical force is $B = V\Delta\rho g \approx (d/\sin \alpha)l\Delta\rho g$, where V is the volume per unit length along the strike, d is the depth of the head of the slab, l is its thickness, and α is the dip of the slab measured from horizontal. If the sinking slab is uncoupled from the horizontal lithosphere by vertical faults, as has been suggested by *Lliboutry* [1969], no tensile 'pulling' force could be exerted directly on the horizontal plate from the sinking one. If the slab is able to transmit tensile stresses or if the tension is compensated by compression from the ridge, the component $F = B \sin \alpha$ parallel to the inclined slab may be in static equilibrium with a vertical guiding force G (arising, e.g., from the depression of the slab below its local equilibrium level) and a horizontal tensile force H parallel to the lithosphere beyond the bend (Figure 4). In this case: $H \approx F \cos \alpha = dl\Delta\rho g \cos \alpha$. With $d \approx 700$ km, $l \approx 100$ km, $\Delta\rho \approx 0.05$ g/cm³, $g \approx 10$ m/sec², and $\alpha \approx 45°$, we obtain $H \approx 2.5 \times 10^{13}$ N/m.

This is nearly an order of magnitude less than the estimated maximum force H from the ridge.

The sinking of the lithosphere is coupled to the whole convection also in a different way. As mantle material is forcefully displaced by the sinking slab, it is 'pumped' through the channel of mobile asthenosphere toward those regions where it can rise again, i.e. under the ridges.

4. Resistance to the Plate Movements

Resistance to the plate movements occurs in two ways: (1) material has to rise against the gravitational force on the path that closes the convection (up to here we have only considered the downward movement of the plate), and (2) differential movements at the edges of the plates and viscous flow of the asthenosphere resist the movements.

1. Let us assume that the mass on the path up is smaller than the mass on the path down, the difference being proportional to $\Delta\rho$. This has implicitly been considered for the sinking slab. At the ridge most of the force H_{ridge} is balanced by a force against the rising of the asthenospheric diapir. The difference between both forces is approximately $(\Delta\rho/\rho)H_{ridge} \approx 3 \times 10^{12}$ N/m in the direction of the movements.

2. The resistive forces evident in the world's great fault systems are difficult to estimate. However, earthquakes and fault creep clearly show that these forces are inadequate to stop the plate movements.

The drift of the lithosphere over the viscous asthenosphere causes a drag against the motion. We have to assume a velocity distribution in the asthenosphere that depends on its flow. Let us assume two cases:

a. The plate is drifting over a 'passive' viscous layer that is dragged along over a bottom that is at rest. In the steady state the drag F_1 for the two-dimensional model is (Figure 5(1)):

$$F_1 \approx \eta b(u_l/a) \tag{5}$$

where b is the length of the plate parallel to its movement, u_l is its velocity, a is the thickness of the asthenosphere, and η is its average viscosity.

b. There is no net transport through a vertical section across the direction of motion, as is shown in Figure 5(2). The velocity function in the asthenosphere has the approximate form

$$u = u_0 + C(z - z_0)^2 \tag{6}$$

and the constants u_0, z_0, and C are determined by the conditions:

$$\text{I.} \quad u_l l + \int_{z=0}^{a} u \, dz = 0$$

(no net transport; for simplicity it is assumed

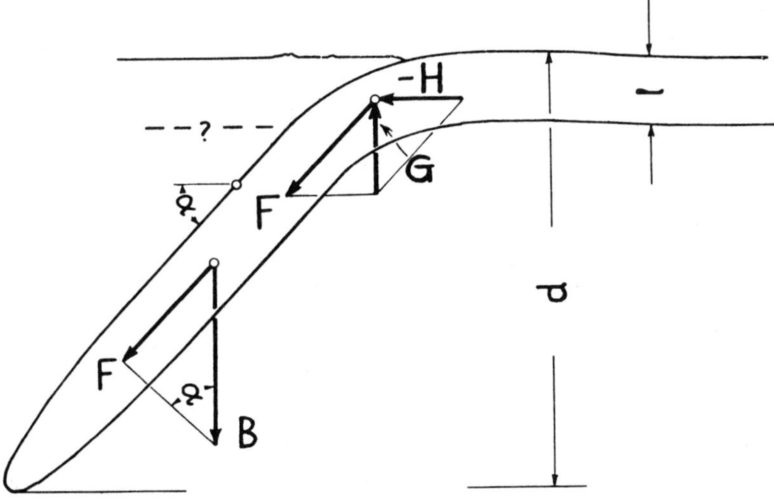

Fig. 4. The slab sinking into the mantle.

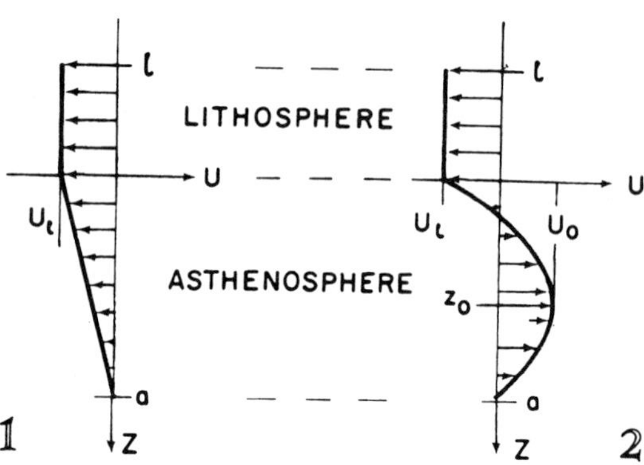

Fig. 5. The velocity of flow in the asthenosphere.

here that the density is constant throughout the vertical section)

II. $u = u_l$ for $z = 0$

III. $u = 0$ for $z = a$

(continuous velocity at the top and bottom of the asthenosphere). With the abbreviation $q = (l + \tfrac{2}{3}a)/(2l + a)$, the constants are:

$$u_0 = u_l\left(1 - \frac{q^2}{2q - 1}\right)$$

$$z_0 = aq \qquad (7)$$

$$C = \frac{1}{2q - 1}\frac{u_l}{a^2}$$

The drag F_2 that the flow through the asthenospheric channel encounters at the top and bottom together is

$$F_2 \approx \eta \cdot b(|u'(0)| + |u'(a)|)$$

where

$$u'(0) = \left.\frac{du}{dz}\right|_{z=0} = -2Cz_0$$

$$u'(a) = \left.\frac{du}{dz}\right|_{z=a} = 2C(a - z_0)$$

Hence

$$F_2 \approx 2\eta bCa = \frac{2}{2q - 1}\,\eta b\,\frac{u_l}{a} = \frac{2}{2q - 1}\,F_1 \qquad (8)$$

With $b \approx 10{,}000$ km, $l \approx 100$ km, $a \approx 200$ km, $\eta \approx 10^{19}$ N sec/m² (10^{20} poises) and $u_l \approx 3$

cm/year or 10^{-9} m/sec, we obtain a numerical estimate of the forces F_1 and F_2:

$$F_1 \approx 0.5 \times 10^{12} \quad \text{N/m}$$

$$F_2 \approx 6 \times 10^{12} \quad \text{N/m}$$

Variation of the geometric parameters within reasonable limits with the viscosity and plate velocity remaining constant would cause F_2 to vary within $\pm\tfrac{1}{2}$ order of magnitude. Furthermore, it is likely that the mantle is somewhat involved in the movements to greater depths than 300 km, since the plates dip to about 700 km depth. This would tend to decrease F_2, particularly for the large plates.

The above estimate is not yet complete. Where the plate is dipping into the more viscous mesosphere (below the asthenosphere), the resistance is likely to grow rapidly, and the leading edge of the plate may therefore encounter a resistance greater than F_2.

A very large special resistance against the plate motion occurs when continental crust is pushed into another plate. The light continental material cannot be taken down with the dipping slab; the buoyant forces against the motion would quickly dominate all other forces [*McKenzie*, 1969].

5. BALANCE OF THE FORCES

It is correct to compare the driving gravitational forces, estimated for the static case, with the forces of reaction against the movements, if the motion is steady. The gravitational forces on the whole convective system are the sum of the net force from the ridge (0.3×10^{13} N/m) and the force from the sinking slab (2.5×10^{13} N/m), hence approximately 3×10^{13} N/m. The estimated resistance from viscous flow of the asthenosphere is between 10^{12} and 10^{13} N/m for a plate of 10,000 km length, moving 3 cm/year.

Since many uncertain parameters have entered into the calculations, it is difficult to say more than that the driving and resistive forces appear to be approximately balanced. Let us discuss two cases:

1. The effects of the errors in the various assumptions cancel in the result or tend to make F_2 too large. The dominating resistance, balancing the driving force and therefore determining the rate of drift, should then act on the edges of the plate: the transform faults, the thrust

fault within the lithosphere below island arcs, the corresponding zone of bending of the plate, and especially the leading edge that bottoms (and, as a special case, a continental margin colliding with an island arc). As we have at present no quantitative estimate of these resistances, we cannot yet use our model to predict the drift rate, e.g., in relation to the plate size. In other words, we cannot yet test this aspect of the model against observations.

2. The effect of the errors is a considerable underestimation of the viscous drag. F_2 would then represent the main resistance, and the proposed model would predict the drift rate to be inversely proportional to the length b of the plate, parallel to the movement, and to be rather small for the largest plates. This appears to contradict the few observations.

Hence the proposed model is more likely to be basically correct if the resistance F_2 is not the dominating one. This would require that the viscosity of the asthenosphere must not be much larger than assumed.

The estimated net driving force from the ridge is only one tenth that from the sinking slab. This may partly explain why, for example, the East Pacific rise spreads faster than the mid-Atlantic ridge; both Pacific plates, but none of the Atlantic plates, sink below island arcs.

The gravitational force acting on the plate from the inclined part, sliding down the ridge (up to 2.5×10^{14} N/m), can be larger than the net gravitational forces acting in the convecting system (3×10^{13} N/m). If so, the lithospheric plate is under compression, perpendicular to the ridge (except near its crest). The compression may be so great that, if building up fast, it may lead to failure along a thrust fault and to the generation of a new island arc. If, on the other hand, the plate, as a Maxwell solid, is subjected slowly to the force along the length of the ridge slope, it should be able to transmit the compressive stress throughout its entire length.

6. Source of Energy

The estimated forces act in the assumed model of plate drift over a long time only if energy is available to maintain the volume and the low viscosity of the asthenosphere. The most likely energy source is heat. Under the assumption that the total heat flow through the earth's surface and the plate drift are in the steady state, the rate at which the mechanical work is done ($\dot{E}$) must not be more than a fraction of the heat flow, simply because any heat engine has an efficiency smaller than 1. $\dot{E}$ can be estimated from the 'driving' force D and the velocity u_l:

$$E \approx D \cdot u_l$$

With $D \approx 3 \times 10^{13}$ N/m and $u_l \approx 10^{-9}$ m/sec, $\dot{E} \approx 3 \times 10^4$ w/m; multiplied with the (double) length of the ridge system (approx. 100,000 km), this yields $\dot{E}_{\text{total}} \approx 3 \times 10^{12}$ w, which is a small fraction of the total heat flow through the earth's surface of approximately 3×10^{13} w.

The rate of global energy release by earthquakes is about 10^{25} ergs/year or 3×10^{10} w, a plausible fraction of the estimated power of the plate drift.

The kinetic energy of the moving plate (with the dimensions assumed above) is negligible ($E_{\text{kin}} = \frac{1}{2} V \rho u^2 \approx 10^{-5}$ joule/m). Therefore the inertial terms of the assumed convection can be neglected in the present analysis.

It is important for the plausibility of the assumed model that a great amount of potential energy is stored. The ridge alone stores about 10^{18} joules/m, which theoretically could feed the process for several million years. The dipping slab of Figure 4 stores even more, approximately 1.5×10^{19} joules/m. (The calculation of these values is omitted here).

7. Possible Tests of the Proposed Mechansim

Some aspects of the present model of the plate drift can be tested experimentally: e.g., the required structure of the ridges, the state of stress of the plates, and the gravity field related to the model.

The upward-pointing wedge of the asthenosphere under the ridges may be detectable by the low velocities and strong attenuation especially of S waves, characteristic for the asthenosphere in general. *Molnar and Oliver* [1969] have shown that the S waves traveling in the uppermost mantle across or along ridges are indeed strongly attenuated. Seismic stations on ridges (e.g., Iceland) generally observe positive travel-time anomalies, indicating a thick low-velocity layer under these stations. Also the geometry of the ridges with a well-defined crest, offset by transform faults, the seismically determined structure of the crust at the ridges [*Ewing*, 1969], and the migration of the ridges and trenches do not conflict with the mechanism.

It has been argued that the broad ridges are implausible unless the asthenosphere can maintain larger shear stresses than are implied by the present model. This argument appears to be invalid. The ridge is lifted up by vertical buoyant forces due to the low density of the asthenosphere maintained by thermal energy. Even if the energy becomes insufficient, the ridge cannot subside completely by outward flow, because this would cause a large mass deficiency, which is nowhere observed on earth, but it

would slowly subside by cooling of the asthenospheric wedge.

The frequent occurrence of chains of marine volcanoes, approximately perpendicular to the ridges, indicates a compressive stress field perpendicular to the ridge, except near the crest as predicted; such chains should form or grow in the direction of the maximum stress [*Walcott*, 1970a]. Observations of the seismic velocity anisotropy on the Pacific plate so far have shown the maximum velocity more or less perpendicular to the East Pacific rise as we would predict if stress were the only cause. The influence of preferred crystal orientation, however, weakens this argument. Tensional earth-

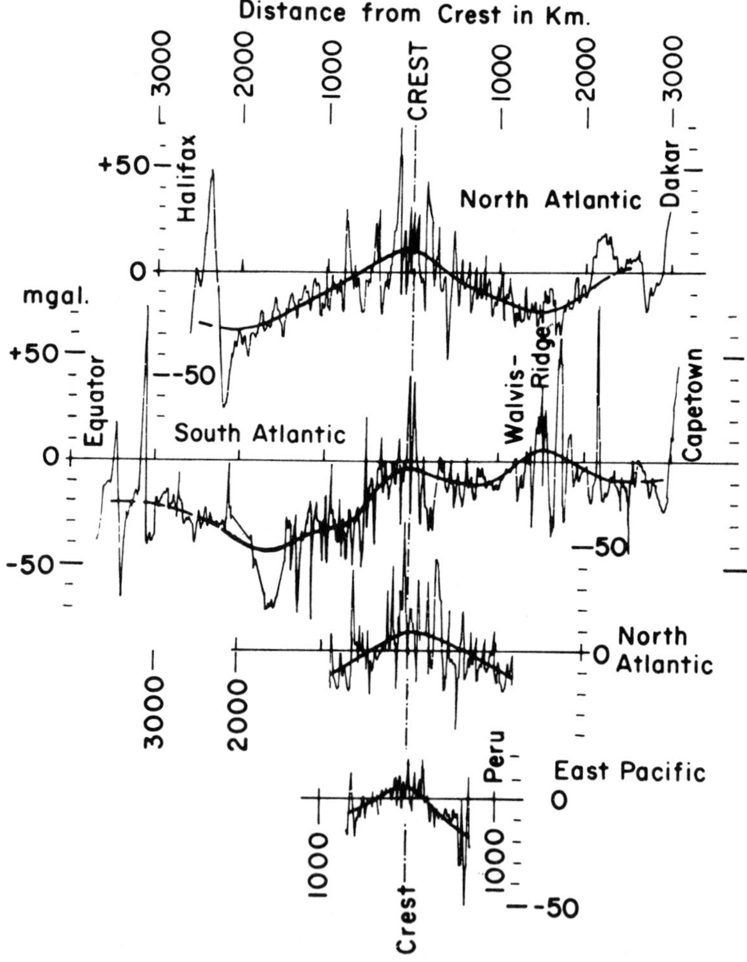

Fig. 6. Free-air gravity anomalies across the mid-ocean ridges, observed and smoothed (first two profiles after *Talwani and LePichon* [1969] ; last two profiles after *Vogt et al.* [1969].

quakes along the ridge crest do not, of course, contradict the proposed mechanism; they are predicted to occur where the plates break apart.

The free-air gravity anomalies over the ridges are not exactly zero, but are always slightly positive near the crest, at least relative to the environment. This is shown in Figure 6 with four profiles taken from different publications. The profiles were brought to the same scale and were strongly smoothed. Until recently, satellite gravity did not have the resolution to show the small positive anomalies related to the ridges [*Kaula*, 1969], but they appear now to be established with significance [*Kaula*, 1970]. The slightly positive anomaly with negative flanks is not necessarily due to deep compensation of the ridge, but may indicate a small mass surplus around the crest. This is indeed inferred from the latest satellite results. It is also the result of a least-squares fit of the observed surface gravity to the computed effect of the topographic ridge and the assumed wedge of asthenosphere, by which the best fitting density contrasts are determined. Using average gravity values (of the four profiles in Figure 6) and an asthenospheric wedge of 1000-km half-width and 90-km height (from the bottom of the lithosphere to 10-km depth under the crest), we obtain a positive mass anomaly above 100 km depth of 2 to 3 $\times$ 10^5 kg/m^2 near the crest. This would favor an active role of the ridge, as mass transport away from it would then lag behind complete equilibrium. In the case of passive upwelling of matter into space made available, when the plates are pulled apart, we should expect a lag in the opposite direction, i.e., a slight mass deficiency. Purely thermal expansion of the lithosphere near the ridges [as suggested by *McKenzie and Sclater*, 1969] also appears to be an unsatisfactory interpretation of the recent gravity data.

A discussion of heat flow is omitted here, although it is of great importance to any theory of the plate motions. The present data appear to be compatible with our model, but may not be discriminating against other models.

Any single discipline of geophysics and geology will probably not provide evidence for or against this and other theories of the plate movements. The answer may come from the increase and combination of knowledge in many fields, such as all aspects of upper mantle physics, the exact nature of the plate movements in the past and at present, the geological record, and others.

8. Conclusions

The model is greatly simplified, and there are large uncertainties in the assumptions. A more rigorous theory would have to consider the earth's spherical shape and the complicated geometry of the individual plates. Moreover, a numerical solution for the flow in the upper mantle related to the plate motions is required. In principle, however, it appears to be consistent with the present ideas of sea-floor spreading and the assumed structure of the upper mantle to envisage the rigid and dense plates sliding from the elevated ridges toward the trenches and sinking down into the mantle, and mobile and light asthenosphere flowing back and rising as a diapir under the ridges. Order-of-magnitude estimates of the inherent gravitational and resistive forces, as well as of the required energy, show that the model is physically conceivable. The upper mantle structure of the ridge and island arc regions, as we know it today, the apparent stress field in the plates, gravity over ridges and island arcs, and other geophysical data appear to support the present hypothesis.

Acknowledgments. I am grateful to many colleagues at the Earth Physics Branch for discussions, without which some ideas never would have formed. K. G. Barr, M. J. Berry, and R. I. Walcott read the manuscript, and helped me improve my English style and the organization of the material presented in this paper. Their comments have been invaluable.

References

Anderson, D. L., Physical properties of the mantle (abstract), *Trans. AGU, 50,* 671, 1969.

Elsasser, W. M., Convection and stress propagation in the upper mantle, *Princeton Univ. Tech. Rep. 5,* June 15, 1967.

Ewing, J., Seismic model of the Atlantic Ocean, in *The Earth's Crust and Upper Mantle, Geophys. Monograph 13,* edited by P. J. Hart, pp. 220–225, AGU, Washington, D. C., 1969.

Hales, A. L., Gravitational sliding and continental drift, *Earth Planet. Sci. Lett., 6,* 31–34, 1969.

Hatherton, T., Gravity and seismicity of asymmetric active regions, *Nature, 221,* 353–355, 1969a.

Hatherton, T., Similarity of gravity anomaly patterns in asymmetric active regions, *Nature, 224,* 357–358, 1969b.

Isacks, B.. and P. Molnar, Mantle earthquake mechanisms and the sinking of the lithosphere, *Nature, 223,* 1121–1124, 1969.

Isacks, B., J. Oliver, and L. R. Sykes, Seismology and the new global tectonics, *J. Geophys. Res., 73,* 5855–5899, 1968.

Jacoby, W. R., Active diapirism under mid-oceanic ridges (abstract), *Trans. AGU, 51,* 204, 1970.

Jeffreys, Sir H., *The Earth, Its Origin, History and Physical Constitution,* fourth ed., Cambridge University Press, London, 1959.

Kanamori, H., and F. Press, How thick is the lithosphere?, *Nature, 226,* 330–331, 1970.

Kaula, W. M., A tectonic classification of the main features of the earth's gravity field, *J. Geophys. Res., 74,* 4807–4826, 1969.

Kaula, W. M., Global gravity and tectonics, paper presented at the Francis Birch Symposium on the Nature of the Solid Earth, Harvard University, Cambridge, Mass., April 16–18, 1970.

Knopoff, L., Continental drift and convection, in *The Earth's Crust and Upper Mantle, Geophys. Monograph 13,* edited by P. J. Hart, pp. 683–689, AGU, Washington, D. C., 1969.

Le Pichon, X., Sea-floor spreading and continental drift, *J. Geophys. Res., 73,* 3661–3697, 1968.

Lliboutry, L., Sea-floor spreading, continental drift, and lithosphere sinking with an asthenosphere at melting point, *J. Geophys. Res., 74,* 6525–6540, 1969.

McKenzie, D. P., The geophysical importance of high-temperature creep, in *History of the Earth's Crust,* edited by R. A. Phinney, pp. 28–44, Princeton University Press, Princeton, N. J., 1968.

McKenzie, D. P., Speculations on the consequences and causes of plate motions, *Geophys. J., 18,* 1–32, 1969.

McKenzie, D. P., and J. G. Sclater, Heat flow in the eastern Pacific and sea floor spreading, *Bull. Volcanol., 33*(1), 101–117, 1969.

Molnar. P., and J. Oliver, Lateral variations of attenuation in the upper mantle and discontinuities in the lithosphere, *J. Geophys. Res., 74,* 2648–2682, 1969.

Morgan, W. J., Rises, trenches, great faults, and crustal blocks, *J. Geophys. Res., 73,* 1959–1982, 1968.

Orowan, E., The origin of the oceanic ridges, *Sci. Amer., 214,* 103–119, 1969.

Press, F., Models of the earth's interior (abstract), *Trans. AGU, 50,* 244, 1969.

Press, F., Constraints on models of the mantle from geophysical data, paper presented at Francis Birch Symposium on the Nature of the Solid Earth, Harvard University, Cambridge, Mass., April 16–18, 1970.

Talwani, M., and X. Le Pichon, Gravity field over the Atlantic Ocean. in *The Crust and Upper Mantle of the Earth, Geophys. Monograph 13,* edited by P. J. Hart, pp. 341–351, AGU, Washington, D. C., 1969.

Talwani, M., X. Le Pichon, and J. R. Heirtzler, East Pacific rise: The magnetic pattern in the fracture zones, *Science, 150,* 1109–1115. 1965.

Vogt, P. R., E. D. Schneider, and G. L. Johnson, The crust and upper mantle beneath the sea, in *The Crust and Upper Mantle of the Earth, Geophys. Monograph 13,* edited by P. J. Hart, pp. 556–617, AGU, Washington, D.C., 1969.

Walcott, R. I., Flexure of the lithosphere at Hawaii, *Tectonophysics, 9*(4), in press, 1970*a*.

Walcott, R. I., Flexural rigidity, thickness, and viscosity of the lithosphere, *J. Geophys. Res., 75,* 3941–3954, 1970*b*.

(Received March 13, 1970;
revised May 19, 1970.)

19

Intraplate Earthquakes, Lithospheric Stresses and the Driving Mechanism of Plate Tectonics

LYNN R. SYKES & MARC L. SBAR
Lamont-Doherty Geological Observatory of Columbia University, Palisades, New York 10964

Focal mechanisms of about eighty intraplate earthquakes and other information on *in situ* stress indicate that the interiors of many lithospheric plates are characterised by large horizontal compressive stresses. These stresses seem to be related to the driving mechanism of plate tectonics.

ABOUT 90% of the world's earthquakes occur along the boundaries of lithospheric plates. Focal mechanism solutions of earthquakes have provided a detailed description of the type of faulting and the nature of the displacements for most of the major plate boundaries[1]. Another class of shocks, intraplate earthquakes, are less numerous than those along plate boundaries and have received relatively little study. Consequently, their cause and tectonic settings have not been well understood.

Here we attempt to determine the type of faulting and to infer the state of stress within lithospheric plates on a world-wide scale from analyses of focal mechanisms of about eighty earthquakes, most of which are located well away from plate boundaries. Our principal result is that a large percentage of these shocks, particularly those not located near mid-ocean ridges or behind subduction zones, are characterised by a predominance of thrust faulting. In the few areas of the world for which they are available, other determinations of the state of stress from overcoring, hydrofracturing and recent crustal movements support our conclusion that large areas in the interiors of many lithospheric plates are characterised by high horizontal compressive stresses. These seem to be uniform in orientation over sizable regions for the few cases where sufficiently dense data are available.

The very existence of earthquakes in the interiors of the plates indicates that the approximation that plates are undeformed except at their edges is not completely correct even though this assumption seems to be a good working hypothesis for many purposes. The intraplate earthquakes discussed in this paper are all of shallow focus and are located in areas where lithospheric plates are horizontal. Here we present evidence which indicates that the most consistent parameter in focal solutions of shallow earthquakes remote from plate boundaries is the orientation of one of the principal stresses. Earthquakes beneath deep-sea trenches, which involve normal faulting, and intermediate and deep shocks[2] are other examples of intraplate events in which the direction of one of the principal stress axes is consistent for various nearby shocks. In contrast, slip vectors, rather than principal stress axes, are consistent for earthquakes along a given plate boundary[3].

While intraplate earthquakes are not numerous compared with shocks along active plate margins, intraplate shocks as large as magnitude 7 have occurred in several populated areas including parts of central and eastern North America, Europe, Australia and South Africa. Examples of destructive and large intraplate earthquakes include New Madrid, Missouri 1811–12; Charleston, South Carolina 1886; Cape Ann, Massachusetts 1755; Grand Banks 1929; Meckering, Australia 1968, and several large earthquakes in the St Lawrence Valley[4].

255

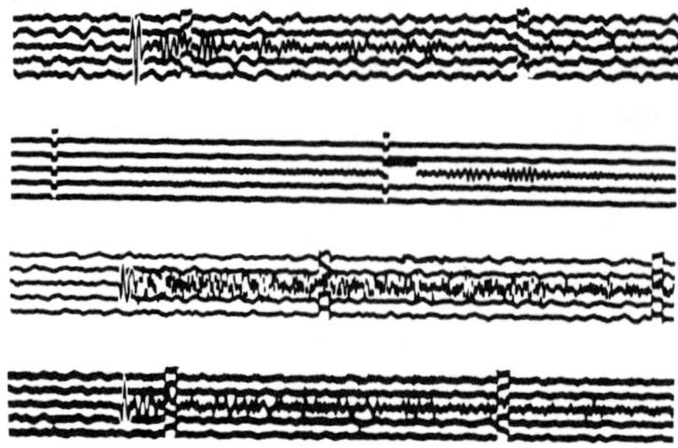

Fig. 1 Short-period vertical seismograms of earthquake near north-east coast of Greenland on November 26, 1971, of magnitude $m_b = 5.2$. Deflexions of trace indicate minute marks, and upward motion of ground is up on records. Station abbreviations, distances and gains are as follows from top to bottom: NUR, 22.8°, 25k; STU, 32.2°, 25k; KBL, 55.7°, 400k; QUE, 59.6°, 200k. The first motion is a clear compression (up) at NUR and a clear dilatational (down) at KBL and QUE. Initiation of P wave and first motion cannot be read at STU, which is inferred to be near a nodal plane. The four stations are labelled on focal mechanism solution in Fig. 2. Arrivals of four P waves are aligned approximately vertically. Light portions of traces are retouched.

These events are characterised by very large areas of perceptibility compared with most shocks of similar magnitudes and energies along plate boundaries. Thus, although they are not numerous, large intraplate shocks constitute an environmental risk that must be taken into account in the design and siting of critical facilities such as nuclear reactors and other large man-made structures. The presence of high compressive stresses in the interiors of many lithospheric plates indicates that the possible triggering of earthquakes by processes that reduce rock strength, such as high-pressure fluid injection and reservoir impounding, must be evaluated carefully.

Large Compressive Stresses

Determinations of stress *in situ*, principally by the overcoring technique, indicate that measured horizontal compressive stresses are much larger than lateral (and even vertical) stresses calculated from the weight of the overburden in Fennoscandia[5], eastern and central North America[6-8], the Colorado Plateau[8-10] and various parts of the USSR[11,12]. Thus, the horizontal component of the stress in these areas must be largely tectonic. Although most of these measurements were made primarily for engineering purposes, Voight[7] recognised that the data may provide important constraints for models of the driving mechanism of large-scale tectonic processes. Nevertheless, very few authors have distinguished between measurements made in what are now recognised as the interiors of plates with those that are pertinent to plate margins.

Sbar and Sykes[8] examined the directions of principal stresses in central and eastern North America as inferred from *in situ* measurements by the overcoring and hydrofracturing techniques, focal mechanisms, and postglacial geological features. They found good agreement among the various techniques and showed that the greatest principal stress trends nearly ENE throughout a large region that includes much of the eastern and central United States. Focal mechanisms of about ten earthquakes in western Europe north of the Alps are characterised by either thrust or strike-slip faulting such that the axes of maximum compressive stress trend consistently north-west[13]. Thus, in two areas where relatively dense observations of stress are available, the maximum compressive stress seems to be horizontal, to be large, and to have a similar trend throughout a broad region. These factors led us to examine the state of stress within lithospheric plates on a global basis.

Focal Mechanism Data

We examined a magnetic tape file of epicentral locations reported by the National Oceanographic and Atmospheric Administration from 1961 to 1971 for obvious intraplate earthquakes. About half of the seventy-nine mechanism solutions used in this article were determined by us. A list of these events and their mechanism parameters is available from us.

Even though intraplate earthquakes are relatively rare, we were able to obtain a large number of solutions by successfully analysing the first motion of compressional waves for events as small as magnitude $m_b = 5.0$ using standard short-period records. Clear, impulsive first motions of P (Fig. 1) and PKP are often observed on short-period records for intraplate events of $5.0 \leq m_b \leq 5.5$. As many intraplate earthquakes involve dip-slip faulting, stations at distances greater than about 40° (Fig. 2) are often remote from a nodal plane, and therefore record P waves with a distinct first motion. The focal mechanism solution shown in Fig. 2 for an event of $m_b = 5.2$ represents about the average quality of the various solutions reported here.

These factors and the simplicity of the waveform of P indicate that the corner frequency for these events is higher than about 1 Hz and that the instruments "see" a point source rather than the initiation of motion from a complex moving source as is generally the case for $m_b > 5.5$. Most intraplate earthquakes have a larger short-period magnitude, m_b, than do events along plate boundaries of the same long-period magnitude, M_s. The enrichment in short-period energy may be caused by greater effective stresses that may be present within plates and/or higher Q paths through the upper mantle, as intraplate events tend to be located in older (and presumably colder and thicker) lithosphere than events along zones of sea-floor spreading and transform faulting.

Stress Distribution

Figure 3 summarises information on the state of stress as inferred from about eighty focal mechanism solutions of intra-

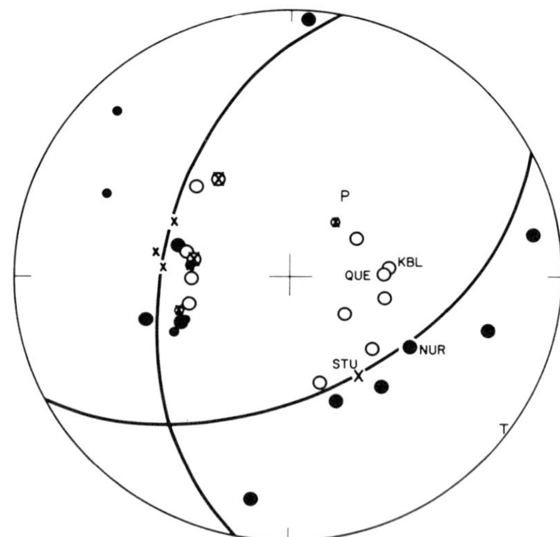

Fig. 2 Focal mechanism (first motion) solution for earthquake near NE coast of Greenland on November 26, 1971. Plot is equal-area projection on lower hemisphere of radiation field. ●, Compressions; ○, dilatations; X, arrivals near nodal plane; P and T are inferred axes of maximum and minimum compressive stress. Large symbols denote more reliable readings. Solution is characterised by a predominance of normal faulting. First motions shown in Fig. 1 are labelled with three-letter codes for stations.

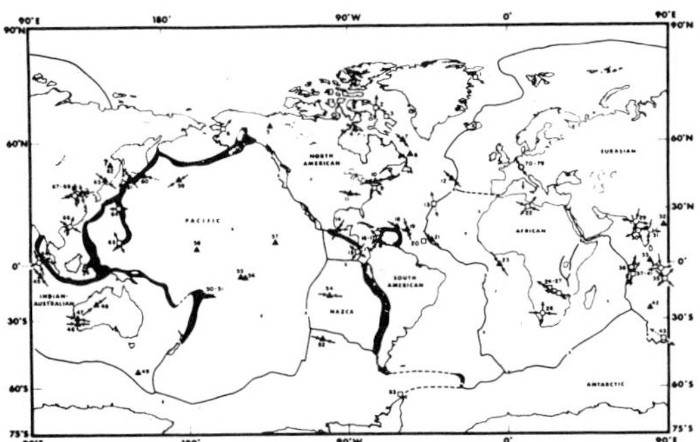

Fig. 3 Worldwide summary of focal mechanism solutions of intraplate earthquakes. ▲, Thrust; ○, strike-slip; □, normal; X, mantle plumes inferred by Morgan[35,36]. Inwardly directed arrows indicate axis of maximum compressive stress (P axis) and outwardly directed arrows denote least compressive stress (T axis). Dashed arrows indicate less reliable directions. Solution in western Europe is representative of ten mechanisms with similar orientation of P axes. Major plates are labelled. Hatching indicates major subduction zones (Mediterranean–Himalayan zone omitted), and solid line represents plate boundaries of extensional and transform type.

plate earthquakes. The solutions are classified as a predominance of either thrust, strike-slip or normal faulting. The first motions of P and PKP are good enough to support this classification for nearly all the events reported. As in other mechanism studies, we equate the P and T axes of the focal solutions with the axes of maximum and minimum compressive stress. For some of the dip-slip events the azimuths of the nodal planes and of the horizontal principal stress are less reliable. Thus, for these events we show only the type of faulting and not the azimuth of the horizontal principal stress. Although the radiation patterns of surface waves can be used to determine this azimuth with greater precision[14-16], we did not employ it here as substantial additional data analysis is involved.

Table 1 Number of Focal Mechanism Solutions as a Function of Predominant Fault Type

	Thrust	Normal	Strike-slip
Total events (1 to 69, Fig. 3)	34	22	13
Total minus regions behind arcs	31	18	8
Total minus regions behind arcs, close to ridges or in East Africa	29	4	7 (6 in Indian plate)

In the Pacific plate each of the solutions in Fig. 3 is characterised by thrust faulting. In the Atlantic, the solutions for earthquakes located well away from the mid-Atlantic ridge also involve thrusting. Table 1 indicates that thirty-four (49%) of the total solutions in Fig. 3 involve thrusting. The percentage of thrusting increases to 73% when events located near ridges, behind subduction zones or in East Africa are deleted.

Each of the three main types of faulting is found behind active subduction zones, although a single stress pattern usually prevails behind a given arc. For example, focal solutions and other geophysical data indicate horizontal compression in the lithosphere behind the Honshu and South American arcs and extension behind the Bonin–Mariana and Tonga–Kermadec arcs[17-21]. The state of stress behind active subduction zones does not seem to be typical of that in horizontal portions of the lithosphere. Evidence for extension in the Basin and Range province of the western United States is omitted from Fig. 3 as the tectonics of that region does not seem to be intraplate and it may reflect processes associated with a former subduction zone or the cessation of subduction near the west coast of the United States[22,23].

Figure 4 shows the type of faulting in oceanic areas as a function of the age of the crust inferred from magnetic anomalies and deep-sea drilling. Thrusting predominates for crust older than 20 m.y.; normal faulting is typical of zones of seafloor spreading at ridge crests and of crust younger than about 10 to 20 m.y. Clearly, the state of stress within the plate has changed from horizontal extension to horizontal compression within the past few tens of millions of years, as it moved away from the ridge crest. The precise age and nature of the boundary between thrusting and normal faulting, however, are not well defined. Six of the eight examples of normal faulting for regions younger than 20 m.y. and not located on ridge crests come from nearly the same location in the Indian Ocean near a large seamount at the southern end of the Chagos–Laccadive ridge[24]. Although solutions characterised by normal faulting were not found near the East Pacific Rise, the one thrusting solution at 10 m.y. indicates that the transition to normal faulting on that ridge system occurs for crust of younger age.

No examples of normal faulting were found for oceanic crust older than 10 to 20 m.y. Normal faulting was detected for an event in New Jersey and for shocks near the coasts of Greenland, the Antarctic Peninsula, and Baffin Island (Fig. 3). All of these events, however, were located on the continental shelf or within continents and not on oceanic crust. We refrain from further speculation about stresses near continental margins, as the number of events is small.

Although not as clear as in the older oceanic lithosphere, many continental areas, interior to plates, are also characterised by large horizontal compressive stresses. Focal mechanism solutions in Fig. 3 and/or measurements of *in situ* stress corroborate this for western Europe and Fennoscandia[5,13], large areas of the USSR[11,12], most of the United States east of the Basin and Range Province[8], western Australia, and peninsular India.

Several examples of normal faulting are found in East Africa (Fig. 3). These events were not located along what are normally regarded as active rifts[27]. Nevertheless, it is difficult to tell if these solutions are better characterised as intraplate events or as earthquakes associated with a wide and complex plate boundary within a continent. A similar problem was encountered in characterising earthquakes in the broad seismic zone, which probably involves continental collision, in the Mediterranean, Central Asia and the Himalayas[26]. The mechanism of the Libyan earthquake in Fig. 3 may be related to this broad plate boundary.

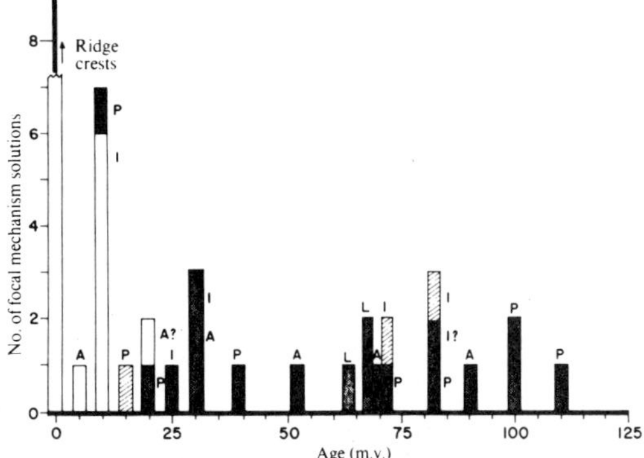

Fig. 4 Type of focal mechanism as a function of age of oceanic crust. Note predominance of normal faulting on, or close to, ridge crests and predominance of thrust faulting for ages greater than 20 m.y. □, Normal; ■, thrust; ▨, strike-slip type faulting. A, Mid-Atlantic ridge; P, Pacific ridges; L, mid-Labrador sea ridge; I, mid-Indian ridges.

Driving Mechanism

From a compilation of focal mechanism solutions Fitch *et al.*[14] concluded that the orientation of the maximum compressive stress (P axis) varied smoothly throughout the Indian–Australian plate. As the axes of least compression (T axes) for their solutions pointed toward the Java trench, they concluded that gravitational sinking of a lithospheric slab at the subduction zone was an important force driving this plate. All of their solutions in the north-east Indian Ocean, however, involved strike-slip faulting. Additional mechanism solutions for this plate (Fig. 3) indicate thrust as well as strike-slip faulting. The two types of solutions show a similar orientation of the maximum compressive stress for nearby events 44 and 45. An explanation of these data is that the minimum and intermediate principal stresses are of similar size, making possible either thrust or strike-slip faulting. As either one of the minimum and intermediate principal stresses is vertical, they both must be similar in magnitude to the stress resulting from the weight of the overburden. This implies that a significant horizontal tensional component is not being applied to the stress field in this region as a result of gravitational sinking. Instead, the mechanism solutions indicate a large horizontal compressive stress, apparently of tectonic origin.

These conclusions are further supported by thrusting mechanisms (Fig. 3) of shocks located seaward of the Aleutian, Kurile, Tonga and Lesser Antillean arcs. If gravitational sinking were primarily responsible for the stress distribution in plates, normal rather than thrust faulting would be typical of intraplate deformation (Fig. 5a). Earthquakes beneath deep-sea trenches[1,21,28–30] are typified by normal faulting, but the causative stress seems to be related mainly to the flexure of the lithosphere[1]. Normal faulting reported for an event on the outer wall of the Japan trench also seems to be related to flexure rather than to a gravitational sinking mechanism as proposed by Shimazaki[31]. Evidence from topographic and gravity anomalies also indicates great horizontal compressive stresses seaward of several Pacific trenches[32,33]. The opening of the South Atlantic is also difficult to explain solely by gravitational sinking as the African and Americas plates are not being subducted except in very local areas[34]. These observations, of course, do not rule out gravitational sinking as a contributory driving force even if it is not the primary driving mechanism. It may well be responsible for the distribution of stress in downgoing plates in island arcs[2].

Several events in Fig. 3 are located so close to plate boundaries that it is not certain if their mechanisms reflect intraplate stresses or boundary effects. Even though events 50, 51 and 60 are located just seaward of deep-sea trenches, their thrusting mechanisms are distinctly different from normal faulting that characterises events within trenches. As thrusting is also found elsewhere in the Pacific plate, we interpret these three solutions as intraplate events and not as boundary effects. Events 61, 64 and 65, however, are located behind subduction zones near volcanic arcs. Their mechanisms may be related to the flexure of the lithosphere, other boundary effects, or, in the case of events 64 and 65, to inter-arc spreading[20]. Event 13 was located about 50 km west of the rift valley and the main seismic zone of the mid-Atlantic ridge[25]. It is difficult to characterise this event as a plate boundary or an intraplate phenomenon.

Several other mechanisms for driving plates and for stressing their interiors are shown schematically in Fig. 5. Observations of normal faulting close to active ridges and thrusting at greater distances are compatible with plates being driven from below by flow in the asthenosphere or by gravitational sliding or pushing from the general area of ridge crests. The easterly trends of the maximum compressive stress in the south-east Pacific provide evidence for these two mechanisms, as they are nearly perpendicular to the crest of the East Pacific Rise. The trends for solutions 18, 19, 21, and 23 in the Atlantic, however, are nearly parallel to the axis of the mid-Atlantic ridge. Although the trends for these three solutions are not precisely constrained by first-motion data, the uncertainty probably does not exceed 25° for events 19 and 23. Likewise, the maximum compressive stress for events in western Australia, at least one of which is well constrained to be nearly east–west from the radiation pattern of surface waves and from observations of surface faulting[14], is nearly parallel to the axis of the nearby branch of the mid-Indian Ocean ridge. Thus, several solutions do not fit a simple model in which gravitational sliding or pushing from a line source located at the ridge crest is the primary motive force of plate tectonics.

Another possible source of compressive stresses within plates is not related to the driving mechanism, but merely to the gradual cooling and thickening of lithospheric plates (Fig. 5) in that stresses may be frozen in as in the cooling of glass. In oceanic areas the transition from normal to thrust faulting occurs for crust about 10 to 20 m.y. old. This age is nearly the same as the thermal time constant for the formation and destruction of the lithosphere[1,34]. Nevertheless, we would expect that compressive stresses frozen into the oceanic lithosphere by this mechanism would exhibit some systematic alignment with respect to the direction of lithospheric cooling and thickening. As in the case of gravitational sliding, the P axes of focal mechanisms do not seem to have such a systematic orientation. Near solutions 18, 19 and 23, however, the pattern of sea-floor evolution is not well known and we cannot evaluate this mechanism definitively.

Morgan[35,36] suggests that mantle plumes are the main driving mechanism of plates. We attempted to test the simple case where each of the plumes he proposes (Fig. 3, X) was of equal strength and the maximum compressive stress was orientated radially about each plume. The orientations of the P axes from focal mechanisms do not obviously correlate with those predicted by the simple model. Also, the directions of principal stresses in Fig. 3 do not seem to correlate with directions of absolute plate motions inferred by Morgan[35,36] from tracks of hot spots.

Although we cannot isolate a single mechanism as the driving force of plate tectonics, data on the distribution of stress on a worldwide basis provide valuable constraints on proposed mechanisms. We refrain from looking for patterns in the directions of principal stresses in Fig. 3 as the data points are widely spaced except for those in eastern North America and northern Europe where the stress fields do, in fact, appear to be uniform over large areas. Additional focal mechanism solutions and *in situ* stress measurements by the overcoring and hydrofracturing techniques should help to clarify if stress fields generally vary rapidly or are nearly uniform in extent over broad areas in the interiors of most plates. The density of stress determinations in oceanic areas could be greatly

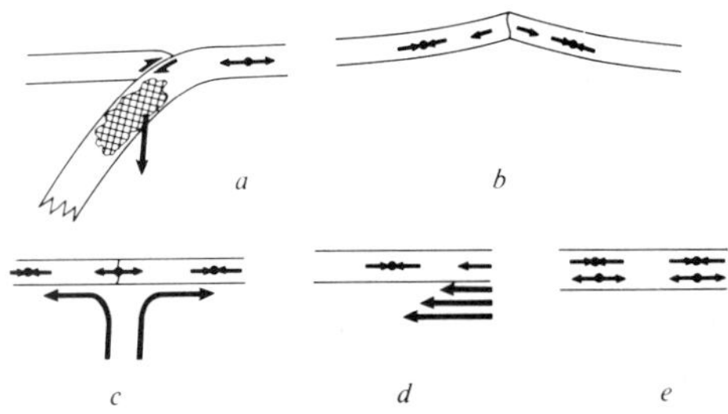

Fig. 5 Schematic diagrams of various mechanisms for generating stresses within lithospheric plates. Single arrows denote relative motions and double arrows denote either compressional (inward) or tensional (outward) deviatoric stresses. *a*, Gravitational sinking in island arc; *b*, gravitational sliding or pushing from ridge; *c*, mantle plume; *d*, rapid flow in asthenosphere; *e*, stresses related to cooling of lithosphere. Hatching in *a* denotes mass excess caused by either cooler material or increased elevation of phase boundaries in sinking slab.

increased by using ocean-bottom seismographs for first-motion studies and by making hydrofracturing measurements in holes drilled into layer 2 of the oceanic crust.

This work was supported by the National Science Foundation, the US Geological Survey and the Advanced Research Projects Agency of the Department of Defense through the Air Force Cambridge Research Laboratories. We thank Paul Pomeroy and Walter Pitman for reviewing the manuscript and D. McKenzie for discussions.

Received July 7; revised August 6, 1973.

[1] Isacks, B. L., Oliver, J., and Sykes, L. R., *J. geophys. Res.*, **73**, 5855 (1968).
[2] Isacks, B., and Molnar, P., *Rev. geophys. Space Phys.*, **9**, 103 (1971).
[3] McKenzie, D., and Parker, R. L., *Nature*, **216**, 1276 (1967).
[4] Smith, W. E. T., *Dom. Obs. Pamph.*, **26**, 271 (1962).
[5] Hast, N., *Tectonophysics*, **8**, 169 (1969).
[6] Hooker, V. E., and Johnson, C. F., in *Proc. Fourth Rock Mech. Symp.* (Department of Energy, Mines and Resources, Ottawa, Canada, 1967).
[7] Voight, B., *Am. Assoc. Petrol. Geol. Mem.*, **12**, 955 (1969).
[8] Sbar, M. L., and Sykes, L. R., *Bull. geol. Soc. Am.*, **84**, 1861 (1973).
[9] Raleigh, C. B., Healey, J. H., and Bredehoeft, J. D., *Geophys. Monog.*, **16**, 275 (1972).
[10] Haimson, B. C., in *Proc. Fourteenth Symp. Rock Mech.*, 689 (American Society of Civil Engineers, 1973).
[11] Kropotkin, P. N., *Phys. Earth planet. Int.*, **6**, 214 (1972).
[12] Turchanivov, I. A., Markov, G. A., Gzovsky, M. V., Kazikayer, D. M., Frenze, U. K., Batugin, S. A., and Chabdarova, U. I., *Phys. Earth planet. Int.*, **6**, 229 (1972).
[13] Ahorner, L., *Geol. Rdsch.*, **61**, 915 (1972).
[14] Fitch, T., Worthington, M. H., and Everingham, I. B., *Earth planet. Sci. Lett.*, **18**, 345 (1973).
[15] Mendiguren, J. A., *J. geophys. Res.*, **76**, 3861 (1971).
[16] Forsyth, D., *Nature*, **243**, 78 (1973).
[17] Ichikawa, M., *Pap. Meteor. Geophys.*, **16**, 104 (1965).
[18] Ichikawa, M., *Pap. Meteor. Geophys.*, **16**, 201 (1966).
[19] Isacks, B., *Abstr. Prog. Seismol. Soc. am. Annual Meeting* (1969).
[20] Karig, D. E., *J. geophys. Res.*, **76**, 2542 (1971).
[21] Katsumata, M., and Sykes, L. R., *J. geophys. Res.*, **74**, 5923 (1969).
[22] Scholz, C., Barazangi, M., and Sbar, M. L., *Bull. geol. Soc. Am.*, **82**, 2979 (1971).
[23] Smith, R., and Sbar, M. L., *Bull. geol. Soc. Am.* (in the press).
[24] Heezen, B. C., and Tharp, M., *Indian Ocean Floor* (map) (National Geographic Society, Washington DC, 1967).
[25] Sykes, L. R., *J. geophys. Res.*, **75**, 6598 (1970).
[26] Molnar, P., Fitch, T. J., and Wu, F. T., *Earth planet. Sci. Lett.* (in the press).
[27] Maasha, N., and Molnar, P., *J. geophys. Res.*, **77**, 5731 (1972).
[28] Stauder, W., *J. geophys. Res.*, **73**, 7693 (1968).
[29] Johnson, T., and Molnar, P., *J. geophys. Res.*, **77**, 5000 (1972).
[30] Molnar, P., and Sykes, L. R., *Bull. geol. Soc. Am.*, **80**, 1639 (1969).
[31] Shimazaki, K., *Phys. Earth planet. Int.*, **6**, 397 (1972).
[32] Hanks, T. C., *Geophys. J.*, **23**, 173 (1971).
[33] Watts, A. B., and Talwani, M., *Geophys. J.* (in the press).
[34] McKenzie, D. P., *Geophys. J.*, **18**, 1 (1969).
[35] Morgan, W. J., *Bull. Am. Assoc. Petrol. Geol.*, **56**, 203 (1972).
[36] Morgan, W. J., *Geol. Soc. Am., Mem.*, **132**, 7 (1973).

20

Some Simple Physical Models for Absolute Plate Motions

SEAN C. SOLOMON AND NORMAN H. SLEEP

Although the relative angular velocities of the earth's plates are well known, the velocities relative to the underlying mantle and the nature of the driving forces are not. We calculate several solutions to the 'absolute' velocity field from the hypothesis that no net torque is exerted on the lithosphere as a whole and a series of assumptions about the forces driving plate motions. The force models include viscous drag at the base of all plates (plus arbitrary forces at ridges and trenches that exert torques of equal magnitude but opposite sign on adjacent plates), drag concentrated beneath continents, drag concentrated to oppose the horizontal translation of sinking slabs, gravitational pull by sinking slabs, and linear combinations of these. These models generally give absolute plate velocities that are very similar to those calculated on the premise that a set of 'hot spots' provides a fixed reference frame. This removes some of the rationale for attributing to these hot spots any significant contribution to the driving forces. The similarity of plate motions inferred from markedly different driving mechanisms precludes using the velocity fields to discriminate among proposed models. Many of the hot spots may be simply due to intraplate tensional stress associated with the forces acting on present plate boundaries. This suggests that calculation of intraplate stress can ultimately serve to choose among force models.

The theory of plate tectonics, which dictates that the earth's surface is made up of a relatively small number of rigid spherical shells or plates of lithosphere that move with respect to each other, has synthesized a wide range of observations from various fields of geology and geophysics. Although the relative motion of the larger plates, about 10 in number, is reasonably well determined [*Morgan,* 1968; *Le Pichon,* 1968; *Chase,* 1972; *Minster et al.,* 1974], the 'absolute' motion of the plates relative to the underlying mantle and the physical mechanisms governing plate motions are not. In this paper we will examine the implications for absolute plate motions of several possible simplified driving mechanisms and relate the results to the origin of seamount chains and midplate stress.

Previous estimates of the absolute velocities of plates were based on the premise that one plate is stationary [*Knopoff and Leeds,* 1972; *Burke and Wilson,* 1972] or that some set of 'hot spots' provides a reference frame fixed with respect to the lower mantle [*Wilson,* 1965; *Morgan,* 1971, 1972, 1973]. Our philosophy in this paper will be first to postulate the nature of the forces driving plate motions and then to examine whether any particular plate might be more or less fixed or whether the 'hot spot poles' have an underlying physical explanation. We shall find that a number of independent postulates for the driving forces produce absolute velocity fields that are very similar to those that have been calculated with respect to a set of hot spots.

DRIVING FORCES

The absolute velocities of the earth's plates may be calculated from simple physics and three main assumptions: (1) the boundaries and relative velocities of the plates are known, (2) there is no net torque on the lithosphere as a whole, and (3) motions in the underlying mantle may be neglected in comparison with plate motions. The first assumption is quite good. We show below that changes in the catalog of plates or in relative rotation vectors within current uncertainties do not appreciably affect absolute velocities. The second assumption is also quite reasonable. It follows as a condition for mechanical equilibrium from the postulate that plates of lithosphere are rigid and transmit horizontal forces, and it is equivalent to the statement that the lithosphere is not accelerating. Torques arising from sources external to the earth (e.g., tides) have not been included. The third assumption may not be correct. It is included for want of better information to give meaning to the motion of absolute velocity. If mantle velocities were known, the calculations below could be repeated, since all forces considered are either independent of or linearly proportional to the relative velocity between plate and mantle.

Possible mechanisms for exerting torque on plates include viscous drag on the base of plates and buoyancy forces at midocean ridges and subduction zones. We treat ridges and subduction zones as line forces, though strictly, these features have finite width.

It is more convenient to treat the lithosphere as a whole than to consider the balance of torques on individual plates for at least two reasons. (1) The uncertain transfer of forces between plates at transform and thrust faults then does not enter into the problem. As an example of such potential difficulties, normal forces across transform faults are not a general property of this type of plate boundary but depend on the general state of stress in the lithosphere. (2) Bilaterally symmetric features such as ridges do not need to be considered explicitly in the torque balance, since they exert no net torque on the lithosphere. An obvious corollary of this approach is that our models for velocity cannot be used to constrain the forces associated with such symmetric structures.

Drag at the base of plates due to the viscosity of the asthenosphere is an important element in the equations balancing torques for any force model. We assume throughout that drag obeys a simple viscous law, so that the drag force is linearly proportional to the velocity of the lithosphere relative to the presumedly fixed underlying mantle. In the simplest model a uniform drag coefficient beneath all plates is adopted. There are grounds for believing that the drag law is complicated by frictional heating [*Schubert and Turcotte,* 1972] and a power law dependence of viscosity on strain rate

260

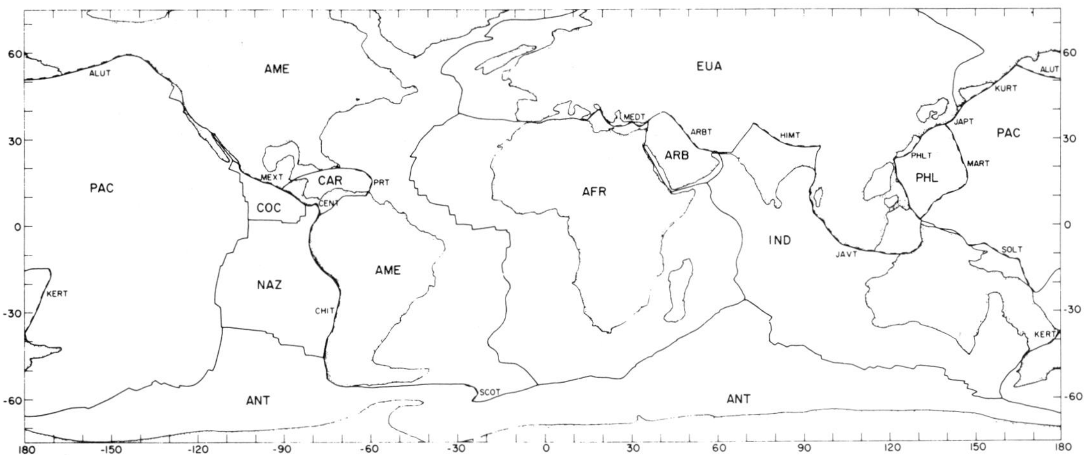

Fig. 1.   Outline of the plates, continental regions (shaded borders), and subduction zones (heavy dashed lines) considered in this study. Cylindrical equidistant projection.

[*Weertman*, 1970]. Incorporation of these effects is probably premature at this time.

The drag coefficient may be regionally variable. Continents, which generally have less well-developed low-velocity zones than oceans, may experience greater drag [*Knopoff*, 1972; *Minster et al.*, 1974; *Alexander and Sherburne*, 1972]. It has also been proposed that drag opposing the horizontal motion of downgoing slabs may provide the major resistance to plate motion, although the physical mechanism for this is unclear [*Talwani*, 1969; *Tullis*, 1972].

Pull on the lithosphere due to negative buoyancy of the downgoing slab is an important element of most dynamical models for plate motions [*Elsasser*, 1969; *Jacoby*, 1970; *Isacks and Molnar*, 1971; *Richter*, 1973a]. According to *Morgan*'s [1972] model the absolute velocity of each of the abutting plates at most subduction zones has a component toward the trench, though the trenchward component of velocity of the subducted plate is generally larger than that of the overthrust plate. This would imply that whereas the subducted slab appears to exert a torque on both plates, the torque exerted

on the downgoing plate may be larger in magnitude. Such a net torque would probably be independent of absolute plate velocity, since no coupling to the lower mantle is explicitly involved. The direction of the force is likely to be normal to the trench by reason of symmetry.

COMPUTATION OF ABSOLUTE VELOCITIES

We compute absolute velocity fields for various possible driving mechanisms by balancing the torques on the plates. As the relative velocities of all plates are reasonably well known, the principal unknown is the absolute velocity of any chosen plate. A three by three system of equations in the vector components of that velocity results when torques are summed (appendix).

The boundaries of the plates that we considered in these calculations are shown in Figure 1. The six major plates (PAC, AME, AFR, EUA, IND, and ANT) compose most of the earth's surface area. We included five additional smaller plates (NAZ, PHL, ARB, CAR, and COC) for which the boundaries and the motions relative to a larger plate are

TABLE 1.   Adopted Values for Relative Plate Motions

Plate	Area*	Continental Area*	Latitude, deg	Longitude, deg	10^{-7} $\omega,$ deg/yr	ω_x	ω_y	ω_z
					Relative Rotation Vector[†]			
PAC, Pacific	2.664	0.046			0.			
EUA, Eurasian	1.675	1.463	63.5	-90.1	8.88	-0.01	-3.96	7.95
AME, American	2.486	1.494	52.0	-73.0	7.86	1.41	-4.63	6.19
IND, Indian	1.503	0.531	58.3	-6.5	12.90	6.73	-0.76	10.97
AFR, African	1.931	0.872	57.6	-63.6	11.06	2.64	-5.31	9.34
ANT, Antarctic	1.477	0.442	69.4	-72.8	9.81	1.02	-3.30	9.18
NAZ, Nazca	0.405	0	59.0	-97.0	14.60	-0.92	-7.46	12.51
COC, Cocos	0.076	0	36.3	-108.5	23.71	-6.06	-18.11	14.05
PHL, Philippine	0.141	0	5.8	-21.2	4.87	4.51	-1.75	0.49
CAR, Caribbean	0.087	0.034	65.3	-75.0	7.76	0.84	-3.14	7.04
ARB, Arabian	0.121	0.104	56.5	-32.4	13.23	6.17	-3.91	11.03

*Earth's radius is set equal to 1.

†The listed rotation vector for a plate is appropriate for a coordinate frame in which PAC is stationary. A right-hand rule is used for the vector $\vec{\omega}$. The Cartesian coordinates x, y, and z are measured along the radius vectors through latitude-longitude pairs (0°, 0°), (0°, 90°), and (90°, 0°). North and east are positive.

reasonably well known. The plate boundaries that we adopted were taken primarily from seismicity maps and are those generally accepted except that the boundary between the American and Eurasian plates from the Nansen ridge to the Japan trench was taken from *Chapman and Solomon* [1973].

The relative plate velocities that we adopted are given in Table 1. The rotation poles and angular velocities are appropriate for a coordinate frame in which the Pacific plate is fixed. Rotation vectors for the major plates and for the Nazca and Cocos plates are from *Chase* [1972]; those for the Philippine and Arabian plates are from *Fitch* [1972] and *McKenzie and Sclater* [1971], respectively. We estimated the rotation vector for the Caribbean plate. The rotation pole for motion between the Caribbean and American plates that best fits the strikes of the El Pilar fault, the Puerto Rico trench, and three segments of the Cayman trough is at 28°S, 69°W. The angular velocity of the American plate with respect to the Caribbean plate is 1.8×10^{-7} deg/yr about this pole if the rate (0.66 cm/yr) of postglacial right lateral movement on the Boconó fault [*Schubert and Sifontes*, 1970] can be used as an index. This figure is in approximate accord with inferred post-Eocene displacement (0.5 cm/yr) along the northern boundary of the Caribbean plate [*Malfait and Dinkelman*, 1972] and with the rate of seismic activity along the Antilles island arc [*Molnar and Sykes*, 1969].

The boundaries of the continental regions that we used for velocity models in which drag is concentrated beneath continents are also shown in Figure 1. These boundaries include continental shelves as well as land areas. The limit of the continental shelf was generally taken to be the 2000-m isobath. The subduction zones included for models in which drag on or pull by the downgoing slab was an essential element are listed in Table 2 and shown in Figure 1. Some regions of very slow convergence (e.g., Azores-Gilbraltar and Macquarie-New Zealand) were not considered.

A couple of points are worth noting in passing. First, the conservation of energy equation in general gives little additional information, since symmetrical forces resisting motion at thrust and transform faults and symmetrical driving forces at ridges and perhaps also trenches are unknown and can be added in equal amounts. An exception to this is noted below: conservation of energy is useful for limiting the ratio of forces due to trench pull and to plate drag. Second, if stresses were computed from the driving mechanism models, ridges and faults would need to be included explicitly. More will be learned once this computation is done.

Absolute velocity fields were calculated for several proposed driving mechanisms that could be modeled quantitatively and that affected absolute velocity. Drag forces beneath the lithosphere, higher drag beneath continents, drag at island arcs, and driving forces due to slabs were considered. The resulting absolute velocities are given in Tables 3–5 and Figures 2–5.

DISCUSSION OF ABSOLUTE VELOCITIES

The simplest force model is one in which the drag coefficient is everywhere uniform and torques exerted on adjacent plates at trenches and ridges are equal in magnitude and opposite in sign. The absolute plate motions for such a model, which we designate model A, are illustrated in Figure 2. This model is one in which there is no net rotation of the lithosphere and for which the rms velocity of the lithosphere as a whole is a minimum (for the assumed plate boundaries and relative velocities).

A most interesting feature of Figure 2 is the striking similarity of the absolute velocity field for this simple model to the velocities calculated from the hypothesis of fixed hot spots [*Morgan*, 1971, 1972, 1973; *Minster et al.*, 1974]. *Minster et al.* [1974] commented, in fact, that a velocity model with no net rotation of the lithosphere would be very similar to their velocity model that best fit the traces of proposed hot spots. Although the fit of model A is close to that of the models of Morgan and Minster et al., it is not perfect: PAC in model A is moving slower than it is in the fixed hot spot models (Table 3), for instance, and AFR is moving faster (Table 4). There is also some disagreement in the poles of the more slowly moving plates, a consequence of the sensitivity of the small angular velocity vectors of these plates to small

TABLE 2. Subduction Zones Included in Calculations

Trench or Subduction Zone	Landward Plate	Length*	Depth,† km	Torque*§			
				T	T_x	T_y	T_z
ALUT, Aleutian	AME	0.601	1.6	0.487	-0.153	0.462	-0.033
MEXT, Mexican	AME	0.346	1.3	0.325	-0.263	0.094	0.165
CENT, Middle American	CAR	0.300	1.3	0.255	-0.207	-0.001	0.149
CHIT, Peru-Chile	AME	1.049	1.8	0.925	0.049	-0.401	0.833
PRT, Puerto Rico	CAR	0.209	2.0	0.161	0.077	-0.006	-0.142
SCOT, South Sandwich	ANT	0.100	3.1	0.085	-0.049	0.054	-0.044
MEDT, Mediterranean	EUA	0.437	0.8	0.318	0.127	-0.291	0.011
ARBT, Arabian	EUA	0.352	···	0.327	0.118	-0.237	0.191
HIMT, Himalayan	EUA	0.406	···	0.372	0.325	-0.115	0.140
JAVT, Java-Banda Sea	EUA	0.898	1.0	0.706	0.580	0.221	0.335
SOLT, New Britain-Solomon-New Hebrides	PAC	0.435	2.1	0.399	0.014	0.288	0.275
KERT, Tonga-Kermadec-New Zealand	IND	0.473	2.4	0.461	0.184	0.192	-0.377
KURT, Kuril	AME	0.389	2.2	0.370	-0.093	0.299	-0.197
JAPT, Japan	EUA	0.062	1.6	0.061	0.002	0.048	-0.037
PHLT, Ryukyu-Philippine	EUA	0.748	1.7	0.575	-0.039	0.219	-0.530
MART, Izu-Bonin-Mariana	PHL	0.703	2.2	0.548	-0.042	0.217	-0.502

*Earth's radius is set equal to 1.
†Average depth of trench below adjacent basin floor is estimated from U.S. Hydrographic Office charts.
§C_t = 1 (see (16) in appendix).

TABLE 3. Absolute Velocity of the Pacific Plate

Pole				
Latitude, deg	Longitude, deg	ω, 10^{-7} deg/yr	v at Hawaii, cm/yr	Explanation
Models Discussed in Text				
-64.3	114.3	7.21	7.65	A (uniform drag beneath all plates)
-62.2	110.4	9.19	9.66	B (drag beneath continents only)
-63.9	107.1	6.88	7.16	C (drag opposing horizontal motion of slabs without ARBT and HIMT)
-56.4	104.5	7.92	8.18	D (maximum pull by slabs plus plate drag)
-66.5	121.7	7.09	7.61	A1 (model A with relative velocities from *Minster et al.* [1974]
-64.3	113.6	7.15	7.57	A2 (model A without PHL, ARB, and CAR)
-63.1	111.7	8.06	8.50	B1 (continents have 3 times more drag than oceans)
-64.2	104.6	7.12	7.35	C1 (model C with ARBT and HIMT)
Fixed Hot Spot Hypothesis				
-67	107	10.5	10.9	*Morgan* [1972,1973]*†
-61.8	117.4	7.96	8.52	*Morgan* [1973]§
-67	135	11	12	*Winterer* [1973]*
-72	97	13	13	*Clague and Jarrard* [1973]*
-67.3	120.6	8.3	8.9	*Minster et al.* [1974]

*Hot spots beneath PAC only were considered.
†Discussed in his text
§Discussed in his table.

changes in the net rotation of the lithosphere.

To test the dependence of model A on the assumed parameters of the plates (Table 1), we considered several minor modifications to the model. We treated a model identical to A except that the relative rotation vectors of *Minster et al.* [1974] instead of those of *Chase* [1972] and *McKenzie and Sclater* [1971] were adopted. (We used the average of the North American-Pacific and South American-Pacific rotation vectors of Minster et al. for our AME-PAC pole and rate, since we did not divide AME into two plates.) The absolute velocities for this model (A1) are in essential agreement with those for model A (Table 3). We also treated a model identical to A except that three of the smaller and less well-defined plates were omitted: PHL was incorporated into PAC, ARB into IND, and CAR into AME. The absolute velocities of the major plates in this model (A2) are almost identical to those in model A (Table 3). Thus the present uncertainties in the boundaries and relative rotations of the plates of the world are sufficiently small not to affect significantly the results of this study.

When all of the drag at the base of the lithosphere is concentrated beneath continental regions, the absolute plate velocities are as shown in Figure 3. This model, designated model B, corresponds to no net rotation of the continents; the lithosphere has a net rotation of 2.0×10^{-7} deg/yr about the pole ($-53.8°$, $100.2°$). The velocities in model B are quite similar both to those for model A and to those in models based on fixed hot spots (Tables 3–5). There are some small distinctions: PAC in model B moves faster than it does in either model A or models in which plates move over a suite of fixed hot spots (Table 3), and AFR moves more slowly in B than in A or than in *Morgan*'s [1971, 1972, 1973] absolute velocity models, though even slower rates for AFR have been proposed (Table 4).

Clearly, a linear combination of models A and B is also permissible. A closer fit of the absolute rotation rate of the Pacific plate to the rate past hot spots is obtained by a combination of the two models than by either end-member. The fit is closest if the drag coefficient beneath continents is 3 to 4 times that beneath oceans (model B1, Table 3). This is a very reasonable ratio, since the predominantly continental plates are moving 3 to 4 times slower than the predominantly oceanic plates in these models [*Minster et al.*, 1974].

If the principal drag resisting plate motions is concentrated to oppose the horizontal motion of sinking slabs, the absolute velocities are as pictured in Figure 4. This model, labeled model C, corresponds to no net rotation of island arcs; there is a small (0.5×10^{-7} deg/yr) net rotation of the lithosphere. ARBT and HIMT are not included in the catalog of subduc-

TABLE 4. Absolute Velocity of the African Plate

Pole			
Latitude, deg	Longitude, deg	ω, 10^{-7} deg/yr	Explanation
Models Discussed in Text			
45.3	-61.2	3.99	A (uniform drag beneath all plates)
35.2	-48.4	2.10	B (drag beneath continents only)
46.7	-54.1	4.34	C (drag opposing horizontal motion of slabs)
55.7	-34.5	3.32	D (maximum pull by slabs plus plate drag)
Fixed Hot Spot Hypothesis			
25	-55	2.45	*Morgan* [1973]*†
38.3	-36.9	2.37	*Morgan* [1973]§
		0	*Burke and Wilson* [1972]*
42.2	-65.2	1.9	*Minster et al.* [1974]

*Hot spots beneath AFR only were considered.
†Discussed in his text.
§Discussed in his table.

TABLE 5. Absolute Velocity of the Eurasian Plate

Pole			
Latitude, deg	Longitude, deg	ω, 10^{-7} deg/yr	Explanation
		Models Discussed in Text	
40.4	-139.5	2.24	A (uniform drag beneath all plates)
-6.6	177.6	1.52	B (drag beneath continents only)
51.8	-130.3	2.26	C (drag opposing horizontal motion of slabs)
49.8	165.3	1.77	D (maximum pull by slabs plus plate drag)
		Fixed Hot Spot Hypothesis	
28.4	133.9	0.82	*Morgan* [1973]*
72	10	4.1	*Duncan et al.* [1972]†
-18	132		*Burke et al.* [1973a]†
38.1	-110.5	1.2	*Minster et al.* [1974]

*Discussed in his table.
†Hot spots beneath EUA only were considered.

tion zones for model C, though a model (C1) in which these regions of continent-continent convergence are included in the subduction zone list gives nearly identical velocities (Table 3). The velocities in model C are similar to those of models A and B, though the agreement with velocities determined with respect to fixed hot spots is not quite as good as it is for the other two models; for instance, much of the Indian plate is moving faster than the Pacific plate. A linear combination of velocities from model C with those from models A and/or B can, of course, also be an acceptable solution. Thus the suggestion of *Talwani* [1969] and *Tullis* [1972] that slabs provide a major source of resistance to plate motion is quite consistent with the absolute velocities of the plates.

A number of velocity models in which torques on subducted plates were an essential component of the driving mechanism were considered. The torques exerted by slabs are balanced by drag beneath the plates (see the final two paragraphs in the appendix). Adding a velocity-independent torque T_0 to the vector equation for the absolute plate velocities amounts to moving all absolute rotation poles toward the 'pole' defined by T_0 and increasing the rms velocity of the earth's plates. Both the magnitude and the pole of the vector T_0 defined by the sum of the torques exerted at the individual subduction zones are obviously dependent on the assumptions about the nature of the slab forces. If the pulling force per unit length of trench is for simplicity taken to be constant and if the 15 oceanic subduction zones in Table 2 are included in the force model, then the pole for T_0 is at latitude and longitude $-3.8°$, $82.3°$. The individual torques (arbitrarily scaled) are listed in Table 2. Note that because the force exerted by a slab is a vector, the torques are not proportional to trench length; the ratio of length to torque is always between 1.0 and 1.4 (in the units of Table 2). If the two continent-continent convergence zones (ABRT and HIMT) are also included, the pole for T_0 is at 11.1°, 58.8°. The force per unit length of trench, of course, is probably not a uniform constant. If this force is proportional, say, to the average difference between the depth of the trench and the depth of adjacent ocean basin floor (Table 2), then a force model including 15 oceanic trenches would have a pole for T_0 at $-19.7°$, 89.4°. Other possibilities for scaling (e.g., force per unit length of trench proportional to total length of subducted slab) might be envisioned.

One important conclusion may be made about all such force models for subduction zones: most of the torques exerted at individual trenches tend to be canceled out by torques at other subduction zones when torque balance is considered for the lithosphere as a whole. The magnitude of the torque sum T_0 for the models discussed above is only 20–30% of the sum of the individual magnitudes of torques at subduction zones. Thus whatever importance the sinking slab may have for driving individual plates, that importance is lessened for the modeling of absolute plate velocities in the manner of this paper. A corollary to this statement is that uncertainties in the relative force per unit length acting at subduction zones are also less important by factors of 3–5 than might initially be imagined.

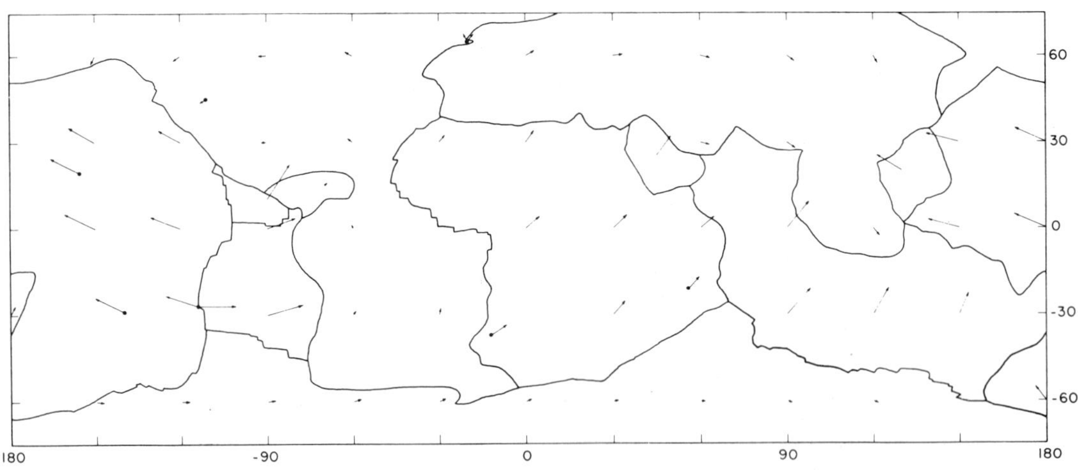

Fig. 2. Absolute velocities of the plates for force model A (uniform drag coefficient beneath all plates). Plate boundaries are from Figure 1. Arrows are proportional to the local plate velocity at the tail of the arrow; a length equal to the distance between tick marks on either the horizontal or vertical axis corresponds to a velocity of 20 cm/yr. Velocities are shown at convenient intervals of latitude and longitude and at several proposed hot spots (dots): Hawaii, Macdonald, Easter, Yellowstone, Iceland, Tristan, Reunion [see *Morgan,* 1973]. Cylindrical equidistant projection.

B. CONTINENT DRAG

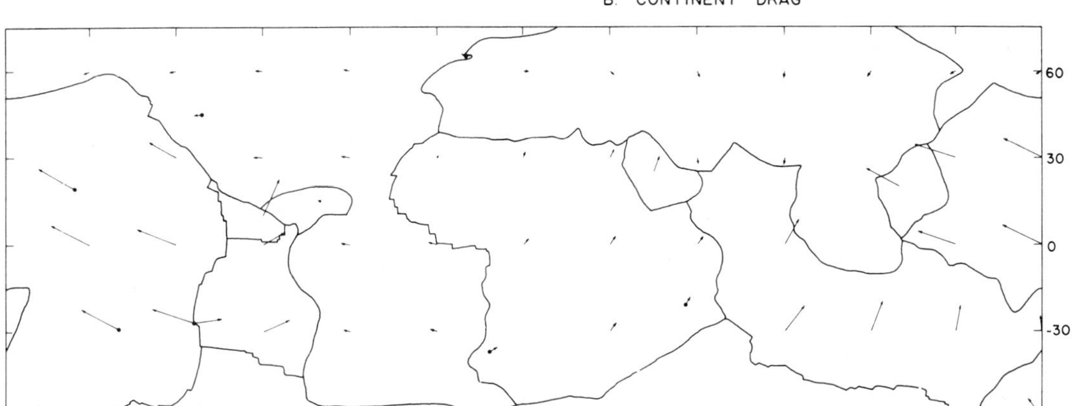

Fig. 3.   Absolute velocities of the plates for force model B (drag concentrated beneath continents). All conventions are as shown in Figure 2.

Even when the relative force per unit length at trenches and the relative drag coefficient beneath plates are specified, there is still an arbitrary parameter reflecting the scaling of pull at trenches to drag (see (18) in the appendix). This results in a family of absolute velocity models for a given torque sum T_0 and a given drag model.

A limit may be placed on the effect of driving forces at trenches on plate motions by simple application of conservation of energy. Clearly, if all of the energy dissipated per unit time by plate drag is equated to the power gained by subduction, then we will obtain an upper bound to the magnitude of the forces at trenches. That this estimate is indeed an upper bound follows from the omission in the energy balance of contributions from ridges or from forces on the overthrust plate at island arcs, since both of these terms likely are opposite in sign to the energy dissipated by drag.

A velocity model including the maximum possible effect of pulling by sinking slabs on subducting plates is shown in Figure 5. The model, designated model D, has been constructed by assuming a uniform drag coefficient beneath all plates and a uniform force per unit length at the 15 oceanic trenches listed in Table 2. (The model is developed in the appendix, (18)–(20), $C/D = 8.4$.) The velocities in this model are again quite similar to those in the force models previously discussed and to those in models constructed from a set of fixed hot spots (Tables 3–5). There is a net rotation of the lithosphere of 1.4×10^{-7} deg/yr about the pole $(-3.8°, 82.3°)$ in model D.

There are two important conclusions that may be made from these various velocity models. The first is that the absolute velocities of the plates cannot serve to discriminate among models for the forces driving plates. A wide assort-

C. SLAB DRAG

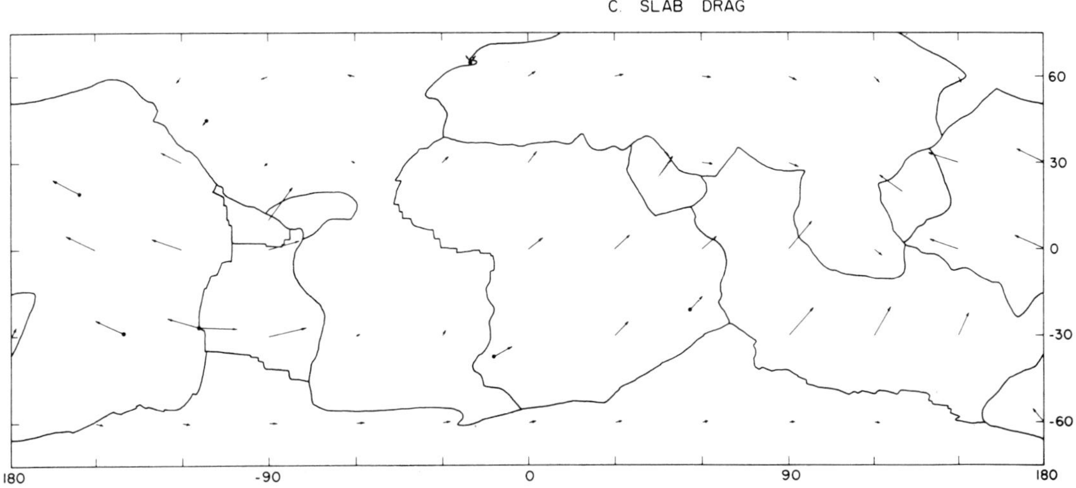

Fig. 4.   Absolute velocities of the plates for force model C (drag concentrated to resist horizontal translation of sinking slabs). All conventions are as shown in Figure 2.

265

ment of assumptions about drag forces and about forces at subduction zones all yielded remarkably consistent patterns of absolute plate velocity. The similarity of velocities in different drag models presumably reflects the fact that either the earth's continental regions or the island arcs or probably the Precambrian shields provide an adequate sampling of the slower-moving plates. The similarity of velocities in different models for subduction zones is principally a consequence of the efficient cancellation of most of the torque exerted by slabs, especially torque about the z axis (through the north pole). It should be noted, however, that there are small differences among the various velocity models that might be used to cast one or more of them into disfavor. Models with net lithosphere rotation greater than perhaps 0.5×10^{-7} deg/yr (e.g., B and D) about an axis significantly different from the earth's spin axis, for instance, can be questioned on paleomagnetic grounds [*McElhinny*, 1973].

To be sure, we assumed a priori the relative plate velocities, which may also contain information about the driving forces. This information may be no easier to unravel, however. Are the predominantly continental plates moving more slowly than the predominantly oceanic plates because there is greater drag beneath continents [*Minster et al.*, 1974] or because the predominantly continental plates are not generally being subducted at island arcs, presumably because of the low density of continental and island arc lithosphere?

The second conclusion is that all of the absolute velocity fields from the various force models are very similar to plate velocities calculated with respect to a global set of fixed hot spots. This result adds a physical basis for the present motions of plates over such hot spots but removes some of the rationale for attributing to hot spots any important contribution to the driving mechanism for plates, in particular, for identifying hot spots with 'plumes' of hot material ascending from the lower mantle [*Morgan,* 1971, 1972, 1973; *Vogt,* 1971].

The internal consistency of the plate motions taken with respect to hot spots precludes an appeal to chance to explain the orientation of seamount chains and other proposed hot spot traces. *Minster et al.* [1974] have shown that the hypothesis that all hot spots have remained fixed with respect to one another for the last 10 m.y. is entirely consistent with relative plate velocities and hot spot traces, though the hot spots appear to show relative motion when longer time periods are considered [*Burke et al.,* 1973b; *Molnar and Atwater,* 1973]. The notion of plumes, however, has several serious difficulties. The greatest of these is that such plumes apparently cannot be modeled physically and can be defined only vaguely. No physically relevant condition has been found for a plume, with conduit radius much narrower than other dimensions of convection in the earth, to form or to be stable. The lack of interference on proposed plumes by plate-induced flow and flow to replenish the lower mantle is hard to explain, since shear would convert plumes into rolls. A shearing flow associated with plate motions would offset the top and bottom of a plume and cause the plume to tilt. Vertical buoyant forces would cause extension of the hot region into a sheet. The effect of shear on rolls has been discussed at length by *Richter* [1973b]. Plumes were not modeled by Richter presumably because conditions for their stability without shear are unknown.

Instead of attempting to refute further a hypothesis that we cannot model we show in the next section that the simple assumptions about driving forces, which give absolute velocities in good agreement with those derived from the hypothesis of fixed hot spots, may also give rise to stresses that can produce seamount chains as a secondary effect without any need for mantle plumes.

SOME COMMENTS ON DRIVING FORCES AND INTRAPLATE STRESSES

A qualitative discussion of forces on plate boundaries and the resulting stress field within plates may be made strictly from plane geometry. Forces perpendicular to the absolute direction of plate motion must originate at the plate boundaries and integrate to zero for each plate, since by definition they cannot be resisted by drag on the base of the plate. Zones of compression or tension perpendicular to absolute motion would occur from opposing forces on opposite sides of the plate. Tension is likely in an inside corner formed by subduc-

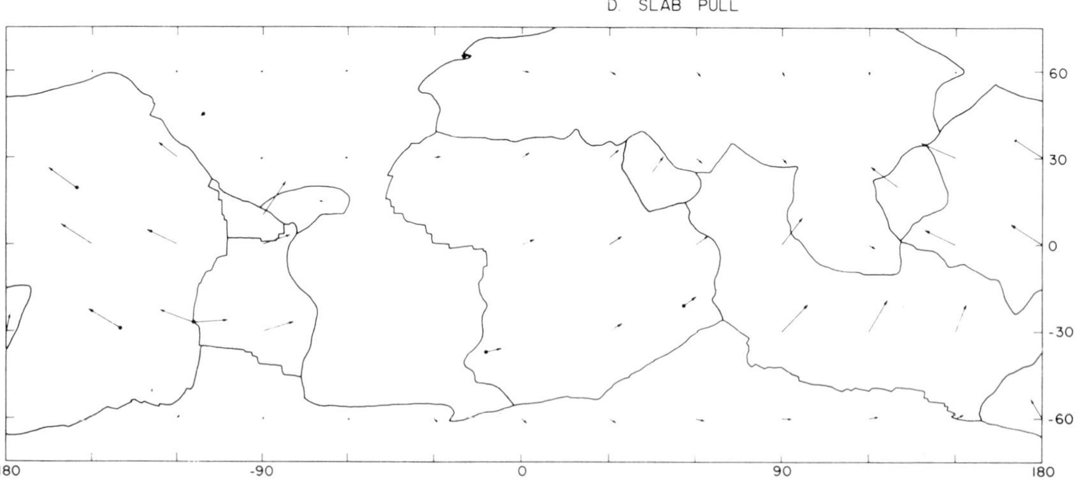

Fig. 5. Absolute velocities of the plates for force model D (maximum pull by slabs on subducting plates plus uniform drag law beneath all plates). All conventions are as shown in Figure 2.

tion zones, which probably pull on the plate. Conversely, plates are probably pushed by forces exerted at spreading centers [*Sykes and Sbar*, 1973], causing compression perpendicular to absolute motion (Figure 6). Gravitational sliding or spreading of the hot buoyant material at midocean ridges is the likely cause of this compressive force [*Lachenbruch*, 1973; *Artyushkov*, 1973].

Stress fields formed in this way are a tempting explanation for seamount volcanism, since tension on the plate can permit volcanic material to upwell passively from the asthenosphere. The heat source for seamounts, which represent only a minor fraction of the magmatic material erupted at midocean ridges, would be according to this hypothesis simply the ambient heat in the interior of the earth. Unlike plume theory the stress field in the interior of plates is amenable to calculation, given a set of forces acting on plate boundaries. Stress models, in addition, are contrained by data, such as midplate earthquake mechanisms [*Sykes and Sbar*, 1973], which are independent of the geometry of seamount chains.

The particular location of a seamount chain in a plate under tension may follow from *Richter*'s [1973b] proposal that secondary convention cells in the mantle should line up with absolute spreading directions to minimize destructive interference with the shear at the base of the plates. Although such cells would have no direct effect on the dynamics of a plate because they produce forces that sum to zero, the stress field would be modulated such that the upwelling axes of the cell would be the foci of maximum tension. These cells provide an explanation for the discrete spacing between seamount chains in the Pacific [*Richter*, 1973b]. Beneath slow-moving plates, such as Africa, a more random distribution of hot spots would be expected, since the small absolute velocity there would be less effective in aligning cells.

Continual motion of the plate with respect to the force field at its boundaries and secondary convective cells beneath could produce chains of volcanoes as new parts of the plate enter zones of tension. Such a seamount chain would be aligned along the absolute spreading direction, but the rate of seamount propagation could be somewhat different from the absolute spreading rate especially if the tensional crack giving rise to the chain modifies the stress field. When a major change in the plate kinematics occurs, both the absolute motion and the stress field within the plate change. A new direction of seamount tracks would then occur.

A difficulty with this hypothesis is that an explanation for the extinction of volcanoes is required. It may be that the local stresses associated with a mature volcano close the vents and preclude further eruption. The stress field also may be such that most of the strain occurs near the active volcanoes. A better understanding of this point will come once quantitative models for intraplate stresses are constructed.

Turcotte and Oxburgh [1973] have proposed that translation of plates over an ellipsoidal earth and lateral thermoelastic stresses, rather than the mechanism discussed above, cause tensional cracks that produce seamount chains and rift structures. Although these processes may provide an explanation for some tensional features, their failure to explain the spatial and temporal distribution of seamounts indicates that other mechanisms must be operating as well.

The greatest stresses from ellipticity are clearly at midlatitudes and on north-south moving plates. Seamount chains in the Pacific, however, are found both far from and near to the equator, and their trend is nearly east-west.

The thermoelastic mechanism of *Turcotte and Oxburgh* [1973] implies that seamount chains should be aligned subparallel to neighboring fracture zones and other fossil lineations rather than along the present direction of absolute motion. It would be hard to dismiss the self-consistent trends of seamount chains at presently active hot spots as coincidence. This hypothesis also cannot readily explain an intersection of seamount chains of different ages, such as the Hawaiian and Marcus-Necker chains, nor changes in the strike of a chain, such as at the Hawaiian-Emperor intersection.

SUMMARY REMARKS

The velocities of the earth's plates with respect to the underlying mantle have been calculated from a number of simple force models for driving forces. Drag beneath plates was in turn supposed to be characterized by a uniform law at

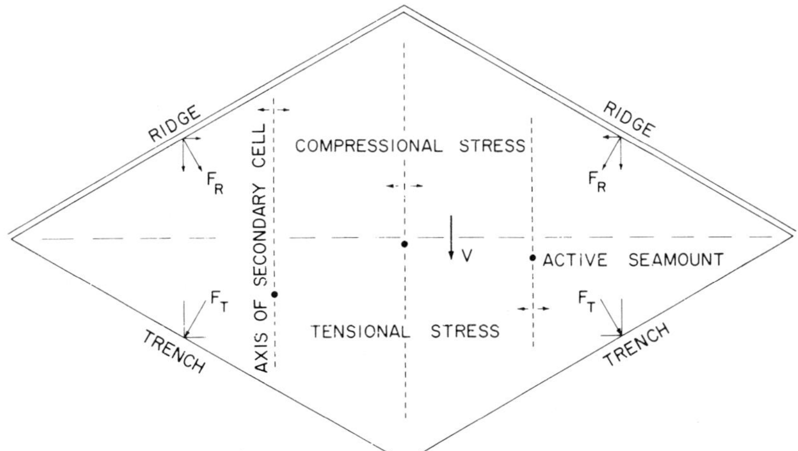

Fig. 6. Notional sketch of forces at plate boundaries (thin arrows), plate motion (heavy arrow), and intraplate stress. The possible relationship among seamount chains, intraplate stress, and secondary convection cells [*Richter*, 1973b] are discussed in the text.

the base of all plates, to be concentrated beneath continents, and to be concentrated to resist horizontal translation of subducting slabs. The forces exerted at ridges were always taken to be symmetric about the ridge axis. The forces exerted at trenches were treated both as being symmetric about the trench axis and as being concentrated primarily on the subducting plate.

The global pattern of absolute velocities is nearly the same for every model considered and is similar to that of the plate velocities calculated with respect to a set of presumedly fixed hot spots. This agreement of velocities in the models to motions estimated with respect to hot spots is best when the comparison is made with studies in which a large number of hot spots are simultaneously considered [*Morgan*, 1973; *Minster et al.*, 1974], rather than with studies of individual hot spots (Tables 3-5). These calculations thus lend a physical basis for absolute plate motions previously proposed but provide absolutely no support for the concept that thermal plumes in the lower mantle help to drive the plates. A good case can be made that intraplate hot spots are the result of tensional stresses arising from forces acting on the plate boundaries.

The absolute plate velocities therefore provide precious little handle on the forces moving the plates. Clearly, the next important calculation to perform is the quantitative modeling of intraplate stress. This stress field should be markedly different for some of the force models considered in this paper, even though the absolute velocities derived from such models are similar. Such indicators of intraplate stress as midplate earthquakes and the seamount chains discussed above should serve to eliminate some of the models proposed for plate driving forces.

Appendix

The relative motion of two rigid plates p and q on a spherical surface can be defined by an angular velocity vector ω_{pq}. The local separation velocity $\mathbf{v}_{pq}$ of the plates at a point with radius vector $\mathbf{r}$ is then

$$\mathbf{v}_{pq} = \omega_{pq} \times \mathbf{r} \qquad (1)$$

The rotation vectors are additive, e.g.,

$$\omega_{pq} + \omega_{qr} = \omega_{pr} \qquad (2)$$

Let 'plate' m be the lower mantle relative to which absolute velocities are calculated. If a linear drag law is assumed, the drag force $\mathbf{F}$ per unit area on any part of the base of the lithosphere is

$$\mathbf{F} = -D\omega_{pm} \times \mathbf{r} \qquad (3)$$

where D is the drag coefficient and may vary spatially.

The condition for equilibrium is that there be no net torque on the plates. By noting that the torque associated with drag is given locally by

$$\mathbf{T} = \mathbf{r} \times \mathbf{F} = -D\mathbf{r} \times (\omega \times \mathbf{r}) \qquad (4)$$

the torque due to drag beneath an entire plate p is

$$\mathbf{T} = \int_{\text{plate } p} dA \, [D\mathbf{r} \times (\omega_{pm} \times \mathbf{r})] \qquad (5)$$

where the integration is carried out over the surface of the plate.

If D is constant on individual plates (one plate can be divided into several plates with different values of D but

each having the same rotation), then the condition for equilibrium on the global lithosphere is

$$\sum_p D_p \int_{\text{plate } p} dA \, \mathbf{r} \times (\omega_{pm} \times \mathbf{r}) = 0 \qquad (6)$$

To evaluate this integral, we use a vector identity

$$\mathbf{L} \equiv \int dA \, (\mathbf{r} \times \omega \times \mathbf{r}) = \int dA \, (\mathbf{r} \cdot \mathbf{r})\omega - \int dA \, (\mathbf{r} \cdot \omega)\mathbf{r} \qquad (7)$$

Letting $\mathbf{r} \cdot \mathbf{r} = 1$ (the radius is constant if we ignore variations in plate thicknesses), we obtain

$$\mathbf{L}_p = A_p \omega_{pm} - \int dA \, (\mathbf{r} \cdot \omega_{pm})\mathbf{r} \qquad (8)$$

where A_p is the area of plate p. We can also write this as

$$\mathbf{L}_p = A_p \omega_{pm} - \mathbf{S}_p \omega_{pm} \qquad (9)$$

Here $\mathbf{S}_p$ is a symmetric matrix defined by

$$S_{ij} = \int_{\text{plate } p} d\phi \int d\theta \, x_i x_j \sin \theta \qquad (10)$$

where $i, j = 1, 2, 3$ are Cartesian indices and θ and ϕ are the spherical coordinates, colatitude and longitude, respectively. Defining

$$\mathbf{Q}_p = (A_p \mathbf{I} - \mathbf{S}_p) \qquad (11)$$

where $\mathbf{I}$ is the three by three identity matrix, we can write (6) as

$$\sum_p D_p \mathbf{Q}_p \omega_{pm} = 0 \qquad (12)$$

Note that $\mathbf{Q}_p$ depends entirely on the geometry of plate p.

In general, any velocity-independent torque $\mathbf{T}_0$ on the plates may be added to the right-hand side of (12). Further, since from (2) ω_{pm} may be written as the sum $\omega_{po} + \omega_{om}$ for some arbitrary plate o, (12) may be regarded as a three by three equation in the absolute velocity ω_{om} of plate o, i.e.,

$$\omega_{om} \sum_p D_p \mathbf{Q}_p = \mathbf{T}_0 - \sum_p D_p \mathbf{Q}_p \omega_{po} \qquad (13)$$

which can be easily solved. A further simplification is possible if D_p is uniform for all plates, since

$$\sum_p \mathbf{Q}_p = \frac{8\pi}{3} \mathbf{I} \qquad (14)$$

For line segments having drag, such as island arcs, (13) may still be used except that $\mathbf{Q}_p$ is defined by

$$Q_{ij} = \int_{\text{line}} dl \, D_p(\delta_{ij} - x_i x_j) \qquad (15)$$

where δ_{ij} is an element of the identity matrix.

Torques due to gravitational pull by subducted slabs are given by

$$\mathbf{T}_t = C_t \int_{\text{line } t} \mathbf{r} \times (d\mathbf{l} \times \mathbf{r}) \qquad (16)$$

where C_t is a constant (assumed to be independent of local subduction rate) and where the line integral is taken counterclockwise about the subducted plate. By analogy to (7), in taking $\mathbf{r} \cdot \mathbf{r} = 1$,

$$\mathbf{T}_t = C_t \int_{\text{line } t} d\mathbf{l} = C_t(\mathbf{r}_2 - \mathbf{r}_1)_t \qquad (17)$$

where $\mathbf{r}_1$ and $\mathbf{r}_2$ are radius vectors to the two ends of the trench. The force exerted by a sinking slab is assumed in writing (16) to act in a direction normal to the strike of the trench.

When both torque exerted on the sinking plates by the subducted slabs and drag beneath all plates are included, then (13) becomes

$$\boldsymbol{\omega}_{om} \sum_p D_p \mathbf{Q}_p = \sum_t C_t(\mathbf{r}_2 - \mathbf{r}_1)_t - \sum_p D_p \mathbf{Q}_p \boldsymbol{\omega}_{po} \qquad (18)$$

Even in the simplified version of (18) in which all D_p are equal (to D, say) and all C_t are equal (to C, say), there is an arbitrary scaling of C to D that will affect $\boldsymbol{\omega}_{om}$. The maximum value of C/D (see below) can be fixed by equating the energy dissipated per time by drag forces to the power gained by subduction. The former quantity is

$$\int_{\text{all earth}} \boldsymbol{\omega} \cdot \mathbf{T}\, dA = D \sum_p \boldsymbol{\omega}_{pm} \cdot (\mathbf{Q}_p \boldsymbol{\omega}_{pm}) \qquad (19)$$

The latter is

$$\sum_t \boldsymbol{\omega}_{tm} \cdot \mathbf{T}_t = C \sum_t \boldsymbol{\omega}_{tm} \cdot (\mathbf{r}_2 - \mathbf{r}_1)_t \qquad (20)$$

where $\boldsymbol{\omega}_{tm}$ is the absolute angular velocity of the subducted plate at trench t. Both (19) and (20) are quadratics in C/D, and both have identical leading terms in $(C/D)^2$ for this simple case of uniform D_p and uniform C_t. Thus by equating the two expressions the desired limit on C/D follows directly. To derive this equality, it was assumed that no work was done at boundaries with symmetric forces such as ridges. If these had been included, an additional term, probably positive, would have appeared on both sides of (20). The value of C/D obtained from equating (19) and (20) would then be reduced. Thus our value of C/D is an upper bound if positive net work is done by symmetric boundaries. In the limit $D \to 0$ a very large absolute velocity about a pole defined by the torque would be necessary to balance (18), and (20) would be set equal to the work done at symmetric resistive boundaries.

It should be noted that the case of a net torque on the plates resulting from trenches accrues from our division of the earth into horizontal plates, sinking slabs, and the asthenosphere. Clearly, radially directed buoyant forces resulting from gravity can produce no net torque on the earth as a whole. There can also be no net torque supplied to the interior of the earth by plates and slabs, since there is no balancing source of torque at the base of the mantle. The slabs and the plates thus must exert equal and opposite torques on the interior. The magnitudes of the torque due to drag at the base of the plates, the torque exerted on the plates by slabs, and the torque exerted on the asthenosphere by slabs must therefore be numerically equal. The case of very large C/D is not likely to occur, since both C and D involve coupling of plates to the asthenosphere.

Acknowledgments. We thank Randall Richardson for assistance in digitizing the plate boundaries and Jeffrey Grossman for considerable programing work. The text of this paper was improved by several suggestions of Clem Chase, Frank Richter, Frank Press, and two anonymous reviewers, to all of whom we are grateful. This research was supported by the Earth Sciences Section, National Science Foundation, NFS grant GA-36132X.

REFERENCES

Alexander, S. S., and R. W. Sherburne, Crust mantle structure of shields and their role in global tectonics (abstract), *Eos Trans. AGU, 53,* 1043, 1972.

Artyushkov, E. V., Stresses in the lithosphere caused by crustal thickness inhomogeneities, *J. Geophys. Res., 78,* 7675, 1973.

Burke, K., and J. T. Wilson, Is the African plate stationary?, *Nature, 239,* 387, 1972.

Burke, K., W. S. F. Kidd, and J. T. Wilson, Plumes and concentric plume traces of the Eurasian plate, *Nature Phys. Sci., 241,* 128, 1973a.

Burke, K., W. S. F. Kidd, and J. T. Wilson, Relative and latitudinal motion of Atlantic hot spots, *Nature, 245,* 133, 1973b.

Chapman, E. D., and S. C. Solomon, American-Eurasian plate boundary in northeast Asia (abstract), *Eos Trans. AGU, 54,* 1204, 1973.

Chase, C. G., The N plate problem of plate tectonics, *Geophys. J. Roy. Astron. Soc., 29,* 117, 1972.

Clague, D. A., and R. D. Jarrard, Tertiary Pacific plate motion deduced from the Hawaiian-Emperor chain, *Geol. Soc. Amer. Bull., 84,* 1135, 1973.

Duncan, R. A., N. Petersen, and R. B. Hargraves, Mantle plumes, movement of the European plate, and polar wandering, *Nature, 239,* 82, 1972.

Elsasser, W. M., Convection and stress propagation in the upper mantle, in *The Application of Modern Physics to the Earth and Planetary Interiors,* edited by S. K. Runcorn, p. 223, Interscience, New York, 1969.

Fitch, T. J., Plate convergence, transcurrent faults, and internal deformation adjacent to southeast Asia and the western Pacific, *J. Geophys. Res., 77,* 4432, 1972.

Isacks, B., and P. Molnar, Distribution of stresses in the descending lithosphere from a global survey of focal mechanism solutions of mantle earthquakes, *Rev. Geophys. Space Phys., 9,* 103, 1971.

Jacoby, W. B., Instability in the upper mantle and global plate movements, *J. Geophys. Res., 75,* 5671, 1970.

Knopoff, L., Observation and inversion of surface-wave dispersion, *Tectonophysics, 13,* 497, 1972.

Knopoff, L., and A. Leeds, Lithospheric momenta and the deceleration of the earth, *Nature, 237,* 93, 1972.

Lachenbruch, A. H., A simple mechanical model for oceanic spreading centers, *J. Geophys. Res., 78,* 3395, 1973.

Le Pichon, X., Sea floor spreading and continental drift, *J. Geophys. Res., 73,* 3661, 1968.

Malfait, B. T., and M. G. Dinkelman, Circum-Caribbean tectonic and igneous activity and the evolution of the Caribbean plate, *Geol. Soc. Amer. Bull., 83,* 251, 1972.

McElhinny, M. W., Mantle plumes, paleomagnetism and polar wandering, *Nature, 241,* 523, 1973.

McKenzie, D., and J. G. Sclater, The evolution of the Indian Ocean since the late Cretaceous, *Geophys. J. Roy. Astron. Soc., 25,* 437, 1971.

Minster, J. B., T. H. Jordon, P. Molnar, and E. Haines, Numerical modeling of instantaneous plate tectonics, *Geophys. J. Roy. Astron. Soc.,* in press, 1974.

Molnar, P., and T. Atwater, Relative motion of hot spots in the mantle, *Nature, 246,* 288, 1973.

Molnar, P., and L. R. Sykes, Tectonics of the Caribbean and middle America regions from focal mechanisms and seismicity, *Geol. Soc. Amer. Bull., 80,* 1639, 1969.

Morgan, W. J., Rises, trenches, great faults, and crustal blocks, *J. Geophys. Res., 73,* 1959, 1968.

Morgan, W. J., Convection plumes in the lower mantle, *Nature, 230,* 42, 1971.

Morgan, W. J., Deep mantle convection plumes and plate motions, *Amer. Ass. Petrol. Geol. Bull., 56,* 203, 1972.

Morgan, W. J., Plate motions and deep mantle convection, Studies in Earth and Space Sciences, *Geol. Soc. Amer. Mem., 132,* 7, 1973.

Richter, F. M., Dynamical models for sea floor spreading, *Rev. Geophys. Space Phys., 11,* 223, 1973a.

Richter, F. M., Convection and large-scale circulation of the mantle, *J. Geophys. Res., 78,* 8735, 1973*b*.

Schubert, C., and R. S. Sifontes, Boconó fault, Venezuelan Andes: Evidence of postglacial movement, *Science, 170,* 66, 1970.

Schubert, G., and D. L. Turcotte, One-dimensional model of shallow mantle convection, *J. Geophys. Res., 77,* 945, 1972.

Sykes, L. R., and M. L. Sbar, Intraplate earthquakes, lithospheric stresses and the driving mechanism of plate tectonics, *Nature, 245,* 298, 1973.

Talwani, M., Plate tectonics and deep-sea trenches (abstract), *Eos Trans. AGU, 50,* 180, 1969.

Tullis, T. E., Evidence that lithospheric slabs act as anchors (abstract), *Eos Trans. AGU, 53,* 522, 1972.

Turcotte, D. L., and E. R. Oxburgh, Mid-plate tectonics, *Nature, 244,* 337, 1973.

Vogt, P. R., Asthenosphere motion recorded by the ocean floor south of Iceland, *Earth Planet. Sci. Lett., 13,* 153, 1971.

Weertman, J., The creep strength of the earth's mantle, *Rev. Geophys. Space Phys., 8,* 145, 1970.

Wilson, J. T., Evidence from ocean islands suggesting movement in the earth, *Phil. Trans. Roy. Soc. London, Ser. A, 258,* 145, 1965.

Winterer, E. L., Sedimentary facies and plate tectonics of equatorial Pacific, *Amer. Ass. Petrol. Geol. Bull., 57,* 265, 1973.

(Received January 5, 1974;
revised March 4, 1974.)

Part VI

THE PATTERN OF FLOW IN THE MANTLE AND ITS EXPRESSIONS

Editor's Comments
on Papers 21 Through 28

Previous sections discussed various aspects of mantle flow and
its relation to plate motions. However, in order to gain a good understanding of the pattern of mantle flow and its influence on the earth's
history, all of these studies should be integrated. Of prime importance is to relate the flow in the mantle to observable features both
from the present and from the more distant geologic past, such as the
earth's gravity field, the planet's chemical differentiation and evolution,
as well as the motions and history of the plates. All these bear on

features of the deep flow and provide information on its pattern and evolution. Still, complete understanding of the geometry of flow in the mantle is a matter for the future.

Richter (1973) proposed two horizontal scales of convection in the mantle: a large-scale flow comprising the observed overturn of lithosphere, and another consisting of small-scale regular cells (actually rolls), having small aspect ratios and strongly modulated by the motions of the overlying plates. This model reconciles the geometry of the plates with theoretical predictions. These ideas were further developed by Richter and Parsons (Paper 21), who supplemented the theoretical study with laboratory experiments. They showed that the surficial plates can suppress rolls elongated perpendicular to their motions, while rolls with axes parallel to plate motion are promoted, and can grow within a few ten million years under plates moving at 5–10 cm/y (with convection extending to a depth of 650 km). Longitudinal rolls are formed even when the unmodulated flow consists of rectangular or spoke-like cells that develop when the Rayleigh numbers are high, as is the case in the mantle. Under slow-moving plates, longitudinal rolls develop only after long periods, so the underlying small-scale flow may retain a complicated pattern. The small-scale flow is supposed to supply the background heat flow in oceans, which cannot be maintained by conduction without assuming unacceptably high temperatures. If convection does not extend through the 650-km phase boundary, many small-scale cells should be present beneath the much larger lithospheric plates; hence the torques exerted on the plates by these cells will nearly cancel out and will have no net influence on the plate motions. McKenzie and Weiss (1975) proposed a similar hypothesis of two scales of mantle convection. They advocate that the 650-km phase boundary is a barrier to convection, and that a separate circulation may occur in the lower mantle. The two-scale flow was further discussed by Parsons and McKenzie (1978), who related it to instability of the thermal boundary layer under mature lithosphere. To date, there is no conclusive observational evidence for the small-scale flow.

The large-scale flow is further discussed by Garfunkel (Paper 22). The rates of plate generation and consumption are not locally balanced, so the shapes of the plates change. The return flow can be balanced on a global scale only, and hence the overall circulation cannot be two-dimensional, and cannot consist of cells in which particles follow closed circuits. In addition, the retrograde subduction (cf. Paper 16) and the associated movement of the descending slabs through the mantle require a flow of the same magnitude as that involved in the overturn of lithosphere. This is, then, an important component of the large-scale flow and must modify parts of any small-scale flow that may exist. Due to the evolution of the plates, the pattern of mid-oceanic ridges and subduction zones changes significantly over a time scale of 100 m.y., so the return flow must also change and is time dependent. In addition there are occasional catastrophic changes, such as continental breakup, jumping of ridges, which is tantamount to breakup of oceanic lithosphere and production of new Benioff zones. The irregular configuration and temporal changes of the pattern of subduction zones, where cold material return to the mantle, act as a complex thermal feedback from the surface. This, and chemical heterogeneity, are expected to produce instabilities. It is suggested that occasional development of new ascending plumes or currents is an important agent which modifies the flow, while breaking plates and changing their motion.

Some workers attempted to calculate the return flow pattern required to balance plate creation and consumption (Chase, 1979a; Parmentier and Oliver, 1979), on the assumption that the flow is confined to the outer 650 km of the mantle. As expected (Paper 22), the flow is three dimensional, and without cells consisting of closed mass circuits. Hager and O'Connell (1978, 1979) calculated the flow induced by the drag of the plates on the underlying mantle. Again the flow is three dimensional and is not anti-parallel to the plate motions. They preferred whole-mantle convection over shallow (in upper 650 km) flow.

The problem of whether convection is mantle-wide or is confined to its outer 650 km is of basic importance, but it is still controversial. Several arguments were cited to show that the 650-km phase transition is a barrier to convection. It was believed that at that depth the mantle minerals break down into oxides, and that the Clapeyron slope, dT/dp, of this trsansition may be negative (Ahrens and Syono, 1967; Anderson, 1967). Even if this were the case, this is not necessarily an obstacle to mantle-wide convection (Schubert, Yuen, and Turcotte, 1975). However, the transformations at 650-km depth are different (Ringwood, 1972; Liu, 1975), so this objection may not apply. Another condition that could inhibit whole-mantle convection would be if the effective viscosity of the deeper phases were very high. However, Davies (1977) demonstrated that a viscosity contrast of about 10^4 is required to exclude any flow from the lower mantle, provided the mantle is compositionally homogenous. If the viscosity contrast is less (which is more probable, as summarized in Part IV), the flow will penetrate into the lower mantle, but there it may be less vigorous. Another argument for shallow convection is provided by the nature of the stresses, as derived from focal mechanisms, in the descending slabs (Isacks and Molnar, Paper 17, 1971), which are most simply explained as showing that the slabs encounter great resistance at a depth of 650-700 km and do not penetrate to greater depths. This is not a compelling argument, however, and alternative interpretations were also given (e.g., discussed by O'Connell, 1977). This problem remains unsettled; Richter (1979) prefers the interpretation that convective flow does not extend across the 650-km transition while Jordan (1977) suggests that there is evidence that some slabs actually extend deeper, though they are not seismically active.

The most important problem regarding the possibility of mantle-wide convection is whether the 650-km transition does or does not signify a chemical stratification, a downward increase in the Fe/Mg ratio, for instance. If it does (Anderson, Sammis, and Jordan, 1971), then convective flow across this boundary is effectively excluded (Richter and Johnson, 1974). On the basis of seismic data, such chemical heterogeneity was found not to be justified by Davies (1974, 1977), Watt, Shankland, and Mao (1977) and Lieberman, Jackson, and Ringwood (1977). However, Liu (1979a, 1979b) compared the properties of high-pressure phases with seismic data and argued in favor of chemical stratification. Isotopic studies were also interpreted as favoring a lower mantle that is not mixed with its outer part, thus implying that the shallow flow does not extend across the 650 km boundary (O'Nions, Evensen, and Hamilton, 1979; Wasserburg and DePaolo, 1979). In this case, two separate convective systems, above and below the 650-km transition, must be envisaged (McKenzie and Weiss, 1975). In any event, evidence for lateral heterogeneity in the lower mantle (Dziewonski, Hager, and O'Connell, 1977), evidence for a thermal boundary layer at its base and an adiabatic temperature gradient in the lower mantle (Jones, 1977; Jeanloz and Richter, 1979) indicate that convective circulation takes place in the lower mantle. Such a deep flow is also required to remove radiogenic heat that is generated in the lower mantle, and to remove the heat which is necessary to drive the dynamo in the earth's core (Gubbins, Masters, and Jacobs, 1979; Elsasser, Olson, and Marsh, 1979). It is the latter effect that produces the thermal boundary layer at the base of the mantle. If two separate convection systems occur below and above the 650-km phase transition, then an additional boundary layer should exist at this depth.

The pattern of flow is, as noted above, time dependent. It can be changed by growth of instabilities that arise occasionally, for example, plumes or mantle diapirs related to continental breakup. Anderson and Perkins (Paper 23) proposed another mechanism that can cause occasional changes in the flow. Under certain cricumstances viscous dissipation can cause "thermal runaway," so that the flow may be locally and temporarily accelerated. The instability can develop within a geologically short period. Thus, an irregular, eddy-like flow pattern may result. Rice and Fairbridge (1975) proposed that this mechanism can account for many episodic events, such as transgressions and regressions, tectonic phases, and neotectonic uplifting, which are so characteristic of the geologic record.

An important expression of recent mantle flow is the earth's gravity field. Analysis of satellite orbits shows that the earth is not in a hydrostatic state, the shape of the geoid being quite irregular. On the other hand, the record of postglacial readjustment shows that mass imbalances of large dimensions will decay within a few ten thousand years. Hence the conclusion that the irregularities of the geoid are dynamically supported. In fact, the lateral density variations inherent in convection will produce gravity anomalies.

This interpretation was explored by Kaula (Paper 24). He attempted to classify the main features of the earth's gravity field in relation to active tectonic processes and to plate tectonic environment. The different situations are then interpreted in terms of mantle flow, mainly at shallow levels. Positive anomalies centered on mid-oceanic ridges can be interpreted as originating from upwelling currents with a free upper boundary, whereas the positive anomalies over subduction zones record the excess mass of the cold descending slabs. Kaula also noted that the large anomaly over west Antarctica could not be interpreted by using this approach, nor as a result of recent deglaciation (cf. Kaula, 1972), so it probably orginates from deep flow, which is not obviously related to surficial features.

In the numerical experiments of McKenzie, Roberts, and Weiss (1974, Paper 6), rising currents are associated with positive gravity anomalies and an upward deformation of the upper (free) surface, but later studies (McKenzie, 1977) showed that this is not always the case. If rising currents exist beneath ridge crests and deform the geoid, then a positive correlation between topography and gravity over ridges may exist. Anderson, McKenzie, and Sclater (1973) suggested evidence for such a correlation, but Cochran and Talwani (1977) found only a correlation between the gravity anomalies and the elevation of sites of hot spot volcanism. Vogt (1976) also noted that the gravity and topographic highs along the ridges are mostly associated with hot spots. As the igneous rocks extruded at hot spots were derived from a source that is distinct from the common basalts of the mid-ocean ridges, Vogt suggests that the volcanic centers are underlain by plumes from which material flows laterally beneath the ridge crests (his "cracked sewer pipe" model). This is an important attempt to discuss features of the mid-oceanic ridges that are not predicted by simple models of plate generation. The importance of localized plumes should be stressed because of their association with continental breakup and changes of plate motions.

Gravity and geoid anomalies, and residual topographic anomalies (i.e., departures from age-depth relations in oceans), being important expressions of conditions below the lithosphere, contain much information about the deep flow. More recent gravity models (e.g., Gaposchkin, 1974; Wagner et al., 1977; Lerch et al., 1979) confirm that some major long-wavelength anomalies are not related to plate boundaries. Such anomalies probably are caused by the deep flow (e.g., Chase, 1979b). In the future, improved gravity data and laser altimetry from satellites will provide more data and lead to more definite conclusions regarding the deep mantle flow.

A most important problem is whether convection occurred throughout the earth's history, and if it did, what was its character? In particular, did a plate tectonic regime, similar to the present one, exist in the geologically ancient past? The Phanerozoic record clearly shows the operation of the plate tectonic regime. The older Precambrian record can probably also be interpreted in terms of plate tectonics or a variant of it, but it is quite clear that the behavior of the earth's outer layers changed with time (Burke et al., 1976; Windley, 1973, 1977). Examination of this problem is not intended here, but a few topics directly related to mantle flow will be mentioned.

One approach is to study the evolution of the mantle as recorded by isotopic and geochemical evidence that can reveal the nature and timing of past differentiation events that can be related to plate tectonic processes such as segregation of oceanic crust at ridges. Church and Tatsumoto (Paper 25) find that the uranium-lead system in mid-oceanic ridge basalts shows that the source rocks of these magmas were affected by a fractionation event about 1.7 b.y. ago, and were not mixed with other parts of the mantle. Similar events are inferred also from the rubidium-strontium systematics of many igneous provinces, both in the oceans and on continents (Oversby and Gast, 1970; O'Nions and Pankhurst, 1974; Brooks et al., 1976). These observations may be interpreted in terms of ancient overturn and differentiation of the mantle that produced distinct chemical reservoirs. This, and the segregation of continental crust imply considerable mass transfer and differentiation within the earth (O'Nions, Evensen, and Hamilton, 1979; Jacobsen and Wasserburg, 1979; Wasserburg and DePaolo, 1979). Modeling of such processes can constrain the sizes of mantle reservoirs and the mass flux between them and thus can be related to the mobility of the mantle, but the actual mechanism and the role of the plate-tectonic processes still need more study.

Concerning the significance of the geologic, and especially the tectonic, record, perhaps the most important problem is to assess the importance of relative horizontal movements of crustal segments. Paleomagnetic data can reveal such motions. Two effects must be separated: polar wandering and relative displacements of individual blocks. Goldreich and Toomre (1969) showed that large shifts of the earth's pole of rotation, and hence also of the paleomagnetic pole, can result from much smaller displacements of continents or other masses, which change the earth's moment of inertia tensor. Thus polar wandering in itself proves a dynamic earth. Irving and Park (Paper 26) examine the paleomagnetic record of the Canadian shield since 2700 m.y. ago. Very large shifts of the paleopole are evident. At certain times the wandering of the pole changed drastically, so its track outlines "hairpins." These were times of enhanced igneous and tectonic activity, which suggests a strong relation between tectonism, igneous activity, and shifting of the crust. They also stress the analogy between the Precambrian and Cretaceous events.

While great shifts of the paleomagnetic pole are well established, it is more difficult to document in detail motions between individual crustal units. The paleomagnetic data from Phanerozoic and late Precambrian rocks indicate considerable relative motions between continental blocks (McElhinny, 1973; Irving, 1977; Morel and Irving, 1978). The continents seem to have aggregated and dispersed several times, as suggested by Wilson (Paper 3). The Precambrian record raises problems, however. Bridern (1976) and Piper (1976) interpreted the world-wide record as being consistent with a model in which a single supercontinent, including all the shields but different from the late Paleozoic Pangea, existed throughout the Proterozoic, but there seems to be good evidence against such a concept. In particular there is evidence for significant motions between Africa and the Canadian shield in the Proterozoic (McGlynn et al., 1975; Irving and McGlynn, 1976; McElhinny and McWilliams, 1977). Relative motions between smaller units were deduced from the paleomagnetic data by some authors (e.g. Irving, Emslie, and Veno, 1974; Cavanaugh and Seyfert, 1977), but there is no general agreement on this matter. McElhinny and McWilliams (1977) thought that the paleomagnetic data showed that there was little relative motion between the ancient cratons in the African and Australian continent. This implies extensive ensialic orogenic activity in the Proterozoic, unlike the younger orogens which formed along subduction zones and between colliding continents.

When using these lines of evidence, as well as other ones, in order to understand the broad aspects of the earth's development, two questions should be discussed separately: First, was the earth a convecting planet throughout its history? Second, if it was, how were the internal movements related to the history of the earth's surface as revealed by the geologic record, and to the earth's geochemical evolution? The existence of a magnetic field 2700 m.y. ago shows that by then the earth possessed a liquid core, so the temperature difference across the mantle was comparable to that of today. This inference, and the large-scale polar wandering, strongly support the case for convection during the early period. Moreover, in the remote geologic past, radiogenic heat was generated at higher rates than now, so that the driving power of convection was stronger and the viscosity was probably lower (because temperatures were higher), so convection must be inferred. The long record of igneous activity in itself also requires an internally dynamic earth (cf. Paper 9). Thus the evidence for mass motions within the earth is strong. Tozer (Paper 27) showed that convection is actually inevitable in a body of the size of the earth if it was formed with sufficient internal heat, which is most likely. Once convection sets in, it will govern the earth's internal temperature and its thermal evolution. He also argued (Tozer, 1974) that convection should occur as an inevitable stage in the history of all terrestrial planets, the Moon, and even smaller bodies, and it will dominate their thermal development as was also found by others (cf. Schubert, 1979).

The answer to the second problem is less clear. Rocks as old as Archean already show evidence and strong deformation, metamorphism, igneous activity which produced rocks such as tonalites and even more acid types, and also indications of a heterogenous mantle. This indicates that the earth was a very active planet during most, if not all, of its history. The Proterozoic record and paleomagnetic data further substantiate this conclusion, but it seems that the behavior of the earth's external layers changed with time (summarized by Windley, 1977), perhaps in response to the earth's declining heat. Combining these considerations, it is reasonable to infer that the history of the earth's surface was controlled by convection in the mantle. The mechanisms are not clear, but operation of a variant of plate tectonics is likely. The appearance of the other terrestrial planets calls for caution, however. Although it is probable that convection occurs in their interiors, this did not affect much their surfaces. Thus, in order to understand the possible relations between internal dynamics and the behavior of their lithospheres, an integrated view of all terrestrial planets is required. This was attempted by Kaula (Paper 28). He showed that the planets' history is determined by the amount of their internal energy, which is due to primordial heating, liberation of gravitational energy by core formation, and radioactive heating. If sufficient, these sources of heat will start convection. The Moon, Mercury, and Mars seem to have passed the peak of their activity. The great activity of the Earth, both present and past, is due to the large amount of its internal energy. Its great size makes cooling slow and assures a long period of continuing activity in the future.

The works presented in this section show the variety of research directions that contribute to the understanding of the nature, geometry, and history of mantle flow and of the many phenomena that result from this flow. Future developments in each direction and synthesis of the results will eventually enable to better understand the way in which the Earth works.

REFERENCES

Ahrens, T. J., and Y. Syono, 1967, Calculated Mineral Reactions in the Earth's Mantle, *Jour. Geophys. Research* **72:**4181–4188.

Anderson, D. L., 1967, Phase Changes in the Upper Mantle, *Science* **157:**1165–1173.

Anderson, D. L., C. G., Sammis, and T. Jordan, 1971, Composition and Evolution of the Mantle and Core, *Science* **171:**1103–1112.

Anderson, R. N., D. P. McKenzie, and J. G. Sclater, 1973, Gravity, Bathymetry and Convection in the Earth, *Earth and Planetary Sci. Letters,* **18:**391–407.

Briden, J. C., 1976, Application of Palaeomagnetism to Proterozoic Tectonics, *Royal Soc. London Philos. Trans.,* ser. A, **280:**405–416.

Brooks, C., S. R. Hart, A. Hofmann, and D. E. James, 1976, Rb-Sr Mantle Isochrons from Oceanic Regions, *Earth and Planetary Sci. Letters* **32:**51–61.

Burke, D., J. F. Dewey, and W. S. F. Kidd, 1976, Dominance of Horizontal Movements Arc and Microcontinental Collisions in the Later Permobile Regime, in *The Early History of the Earth,* B. F. Windley, ed., Wiley, New York, pp. 113–129.

Cavanaugh, M. D., and C. K. Seyfert, 1977, Apparent Polar Wander Paths and the Joining of the Superior and Slave Provinces during Early Proterozoic time, *Geology* **5:**207–211.

Chase, C. G., 1979a, Asthenospheric Counterflow: A Kinematic Model, *Royal Astron. Soc. Geophys. Jour.* **56:**1–18.

Chase, C. G., 1979b, Subduction, the Geoid, and Lower Mantle Convection, *Nature* **282:**464–468.

Cochran, J. R. and M. Talwani, 1977, Free Air Gravity Anomalies in the World's Oceans and Their Relationship to Residual Elevation, *Royal Astron. Soc. Geophys. Jour.* **50:**494–552.

Davies, G. F., 1974, Limits on the Constitution of the Lower Mantle, *Royal Astron. Soc. Geophys. Jour.* **38:**479–503.

Davies, G. F., 1977, Whole Mantle Convection and Plate Tectonics, *Royal Astron. Soc. Geophys. Jour.* **49:**459–486.

Dziewonski, A. M., B. H. Hager, and R. J. O'Connell, 1977, Large-Scale Heterogeneities in the Lower Mantle, *Jour. Geophys. Research* **82:**239–255.

Elsasser, W. M., Olson, P., and Marsh, B. D., 1979, The Depth of Mantle Convection *Jour. Geophys. Research* **84:**147–155.

Gaposchkin, E. M., 1974, Earth's Gravity Field to the Eighteenth Degree and Geocentric Coordinates for 104 Stations from Satellite and Terrestrial Data, *Jour. Geophys. Research* **79:**5377–5411.

Goldreich, P., and A. Toomre, 1969, Some Remarks on Polar Wandering, *Jour. Geophys. Research* **74:**2555–2567.

Gubbins, D., T. G. Masters, and J. A Jacobs, 1979, Thermal Evolution of the Earth's Core, *Royal Astron. Soc. Geophys. Jour.* **59:**57–99.

Hager, B. A., and R. J. O'Connell, 1978, Subduction Zone Dip Angles and Flow Driven by Plate Motions, *Tectonophysics* **50:**111–133.

Hager, B. A., and O'Connell, R. J., 1979, Kinematic Models of Large-Scale Flow in the Earth's Mantle, *Jour. Geophys. Research* **84:**1031–1048.

Irving, E., 1977, Continental Drift since the Devonian, *Nature* **270:**304–309.

Irving, E., R. F. Emslie, and H. Ueno, 1974, Upper Proterozoic Paleomagnetic Poles from Laurentia and the History of the Grenville Structural Province, *Jour. Geophys. Research* **79:**5491–5502.

Irving, E., and McGlynn, J. C., 1976, Proterozoic Magnetostratigraphy and the Tectonic Evolution of Laurentia, *Royal Soc. London Philos. Trans.* ser. A, **280:**433–468.

Isacks, B., and P. Molnar, 1971, Distribution of Stresses in the Descending Lithosphere from a Global Survey of Focal Mechanism Solution of Mantle Earthquakes, *Rev. Geophys. Space. Phys.* **9:**103–174.

Jacobson, S. B., and Wasserburg, G. J., 1979, The Mean Age of Mantle and Crust Reservoirs, *Jour. Geophys. Res.* **84:**7411–7427.

Jeanloz, R., and F. M. Richter, 1979, Convection, Composition, and the Thermal State of the Lower Mantle, *Jour. Geophys. Res.* **84:**5497–5504.

Jones, G. M., 1977, Thermal Interaction of the Core and the Mantle and Long-term Behavior of the Geomagnetic Field, *Jour. Geophys. Res.* **82:**1703–1709.

Jordan, T. H., 1977, Lithospheric Slab Penetration into Lower Mantle Beneath the Sea of Okhotsk, *Jour. Geophys.* **43:**473–496.

Kaula, W. M., 1972, Global Gravity and Mantle Convection, *Tectonophysics,* **13:**341–359.

Lerch, F. J., S. M. Klosko, R. E. Laubscher, and C. A. Wagner, 1979, Gravity Model Improvements Using Geos 3 (GEM 9 and 10), *Jour. Geophys. Research* **84:**3897–3916.

Lieberman, R. C., I. Jackson, and A. E. Ringwood, 1977, Elasticity and Phase Equilibria of Spinel Disproportionation Reactions, *Royal Astron. Soc. Geophys. Jour.* **50:**555–586.

Liu, L. G., 1975, Post-Oxide Phases of Forsterite and Enstatite, *Geophys. Res. Lett.* **2:**417–419.

Liu, L. G., 1979a, On the 650 km Seismic Discontinuity, *Earth and Planetary Sci. Letters* **42:**202–208.

Liu, L. G., 1979b, Calculations of High-Pressure Phase Transitions in the System MgO-SiO$_2$ and Implications for Mantle Discontinuities, *Physics Earth and Planetary Interiors* **19:**319–330.

McElhinny, M. W., 1973, *Palaeomagnetism and Plate Tectonics,* Cambridge University Press, Cambridge, 358p.

McElhinny, M. W., and M. O. McWilliams, 1977, Precambrian Geodynamics—A Palaeomagnetic View *Tectonophysics* **40:**137–159.

McGlynn, J. C., E. Irving, K. Bell, and G. Pullaiah, 1975, Palaeomagnetic Poles and Proterozoic Supercontinent, *Nature* **255:**318–319.

McKenzie, D. P., 1977, Surface Deformation, Gravity and Convection, *Royal Astron. Soc. Geophys. Jour.* **48:**211–238.

McKenzie, D. P., Roberts, J. M., and Weiss, N. O., 1974, Convection in the

Earth's Mantle: Towards a Numerical Simulation, *Jour. Fluid Mechanics* **42:**465–538.

McKenzie, D. P., and N. O. Weiss, 1975, Speculations on the Thermal and Tectonic History of the Earth, *Royal Astron. Soc. Geophys. Jour.* **42:**131–174.

Morel, P., and E. Irving, 1978, Tentative Paleocontinental Maps for the Early Phanerozoic and Proterozoic, *Jour. Geology* **86:**535–561.

O'Nions, R. K., Evensen, N. M., and Hamilton, P. J., 1979, Geochemical Modeling of Mantle Differentiation and Crustal Growth, *Jour. Geophys. Research* **84:**6091–6101.

O'Nions, R. K., and Pankhurst, R. J., 1974, Petrogenetic Significance of Isotope and Trace Element Variations in Volcanic Rocks from the Mid-Atlantic, *Jour. Petrology* **15:**603–634.

Oversby, V. M., and Gast, P. W., 1971, Isotopic Compositions of Lead from Oceanic Islands, *Jour. Geophys. Research* **75:**2097–2114.

Parmentier, E. M., and J. E. Oliver, 1979, A Study of Shallow Global Mantle Flow Due to the Accretion and Subduction of Lithospheric Plates, *Royal Astron. Soc. Geophys. Jour.* **57:**1–22.

Parsons, B., and D. McKenzie, 1978, Mantle Convection and the Thermal Structure of Plates,:*Jour. Geophys. Research* **83:**4485–4496.

Piper, J. D. A., 1976, Palaeomagnetic Evidence for a Proterozoic Supercontinent, *Royal Soc. London, Philos. Trans.* ser. A, **280:**469–483.

Rice, A., and R. W. Fairbridge, 1975, Thermal Runaway in the Mantle and Neotectonics, *Tectonophysics* **29:**59–72.

Richter, F. M., 1973, Dynamical Models for Sea-floor Spreading, *Rev. Geophys. Space Phys.* **11:**223–287.

Richter, F. M., 1979, Focal Mechanisms and Seismic Energy Release of Deep and Intermediate Earthquakes in the Tonga-Kermadec Region and Their Bearing on the Depth Extent of Mantle Flow, *Jour. Geophys. Res.* **84:**6783–6795.

Richter, F. M., and C. E. Johnson, 1974, Stability of a Chemically Layered Mantle, *Jour. Geophys. Res.* **79:**1635–1639.

Ringwood, A. E., 1972, Phase Transformation and Mantle Dynamics, *Earth and Planetary Sci. Letters* **14:**233–241.

Schubert, G., 1979, Subsolidus Convection in the Mantles of the Terrestrial Planets *Ann. Rev. Earth Planetary Sci.* **7:**289–342.

Schubert, G., D. A. Yuen, and D. L. Turcotte, 1975, Role of Phase Transitions in a Dynamic Mantle, *Royal Astron. Soc. Geophys. Jour.* **42:**705–735.

Tozer, D. C., 1974, The Internal, Evolution of Planetary Sized Objects, *Moon* **9:**167–182.

Vogt, P. R., 1976, Plumes, Subaxial Flow and Topography along the Mid-Oceanic Ridge, *Earth and Planetary Sci. Letters* **29:**309–325.

Wagner, C. A., F. J. Lerch, D. Brown, and J. A. Richardson, 1977, Improvement in the Geopotential Derived from Satellite and Surface Data (GEM 7 and 8), *Jour. Geophys. Research* **82:**901–914.

Watt, J. P., Shankland, T. J., and Mao, N. H., 1975, Uniformity of Mantle Composition, *Geology* **3:**91–94.

Windley, B. F., 1973, Crustal Development in the Precambrian, *Royal Soc. London Philos. Trans.,* ser. A, **273:**321–341.

Windley, B. F., 1977, *The Evolving Continents,* Wiley, New York, 385p.

Wasserburg, G. J., and D. J. DePaolo, 1979, Models of earth structure inferred from Neodymium and Strontium isotopic abundances, Proceed. National Academy Sci. **76:**3594–3598.

21

On the Interaction of Two Scales of Convection in the Mantle

FRANK M. RICHTER AND BARRY PARSONS

INTRODUCTION

The theory of plate tectonics successfully provides the basis for an explanation of a wide range of disparate geophysical observations. Its success lies in the simplicity of the basic hypothesis that the outer layer of the earth can be divided into a number of quasi-rigid spherical caps, or plates, in relative motion and in the fact that the theory is a purely kinematic one. In other words, the surface features that the theory explains, recently reviewed by *McKenzie* [1972], depend only on the rigidity of the plates away from their boundaries and on their relative motion and not on whatever mechanisms maintain that state. Hence the next step of providing a dynamic basis for the kinematic theory, a step that is necessary in considering how the system of plates evolves with time, is made that much more difficult. If it is assumed that the most likely driving mechanism is a form of thermal convection [*McKenzie et al.*, 1974], then few surface observations remain that can be related to convective motions in the mantle. Here we shall attempt to use such remaining observations to construct a hypothesis in which convection occurs in the upper mantle on two distinct horizontal length scales and to suggest what further observations could be made to test its consequences.

The existence of the plates themselves and of their relative motion is the most self-evident observation. As plates move apart at midocean ridges, intruded material from the upper mantle cools and is added to the plates. The creation of new plate is balanced by the subduction of surface material in the vicinity of ocean trenches. This circulation must be closed by a return flow from the trenches to the ridges in order to conserve mass. The circulation involving the plates and the return flow will be called the large-scale circulation, in that it necessarily has horizontal length scales of the order of plate dimensions. These can be as large as 10,000 km in the case of the Pacific.

One of the major accomplishments of plate tectonics has been to provide an explanation of the variation of heat flow and ocean depth when moving away from midocean ridges [*McKenzie*, 1967; *Sclater and Francheteau*, 1970; *Sclater et al.*, 1971]. Figure 1 shows a typical plot of heat flow versus the age of the plate at the location of the measurements. The heat flow anomaly associated with the ridge crest extends out to about 50 m.y. and can be thought of as resulting from heat transported by the large-scale flow. In ocean floor older than 50 m.y., heat flow values become approximately constant. The constant background value is consistent with the idea of a plate of constant thickness having reached thermal equilibrium [*McKenzie*, 1967]. This, in turn, implies that there is a relatively uniform supply of heat to the base of the plate. Since heat production by radioactive heat sources within the plate itself is generally accepted to be small in comparison with the observed heat flow, the upper mantle must be transporting about 1 μcal cm^{-2} s^{-1}. The evidence from the heat flow data is supported by an extension of the empirical depth-age curve [*Sclater et al.*, 1975], which shows that the ocean depth also tends to flatten out with age of the ocean floor. This is again consistent with the above argument; however, the depth data in older regions are not nearly as extensive or reliable as the heat flow measurements, a reversal of their relative reliabilities near the ridge crest.

The question then becomes, What is the mechanism by which the upper mantle transports the large background value of heat flow? It does not seem possible to provide this value of heat flux by conduction alone. Recent estimates of the total (lattice and radiative) thermal diffusivity of typical earth materials suggest a value of about $1.5 \cdot 10^{-2}$ cm^2 s^{-1} (see the discussion below). This value is almost an order of magnitude too small for the upper mantle to transport the required heat by conduction if one rejects widespread melting below 100 km. Some form of 'local' convection is the most likely heat transport mechanism not requiring an unacceptably large temperature increase below 100 km. By local convection we envisage convection under the plates having a horizontal scale

283

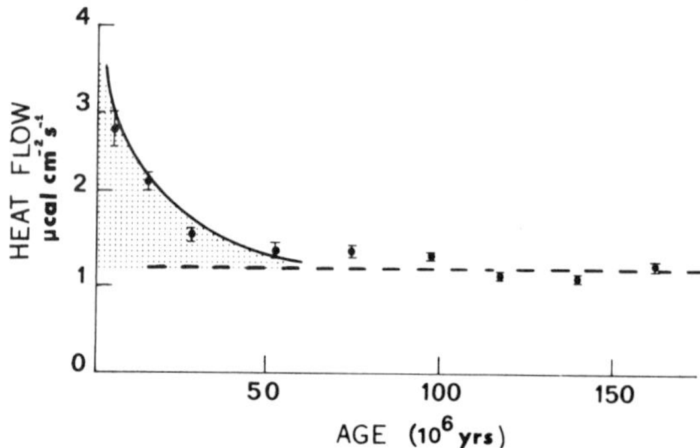

Fig. 1. Mean heat flow against age of province for the North Pacific (data from *Sclater and Francheteau* [1970, Table 1]). The length of the bar gives the standard error. We interpret the shaded area as resulting from heat transported by the large-scale flow, while the background value of about 1.1 μcal cm^{-2} s^{-1} requires local small-scale convection.

of the order of the depth of the fluid, which we assume is small in comparison with typical plate dimensions. As we shall attempt to justify in the discussion below, we shall assume that this depth is down to the 650-km-deep seismic discontinuity. In contrast to the large-scale flow, the convection that we are now suggesting represents a distinct second scale of motion, the small-scale convection. This suggestion is consistent with laboratory observations and analytical calculations on convection [*Busse and Whitehead*, 1971; *Schlüter et al.*, 1965; *Busse*, 1967], which show that the horizontal scales of the convective planform in Rayleigh-Benard type convection are of the order of the depth of the layer. Figure 12, shown later in the paper, illustrates the two-scale concept by showing Rayleigh-Benard cells (the small-scale flow) within the large-scale flow consisting of the plates and return flow at depth. The large-scale flow is assumed to extend to the same depth as the Rayleigh-Benard cells. In our experiments on the interaction of these two scales of motion we represent the small-scale flow by classical Rayleigh-Benard convection. This should be regarded as an idealization in that convection with some internal heat generation [*McKenzie et al.*, 1974] is probably a more appropriate model for any form of mantle convection. The large-scale flow is modeled by the shear flow produced by uniformly moving the upper boundary of the convecting layer.

The idea of two scales of motion in the upper mantle, one required by the moving plates and the other suggested by heat flow observations in old ocean floor, has been considered by *Richter* [1973b], using both analytic and numerical models. These models were meant to help understand the interaction of convection with the large-scale flow. The principal result is that the large-scale flow suppresses convective rolls with horizontal scales of the order of the depth and with their axes oriented perpendicular to the direction of the large-scale flow (transverse rolls). Given sufficient time, the decaying transverse rolls are replaced by a new set of rolls with their axes aligned in the direction of large-scale flow (longitudinal rolls). This earlier work has the limitation that it is formally valid for Rayleigh numbers only slightly in excess of critical, whereas the heat transport in the upper mantle suggests that the Rayleigh number is likely to be several hundred times greater than critical there. The techniques used by Richter cannot be easily extended to more realistic Rayleigh numbers for two principal reasons. First, the analytic technique used can-

not cope with strong nonlinearities, i.e., with large transport of heat by the flow. The numerical techniques are already costly and would prove prohibitive if three space dimensions were to be considered. Laboratory experiments become a logical approach, and in this paper we report on the results of experiments on convection in a large-scale flow produced by a moving boundary, which corroborate and extend the earlier theoretical results.

It is worthwhile to emphasize the three-dimensional nature of convection, since it is one of the reasons why experiments are required. *Busse* [1967] has shown that two-dimensional rolls are a stable form of convection in a fluid layer bounded by rigid nonmoving boundaries only in the limited parameter range of Rayleigh numbers less than 13 times critical. At Rayleigh numbers greater than this value a second set of rolls perpendicular to the original one grows, and the new planform, is rectangular. This three-dimensional form of convection is called bimodal, and excellent photographs are shown by *Busse and Whitehead* [1971]. At even larger Rayleigh numbers, about 100 times critical, a new planform develops through a collective instability. The corners of the bimodal rectangles join, and the resulting planform resembles spokes radiating out of central upwellings or downwellings. We will call this final planform multimodal. Roll, bimodal, and multimodal convection will be seen in our experiments as the initial state that we subject to a large-scale shear.

The existing theory and experiments showing the limited stability of two-dimensional solutions cannot be ignored when convection models that have been applied to the mantle are evaluated. Most mantle convection models are two-dimensional, but the calculations are performed at Rayleigh numbers where present understanding suggests that two-dimensional solutions are unstable. The point is that when a model is governed by nonlinear dynamics, it is not sufficient to find a solution that satisfies the equations and boundary conditions. A physically realizable solution must satisfy these constraints, but it must also be shown to be stable. For example, a state of no motion is an exact solution to the Rayleigh convection equations for all Rayleigh numbers, but it is a physically realizable solution only for Rayleigh numbers less than critical. The burden of discussing the stability of a solution is often ignored in geophysical applications and may in part be responsible for the proliferation of mantle convection models that bear little resemblance to one another. Laboratory experiments have the advantage that they provide the physically realizable solution.

Experiments on convection under moving boundaries have been done by *Chandra* [1938] and *Ingersoll* [1966]. These experiments generally suggest that given sufficient shear, longitudinal rolls are the preferred form of convection, but the fluids used were of low viscosity, and the results are complicated by instabilities other than those of thermal origin. Furthermore, these studies do not consider the question of the time required for longitudinal rolls to develop. In order for longitudinal rolls to be important under present plates, we must show that they can develop on a time scale not greater than that over which plate motions significantly change. Our experiments, which cover the Rayleigh number range from critical to about 10^6, agree with the earlier experiments in that longitudinal rolls are observed for all Rayleigh numbers and plate velocities used. When we apply the results to the earth, we find that longitudinal rolls can develop in less than about 200 m.y. if the Rayleigh number is greater than 10^6 and the plate velocity is greater than 2 cm yr^{-1}.

Model, Apparatus, and Procedure

To investigate the effect of a moving horizontal boundary on Rayleigh-Benard convection, we consider the model shown in Figure 2, which is governed by equations (6)–(8) given below. The model consists of a fluid layer, across which is applied a vertical temperature increase ΔT, contained in a box whose upper boundary can be moved at a constant rate U. The variables are nondimensionalized by using the thermal diffusion scales. Letting asterisks denote dimensional variables, we write

$$\begin{bmatrix} x^* \\ y^* \\ z^* \end{bmatrix} = D \begin{bmatrix} x \\ y \\ z \end{bmatrix} \tag{1}$$

for distance,

$$\mathbf{q}^* = \begin{bmatrix} u^* \\ v^* \\ w^* \end{bmatrix} = \frac{\kappa}{D}\,\mathbf{q} = \frac{\kappa}{D} \begin{bmatrix} u \\ v \\ w \end{bmatrix} \tag{2}$$

for velocity,

$$T^* = T_0^* + \Delta T \theta \tag{3}$$

for temperature,

$$p^* = \rho_0 g D (1 - z) + (\rho_0 \kappa \nu / D^2) p \tag{4}$$

for pressure, and

$$t^* = (D^2/\kappa) t \tag{5}$$

for time, where D is the depth, κ is the thermometric diffusivity, T_0^* is the temperature at $z^* = D$, ΔT is the temperature change across the layer, ρ_0 is the density at temperature T_0^*, g is the acceleration of gravity, and ν is the kinematic viscosity.

The nondimensional equations governing a nonrotating Boussinesq Newtonian fluid are

$$\frac{1}{\sigma}\frac{D\mathbf{q}}{Dt} = -\nabla p + \nabla^2 \mathbf{q} + Ra\,\theta \mathbf{k} \tag{6}$$

$$\frac{D\theta}{Dt} = \nabla^2 \theta \tag{7}$$

$$\nabla \cdot \mathbf{q} = 0 \tag{8}$$

where

$$\frac{D}{Dt} = \frac{\partial}{\partial t} + \mathbf{q} \cdot \nabla$$

$$\nabla = \mathbf{i}\frac{\partial}{\partial x} + \mathbf{j}\frac{\partial}{\partial y} + \mathbf{k}\frac{\partial}{\partial z}$$

$$\nabla^2 = \frac{\partial^2}{\partial x^2} + \frac{\partial^2}{\partial y^2} + \frac{\partial^2}{\partial z^2}$$

$\sigma = \nu/\kappa$ is the Prandtl number, and $Ra = g\alpha\Delta T D^3/\kappa\nu$ is the Rayleigh number, where α is the coefficient of thermal expansion.

The boundary conditions are as follows:

$$\theta = 1 \qquad u = v = w = 0 \qquad \text{at} \quad z = 0 \tag{9}$$

$$\theta = 0 \qquad u = \frac{D}{\kappa} U \qquad v = w = 0 \tag{10}$$

$$\text{at} \quad z = 1$$

$$\theta \ \text{unspecified} \qquad u = v = w = 0 \tag{11}$$

$$\text{on} \quad x = 0,\, H/D \ \text{and} \ y = 0,\, H/D$$

The problem is specified by four nondimensional parameters: the Rayleigh number Ra, the Prandtl number σ, the Peclet number UD/κ, and the aspect ratio H/D. The model is to be investigated experimentally, and the range of interest of the above parameters dictates the choice of fluid and apparatus specifications. The simplest choice involves the Prandtl number; we want it to be as large as possible since in the earth it is about 10^{24}. The effect of such a large Prandtl number is to make the inertial terms negligible in the momentum equation (6). We use as an experimental fluid Dow Corning 200 silicone oil that has a Prandtl number of 8600. The complete properties of this oil are taken or extrapolated from data in *Dow Corning Bulletin 22-092a* and are as follows: $\rho_0 = 0.971$ g cm^{-3}, $\nu = 10$ cm^2 s^{-1}, $(1/\nu)(\partial\nu/\partial T) = 0.01\,°C^{-1}$, $\kappa = 1.16 \times 10^{-3}$ cm^2 s^{-1}, and $\alpha = 9.6 \times 10^{-4}\,°C^{-1}$. Oils with larger Prandtl numbers are available, but their viscosities are so large that the Rayleigh number range that one can cover in

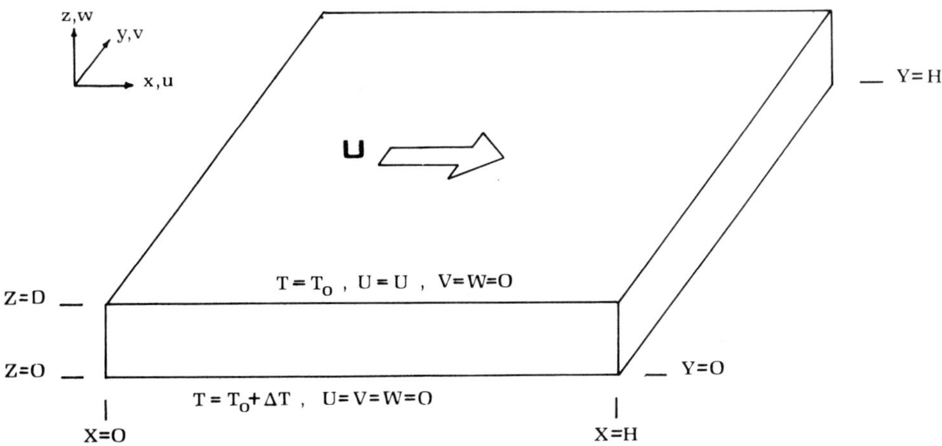

Fig. 2. Model for convection under a moving boundary.

a reasonable size apparatus is too limited. A Prandtl number of 8600 is already large enough to make the advection of momentum negligible in our experiments. The Rayleigh number range that we cover is from 1.7×10^3 to 1.0×10^6, as most estimates of mantle Rayleigh numbers are in this range. Since a reasonable temperature change in an experiment is of the order of 10°C, this Rayleigh number range requires depths of from 1 to 7 cm. The largest depth determines the horizontal dimensions of the convecting layer. We want the aspect ratio to be large $(H/D \gtrsim 15)$, because we are specifically interested in convection in a large-scale flow and also because we want to avoid the influence of vertical boundaries on which the temperature cannot be controlled or meaningfully specified. Finally, a plate spreading velocity of 10 cm yr^{-1} in the earth system results in a Peclet number of about 1500, which in turn requires experimental boundary velocities up to 0.03 cm s^{-1}.

The experimental apparatus was specifically designed to cover the parameter ranges discussed above, with the added consideration that the primary interest is in the visual observation of convective planforms. The apparatus is in principle based on a convection tank described by *Chen and Whitehead* [1968], and a side view is shown in Figure 3. It consists of a layer of silicone oil bounded above and below by double glass plates forming a channel through which water was circulated via a thermostatically controlled constant temperature bath. The depth of the silicone oil can be varied by changing the machined spacers that support the upper boundary. The depths used are 1.591, 3.495, and 6.936 cm (all ±0.005 cm). The horizontal dimensions of the convecting region are slightly greater than 100 × 100 cm, and thus the smallest aspect ratio is 15:1. A uniformly moving upper boundary is provided by a 0.025-cm-thick cellulose acetate sheet that is flush against the upper boundary and attached to variable speed motor-driven rollers at each end of the tank. The temperature on the boundaries was determined as the average of the inflow and outflow temperatures measured in the constant-level tanks at each end of the water channels. There is always a slight difference between inflow and outflow water temperatures, which is a function of total imposed vertical temperature change ($\sim 0.02\Delta T$). By circulating the water in opposite directions in the two channels, this effect is minimized, and ΔT is kept more nearly constant.

The operating procedure begins by controlling the initial conditions for the onset of convection. We shine light from a heat lamp through a grid placed on the upper water channel and consisting of regularly spaced tapes (Figure 4a). The absorption of radiation by the oil results in a small (~ 0.1°C) periodic temperature perturbation. After about one thermal time scale (D^2/κ) the thermostats are turned on, and the desired vertical temperature change is imposed. Once the water baths reach their steady temperature, the heat lamp and grid are removed. The small temperature perturbations induce a perfect set of rolls. This inducing technique, first used by *Chen and Whitehead* [1968], provides steady, reproducible initial conditions for experiments that start as transverse rolls or bimodal convection. In cases beginning as multimodal convection, no induction was used to avoid any transient preferred orientation in the initial conditions.

After inducing, the convection pattern is visualized by using a shadowgraph projected onto a screen (Figure 4b). A slightly divergent beam of light passes through the convecting layer and is then intercepted by a screen. The temperature dependence of the index of refraction of the silicone oil causes the light to bend, so that cold fluid (downwelling) is seen as light areas on the screen, and warm fluid (upwelling) appears as dark areas. The observing system integrates across the total depth of the fluid and therefore provides no information about vertical structure, but since we are primarily interested in the planform, it is well suited to our goals. Each experiment consists of first establishing a reproducible initial condition and observing this planform long enough to establish its stability. Then at time $t = 0$ the boundary begins to move, and changes in the planform due to the shear are recorded by photographs taken at regular intervals.

RESULTS

Before the experiments are discussed, it is useful to recall the theory on convection in a shear flow near the critical Rayleigh number. Figure 5 shows the amplitude of transverse rolls (A) and longitudinal rolls (B) as a function of time for four values of shear. The structure of the shear is $U \cos \pi z$,

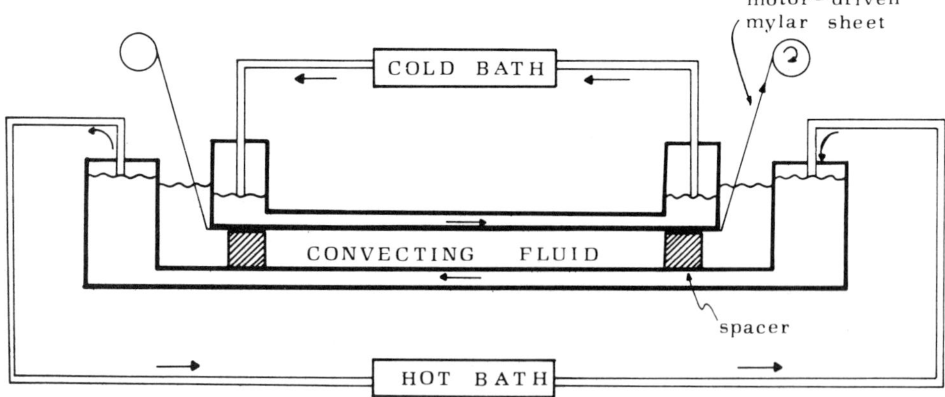

Fig. 3.   Schematic diagram showing a side view of the experimental apparatus. The arrows indicate the flow of thermostatically controlled water. The motor-driven mylar sheet provides a moving upper boundary to the convecting fluid layer.

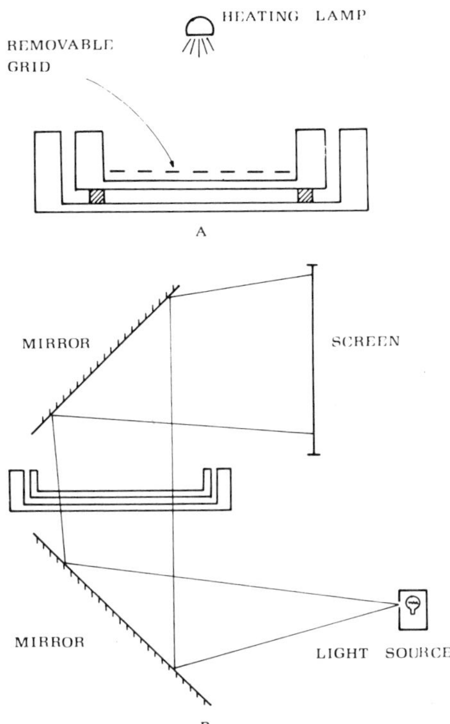

Fig. 4. Schematic diagram illustrating operating procedure. (a) Inducing regular transverse rolls by shining light from a heat lamp through a grid and thereby causing periodic temperature perturbations. (b) Method of observing the convecting pattern by projecting slightly divergent light through the convecting layer and onto a screen. The dependence on temperature of the index of refraction of the convecting fluid makes warm upwelling regions appear as dark areas and cold downwelling regions appear as light areas on the screen.

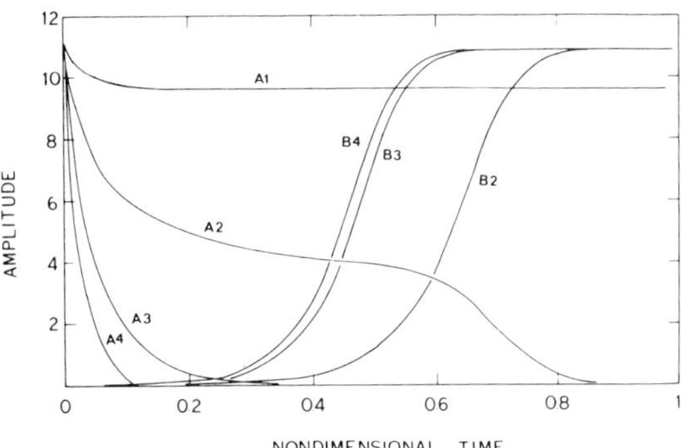

Fig. 5. Amplitude as a function of time for transverse rolls (A) and longitudinal rolls (B) for four values of the shear flow, the nondimensional horizontal velocity varying with depth as given for each example. For case A1, shear is 10.88 cos πz; for case A2-B2, shear is 21.76 cos πz; for case A3-B3, shear is 32.64 cos πz; and for case A4-B4, shear is 43.52 cos πz [from *Richter*, 1973b].

where U is measured in units of A_0, the initial amplitude of the transverse rolls. When U is sufficiently large, as it is in cases A3-B3 and A4-B4, the problem becomes pseudo–two-dimensional in that the transverse rolls decay to insignificant amplitude before the longitudinal rolls begin to grow. In an experiment we would expect to see the decay of the transverse rolls on a relatively fast time scale, a period of no convection, and finally the growth of the longitudinal rolls. The time required for the longitudinal rolls to reach a steady amplitude becomes independent of U for large U. For smaller values of U, such as those in case A2-B2, both transverse and longitudinal rolls will coexist for a short period of time, and in an experiment we would expect to see a transient rectangular planform. Case A1 has the same behavior as case A2-B2, except that the longitudinal mode B1 does not reach significant amplitude until 2.5 time units. The time required for the longitudinal rolls to reach steady amplitude is thus very dependent on U for small values of U.

The amplitude behavior described is based on a model that differs from the experiments in two principal ways. First, the theory is for free (no stress) boundary conditions, whereas the experiments have rigid (no slip) boundaries. Second, the structure of the shear is $U \cos \pi z$ in the theory, whereas it is $U(3z^2 - 2z)$ (Figure 6) in the experiments. Despite these differences, the experiments behave qualitatively as the theory suggests. For

experiments having transverse rolls as initial conditions, we see pseudo–two-dimensional behavior for large values of shear and a transient rectangular planform for smaller values of shear. In all experiments there is a tendency for the time required for the longitudinal rolls to reach steady amplitude to become independent of the magnitude of the shear at large values of the shear.

Table 1 summarizes the experiments and shows the Rayleigh number, the Peclet number (which is a measure of the boundary velocity), the nondimensional time required for longitudinal rolls to reach a steady amplitude, the equivalent dimensional time applied to the upper mantle, the dimensional plate velocity, and the initial conditions. Depending on the initial conditions, the results fall into three classes. First, we have the experiments that start as transverse rolls (1–12 and 23). Figure 7 shows four photographs taken from experiment 1 and illustrates the initial fast decay of the transverse rolls to a new reduced amplitude, followed by a brief period of both transverse and longitudinal rolls and finally only steady longitudinal rolls. This sequence is like case A2-B2 of Figure 5 and is also typical of experiments 5, 6, and 23. At larger rates of shear (experiments 2–4 and 7–12) the behavior is like that in cases A3-B3 and A4-B4, in which the transverse rolls become insignificant before the longitudinal rolls begin to grow. A se-

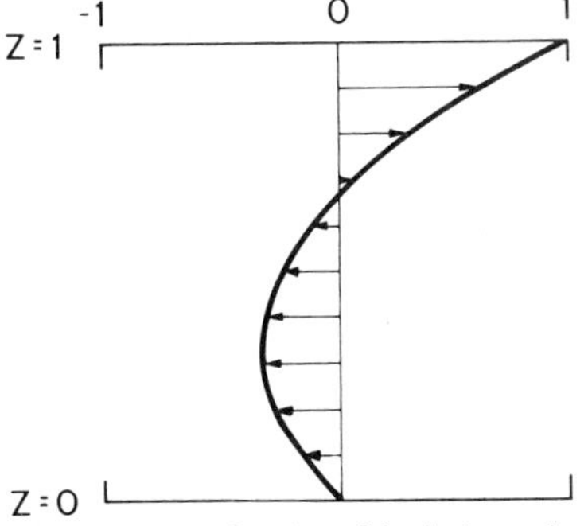

Fig. 6. Velocity as a function of depth due to the moving upper boundary. This profile is observed when the Rayleigh number is below critical.

287

TABLE 1. Summary of Experiments

Experiment	Rayleigh Number	Peclet Number	Nondimensional Time*	Time,* m.y.	Plate Velocity, cm/yr	Initial Conditions†
1	3.3×10^3	31	1.07	7861	0.24	T
2	3.0×10^3	86	0.63	4628	0.67	T
3	3.5×10^3	96	0.58	4261	0.74	T
4	3.4×10^3	144	0.68	4995	1.12	T
5	1.8×10^4	87	0.34	2497	0.68	T
6	1.8×10^4	188	0.15	1102	1.46	T
7	1.8×10^4	188	0.114	837	1.46	T
8	1.8×10^4	167	0.142	1043	1.30	T
9	1.8×10^4	270	0.097	712	2.10	T
10	1.8×10^4	257	0.114	837	2.00	T
11	1.7×10^4	463	0.090	661	3.61	T
12	1.8×10^4	437	0.085	624	3.40	T
13	5.1×10^4	72	0.295	2174	0.56	B
14	5.0×10^4	183	0.065	477	1.42	B
15	5.3×10^4	298	0.045	330	2.32	B
16	5.2×10^4	396	0.037	271	3.09	B
17	5.1×10^4	416	0.035	257	3.24	B
18	1.4×10^5	271	0.034	249	2.12	M
19	1.4×10^5	456	0.026	191	3.55	M
20	1.4×10^5	717	0.023	168	5.59	M
21	1.4×10^5	1178	0.026	191	9.18	M
22	1.4×10^5	675	0.021	154	5.26	M
23	2.9×10^5	767	0.011	80	5.97	TT
24	4.1×10^5	287	0.042	308	2.23	M
25	4.1×10^5	478	0.018	132	3.73	M
26	4.1×10^5	688	0.018	132	5.36	M
27	4.1×10^5	997	0.020	146	7.77	M
28	4.1×10^5	870	0.014	102	6.80	M
29	6.9×10^5	454	0.024	176	3.55	M
30	6.7×10^5	830	0.014	102	6.47	M
31	6.8×10^5	1659	0.007	51	12.9	M

*Values given are the times required for the complete development of longitudinal rolls, measured from the initiation of the large-scale shear flow.

†T, transverse rolls; B, bimodal; M, multimodal; TT, transient transverse rolls.

quence of photographs would be similar to Figure 7, except that the lower-left picture would show no convection at all.

Experiments 13–17 (Table 1) are at a Rayleigh number ($\sim 5 \times 10^4$) where the stable form of convection in the absence of shear is bimodal, characterized by two sets of rolls at right angles. This planform is achieved by inducing transverse rolls and allowing sufficient time for these to evolve into a bimodal pattern. Figure 8 shows photographs from experiment 15, and the sequence of events is typical of all experiments starting as bimodal convection. From Table 1 one can see the tendency of the time required for longitudinal rolls to reach steady amplitude to become independent of the shear.

Experiments 18–31 (except 23) are at Rayleigh numbers for which the initial condition is multimodal. The initial condition in Figure 9 (experiment 20) is a good example of multimodal flow, which is characterized by a spoke pattern of upwellings and downwellings. The vertical structure of the multimodal flow is certainly complicated, and it must be kept in mind that the optical system used to visualize the flow integrates over the entire depth of the fluid. This accounts for the apparent intersection of regions of upwelling (warm) with regions of downwelling (cold). Figure 9 is typical of the experiments that begin as multimodal convection. The effect of the shear was in all cases to produce a convective planform best described as longitudinal rolls. Due to the random initial conditions, the final rolls have a greater tendency to 'pinch out' along their axis than they do in the lower Rayleigh number cases. The bottom two photographs in Figure 9 are at times 0.023 and 0.046, indicating a relatively stable flow despite the pinching out. Figure 9 also illustrates the difficulty in choosing a time for the

development of the longitudinal rolls. We chose 0.023 on the basis that thereafter the pattern changed only very slowly. Our choice of times reported for experiments starting as multimodal flows is thus generally an upper bound. As is the case in Figure 9, there was usually a strong tendency for the convection to align with the shear before the reported time.

The subjectiveness of the time scale is the major source of error in the values shown in Table 1. We emphasize that the times given, especially in the multimodal cases, may be too large, sometimes by as much as 50%. The Rayleigh number may be in error owing to sagging of the mylar sheet away from the upper boundary. The best way to test the Rayleigh number is to. compare the experimental and theoretical Rayleigh numbers for the onset of convection. Such a comparison suggests an error of less than 5%, but it may be slightly larger once the mylar sheet begins to move. The Peclet number, although it is reported as a single value, was in fact slowly increasing with time owing to the changing radius of the pickup roller as mylar was rolled onto it. The total change in the Peclet number was sometimes as large as 10%. The errors in Rayleigh and Peclet numbers reported are small in comparison with the differences between cases and therefore are of little consequence in this study.

Figure 10 summarizes all the experiments by showing the time required for the longitudinal rolls to develop as a function of the Peclet number for seven values of the Rayleigh number. The initial conditions for experiments on curves 1 and 2 are transverse rolls, and those for experiments on curve 3 are bimodal. In these experiments it is relatively easy to decide on the time required for longitudinal rolls to develop, and this is reflected in the good agreement of parametrically similar experiments. Experiments along curves 4–6 begin as multimodal convection, and it is more difficult to choose an unambiguous time scale. As can be seen in Figure 9, after a relatively short time (~50% earlier than the reported time) the convective pattern already consists of segments aligned in the direction of the moving boundary. The time reported in Figure 10 is longer because it depends on the segments' joining into rolls extending the length of the tank and thus represents an upper bound on the time. Curves 4–6 could be shifted to earlier times by as much as 50% if a different criterion were used. The statement that we are reporting an upper bound is supported by experiment 23. This experiment is at a Rayleigh number (2.9 × 10^5) where the stable planform is the spoke pattern. However, the initial conditions for this experiment are transverse rolls, which are transient in that they would have developed into the spoke pattern had the boundary not begun to move. Since the initial conditions were very regular, a clean longitudinal pattern eventually developed for which there was no difficulty choosing an unambiguous time. The time is shown as an asterisk in Figure 10 and is lower by more than 50% than the value that one would predict by interpolating between curves 4 and 5. Furthermore, experiment 23 may be typical of changes in boundary direction, since in such cases the exchange would be from rolls aligned with the old shear to rolls aligned with the new spreading direction.

An alternative way of presenting the results is to show growth time as a function of Rayleigh number for various boundary velocities (Figure 11). The curves are in dimensional quantities using a depth of 575 km (base of the lithosphere to the 650-km discontinuity) and a thermometric diffusivity of 1.5 × 10^{-2} cm² s^{-1}. The 2 cm yr^{-1} curve increases at large Rayleigh numbers because the time scales are a function of both the shear and the Rayleigh number. As the Rayleigh number is increased, this value of shear can be regarded as becoming effectively weaker, and it eventually causes the times to increase. Curves for 5 and 10 cm yr^{-1} are also given. Finally, the dashed curve is an alternative to the 10 cm yr^{-1} solid curve and represents the time required for 'local' alignment of the convection in the direction of shear. This dashed curve represents the time for the development of rolls of a quality intermediate between the qualities of the two middle photographs in Figure 9. It is also the most appropriate time–Rayleigh number curve for cases starting as misoriented rolls at high Rayleigh number.

DISCUSSION

The problem with all suggestions about the form of convection in the mantle is the lack of observational constraints that can be directly related to the convection and are not explicable merely in terms of the rigidity of the surface plates. In the above sections we have put forward the idea that convection in the mantle occurs on two distinct horizontal length scales and then examined some of its consequences. This was done as the best way to reconcile the observations of the existence of the plates themselves, the value of the mean heat flow in older

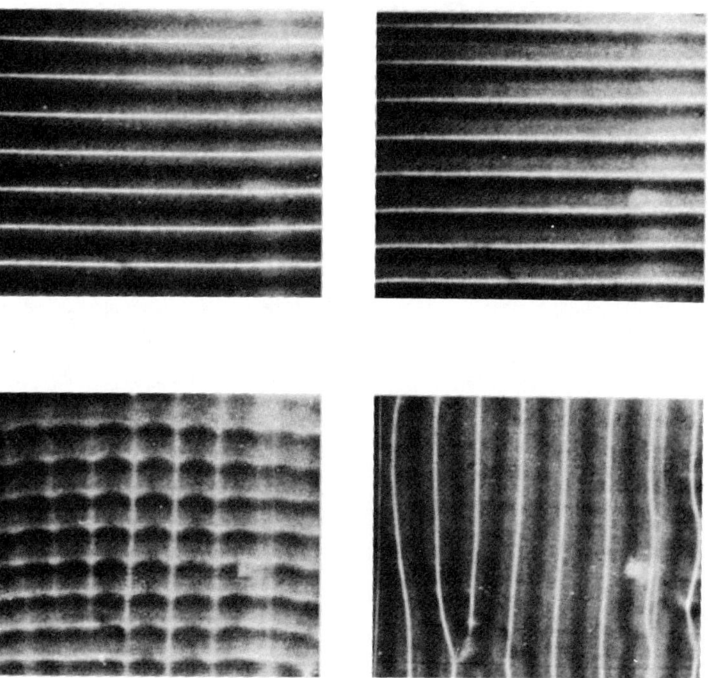

Fig. 7. Photographs from experiment 1; the Rayleigh number is 3.3 × 10^3, and the Peclet number is 31. Light regions represent cold downwelling fluid, while dark regions represent warm upwelling fluid. The top left photograph shows initial transverse rolls at nondimensional time $t = 0$. The top boundary begins to move from top to bottom of the pictures at $t = 0$ and continues to move thereafter. The top right photograph shows decaying transverse rolls at $t = 0.4$. The bottom left photograph shows decaying transverse and growing longitudinal rolls at time $t = 0.7$. The bottom right photograph shows final longitudinal rolls at $t = 1.0$.

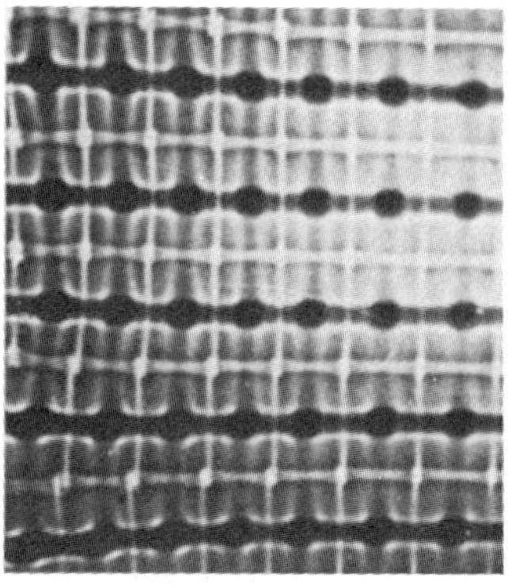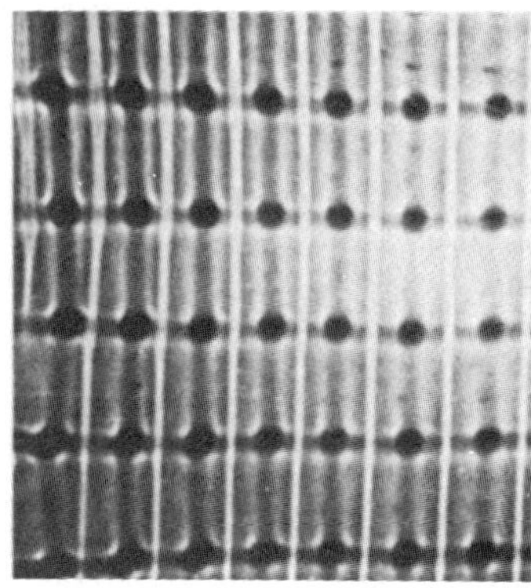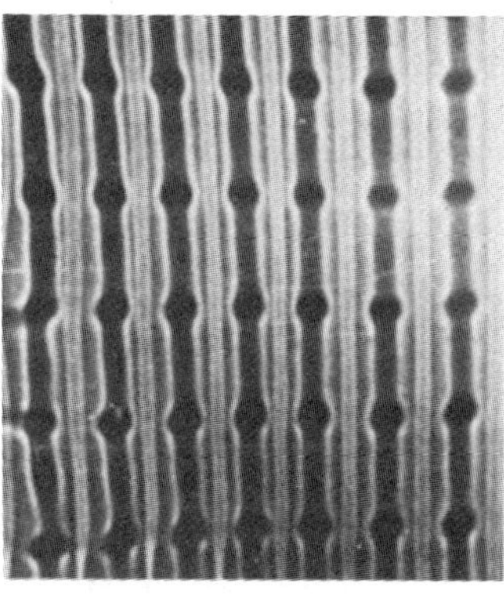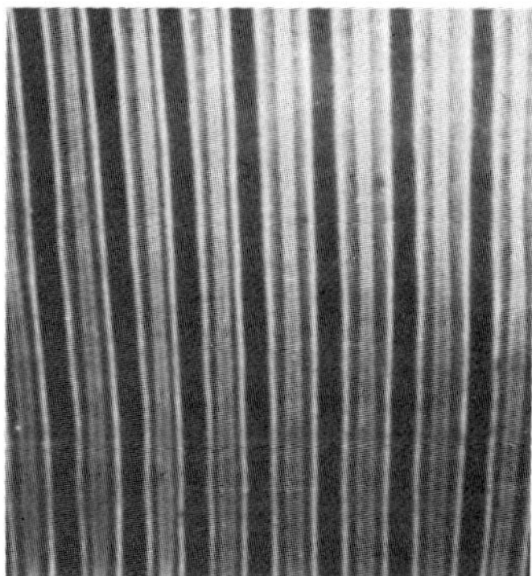

Fig. 8.   Photographs from experiment 15; the Rayleigh number is 5.3 × 10⁴, and the Peclet number is 298. The top left photograph shows initial bimodal convection at $t = 0$. After $t = 0$ the upper boundary is moving from the top toward the bottom of the pictures. For the top right photograph, $t = 0.015$. For the bottom left photograph, $t = 0.030$. For the bottom right photograph, $t = 0.045$.

ocean basins, and the understanding of convection derived from laboratory and numerical experiments. However, because of the paucity of observational constraints, it seems necessary to reexamine points that are fundamental to the hypothesis and to be explicit as to whether they are reliable observations or merely reasonable assumptions where there are no conclusive observations. For the same reason we shall attempt to suggest observations that could be made in the near future to test predictions that follow from the idea of two scales of mantle convection.

The smaller scale of convection was introduced to maintain the relatively constant value of heat flow observed in older ocean basins (Figure 1). This depended on the argument that values of thermal conductivity for mantle materials were too small to allow this heat flow value to be supplied by conduction without requiring large temperature gradients and extensive melting in the upper mantle. Until recently, values of thermal conductivity in the mantle were presumed to be much larger owing to the radiative transfer of heat discussed by *Clark* [1957]. The dominant contribution to the radiative conductivity in olivine comes from wavelengths in a passband from 1.0 to 3.0 μ. The high values of radiative conductivity resulted from using values of the opacity (absorption plus scat-

tering coefficients) at those wavelengths obtained at room temperature. However, measurements at high temperatures [*Fukao et al.*, 1968] showed that the absorption in this passband increased with temperature and hence stopped the radiative conductivity from increasing rapidly with temperature. Combining these results with measurements of total conductivity [*Fujisawa et al.*, 1968; *Kanamori et al.*, 1968] led to an approximately constant value of 0.011 cal cm⁻¹ s⁻¹ deg⁻¹ for the thermal conductivity of olivine in the upper mantle [*Fukao*, 1969]. These conclusions are supported by the results of *Schatz and Simmons* [1972], who made simultaneous determinations of the total conductivity and opacity for a variety of mineral types and gave a maximum value of 0.020 cal cm⁻¹ s⁻¹ deg⁻¹ in the upper 400 km of the mantle. The above determinations have been criticized [*Duba and Nicholls*, 1973; A. Duba, personal communication, 1974] on the grounds that they were performed under temperature and oxygen fugacity conditions favoring oxidation [*Fukao et al.*, 1968] or reduction [*Schatz and Simmons*, 1972] of olivine. Indeed, in both cases color changes in the specimens were reported, and in one case [*Fukao et al.*, 1968] the change was associated with irreversible effects. The oxidation state of the iron could affect the measured absorption coefficients, yet these two sets of experiments were carried out under different conditions and gave similar results. Hence the values of thermal conductivity derived from both are probably reliable. Allowing for the small effects of pressure and the olivine-spinel transition, we have used a constant value of 0.015 cm² s⁻¹ for the thermal diffusivity in the upper mantle.

Having proposed the existence of small-scale convection beneath the lithosphere, we went further and assumed that the convecting region extends to the 650-km-deep seismic discontinuity. Using 75 km as the best estimate of the thickness of the oceanic lithosphere [*Sclater and Francheteau*, 1970; *Kanamori and Press*, 1970], we obtain 575 km for the depth of the convecting region. The only evidence supporting this assumption comes from observations of the focal mechanisms of deep-focus earthquakes [*Isacks and Molnar*, 1969, 1971]. The distribution of earthquake locations is presumed to map out the sinking lithosphere after it has descended beneath the oceanic trenches, and so the existence of earthquakes down to the depth of the phase boundary between 650 and 700 km shows that convective motion can penetrate at least this deep. When the sinking lithosphere has not reached a depth of about 350

km, the earthquakes mainly have focal mechanisms whose least principal stress axis (tension) is aligned with the dip of the slab. This also occurs for descending lithospheric slabs having deeper portions where gaps in the seismicity and the geometrical configuration suggest that the deeper section has become detached. Beneath 350 km the focal mechanisms show that the greatest principal stress axis (compression) is aligned with the dip of the descending lithosphere. In those cases where the lithosphere has reached a depth of 650 km, not only earthquakes below 350 km but also those over the whole length of the slab exhibit downdip compression. Unfortunately, examples of the latter form only a small sample, the slabs beneath the Tonga and Honshu arcs giving the most reliable evidence. These observations can be interpreted in terms of a resistance to motion that begins to increase at about 350 km and increases sharply at 650 km. They also suggest that the convective motion does not penetrate below 650 km. The absence of earthquakes below 700 km is consistent with this conclusion, although such evidence is highly ambiguous without more knowledge of the changes in material properties at this boundary. The lack of knowledge about the properties

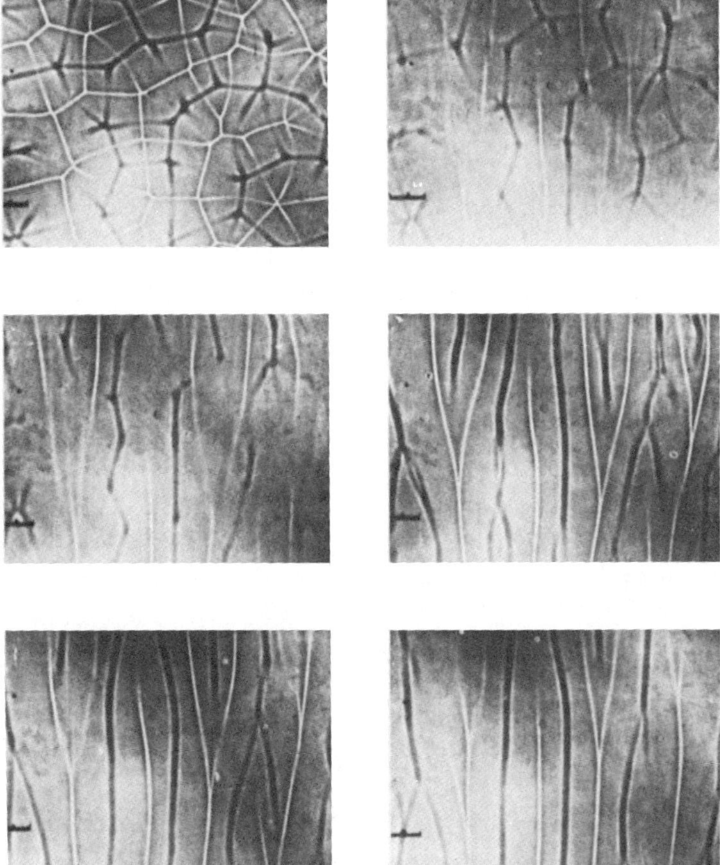

Fig. 9. Photographs from experiment 20; the Rayleigh number is 1.4 × 10⁶, and the Peclet number is 717. The top left photograph shows initial multimodal convection at $t = 0$. After $t = 0$ the upper boundary moves from the bottom of the pictures toward the top. For the top right photograph, $t = 0.002$. For the middle left photograph, $t = 0.007$. For the middle right photograph, $t = 0.015$. For the bottom left photograph, $t = 0.023$. For the bottom right photograph, $t = 0.046$.

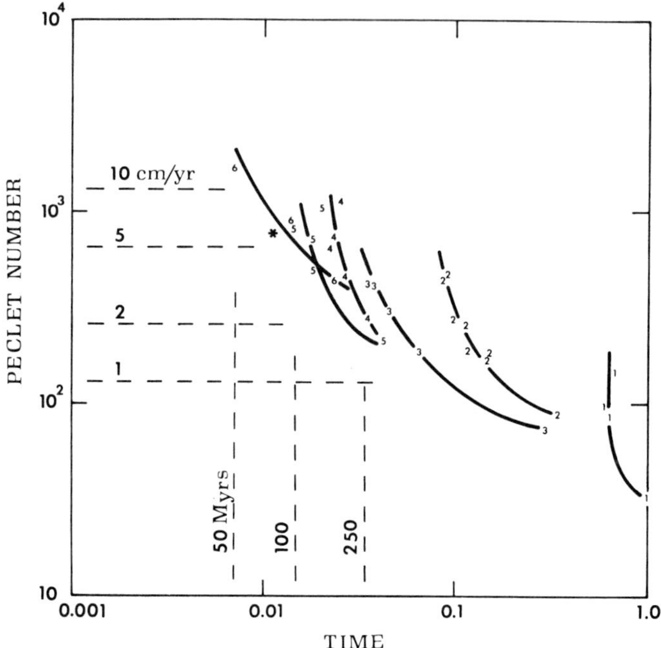

Fig. 10. Graph showing the time required for longitudinal rolls to reach steady amplitude as a function of the boundary velocity for various Rayleigh numbers. The experimental data points (Table 1) are indicated by numbers. Curve 1 is for a Rayleigh number of $\sim 3.3 \times 10^3$; for curve 2, $Ra \sim 1.8 \times 10^4$; for curve 3, $Ra \sim 5.1 \times 10^4$; for curve 4, $Ra \sim 1.4 \times 10^5$; for curve 5, $Ra \sim 4.1 \times 10^5$; and for curve 6, $Ra \sim 6.8 \times 10^5$. The asterisk represents experiment 23, which is a special case, as is discussed in the text. The dimensional quantities along the axis are based on a convective depth of 575 km and a thermometric diffusivity of 1.5×10^{-2} cm² s⁻¹.

of the 650-km phase boundary makes it difficult to understand why it would present a barrier to convection. *Isacks and Molnar* [1971] suggested that if the phase boundary had a negative Clapeyron slope (absorption of heat in going to the denser phase), the phase boundary would be deeper in the sinking lithosphere and the resulting buoyancy forces would prevent penetration. However, the calculations of *Richter* [1973a] showed that the effect of such a phase boundary on convection is only mildly stabilizing relative to convection without a phase boundary and does not provide a barrier against convection. This occurs because the additional cooling effect due to the absorption of heat counteracts the buoyancy effect of the shift in the phase boundary. Another suggestion put forward by *Isacks and Molnar* [1971] was that the successive phases in the upper mantle become more resistant to deformation. Lack of knowledge of material properties and insufficient resolution in the studies of postglacial uplift in North America [*Cathles*, 1971] do not enable us to say whether there is a region around 650–700 km that is more resistant to deformation. Therefore we regard the above evidence as highly suggestive of the assumed depth of convection but not conclusive. Observational tests such as those discussed below will be required as further evidence for the depth of the small-scale convection.

The values of thermal diffusivity κ and depth of convection D having been obtained as described above, the times measured in the experiments could be rescaled in terms of D^2/κ to give times appropriate for the small-scale upper mantle convection. The Rayleigh number appropriate to the small-scale

convection is also needed before we can apply the experimental results. Direct use of the definition is a poor way to determine the Rayleigh number. The viscosity ν is known only to within an order of magnitude, but, more importantly, we do not know what value to use for the temperature difference ΔT. A better method of estimating the Rayleigh number is to take the results of numerical experiments [*McKenzie et al.*, 1974] and use the range of Rayleigh numbers needed to match the mean heat flux out of the convection cells in the numerical experiments with that observed. This method suggests values between 10^5 and 10^6, the most likely estimate lying in the upper part of this range. Applying these values to the experimental results (Figures 10 and 11) leads to the following conclusions. For an absolute plate velocity of 10 cm yr⁻¹ the small-scale convection would become aligned as longitudinal rolls on time scales from 20 to 50 m.y. In this case the resulting arrangement would look like that shown in Figure 12. However, when the plate velocity is lower than 2 cm yr⁻¹, the time scales increase to several hundred million years. Comparing these times with the lengths of time over which no major readjustments in plate motions might be expected to occur suggests that sufficient time is available for rolls to evolve in the former case but not in the latter. We must now examine how and where observations could be made to test the existence of such processes in the upper mantle.

Our experiments were concerned with observations of the horizontal planform of convection. In a similar way, we seek out surface geophysical measurements that, although they are related only in an average way to the detailed mantle structure

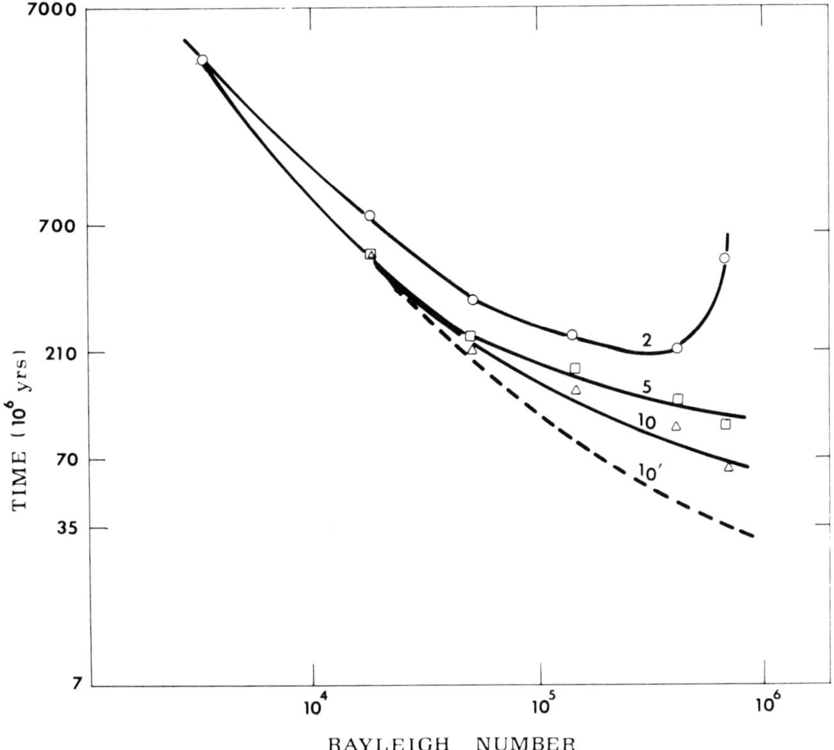

Fig. 11. Graph showing the time required for longitudinal rolls to reach steady amplitude as a function of Rayleigh number for three values of the boundary velocity (2, 5, and 10 cm yr^{-1}). The time and boundary velocity are dimensionalized by using $D = 575$ km and $\kappa = 1.5 \times 10^{-2}$ cm^2 s^{-1}. The data points are taken directly or interpolated from Table 1. The dashed curve is an alternative 10 cm yr^{-1} curve as discussed in the text.

beneath, will enable the horizontal planform of convection to be mapped out. The calculations of *McKenzie et al.* [1974] give the gravity anomalies and surface elevations associated with convection. They found that positive gravity anomalies were associated with upward deformations of the surface, both occurring over upwelling fluid. They also gave a linear relationship between the gravity anomaly and the surface elevation. The size of the gravity anomaly for a given elevation depends on the competition between the gravitational effect of the hot upwelling fluid and that of the upward surface deformation, which are in opposite senses. Hence the magnitude of the slope of the relation between gravity and elevation characterizes the convection. *Anderson et al.* [1973] examined the relationship between free-air gravity anomalies and depth

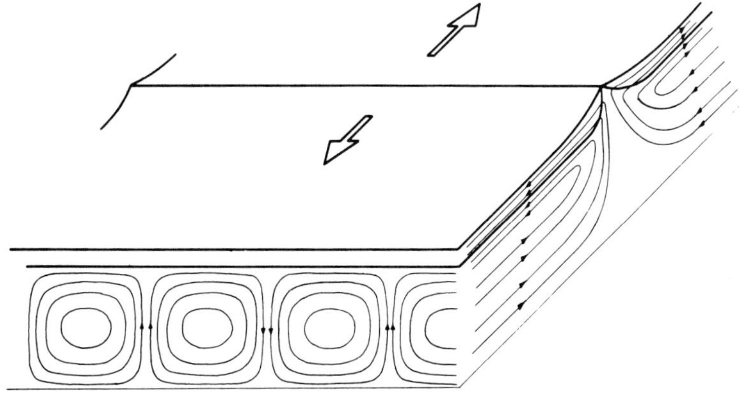

Fig. 12. Illustration of the large-scale flow in the vicinity of a ridge and the superimposed small-scale longitudinal rolls. The form of the cross section of the small-scale convection is schematic only.

along oceanic ridge crests and obtained a linear relationship between the two whose slope agreed with that obtained from the numerical experiments, although admittedly there was a lot of scatter. More recently, *Sclater et al.* [1975] found a similar relationship in two dimensions over the North Atlantic. They defined a residual elevation anomaly to be the difference between the measured depth and the depth to be expected as a result of the cooling of the lithosphere as it moves away from the ridge crest [*Sclater et al.*, 1971]. By constructing contour maps of 5° × 5° average residual elevation and free-air gravity anomalies, they demonstrated a good visual correlation between the two. Moreover, the values also satisfied a linear relationship whose slope agreed with that given by *McKenzie et al.* [1974]. The importance of using both the residual elevation and the gravity anomalies is that the visual correlation between the two, their horizontal extent, and the magnitude of the slope of the correlation can be combined into a convincing argument that the anomalies are produced by convection. The above considerations suggest that determinations of the horizontal variations of this double set of anomalies will provide a good way of mapping out the horizontal planform of upper mantle convection. Preliminary observations of such correlations have also been made in the northeast Pacific [*Menard*, 1973] and the southeast Indian oceans [*Weissel and Hayes*, 1974].

As yet, the observed anomalies do not exhibit any distinctive roll-like features, certainly not in the North Atlantic. The experiments show, however, that we can only reasonably expect to see a roll-like planform beneath a lithospheric plate that is moving sufficiently fast. There would seem to be little argument that the best place to look for such a phenomenon is the Pacific Ocean. Although, as we shall discuss below, there seems little reason to suppose that mantle hot spots [*Morgan*, 1972] are fixed, thereby providing an absolute reference frame, if we suppose for the moment that the relative motion of the hot spots is slow in comparison with the velocity of the Pacific plate, then the results of *Minster et al.* [1974] can still be used as a guide. It is thus suggested that the Pacific may well have absolute velocities of the order of 9 cm yr^{-1}. In order for the rolls to have aligned themselves in the present direction of motion, there must have been sufficient time since the last major readjustment in the motion of the Pacific plate. Evidence from around the Pacific border [*Uyeda and Miyashiro*, 1974; *Atwater*, 1970] and from changes in the direction of Pacific island chains [*Clague and Jarrard*, 1973] indicates that such a change occurred between 40 and 25 m.y. B.P. This change was associated with the beginning of the formation of the Izu-Mariana arc system and the Philippine Sea and the initiation of the Mizuho orogeny in Japan on the western side of the Pacific [*Uyeda and Miyashiro*, 1974]. During the same period the Pacific-Farallon ridge system began to interact with the trench that then existed off the coast of California [*Atwater*, 1970; *Atwater and Molnar*, 1973]. The approximate dates of this activity agree with dates around the bend in the Hawaiian-Emperor chain [*Clague and Jarrard*, 1973; *Dalrymple et al.*, 1974]. If it is assumed that the direction of motion of the Pacific has been constant over the last 25-40 m.y., there has probably been sufficient time for rolls to develop. Certainly, if we look at the sequence in Figure 9, at least some roll-like features will have evolved even at an intermediate stage. Hence a test of the phenomena explored in this paper could be made by constructing contour maps of the residual elevation and free-air gravity anomalies over the Pacific. In the likely case that there are insufficient data to do

this, a few long lines parallel to each other and perpendicular to the direction of motion of the Pacific plate should suffice to reveal the existence of roll-like features. Although we do not know exactly the absolute velocity of the Pacific plate, the rotation poles of *Clague and Jarrard* [1973] probably give the best direction to use consistent with paleomagnetic data and sediment distributions.

The observations discussed above would seem to provide the best test of the hypothesis, and we have not utilized the occurrence of volcanic features in the interior of plates because of their uncertain relation to mantle convection. The linear island chains in the Pacific, for instance, have previously been explained in terms of a plate moving over a hot spot rising from the lower mantle [*Wilson*, 1963; *Morgan*, 1971, 1972]. There seems to be little difficulty in adapting the hypothesis proposed here to explain these features, either by modulating the amplitude of convection along the roll [*Richter*, 1973b] or by invoking the fracture properties of the plate. Also, age gaps or ages that do not match the predicted progression [*Clague and Jarrard*, 1973] are not unexpected. However, the main point to be made is that, regardless of the exact form of upper mantle convection or the true explanation of these surface features, a very wide variety of different features that provide potential explanations arises in such a simple system as a uniformly heated layer of convecting fluid beneath a moving boundary. This is amply illustrated by the sequence in Figure 9. The experiments indicate that beneath a stationary or very slow moving plate the form of convection to be expected is more like the first photograph in this sequence. The characteristic form of convection beneath a stationary boundary at these Rayleigh numbers is dominated by both upwelling and downwelling spouts of fluid. Thus the same system contains a potential explanation of some features of Neogene geology in Africa, such as the basin and swell topography [*Burke and Wilson*, 1972; *Wilson*, 1973]. Regardless of whether the form of convection is roll-like or spout-like, the mapping of the horizontal planform of convection using residual elevation and gravity anomalies will give some information about the depth of convection. As was pointed out earlier, laboratory experiments and analytical calculations show that the dominant horizontal length scale is determined by the depth of the fluid. Hence a characteristic wavelength for the pattern obtained, and any other associated features, would be a strong indication of a given depth of convection. We also see that many assumptions used in previous hypotheses can be discarded as unnecessary. Thus there is no need to locate the source of plumes in the lower mantle. Recent reconstructions have also shown that hot spots are not stationary with respect to one another [*Molnar and Atwater*, 1973; *Burke et al.*, 1973], in agreement with laboratory observations that the pattern of upwelling and downwelling spouts is seen to evolve slowly with time. Therefore there is the possibility of explaining many surface features in terms of a well-defined system of convection whose physical basis can be properly studied.

Throughout this paper we have avoided the question of the maintenance of the large-scale circulation, taking its existence to be given by observations. By assuming two scales of convection, where the horizontal scales are distinct, we could put aside this question and just investigate the interaction of the small-scale flow in the interior of the large-scale flow. Underlying this approach is the assumption that the importance of Rayleigh-Benard convection and its variants that have been studied numerically is in the vertical transport of heat (the small scale of convection) and that an explanation of the large-scale circulation must be sought elsewhere. The experiments demonstrate the plausibility of this assumption and the resulting properties of such a system. Our experiments do make a comment on the maintenance of the large-scale flow by showing that small-scale convection interacting with the overlying plate is suppressed. In other words, a small-scale convective mode that influences the motion of the plates is at a competitive disadvantage with other modes that do not interact with the plate motion. It is thus suggested, especially in the case of oceanic plates, that the large-scale flow is maintained by effects near the boundaries, whereas the available potential energy under the interior of plates goes into driving the small-scale flow.

Acknowledgments. We are grateful to Woods Hole Oceanographic Institution and in particular to J. Whitehead for the use of laboratory facilities. We benefited greatly from Jack Whitehead's long experience with convection experiments. Robert Frazel helped at every stage, from the building of the apparatus to photography. R. Goines and J. Keller of Dow Corning Corporation, Midland, Michigan, were most kind in donating otherwise unavailable silicone adhesives. The research was supported primarily by the Earth Sciences Section, National Science Foundation, grant GA-36132X. Additional support was provided by the Advanced Research Projects Agency, monitored by the Air Force Office of Scientific Research under contract F44620-71-C-0049, and a portion of the laboratory support was provided by National Science Foundation grant GA-35447.

REFERENCES

Anderson, R. N., D. P. McKenzie, and J. G. Sclater, Gravity, bathymetry and convection within the earth, *Earth Planet. Sci. Lett., 18,* 391–407, 1973.

Atwater, T., Implications of plate tectonics for the Cenozoic tectonic evolution of western North America, *Geol. Soc. Amer. Bull., 81,* 3513–3536, 1970.

Atwater, T., and P. Molnar, Relative motion of the Pacific and North American plates deduced from sea-floor spreading in the Atlantic, Indian, and South Pacific Oceans, in *Proceedings of the Conference on the Tectonic Problems of the San Andreas Fault*, pp. 136–148, Stanford University Press, Stanford, Calif., 1973.

Burke, K., and J. T. Wilson, Is the African plate stationary?, *Nature, 239,* 387–389, 1972.

Burke, K., W. S. F. Kidd, and J. T. Wilson, Relative and latitudinal motion of Atlantic hot spots, *Nature, 245,* 133–137, 1973.

Busse, F. H., On the stability of two-dimensional convection in a layer heated from below, *J. Math. Phys., 46,* 140, 1967.

Busse, F. H., and J. Whitehead, Instabilities of convection rolls in a high-Prandtl number fluid, *J. Fluid Mech., 47,* 305, 1971.

Cathles, L. M., The viscosity of the earth's mantle, Ph.D. thesis, Princeton Univ., Princeton, N. J., 1971.

Chandra, K., Instability of fluids heated from below, *Proc. Roy. Soc., Ser. A, 164,* 231, 1938.

Chen, M. M., and J. Whitehead, Evolution of two-dimensional periodic Rayleigh convection cells of arbitrary wavenumbers, *J. Fluid Mech., 31,* 1, 1968.

Clague, D. A., and R. D. Jarrard, Tertiary Pacific plate motion deduced from the Hawaiian-Emperor chain, *Geol. Soc. Amer. Bull., 84,* 1135–1154, 1973.

Clark, S. P., Radiative transfer in the earth's mantle, *Eos Trans. AGU, 38,* 931–938, 1957.

Dalrymple, G. B., M. A. Lanphere, and E. D. Jackson, Contributions to the petrography and geochronology of volcanic rocks from the leeward Hawaiian Islands, *Geol. Soc. Amer. Bull., 85,* 727–738, 1974.

Duba, A., and I. A. Nicholls, The influence of oxidation state on the electrical conductivity of olivine, *Earth Planet. Sci. Lett., 18,* 59–64, 1973.

Fujisawa, H., N. Fujii, H. Mizutani, H. Kanamori, and S. Akimoto, Thermal diffusivity of Mg_2SiO_4, Fe_2SiO_4, and NaCl at high pressures and temperatures, *J. Geophys. Res., 73,* 4727, 1968.

Fukao, Y., On the radiative heat transfer and the thermal conductivity in the upper mantle, *Bull. Earthquake Res. Inst. Tokyo Univ., 47,* 549–569, 1969.

Fukao, Y., H. Mizutani, and S. Uyeda, Optical absorption spectra at high temperatures and radiative thermal conductivity of olivines, *Phys. Earth Planet. Interiors, 1,* 57–62, 1968.

Ingersoll, A. P., Thermal convection with shear at high Rayleigh number, *J. Fluid Mech., 25,* 209, 1966.

Isacks, B., and P. Molnar, Mantle earthquake mechanisms and the sinking of the lithosphere, *Nature, 223,* 1121–1124, 1969.

Isacks, B., and P. Molnar, Distribution of stresses in the descending lithosphere from a global survey of focal mechanism solutions of mantle earthquakes, *Rev. Geophys. Space Phys., 9,* 103–174, 1971.

Kanamori, H., and F. Press, How thick is the lithosphere?, *Nature, 226,* 330–331, 1970.

Kanamori, H., N. Fujii, and H. Mizutani, Thermal diffusivity measurement of rock-forming minerals from 400° to 1100°K, *J. Geophys. Res., 73,* 595–605, 1968.

McKenzie, D. P., Some remarks on heat flow and gravity anomalies, *J. Geophys. Res., 72,* 61, 1967.

McKenzie, D. P., Plate tectonics, in *The Nature of the Solid Earth,* edited by E. C. Robertson, p. 323, McGraw-Hill, New York, 1972.

McKenzie, D. P., J. M. Roberts, and N. O. Weiss, Convection in the earth's mantle: Towards a numerical simulation, *J. Fluid Mech., 62,* 465–538, 1974.

Menard, H. W., Depth anomalies and the bobbing motion of drifting continents, *J. Geophys. Res., 78,* 5128–5137, 1973.

Minster, J. B., T. H. Jordan, P. Molnar, and E. Haines, Numerical modelling of instantaneous plate tectonics, *Geophys. J. Roy. Astron. Soc., 36,* 541–576, 1974.

Molnar, P., and T. Atwater, Relative motion of hot spots in the mantle, *Nature, 246,* 288–291, 1973.

Morgan, W. J., Convection plumes in the lower mantle, *Nature, 230,* 42–43, 1971.

Morgan, W. J., Deep mantle convection plumes and plate motions, *Amer. Ass. Petrol. Geol. Bull., 56,* 203–213, 1972.

Richter, F. M., Dynamical models for sea floor spreading, *Rev. Geophys. Space Phys., 11,* 223–287, 1973a.

Richter, F. M., Convection and the large-scale circulation of the mantle, *J. Geophys. Res., 78,* 8735–8745, 1973b.

Schatz, J., and G. Simmons, Thermal conductivity of earth materials at high temperatures, *J. Geophys. Res., 77,* 6966–6983, 1972.

Schlüter, A., D. Lortz, and F. Busse, On the stability of steady finite amplitude convection, *J. Fluid Mech., 23,* 129–144, 1965.

Sclater, J. G., and J. Francheteau, The implications of terrestrial heat flow observations on current tectonic and geochemical models of the crust and upper mantle of the earth, *Geophys. J. Roy. Astron. Soc., 20,* 509–542, 1970.

Sclater, J. G., R. N. Anderson, and M. L. Bell, The elevation of ridges and the evolution of the central eastern Pacific, *J. Geophys. Res., 76,* 7888–7915, 1971.

Sclater, J. G., L. A. Lawver, and B. Parsons, Comparison of long wavelength residual elevation and free-air gravity anomalies in the North Atlantic and possible implications for the thickness of the lithospheric plate, *J. Geophys. Res., 80,* 1031–1052, 1975.

Uyeda, S., and A. Miyashiro, Plate tectonics and the Japanese Islands: A synthesis, *Geol. Soc. Amer. Bull., 85,* 1159–1170, 1974.

Weissel, J. K., and D. E. Hayes, The Australian-Antarctic discordance: New results and implications, *J. Geophys. Res., 79,* 2579–2587, 1974.

Wilson, J. T., A possible origin of the Hawaiian Islands, *Can. J. Phys., 41,* 863–870, 1963.

Wilson, J. T., Mantle plumes and plate motions, *Tectonophysics, 19,* 149–164, 1973.

(Received September 30, 1974;
revised January 29, 1975;
accepted January 30, 1975.)

22

Reprinted from *Jour. Geophys. Research* **80**:4425–4432 (1975)

Growth, Shrinking, and Long-Term Evolution of Plates and Their Implications for the Flow Pattern in the Mantle

ZVI GARFUNKEL

Surficial manifestations of the long-term evolution of plates are examined and used to derive constraints on the pattern of flow in the mantle. Generation and consumption of plates are not balanced around single plates, so the sizes of plates and the relative positions of ridges and descending slabs change, a complex return flow thus being required. At present a net flow should occur from the sites of much subduction toward the southern hemisphere, where sea floor spreading predominates. To allow the relative motions between descending slabs, masses in the mantle should be displaced at a rate which is several times larger than the rate of generation of lithosphere. The patterns of plate motions and of the return flow change considerably over periods of 100 m.y. The formation of instabilities occurs much faster than the cycling of mantle material. Hence the flow in the mantle cannot consist of regular steady state cells in which material moves in simple circuits and which are linked to plate boundaries. The deep flow is not parallel to the relative motions of overlying plates, so it is not two dimensional. The magnitude and non-steady state of the sublithospheric flow make it doubtful that instabilities originating within the lithosphere alone drive the flow. This fact, together with the abrupt changes of plate motions, including jumping of ridges and continental breakup, suggests that instabilities at depth, e.g., mantle diapirs, are most important. The plate tectonic regime helps generate such instabilities in the mantle by disturbing the temperature distribution there.

INTRODUCTION

The theory of plate tectonics requires considerable mass movement in a large part of the earth: plates move about on the surface, material rises from depth to generate oceanic lithosphere at ridges, and lithospheric slabs descend in subduction zones to depths reaching 700 km. To balance this overturn of lithospheric mass, a return flow should occur to at least a depth of 700 km, i.e., in about a quarter of the mass of the mantle. However, while the motions of the surficial plates are fairly well known, the pattern of the deep return flow remains obscure.

In addition to the continuing overturn of lithosphere the plate tectonic regime leads to characteristic long-term effects. Notably, the sizes and shapes of plates change continuously. In the long run these changes are related to changes in the pattern of plate motions. The purpose of this paper is to discuss such long-term features of plate evolution and to examine their implications for the pattern of mass flow in the mantle.

The discussion is based on surficial phenomena, so the conclusions do not depend on physical models of the flow, though it is accepted that the flow is some kind of thermal convection. The aim is to derive constraints on the principal features of the flow pattern rather than to offer a physical mechanism; successful physical models should satisfy these constraints.

Quantitative data on plate behavior are available only for a short part of the earth's history. However, the gross uniformity of the main traits of the earth's behavior during the Phanerozoic and much of the Precambrian, as recorded on continents, suggests that the same fundamental processes were in operation. It is assumed therefore that the processes which now drive and control the motion and evolution of plates were active during a considerable portion of the earth's history. Observations on the geologically recent past are supposed to represent much older periods, for which direct information about plate behavior is not available. It is also assumed that the size of the earth did not change during this period.

The flow in the mantle is supposed to involve only the outer 700 km or so of the earth, as is suggested by the seismic behavior of the descending slabs [*Isacks and Molnar,* 1971] and by the differences between the composition of the lower mantle and that of its shallower parts [*Anderson and Jordan,* 1970]. However, some conclusions do not depend on this assumption.

GROWTH AND SHRINKING OF PLATES

Rates of production and consumption of surface area along various plate boundaries during the last 5–10 m.y. are summarized in Tables 1 and 2. Plate motions during this period are known quite well, so the errors are probably less than 10%. These values are used to calculate the rates of change of the areas of plates (in square kilometers per year), which are given below:

Antarctican	+0.50
African	+0.26
(including Somali)	
American	+0.30
Eurasian	+0.06
Arabian	−0.11
Indian	−0.37
Pacific	−0.45
Cocos	−0.08
Nazca	−0.11

Clearly, sea floor spreading and subduction are not balanced locally, i.e., around single plates, so the areas and shapes of plates change significantly in geologic time.

The Antarctican plate grows fastest, being almost surrounded by fast spreading ridges. The African plate (including Somali) is also largely surrounded by ridges, but part of this growth is offset by subduction in the Mediterranean Sea region. The American plate grows along its Atlantic margin, while its Pacific margin remains virtually unchanged.

The Pacific, Cocos, and Nazca plates shrink at a total rate of about 0.7 km²/yr. Here the overturn of oceanic lithosphere is faster than it is elsewhere, but the result is a net loss of area, so the Pacific Ocean shrinks. However, in the south the Antarc-

TABLE 1. Rates of Sea Floor Spreading Along Various Plate Boundaries

Plate Boundary	Relative Motion			Distance From Rotation Pole*		Rate of Area Generation, km²/yr	Reference
	Geographic Position		V_{max}, cm/yr	θ_1	θ_2		
				Atlantic Ocean			
Eurasian-American	68°N,	137°E	3.08	0†	72	0.136	*Pitman and Talwani* [1972]
African-American	70°N,	−33°E	4.44	30	127	0.414	*Pitman and Talwani* [1972]
Total						0.550	
				Pacific Ocean			
Pacific-American	55°N,	−65°E	−8.5	41	49	0.050	Author's estimate
Cocos-Nazca	5°N,	−122°E	12.	20	30	0.116	*Herron* [1972]
Pacific-Nazca	58°N,	−93°E	−20.	56	98	0.880	*Herron* [1972]
Pacific-Cocos	40°N,	−108°E	−28.4	23	38	0.246	Calculated from *Herron* [1972]
Antarctican-Pacific	−70°N,	118°E	−12.	17	64	0.398	*Le Pichon* [1968]
Antarctican-Nazca	38°N,	−110°E	9.3	81	89	0.080	
Pacific–Juan de Fuca						0.070	Estimated from *Atwater* [1970]
Total						1.840	
				Indian Ocean			
Arabian-African	36°N,	18°E	3.2	16	33	0.026	*McKenzie et al.* [1970]
Arabian-Somali	26°N,	21°E	3.9	25	37	0.026	*Laughton et al.* [1970]
Somali-Indian	16°N,	48°E	−6.9	10	48	0.140	*McKenzie and Sclater* [1971]
Indian-Antarctican	11°N,	32°E	7.1	53	116	0.470	*McKenzie and Sclater* [1971]
Antartican-African	16°N,	161°E	2.5	98	139	0.098	Calculated from *McKenzie and Sclater* [1971]
Total						0.760	
Grand total						3.150	

* The rate of generation of area between points at angular distances θ_1 and θ_2 from the rotation pole is given by $ds/dt = {}_{\theta_1}\!\int^{\theta_2} V_{max} \cdot \sin \theta \cdot R \cdot d\theta = V_{max} \cdot R \cdot (\cos \theta_1 - \cos \theta_2)$, where R is the earth's radius.

† A fair approximation.

tican plate grows at a rate of 0.24 km²/yr. The Indian plate also loses area.

Many lines of evidence show that the area of continental crust has increased during geological history [*Wilson, 1967; Taylor, 1967; Ringwood, 1969*]. To produce the area shallower than 1 km below sea level, i.e., 190 × 10⁶ km² [*Menard and Smith, 1966*] during 3500 m.y., the average rate of growth had to be 0.054 km²/yr. Compared with this value, the rate of loss of continental area along the margin of the Arabian plate and along the Himalayas (Table 2) is surprisingly large: it is almost 0.3 km²/yr, which is about 5 times the long-term average rate of continental growth. This is a transient circumstance, because it is improbable that large volumes of relatively light and buoyant continental crust can descend into the heavier mantle. Except for this feature the values of the rates of change of plate areas given in the in-text table above probably represent in a general way the geologic past.

The oldest oceanic crust is of Jurassic age, about 180 m.y. old. The possible growth of continents during this period is negligible (amounting to only 10 × 10⁶ km² when the above long-term average is used; even a value several times larger will not seriously affect the following calculation). Thus on an earth of constant size the entire oceanic area, about 310 × 10² km² [*Menard and Smith, 1966*], was generated within the last 180 m.y. In addition, the asymmetric magnetic anomalies in the Pacific Ocean show that an area equal to about three quarters of this ocean was generated during this time interval but has already been subducted [*Larson and Pitman, 1972*]. Hence during the last 180 m.y., oceanic lithosphere was generated at an average rate of about 2.4 km²/yr, which is not very different from the estimate of 3.1 km²/yr (Table 1) for the present rate. Some fluctuations could have occurred.

The shrinking of the Pacific Ocean is a long-term process; the fit of the continents at the beginning of the Mesozoic [*Bullard et al., 1965; Smith and Hallam, 1970*] shows that the area

of the Tethys Ocean was then about 30 × 10⁶ km². This area has been consumed, and concurrently, the Atlantic and Indian oceans were formed, covering areas (deeper than 1 km) of 76 × 10⁶ and 69 × 10⁶ km², respectively [*Menard and Smith, 1966*]. As the area of the continents hardly changed during this period (see discussion above), the area of the remaining ocean, that is, the Pacific, must have decreased by about 115 × 10⁶ km². Hence the average rate of area loss since 180 m.y. ago has been about 0.65 km²/yr, similar to the current rate of 0.7 km²/yr (see in-text table above). While the Pacific Ocean shrank, the Kula plate was entirely consumed, and so were large parts of the Phoenix and Farallon plates [*Atwater, 1970; Larson and Pitman, 1972*]. At the present rate of loss of area, most of the Pacific Ocean can be consumed within about 200 m.y., and the bordering continents will collide.

Other oceans disappeared in the geologic past when the bordering continents collided, a situation indicating that subduction and sea floor spreading were not locally balanced. In addition to the Tethys, one may cite the proto–North Atlantic, which disappeared early in the Paleozoic at the site of the Caledonides [*Wilson, 1966; Harland and Gayer, 1972*], and the Siberian Ocean, which disappeared in the Paleozoic at the site of the Uralides [*Hamilton, 1970*].

The Pacific Ocean could not have been shrinking at the present rate from before the Mesozoic. If it had been, then in the Paleozoic it would have been much larger than the total oceanic area on earth. At that time the Pacific Ocean was probably growing while other oceans were being consumed at the Caledonian and Hercynian orogenic belts. Thus the overall balance between subduction and spreading in the Pacific Ocean must have been reversed.

Neither oceanic plates nor oceans can grow indefinitely to exceed the total area of oceanic crust on earth; otherwise, much continental crust would have to be subducted, an unlikely prospect. If the Atlantic and Indian oceans continue to

TABLE 2. Rates of Subduction Along Various Plate Boundaries

Plate Boundary	Relative Motion			Distance From Rotation Pole		Rate of Area Consumption, km²/yr
	Geographic Position		V_{max}, cm/yr	θ_1	θ_2	
Alpine Orogenic Belt						
African-Eurasian	26°N,	−37°E	3.0	14	63	0.104
Arabian-Eurasian	34°N,	−10°E	5.7	38	66	0.138
Indian-Eurasian (Himalayas)	28°N,	26°E	7.2	38	64	0.155
Indian-Eurasian (Java trench)	28°N,	26°E	7.2	64	102	0.300
Total						0.697
East Pacific						
Cocos-American	31°N,	−119°E	21.6	19	41	0.256
Nazca-American	56°N,	−113°E	12.	55	106	0.652
Antartican-American	−79°N,	−81°E	4.4	34	19	0.033
Pacific-American	55°N,	−65°E	−8.5	38	62	0.175
Juan de Fuca-American						0.075*
Total						1.191
West Pacific						
American-Eurasian	68°N,	137°E	−3.08	0	17	0.008
Pacific-Eurasian†	69°N,	−72°E	−10.5	49	109	0.660
Pacific-Indian‡	−62°N,	174°E	15	22	48	0.246
				60	70	0.151
Pacific-Indian	−62°N,	174°E	15	9	22	0.058
				48	60	0.162
Total						1.285
Grand total						3.173

All relative plate motions are from Table 1 or are calculated from it.
*Estimate.
†Includes Philippine plate.
‡Pacific plate consumed in New Guinea and Tonga-Kermadec trenches; Indian plate consumed along rest of plate boundary.

grow at the present rates, the maximum possible size will be reached in about 250 m.y. Then either sea floor spreading will cease, or new subduction zones will have to develop. The balance between generation and subduction of oceanic lithosphere will change over large areas.

The constantly changing sizes and shapes of plates clearly must be accompanied by modifications of the geometry of plate motions. In addition, when continents collide, the relative motion of the plates containing them should change or stop to avoid subduction of much continental crust.

However, motions of plates also change without apparent relation to changes of their sizes and shapes. Such events, accompanied by jumping of midoceanic ridges, occurred in the oceans [*McKenzie and Sclater*, 1971; *Herron*, 1972; *Pitman and Talwani*, 1972]. Most spectacular changes result from continental breakup, e.g., the dispersal of Laurasia and Gondwanaland. Similar events probably also occurred earlier in the earth's history: the western margin of North America, for instance, bears evidence of Precambrian events of continental splitting [*Monger et al.*, 1972; *Hoffman*, 1973].

At present the absence of local balance between sea floor spreading and subduction is well expressed not only in terms of changing areas of individual plates but also on a broader scale: more than three quarters of all sea floor spreading occurs in the southern hemisphere, or rather in the hemisphere centered on the Scotia Sea, whereas only about one third of all subduction occurs there. An area of about 0.96 km² is annually consumed in a relatively small region around Southeast Asia (Table 2). This is about 30% of the global rate, yet is not obviously related to any nearby region of sea floor spreading, since the neighboring Indian and Pacific plates are shrinking.

Along the Central and Southern American trenches an area of about 0.8 km² is consumed annually (Table 2); in the neighboring South Atlantic Ocean, only half of this area is generated, while in the eastern Pacific Ocean an area twice as large is generated (Table 1). The ridges around the Antarctican and African plates do not alternate with subduction zones, so the spreading along these ridges cannot at all be related directly to subduction in neighboring regions.

The foregoing considerations show that shrinking of some plates and growth of others, coupled with continuous and episodic changes in the pattern of plate motions and interaction, are essential features of the long-term evolution of plates, expressing the absence of local balance between sea floor spreading and subduction. The quantitative relations discussed above show that 100 m.y. or a few hundred million years is a typical period for some fundamental changes in the pattern of plate motions to take place, and this may also be the lifetime of some oceanic plates.

CONSEQUENCES REGARDING FLOW IN THE MANTLE

The features of the plate evolution and motions discussed above have important consequences with regard to the nature of flow in the underlying mantle.

The non–steady state nature of plate motions implies that the underlying mass motions must also be time dependent, because (1) generation and consumption of plates are coupled with a return flow in the mantle which extends to a depth of 700 km at least and (2) the overturn of lithospheric mass relative to the mass of the mantle is significant during geologically long periods. To produce a 70-km-thick lithosphere [*Kanamori and Press*, 1970] at a rate of 2.4 km²/yr, a mass of about 5.5×10^{17} g ($\rho = 3.3$ g/cm³) should rise annually from the mantle; at this 180-m.y. average rate the entire mass shallower than 700 km, which is assumed to participate in the flow, can be recycled in somewhat more than 2000 m.y., and the mass down to 1000 km can be recycled in about 3000 m.y. In the Proterozoic this rate was probably higher than it is now, because more radiogenic heat was then produced in the earth, an action which could induce more vigorous convection; hence a complete overturn could occur in 1500 m.y. This value is similar to the differentiation events recorded by lead isotopes from volcanic islands [*Oversby and Gast*, 1970], which *Morgan* [1972] interpreted as recording previous times when the source rocks of the volcanics were part of the lithosphere. Therefore the mass motions in the mantle associated with the plate tectonic regime cannot be approximated well by a steady state model over periods of a few hundred million years.

The continuing changes of the sizes and shapes of plates require that the relative positions of the midoceanic ridges and of the subduction zones must also change, so at least some must move in relation to the deeper mantle. *Elsasser* [1971] noted that as a result of the shrinking of the Pacific Ocean the subduction zones surrounding it should approach each other; therefore at least some descending slabs must sink in relation to the mantle and not merely descend parallel to themselves. Elsasser called this motion 'retrograde subduction.' Such a behavior is in line with gravitative sinking of the slabs, which are colder and denser than the surrounding mantle [*Elsasser*, 1971; *Turcotte and Schubert*, 1971]. When a descending lithospheric slab sinks in such a manner through the mantle, it displaces some material from its lower side, and an equal volume must be filled on the other side (Figure 1). Since equal volumes are displaced on the two sides of the slabs, these mass motions cannot be simply related to the return flow from the subduction zones to ridges.

The magnitude of the sublithospheric mass motions necessary to allow retrograde subduction can be estimated from the rate of shrinking of the Pacific Ocean. While this ocean shrinks at a rate of about 0.7 km²/yr, the subduction zones which surround most of it sweep out a comparable area between themselves. The descending slabs reach to depths exceeding

500 km, so while their relative positions change, they displace from the Pacific side a mass exceeding 12×10^{17} g/yr ($\rho = 3.5$ g/cm³), which is more than twice the mass that rises in all the ridges. A similar mass is displaced on the other side of the slabs. This is a crude estimate because it does not correct for absence of subduction along some parts of the Pacific Ocean or for possible differences in the depths to which various slabs extend. However, this estimate does show that the sublithospheric mass flow which accommodates retrograde subduction is several times larger than the mass flow necessary to generate oceanic lithosphere.

Two circumstances are essential in the foregoing calculation: (1) The moving subduction zones around the shrinking Pacific Ocean sweep out areas which are not much smaller than the global rate of sea floor spreading. (2) The descending slabs reach a depth as much as 10 times deeper than the base of the lithosphere. These conditions probably obtained also in the geologic past: it may be expected that, similar to the present situation, shrinking of ancient oceans was of the same order of magnitude as plate displacement and rates of sea floor spreading. The existence of deep descending slabs in the geologic past is indicated by the record in old orogenic belts, especially by the record of andesitic volcanism and related plutonism [*Dickinson*, 1972].

Hence plate tectonics is the direct surface expression of, and is directly linked to, mass movements in the mantle that are considerably more extensive than the overturn of oceanic lithosphere. The relation between the extra flow and the retrograde descent of slabs strongly suggests that some features of subduction zones are surficial manifestations of these additional motions and are not related only to the behavior of the descending slabs.

In the absence of local balance between subduction and sea floor spreading the sublithospheric flow should allow for a net flux from the shrinking plates to the growing ones. At present this requires a net flow toward the southern hemisphere, or rather toward the South Atlantic and circum-Antarctica regions, where most plate growth occurs, and away from the shrinking northern and central Pacific Ocean, this flow not being obviously related to the generally north-south trending subduction zones bordering the Pacific. Such a globally asymmetric deep flow is perhaps related to the pear shape of the earth [*King-Hele and Cook*, 1973].

In the Cretaceous and earlier, before the growth of the Antarctican plate became important, the net deep flow was toward the spreading regions in the central and southern Atlantic Ocean and the Indian Ocean and away from the shrinking Pacific Ocean (the Tethys was also shrinking, but this was of little importance at that time). Still earlier, in the Paleozoic, most plate growth probably occurred in the Pacific Ocean while continents aggregated on the opposite hemisphere, where the Tethys, Siberian, and other oceans were shrinking or being closed. Thus global oscillations are indicated.

Hence the long-term evolution of plates indicates that the sublithospheric flow is quite complex and time dependent and the part of it that is directly related to plate evolution is considerably larger than the mass motions involved in the overturn of lithosphere. This flow accounts for the changes of the relative positions of the descending slabs and the ridges and for the absence of local balance between plate generation and consumption, and it also allows a net sublithospheric mass flux from some parts of the earth where much subduction occurs to others where plate growth predominates.

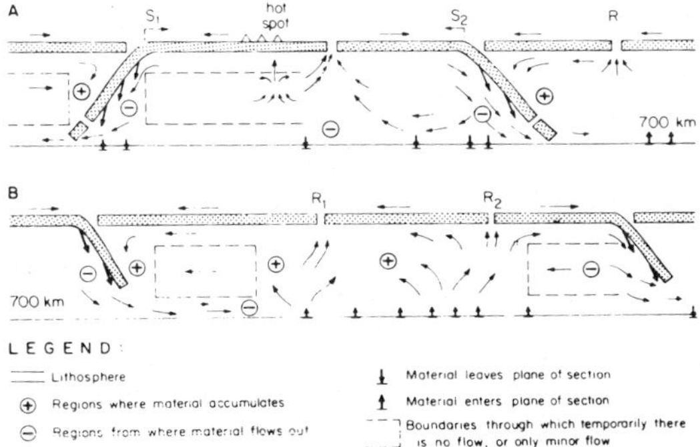

Fig. 1.   Possible patterns of the master flow in the mantle (see text), compatible with the long-term behavior of plates. The flow is not two dimensional, so it is not adequately represented in cross sections. (*a*) Shrinking mantle compartment between approaching lithospheric slabs (Pacific Ocean type). Subduction zones S₁ and S₂ approach each other as a result of retrograde subduction, while ridge R recedes from S₂. The plates between S₁ and S₂ lose area. (*b*) Expanding mantle compartment (Atlantic Ocean type). Ridges R₁ and R₂ recede from each other, while the plates delimited by them grow.

Since the surficial plates control the manner in which the heat generated in the earth is given off, the above conclusions show that the plate tectonic regime introduces irregularity into the temperature distribution in the mantle; the present average heat flow from the oceans is about 1.5 HFU. If it is assumed that the continental lithosphere (which does not participate in the mass overturn) receives a heat flux of 0.5 HFU from the underlying mantle, then the total heat flow from the convecting part of the earth is 1.8×10^{20} cal/yr. The heat flow from the oldest oceanic areas in the western Pacific, west central Atlantic, and northeastern Indian oceans is only 1.1–1.2 HFU [*Langseth and von Herzen*, 1970]. The difference between this value and the average oceanic heat flow results from cooling of the oceanic lithosphere as it becomes older; i.e., this is the heat released as a consequence of the generation of plates. Over the entire oceanic area this amounts to about 0.35×10^{20} cal/yr, or about 20% of the heat given off by the convecting part of the earth. This is a minimum estimate, because some oceanic lithosphere may not reach a steady thermal state prior to subduction, and interaction with seawater may be important at ridges. Thermal models of the lithosphere lead to higher estimates [*Sclater and Francheteau*, 1970].

As plate generation controls the disposal of a significant fraction of the thermal budget of the earth, it follows that the irregular distribution and long-term changes of sea floor spreading should influence the temperature distribution in the mantle. Complementary to this relationship is the uneven descent of cooled lithospheric material into various parts of the mantle. At present much more heat is given off by plate generation in the southern hemisphere, where most sea floor spreading occurs, than in the northern hemisphere, while most cooled lithosphere descends in a few rather limited regions. As similar irregularities existed in the past, important complexities in the temperature distribution, including lateral gradients, should have been generated in the convecting part of the mantle, provided that the plate tectonic regime operated long enough.

It is not known when this regime was first established, but the similarity of Proterozoic structures to more recent ones strongly suggests that the plate tectonic regime has been in operation during the last 2000 m.y. at least [*Salop and Scheinmann*, 1969; *Burke and Dewey*, 1973; *Bridgewater et al.*, 1973; *Hoffman*, 1973]. It was estimated above that the entire outer 700 or 1000 km of the mantle can be recycled in less than 2000 or 3000 m.y., respectively. Thus the plate tectonic regime had ample time to impose its influence on the thermal regime in the convecting part of the mantle.

PATTERN OF FLOW IN THE MANTLE

The foregoing conclusions will now be used to discuss the flow pattern. While numerous possibilities present themselves, some generalizations are possible.

Flow in the mantle was often described in terms of discrete, episodic or steady state, convective cells or rolls in which material moves in a circuit. Following the physical theory of convection and the experimental results [e.g., *Brindley*, 1967], periodic patterns were generally considered. The cells were assumed to be linked to subduction zones or to all nonconservative plate boundaries [e.g., *Hess*, 1962; *Holmes*, 1965; *Turcotte and Oxburgh*, 1972]. Models of convection cells not specifically linked to plate boundaries were also considered [e.g., *Knopoff*, 1964; *Runcorn*, 1965].

Models of a flow consisting of discrete cells, which are linked to plate boundaries, are unsatisfactory irrespective of the depth of the cells, because they predict, contrary to observation, local balance between the material which rises to build new lithosphere and the material which returns to the mantle in subduction zones. Consequently, such models do not allow changes of the sizes and shapes of plates or changes of the relative positions of subduction zones or of midoceanic ridges. To account for these features of plate evolution, cells linked to plate boundaries should have unequal sizes and irregular and continuously changing shapes, and some should be leaking and should lose material to the surrounding regions, while other cells, linked to growing plates, should expand. However, even such irregular and time-dependent cells are incompatible with ridges which do not alternate with subduction zones, like those around Africa and Antarctica. If periodic cells exist in the mantle, they are decoupled from plate boundaries, and they cannot account for the overturn of lithosphere.

Hence the overturn of lithospheric material and the long-term evolution of plates cannot be described by a flow pattern consisting of discrete circuital cells; instead, a worldwide flow must be considered. It can be discussed in terms of two components: (1) a mass flux, to be called the master flow, which is required to account for observable features of plate evolution (this flux provides the return flow from sites of plate consumption to sites of plate generation and accounts for the other mass motions required by the long-term features of plate evolution, which were discussed above) and (2) additional mass circuits underneath the lithosphere superimposed on the master flow. The master flow is the minimal flow required by the known features of plate evolution, whereas the other components of flow, of unknown magnitude (which is perhaps relatively small), do not have a net result in terms of plate evolution, though they may contribute to the mechanism which drives and modifies the motions of plates (see discussion below).

Present knowledge does not define uniquely the master flow, but the previous discussion revealed many of its features, so a simple model with the required properties can be outlined. In this model (Figure 1) the descending lithospheric slabs lose their identity at some depth, probably 600–700 km or somewhat deeper, and their mass is incorporated into a large reservoir, possibly of global extent. Elsewhere material rises directly from this reservoir or from shallower parts of the mantle to build new oceanic lithosphere. A rising plume is obviously formed beneath any ridge segment, but it need not extend deeper than the low-velocity zone. The latter may be replenished in a diffuse manner or by deep plumes whose position and geometry are unrelated to ridges, so migration of ridges relative to underlying mantle is possible.

Additional motions in the mantle are required to accommodate retrograde subduction or any motions through the mantle of the descending slabs not parallel to themselves, as well as the other long-term features of plate evolution. Now such motions cause the mantle compartment under the Pacific Ocean to shrink while lithospheric slabs which border it approach each other. Some material of this compartment is incorporated into plates at ridges and eventually is swept away from the Pacific Ocean as outwardly slanting slabs and may accumulate elsewhere (Figure 1). For some time, no material at all need move into this shrinking mantle compartment. Concurrently, other compartments of the mantle grow by influx of material.

It is noteworthy that all descending slabs slant away from the plates to which they are attached, and the distribution of earthquakes in Benioff zones [*Isacks and Molnar*, 1971] does not reveal any tendency of the slabs to turn backward. On the contrary, some deep detached pieces of lithosphere are displaced farther away from the subduction zones than the shallower parts of the slabs are [*Isacks and Molnar*, 1971; *Pascal et al.*, 1973]. The sublithospheric flow seems to carry these detached pieces away from their parental plates. As most descending slabs are parts of shrinking plates, they probably help evacuate material from the underlying parts of the mantle.

In this model it is possible for large volumes of the mantle to move, one relative to the others, without undergoing changes of their external shapes. For instance, large regions under the Pacific Ocean probably move toward ridges where their marginal parts rise. Groups of mantle diapirs which generate 'hot spots' or localize volcanism may be embedded in such regions and may therefore retain their relative positions for some time.

In this model, various particles follow complicated paths, quite unlike simple circuits of the cellular models, and their velocities may change considerably with time. The entire flow system is balanced only on a global scale and probably cannot be divided into a few partial mass circuits. An essential feature is that motion at depth is generally not parallel to the directions of relative motions of overlying plates. This is necessary to balance the uneven distribution of plate consumption and generation. Thus the flow is not two dimensional and is not adequately represented in planar cross sections.

Various local mass circuits may be superimposed on the master flow, such as systems of periodic cells or rolls, which may even be excited by plate motions [*Richter*, 1973] but cannot contribute to the overall overturn of lithosphere and the long-term evolution of plates. Also, the cells will occasionally be caught between approaching lithospheric slabs which delimit shrinking oceans, or alternatively, they may collide with slabs whose positions relative to the surrounding mantle change. In such events the convection cells should be modified or destroyed, so in the long run (possibly within 500 m.y.) the flow in the cells should be time dependent.

Local overturns of mass underneath the lithosphere may also be caused by ascent of buoyant masses such as diapirs or plumes [*Ramberg*, 1972a, b]. Unlike systems of periodic cells underneath the lithosphere, to which surficial features cannot readily be related, localized diapirs or plumes can be associated with various surficial manifestations. Such manifestations are 'hot spots,' i.e., loci of volcanic activity in plate interiors or loci of exceptional volcanic productivity on constructive plate margins [*Wilson*, 1973; *Morgan*, 1971, 1972], and especially, continental rifts (or aulacogenes) in which volcanism is combined with limited crustal separation and which may pass laterally into midoceanic ridges or develop into them.

It was suggested that the gravitative sinking of lithospheric slabs that have negative buoyancy, possibly supplemented by instabilities at ridges, provides the driving force behind plate motions and the overturn of lithospheric material [*Elsasser*, 1969, 1971; *Jacoby*, 1970; *Turcotte and Schubert*, 1971]. The influence of the descending slabs (but not of ridges) is expressed in the present absolute plate motions [*Morgan*, 1972]: Plates with descending slabs move faster than other plates and move toward the subduction zones. However, there is no clear correlation between the velocity of plate motion, plate size, and the size of the attached descending slab. The motions of the Antarctican and African plates, practically surrounded by

ridges, are not well explained. It is unlikely, therefore, that instabilities within the lithosphere alone drive the plate motions.

Not less important are the following observations which emerge from the above discussion: (1) Motions of plates and their evolution are direct surface manifestations of a complex sublithospheric flow that is much larger than the overturn of lithospheric mass; this flow is not well explained as being driven only by instabilities within the plates. (2) Abrupt changes in the geometry and rates of plate motions are not explained by instabilities which exist all the time within plates. Known instances cannot be related to changes in the configuration of subduction zones. In particular, there is no explanation for jumping of ridges or for changes of plate motions which were accompanied by jumping of ridges. Nor can continental breakup (e.g., dispersal of Gondwanaland) be specifically related to instabilities within plates or to the influence of preexisting or the birth of new descending slabs.

These arguments show that instabilities in the lithosphere probably are not of prime importance in driving the master flow, since they do not account for the long-term features of the plate tectonic regime. These features are best explained as resulting from instabilities in the sublithospheric mantle. It was shown above that the plate tectonic regime generated an irregular temperature distribution in the mantle, amounting to horizontal temperature gradients, which are superimposed on the gradients resulting from the heat generation within the mantle. Thus instabilities arise and should eventually lead to convection. In these circumstances the motions are not expected to have a simple, e.g., periodic, pattern.

In this context, local mass overturns, that is, hot spots or diapirs, may have a special role. Inception of these features and their activity require abrupt disturbances of the thermal and mechanical conditions in a relatively small region and a supply of heat to generate magmas. These disturbances are best explained as a result of the ascent of hot material from below the plates. Association with deep ascending masses is also strongly suggested by the cases in which continental rifting is a precursor of continental breakup and the birth of new midoceanic ridges [*Burke and Dewey*, 1973]. Such disturbances are to be expected in a mantle with a complex temperature distribution. Furthermore, the volcanics produced in such places are much richer in radioactive elements than are the normal igneous products at midoceanic ridges [*Engel and Engel*, 1970], an indication that their source materials were richer in such elements than the normal surrounding mantle material and therefore became heated and buoyant, thus producing local disturbances in the mantle. These disturbances can grow to become active diapirs within geologically brief times [*Ramberg*, 1972a, b]. However, the small areal extent of the surficial expression of many hot spots and their geologically short life span show that the underlying disturbances are small and decay within 10–100 m.y. These are mostly local features which just keep the convecting part of the mantle agitated in an irregular and constantly changing manner. Only a few disturbances, which are large enough and properly located, trigger much larger instabilities in the surrounding mantle. Such disturbances modify the patterns of flow in the mantle and of plate motions while they become integral parts of the master flow.

Occasionally, new subduction zones should form, and this too should modify considerably the master flow in the mantle, but such events are probably rare. There are no well-documented cases of transformation of Atlantic-type coasts to Pacific-type coasts with formation of new subduction zones within the last 350 m.y. (post-Devonian time), though the details of structure of many subduction zones have changed considerably within this period and many have been eliminated. The much more common mantle diapirs appear to be more important as agents modifying the master flow in the mantle.

The pattern of the flow is therefore the cumulative result mainly of the superposition of numerous mantle diapirs, though development of new subduction zones should also contribute to the secular changes of the flow pattern. Instabilities within the lithosphere, especially the negative buoyancy of descending slabs, certainly contribute to the maintenance of the flow and supplement the deep instabilities. Continued activity of the plate tectonic regime maintains an ever-changing temperature distribution in the mantle, so new instabilities develop, and these in turn maintain the complexity of the flow and its time-dependent nature.

The stress on the time-dependent nature of the flow expresses the circumstance that geologic observations are concerned with manifestations of the plate tectonic regime during periods that are comparable to, or much shorter than, overturn periods of the mass in the mantle. As is discussed above, particles of the convecting part of the mantle are incorporated into the lithosphere at intervals of the order of 10^9 yr. Since the sublithospheric mass flux is several times larger than the overturn of lithosphere, the average interval between successive times that any particle comes closest to the surface is of the order of several 10^8 yr, possibly 5×10^8 yr. Relative to the time resolution of geological methods, convection in the mantle is very slow, and time-dependent features are very conspicuous.

Acknowledgments. I wish to thank R. Freund and G. Steinitz for valuable comments and criticism.

REFERENCES

Anderson, D. L., and T. Jordan, The composition of the lower mantle, *Phys. Earth Planet. Interiors, 3,* 23, 1970.

Atwater, T., Implications of plate tectonics for the Cenozoic tectonic evolution of western North America, *Geol. Soc. Amer. Bull., 81,* 3513, 1970.

Bridgewater, D., A. Escher, and J. Watterson, Tectonic displacements and thermal activity in two contrasting Proterozoic mobile belts from Greenland, *Phil. Trans. Roy. Soc. London, Ser. A, 273,* 547, 1973.

Brindley, J., Thermal convection in horizontal fluid layers, *J. Inst. Math. Its Appl., 3,* 313, 1967.

Bullard, E. C. M., J. E. Everett, and A. G. Smith, The fit of the continents around the Atlantic, *Phil. Trans. Roy. Soc. London, Ser. A, 258,* 41, 1965.

Burke, K., and J. F. Dewey, Plume generated triple junctions: Key indicators in applying plate tectonics to old rocks, *J. Geol., 81,* 406, 1973.

Dickinson, R. W., Evidence of plate tectonic regimes in the rock record, *Amer. J. Sci., 272,* 551, 1972.

Elsasser, W. M., Convection and stress propagation in the upper mantle, in *The Application of Modern Physics to the Earth and Planetary Interiors,* edited by S. K. Runcorn, p. 223, Interscience, New York, 1969.

Elsasser, W. M., Sea floor spreading as thermal convection, *J. Geophys. Res., 76,* 1101, 1971.

Engel, A. E. J., and C. G. Engel, Mafic and ultramafic rocks, in *The Sea,* vol. 4, part 1, edited by A. E. Maxwell, p. 465, Interscience, New York, 1970.

Hamilton, W., The Uralides and the motion of the Russian and Siberian platforms, *Geol. Soc. Amer. Bull., 81,* 2553, 1970.

Harland, W. B., and R. A. Gayer, The Arctic Caledonides and earlier oceans, *Geol. Mag., 109,* 289, 1972.

Herron, E. M., Sea-floor spreading and the Cenozoic history of the east-central Pacific, *Geol. Soc. Amer. Bull., 83,* 1671, 1972.

Hess, H. H., History of ocean basins, in *Petrologic Studies: A Volume in Honor of A. F. Buddington,* edited by A. E. J. Engel, H. L. Jones, and B. F. Leonard, p. 599, Geological Society of America, New York, 1962.

Hoffman, P., Evolution of an early Proterozoic continental margin: The Coronation geosyncline and associated aulacogens of the northwestern Canadian Shield, *Phil. Trans. Roy. Soc. London, Ser. A, 273,* 547, 1973.

Holmes, A., *Principles of Physical Geology,* 2nd ed., Nelson, London, 1965.

Isacks, B., and P. Molnar, Distribution of stresses in the descending lithosphere from a global survey of focal mechanism solutions of mantle earthquakes, *Rev. Geophys. Space Phys., 9,* 103, 1971.

Jacoby, W. R., Instability in the upper mantle and global plate movements, *J. Geophys. Res., 75,* 5671, 1970.

Kanamori, H., and F. Press, How thick is the lithosphere?, *Nature, 226,* 330, 1970.

King-Hele, D. G., and G. E. Cook, Refining the earth's pear shape, *Nature, 246,* 86, 1973.

Knopoff, L., The convection current hypothesis, *Rev. Geophys. Space Phys., 2,* 89, 1964.

Langseth, M. J., Jr., and R. P. von Herzen, Heat flow through the floor of the world ocean, in *The Sea,* vol. 4, part 1, edited by A. E. Maxwell, p. 299, Interscience, New York, 1970.

Larson, R. L., and W. C. Pitman III, World-wide correlation of Mesozoic magnetic anomalies and its implications, *Geol. Soc. Amer. Bull., 83,* 3645, 1972.

Laughton, A. S., R. B. Whitmarsh, and M. T. Jones, The evolution of the Gulf of Aden, *Phil. Trans. Roy. Soc. London, Ser. A, 267,* 227, 1970.

Le Pichon, X., Sea floor spreading and continental drift, *J. Geophys. Res., 73,* 3661, 1968.

McKenzie, D. P., and J. G. Sclater, Evolution of the Indian Ocean, *Geophys. J. Roy. Astron. Soc., 24,* 437, 1971.

McKenzie, D. P., P. Molnar, and D. Davies, Plate tectonics of the Red Sea and East Africa, *Nature, 226,* 243, 1970.

Menard, H. W., and S. M. Smith, Hypsometry of ocean basin provinces, *J. Geophys. Res., 71,* 4305, 1966.

Monger, J. W. H., J. G. Souther, and H. Gabrielse, Evolution of the Canadian Cordillera: A plate tectonic model, *Amer. J. Sci., 272,* 577, 1972.

Morgan, W. J., Convective plumes in the lower mantle, *Nature, 230,* 42, 1971.

Morgan, W. J., Plate motions and deep mantle convection plumes and plate motions, *Geol. Soc. Amer. Mem. 132,* 7, 1972.

Oversby, V. M., and P. W. Gast, Isotopic composition of lead from oceanic islands, *J. Geophys. Res., 75,* 2097, 1970.

Pascal, G., J. Dubois, M. Barazangi, B. Isacks, and J. Oliver, Seismic velocity anomalies beneath the New Hebrides island arc: Evidence for a detached slab in the upper mantle, *J. Geophys. Res., 78,* 6998, 1973.

Pitman, W. C., III, and M. Talwani, Sea-floor spreading in the North Atlantic, *Geol. Soc. Amer. Bull., 83,* 619, 1972.

Ramberg, H., Mantle diapirism and its tectonic and magmagenetic consequences, *Phys. Earth Planet. Interiors, 5,* 45, 1972a.

Ramberg, H., Theoretical models of density stratification and diapirism in the earth, *J. Geophys. Res., 77,* 877, 1972b.

Richter, F. M., Convection and large-scale circulation of the mantle, *J. Geophys. Res., 78,* 8735, 1973.

Ringwood, A. E., Composition and evolution of the upper mantle, in *The Earth's Crust and Upper Mantle, Geophys. Monogr. Ser.,* vol. 13, edited by P. J. Hart, p. 1, AGU, Washington, D. C., 1969.

Runcorn, S. K., Changes in the convection pattern in the earth's mantle and continental drift: Evidence for a cold origin of the earth, *Phil. Trans. Roy. Soc. London, Ser. A, 258,* 228, 1965.

Salop, L. I., and Yu. M. Scheinmann, Tectonic history and structures of platforms and shields, *Tectonophysics, 7,* 565, 1969.

Sclater, J. G., and J. Francheteau, The implications of terrestrial heat flow observations on current tectonic and geochemical models of the crust and upper mantle of the earth, *Geophys. J. Roy. Astron. Soc., 20,* 509, 1970.

Smith, A. G., and A. Hallam, The fit of the southern continents, *Nature, 225,* 139, 1970.

Taylor, S. R., The origin and growth of continents, *Tectonophysics, 4,* 17, 1967.

Turcotte, D. L., and E. R. Oxburgh, Mantle convection and the new global tectonics, *Ann. Rev. Fluid Mech., 4,* 33,, 1972.

Turcotte, D. L., and G. Schubert, Structure of the olivine-spinel phase boundary in the descending lithosphere, *J. Geophys. Res., 76,* 7980, 1971.

Wilson, J. T., Did the Atlantic close and then reopen again?, *Nature, 211,* 676, 1966.

Wilson, J. T., Theories of building of continents, in *The Earth's Mantle,* edited by T. F. Gaskell, p. 445, Academic, New York, 1967.

Wilson, J. T., Mantle plumes and plate motions, *Tectonophysics, 19,* 149, 1973.

(Received January 3, 1975;
revised May 27, 1975;
accepted June 10, 1975.)

23

Runaway Temperatures in the Asthenosphere Resulting From Viscous Heating

ORSON L. ANDERSON AND PRISCILLA C. PERKINS

The purpose of this report is to examine some important consequences of the fact that the heat flow equation, in the model of one-dimensional viscous flow at constant stress, is the same one-dimensional equation used to describe thermal explosions in solids [*Bowden and Yoffe*, 1952, 1958]. This conclusion about the state of the asthenosphere arises from the fact that the viscosity decreases exponentially with increasing temperature, obeying the Arrhenius equation for the cases of molten basalt [*Shaw*, 1969] and rocks resembling mantle materials [*Stocker and Ashby*, 1973; *Kirby and Raleigh*, 1973]. One can therefore postulate the existence of instabilities in asthenospheric flow that would be thermal runaways. These thermal instabilities would be analogous to the instabilities that exist in solid explosive material. When it is triggered in such material, the instability manifests itself as a thermal explosion.

The first author to point out the possibility of thermal runaways in the earth's interior due to nonlinear viscous heating exhibited by the heat equations was *Gruntfest* [1963]. He showed that under certain boundary conditions (one-dimensional plane viscous flow under constant stress in a slab) the temperature and strain rate become arbitrarily large in finite time under adiabatic or near-adiabatic conditions. Gruntfest suggested that near-adiabatic conditions ought to be present in the interior of the earth because of the long time scale of geologic processes and the low conductivity of rocks.

Many of Gruntfest's ideas have been applied to geologic processes by *Shaw* [1969, 1970, 1973] and by *Shaw et al.* [1971]. These applications concern (1) the rheological behavior of basalt; (2) the time constants for the thermal feedback processes in materials with viscosities to be expected in the asthenosphere; (3) the relation between earth tides and the magmatic history in the Sierra Nevada; and (4) volcanism in Hawaii. *Fujii and Uyeda* [1974] applied Gruntfest's equations and the parameters of basalt computed by *Shaw* [1969] for Gruntfest's equations to explain the size of intrusive dikes and the eruptive history of volcanoes.

Anderson and Perkins [1974] have suggested that thermal runaways in the asthenosphere could partially account for the wide extent and complex patterns of Cenozoic igneous activity in the southwestern United States. They also noted that thermal instabilities can be initiated in the asthenosphere in spite of the extremely low Reynolds number for viscous flow. *Gilluly* [1970, 1973] and *Noble* [1972] have each remarked that the steady state subduction model of plate tectonics does not adequately explain the igneous activity of the southwestern

United States. For example, *Noble et al.* [1973] observed that the late Tertiary volcanic field of south central Nevada exhibits magmas generated from different materials and/or different levels.

Anderson and Perkins [1974] postulated that the succession of trench and transform fault geometries at the Pacific margin, together with the encounter of the Pacific ridge system with the American plate, disrupted the flow regime and consequently disrupted the thermal gradients in the asthenosphere, thus creating conditions favorable for the triggering of thermal runaways. This hypothesis is similar to *Noble*'s [1972, p. 146] suggestion that 'changes in stress and (or) flow patterns in the asthenosphere caused by changes in relative movement of the North American and Farallon plates . . . caused the activation of gravitational instabilities . . . which in turn allowed the upward movement of large diapirs of mantle material.'

The thermal feedback process producing the runaway would be expected to produce partial melting and to be responsible for magma generation. The total pattern of igneous activity in the southwestern United States otherwise is difficult to explain by plate tectonic models of volcanism. It should be stressed that the Gruntfest thermal instabilities are not proposed to be the driving mechanism for mantle convection and plate motion. We consider them to be transient irregularities superimposed at various scales on existing patterns of laminar asthenospheric flow.

THERMAL EXPLOSIONS AND THERMAL RUNAWAYS

We begin by copying the equation of heat flow used to describe thermal explosions [*Bowden and Yoffe*, 1952, 1958]:

$$C\rho \; \partial T/\partial t = k(\partial^2 T/\partial x^2 + \partial^2 T/\partial y^2 + \partial^2 T/\partial z^2)$$

$$+ A \exp (-E/R)[(1/T) - (1/T_0)] \qquad (1)$$

On the left is the expression for self-heating, which is controlled by specific heat and density. The first term on the right is the heat lost by conduction, controlled by the thermal conductivity k, and the second term is the rate of production of heat by a chemical reaction, where A is a constant, the parameter E is the activation energy, and R is the gas constant. This equation has received extensive treatment in the theory of explosions and has been solved for the linear, cylindrical, and spherical coordinate systems. If a thermal pulse $\Delta T = T - T_0$ of sufficient magnitude is applied, the nonlinearity of the equation produces a burning reaction. There are two important cases. In the adiabatic case the term on the left balances the second term on the right; the first term on the right is zero. The temperature increases exponentially with time. In the case where the term on the left is zero the temperature remains constant at some high value, and the energy of the system is steadily depleted. The only difference between the solutions for the different coordinate systems is that the time to go from the initial to the final temperature state is different. It is smallest for the linear case and largest for the spherical case [*Bowden and Yoffe*, 1958].

In viscous heat dissipation the rate of production of viscous heat is $\sigma \cdot \dot{\epsilon}$, where σ is stress and $\dot{\epsilon}$ is strain rate. When the expressions for $\dot{\epsilon}$ appropriate to the flow of rocks are used, $\sigma \cdot \dot{\epsilon}$ has the same form as the term on the far right in (1) [*Kirby and Raleigh*, 1973], and the heat conduction equation is a differential equation of the same form as the one that controls the temperature-time history in a thermal explosion [see also *Shaw*, 1969]. We propose that the physics of instabilities in the asthenosphere can be guided by what is known about the initiation and burning of solid explosives. In the extreme case of an adiabatic boundary condition at constant stress, an explosion is easy to initiate, and the time for the temperature to become infinite, t_∞, depends on the constants in the strain rate expression such as the activation energy and the stress and the initial viscosity. In the nonadiabatic case under constant stress the unstable condition depends upon whether t_∞ is small in comparison with the time required for a temperature pulse to die away in the absence of viscous heating. The ratio of these two times, called the Gruntfest parameter G, which controls the conditions for the existence of the stability, depends upon the dimensions of the body, the stress, the activation energy, and the initial viscosity as well as upon the thermal conductivity. Large bodies are less stable than small bodies (hence explosives are transported in small separately wrapped packages). Bodies under large shear stress are less stable than those under small shear stress, and bodies whose reactions are controlled by a small activation energy are less stable than those with a large activation energy.

These ideas have been quantified in a simple expression by replacing, for mathematical convenience, the viscosity-temperature relation in the Arrhenius form by an equivalent expression [*Gruntfest*, 1963]:

$$\eta = \eta_0 \exp\left[-a(T - T_0)\right] \qquad (2)$$

where a is proportional to the reciprocal of the activation energy and the sign is reversed to compensate for the reciprocation of the temperature. When (2) is used in the term for viscous heating, in place of the Arrhenius equation, an equation equivalent to (1) is obtained with the following result for the one-dimensional case [*Gruntfest*, 1963]. For the adiabatic case the time required for the temperature to become very large is

$$t_\infty{}^{ad} = C\eta_0/a\sigma^2 \qquad (3)$$

For nonadiabatic cases the value of t_∞ can also be calculated, but this turns out to be controlled by G, the Gruntfest parameter given by

$$G = a\sigma^2 l^2/k\eta_0 \qquad (4)$$

The value of G corresponding to (3) is infinite, since for the adiabatic case $k = 0$ in (4). For some nonadiabatic cases the system becomes unstable. This instability occurs when $G > 1$, according to the model, with the result that $t_\infty > t_\infty{}^{ad}$, t_∞

becoming infinite as G approaches zero. The value of t_γ can be computed from Gruntfest's theory if the values of the physical parameters in (3) are known. If we consider the parameters in turn for a fixed G, then t_∞ becomes longer as the stress decreases, becomes shorter as the length of the body decreases, and becomes shorter as the activation energy E decreases (or a increases).

Shaw [1969] used Gruntfest's theory, which leads to (4), to compute t_γ for basalt with constant $\sigma = 100$ bars, $\eta_0 = 10^{21}$ poises, a shear couple of about 10 km, and the value of a determined from his measurements of basalt. He found that $t_\gamma = 3.25$ m.y. for a temperature runaway for $G = 1$, according to (3). The activation energy for the viscosity of mantle material is probably higher than that for liquid basalt, so that the time would be correspondingly longer, but if one uses the Arrhenius equation for viscosity instead of (2), the time constants would be shorter. However, let us use as a reasonable approximation Shaw's calculation for 3.25 m.y. for the one-dimensional case at a constant stress of 100 bars. We then examine the effects that different boundary conditions would have upon the magnitude of t_γ. In the asthenosphere the viscosity η_0 might be 2 or 3 orders of magnitude less than 10^{21}, but the stress would be less, perhaps a few bars [*Kirby and Raleigh*, 1973] instead of 100 bars, so that these corrections would probably compensate each other in (4).

The assumption of constant stress would be only approximate. *Nitsan* [1973] reported calculations for the case of thermal feedback by using the Arrhenius equation for $\eta(T)$, but he used as a boundary condition the case of constant velocity on the edge of a semi-infinite slab. He found that for a wide range of parameters a low-viscosity layer of narrow dimensions is formed in which all the shear takes place and all the viscous heat is generated. He also found that the temperature effect overwhelmed the stress effect even when the strain rates were assumed proportional to the third power of stress. This stabilized solution is an idealized model for the low-velocity zone.

ASTHENOSPHERIC FLOW

In asthenospheric flow the boundary conditions are not really equivalent to a one-dimensional slab with a constant velocity or to a one-dimensional slab under constant stress. In a flow field of the asthenosphere where there exist a finite but small velocity gradient and a finite but small stress gradient the conditions of instability would be intermediate to the two cases. Gruntfest found that instabilities exist for values of G of about 1 or greater in the case of constant stress. In the asthenosphere with a small stress gradient the conditions for instability would be intermediate to the boundary conditions used by Gruntfest and by Nitsan, and we would therefore expect the instability to occur at larger values of G than in the constant stress case. This is the same as increasing t_∞.

The effect of increased pressure would be to raise the activation energy above the value used by Shaw for 1 atm, and this in turn would increase the value of t_∞ required for the temperature to become very large. All these effects imply that Shaw's estimate of t_∞ is a lower limit. Also, the coordinate system for the three-dimensional case would change the estimate of t_γ because the value of t_∞ computed from a one- or two-dimensional model should be less than that computed from a three-dimensional model. This relation is indicated by the calculations for thermal explosions [*Bowden and Yoffe*, 1958]. Although considerable uncertainty is indicated for the exact value of t_∞ owing to these various effects, it appears that the time for a thermal runaway in the asthenosphere might be a few tens of millions of years after the initiating event. Confidence in this estimate is strengthened by the fact that the one-dimensional case of plane stress is used to guide experiments in thermal explosions that depart from this simple analytical model.

Once the event is initiated, the gravity gradient will make the hotter material rise toward the lithosphere in a movement that we propose to call a surge. This concept is similar to that proposed by *Shaw* [1973] and *Shaw and Jackson* [1973] for mantle melting in the vicinity of Hawaii.

GEOLOGICAL SIGNIFICANCE

If the surge is strong enough to rise to the lithosphere boundary, we expect that it will not penetrate the higher-viscosity material. We postulate that the surging material will tend to spread out horizontally and then sink as it cools, forming structures resembling partial eddies. These are not true eddies with a complete circular path since the time required for the surge to die out is about the same as that required for the surge to grow (tens of millions of years). The geological significance of this mechanism is that conditions of reverse flow, stagnation points, and hot spots are expected to exist near the lithosphere boundary for a few tens of millions of years. Reverse flow at a boundary could occur several times during the lifetime of a single surge, so that at any particular point on the surface the span of time for a particular magmatic epoch could be smaller than the value of t_∞. On an even smaller scale, once magma has accumulated, individual intrusive or eruptive events may be dominated by thermal runaways with correspondingly shorter time scales, as was suggested, for example, by *Fujii and Uyeda* [1974].

The existence of surges disturbing the thermal regime of the asthenosphere offers a wealth of possibilities to explain the ultimate origin of many igneous events. Thermal surges would be expected to favor irregular distribution of volatiles, possibly including local concentrations that would lead to kimberlite eruptions [*Anderson and Perkins,* 1974]. Surges would promote the formation of zones of partial melting in the lower crust and upper mantle. Monotonic sequences in time and space of magmatic events, implied by some steady subduction models, are no longer required. In this respect our model is similar to the 'multiple diapir' model proposed by *Noble* [1972, p. 147] to explain episodic volcanism '. . . with significant temporal variations in the distribution and nature of igneous activity.'

If the potential for thermal runaways exists in the asthenosphere, as it does in an untriggered solid explosive, we can seek the identity of processes that might initiate a thermal runaway. We postulate that important events at plate boundaries, such as the breakup of continents or the replacement of downgoing slabs by transform faults, are of sufficient magnitude to trigger the runaways by changing flow patterns in the asthenosphere [*Anderson and Perkins,* 1974]. Considering mechanisms analogous to those that trigger thermal explosions in solids, we also suggest that instabilities in the

asthenosphere could be initiated by friction at a lithosphere-asthenosphere boundary, a thermal hot spot, or a change in activation energy by dilution with a suitable volatile. In this way plate tectonic theory can be extended to show correlations between large-scale plate boundary adjustments and the initiation of episodes of extensive magmatism at some distance from the plate boundary.

Acknowledgments. We thank S. Uyeda for bringing to our attention the theory of Gruntfest. We are especially grateful to H. R. Shaw and D. C. Noble for helpful discussions and correspondence concerning instabilities in asthenospheric flow and episodic volcanism in the southwestern United States. However, we accept full responsibility for the conclusions presented in this paper. This research was supported by NSF grant GA 35062. Publication 1242, Institute of Geophysics and Planetary Physics.

REFERENCES

Anderson, O. L., and P. C. Perkins, A plate tectonics model involving non-laminar asthenospheric flow to account for irregular patterns of magmatism in the southwestern United States, *Phys. Chem. Earth, 7,* in press, 1974.

Bowden, F. P., and A. D. Yoffe, *Initiation and Growth of Explosion in Liquids and Solids,* p. 8, Cambridge University Press, London, 1952.

Bowden, F. P., and A. D. Yoffe, *Fast Reactions in Solids,* pp. 20–25, Butterworths, London, 1958.

Fujii, N., and S. Uyeda, Thermal instabilities during flow of magma in volcanic conduits, submitted to *J. Geophys. Res.,* 1974.

Gilluly, J., Crustal deformation in the western United States, in *The Megatectonics of Continents and Oceans,* edited by H. Johnson and B. L. Smith, pp. 47–73, Rutgers University Press, New Brunswick, N. J., 1970.

Gilluly, J., Steady plate motion and episodic orogeny and magmatism, *Geol. Soc. Amer. Bull., 84,* 499–514, 1973.

Gruntfest, I. J., Thermal feedback in liquid flow, plane shear at constant stress, *Trans. Soc. Rheol., 7,* 195–207, 1963.

Kirby, S. H., and C. B. Raleigh, Mechanisms of high temperature, solid state flow in minerals and ceramics and their bearing in the creep behavior of the mantle, *Tectonophysics, 19,* 165–194, 1973.

Nitsan, U., Viscous heat production in a slab, *J. Geophys. Res., 78,* 1395–1397, 1973.

Noble, D. C., Some observations on the Cenozoic volcano-tectonic evolution of the Great Basin, western United States, *Earth Planet. Sci. Lett., 17,* 142–150, 1972.

Noble, D. C., C. E. Hedge, E. H. McKee, and M. K. Korringa, Reconnaissance study of the strontium isotopic composition of Cenozoic volcanic rocks in the northwestern Great Basin, *Geol. Soc. Amer. Bull., 84,* 1393–1406, 1973.

Shaw, H. R., Rheology of basalt in the melting range, *J. Petrol., 10,* 510–535, 1969.

Shaw, H. R., Earth tides, global heat flow, and tectonics, *Science, 168,* 1084–1087, 1970.

Shaw, H. R., Mantle convection and volcanic periodicity in the Pacific: Evidence from Hawaii, *Geol. Soc. Amer. Bull., 84,* 1505–1526, 1973.

Shaw, H. R., and E. D. Jackson, Linear island chains in the Pacific: Result of thermal plumes or gravitational anchors, *J. Geophys. Res., 78,* 8634–8652, 1973.

Shaw, H. R., R. W. Kistler, and J. F. Evernden, Sierra Nevada plutonic cycle, 2, Tidal energy and a hypothesis of orogenic-epeirogenic periodicities, *Geol. Soc. Amer. Bull., 82,* 869–896, 1971.

Stocker, R. L., and M. F. Ashby, On the rheology of the upper mantle, *Rev. Geophys. Space Phys., 11,* 391–426, 1973.

(Received September 14, 1973;
revised February 6, 1974.)

24

GLOBAL GRAVITY AND TECTONICS

by William M. Kaula

The improved resolution of the new solution for the earth's gravity field by Gaposhkin and Lambeck (1971) results in a significant change in our knowledge of the relationship of the gravity to tectonics in the southern oceans. Large positive anomaly areas are now located along the ocean rises, the sole exception in the oceans being Hawaii. This correlation of positive anomalies with rises is enhanced for representations that are isostatic and residual to a fifth-degree figure.

The trench and island arc belts are also predominantly positive. The paradox of both tension and compression features being areas of mass excess near the surface may be the consequence of the lithosphere acting primarily as a free boundary to vertical stresses in the former case and a fixed boundary in the latter, as well as both being stagnation points in mantle flow.

The commonest negative anomaly features are the ocean basins, always located to the flanks of the rises. The basins may be regions of maximum horizontal acceleration (Lagrangian) in the asthenospheric flow, or there may be a downflow of a denser component from the flow between the rise to the basin.

Correlation of negative anomalies on land with glaciated areas is marked, particularly isostatic anomalies residual to the fifth degree. The

Antarctic negative is much too large to account for by glacial melting; it appears to require a deficiency of the asthenospheric flow associated with the outward migration of ocean rises. The Himalayas-Turkestan negative seems well explained by great thicknesses of low-density crust carried by stiff lithosphere, which to some extent displaces asthenospheric material.

INTRODUCTION

This paper is a continuation of Kaula (1969), which attempted a tectonic classification of the main features of the earth's gravitational field. On the basis of magnitude and extent of mean gravity anomalies for 5° squares, 19 areas on the earth were selected as markedly positive, 14 as markedly negative, 10 as exceptionally mild. Other geological and geophysical data for each of these 43 areas were examined. On the basis of certain patterns of correlation, 11 types of areas were defined. It was found that, where characteristics of different types appeared, certain characteristics were dominant over others. In general, characteristics associated with positive anomalies were dominant over those associated with mild or negative anomalies, and characteristics associated with recent tectonics dominant over those associated with ancient. The 11 types in order of dominance, with sign and a leading example of each given in parentheses, were: trench and island arc (+, Indonesia-Philippines), Cenozoic oceanic flood basalts (+, Iceland-North Atlantic), Cenozoic orogeny with Quaternary extrusives (+, Caucasus), Quaternary glaciation (−, Canadian Shield), vigorous ocean rise (0, southeast Pacific), current orogeny without extrusives (−, Himalayas), ocean basin (−, Somali-Arabian), continental basin (−, Parnaiba Basin), pre-Cenozoic orogeny (0, eastern United States), continental shield (0, Brazilian Shield), pre-Cenozoic ocean flood basalts (0, Darwin Rise). The strongest correlation found was between positive gravity anomalies and Quaternary volcanism. Positive correlation of gravity anomalies with topography residual to a fifth-degree figure was almost universal. The extent to which the different area types relate to the global tectonics inferred from paleomagnetic and seismic data varied from strong (trench and island arc, vigorous ocean rise) to negligible (Cenozoic oceanic flood basalts, Quaternary glaciation). The lack of systematic correlation between temperature indicators (heat flow, P_n velocities, seismic station delays) and gravity anomalies indicated that horizontal variations in petrology are significant.

Since Kaula (1969), there has been a major improvement in the determination of the global gravity field by Gaposhkin and Lambeck (1971). In this paper, we first examine this improved determination, and attempt to transform it so as to be most useful for geophysical interpretation. Then we hypothesize mechanisms by which the main features of the gravitational field are maintained.

THE DATA

Figure 1 of this paper differs from Fig. 1 of Kaula (1969) in four significant respects:

1. The new determination of the gravity field by Gaposhkin and Lambeck (1971) is used. This analysis is based primarily on the orbits of 21 artificial satellites and secondarily on mean gravity anomalies for 5° by 5° squares covering 56% of the earth.

2. The gravimetry is the same as that used by Kaula (1966), but the manner of combination of data in effect gives higher weight to the satellites than in the 1966 analysis.

3. The results are given in the form of spherical harmonic coefficients of the potential, complete through the 16th degree, (plus 33 coefficients of higher degree to which satellite orbits are sensitive), rather than area means. Hence, the resolution, or shortest half wavelength represented, is about 11° or 1200 km.

4. The free air anomalies in Fig. 1 are referred to the figure of hydrostatic equilibrium, an ellipsoid of flattening 1/299.8, in accord with the explanation of Goldreich and Toomre (1969) for the excess oblateness.

There are two major effects of these changes:

1. The improved resolution results in the breakup of the two largest features in the southern oceans. The large area of mild anomaly in the South Pacific is now resolved into two negative areas with a positive area between, the former over basins and the latter along the East Pacific Rise. In the area between Africa and Antarctica, a single large positive feature centered in the "vee" between the two rises is now divided into two positive features over the rises and an area of mild anomaly between. In general, most of the ocean rises are now positives, rather than "mild" features.

2. The use of the hydrostatic flattening results in the intensification of the negative anomalies in the glaciated areas near the poles: at the South Pole to an extent that is much greater than can be imputed to glacial loading.

Lesser effects are the appearance of the highest Himalayas as a small positive belt; the reduction of an overlap of the southeast Indian Ocean rise by the south Australian basin negative anomaly; the emphasis of the positive belt from the Carpathians to Iran; the reduction of the East Mediterranean negative; and the reduction or removal of positive features in areas of slight recent tectonic activity in northeast USSR, the Central Pacific, and Australia.

Figure 2 is the corresponding isostatic anomaly map, using the spherical harmonic expansion of the Airy-Heiskanen 30-km crust isostatic correction calculated by Uotila (1962). As usual, oceanic maxima and continental minima are enhanced in the isostatic map, but significant change in the pattern would occur only if the compensation were placed at, or below, asthenospheric depths.

As previously pointed out (Kaula, 1967), the correlation of gravity with topography is poor for the fifth and lower degrees. On the other hand, Hide and Malin (1970) have recently shown that the low degree harmonics of the gravity

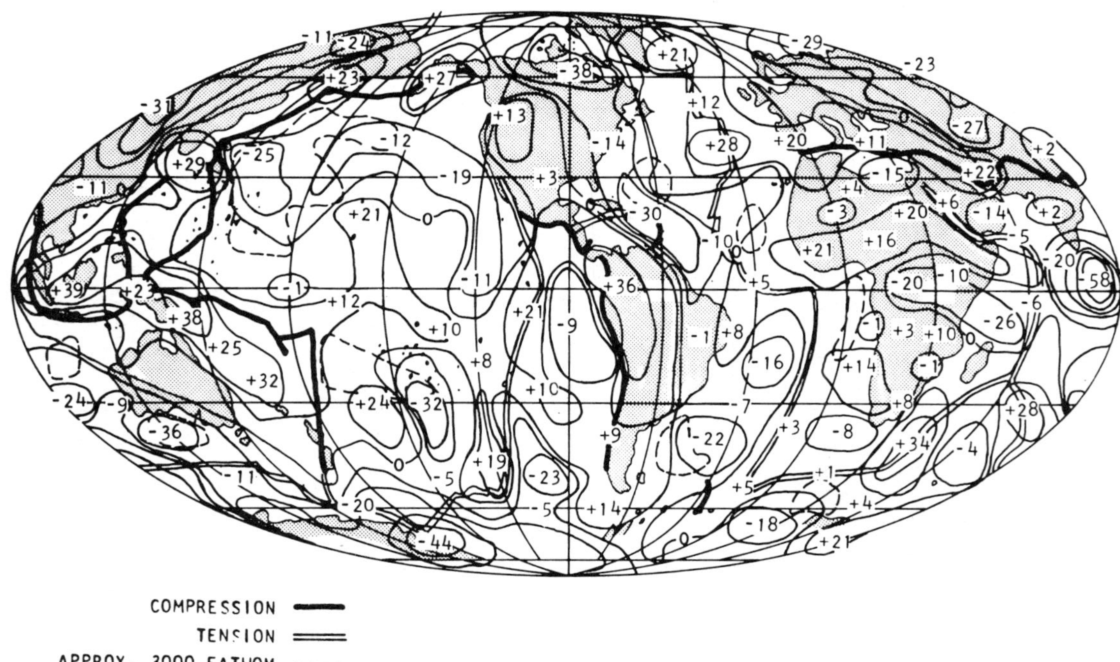

Fig. 1. Free air anomalies in milligals referred to an ellipsoid of flattening 1/299.8. Calculated from the spherical harmonic coefficients of the gravitational field of degrees 2 through 16 of Gaposhkin and Lambeck (1971). (Non-zero contours enclosing only one value have been omitted on all figures.) Global tectonic lines of compression and tension from Isacks et al. (1968), and major basins indicated by approximate 3000-fathom line on all figures.

310

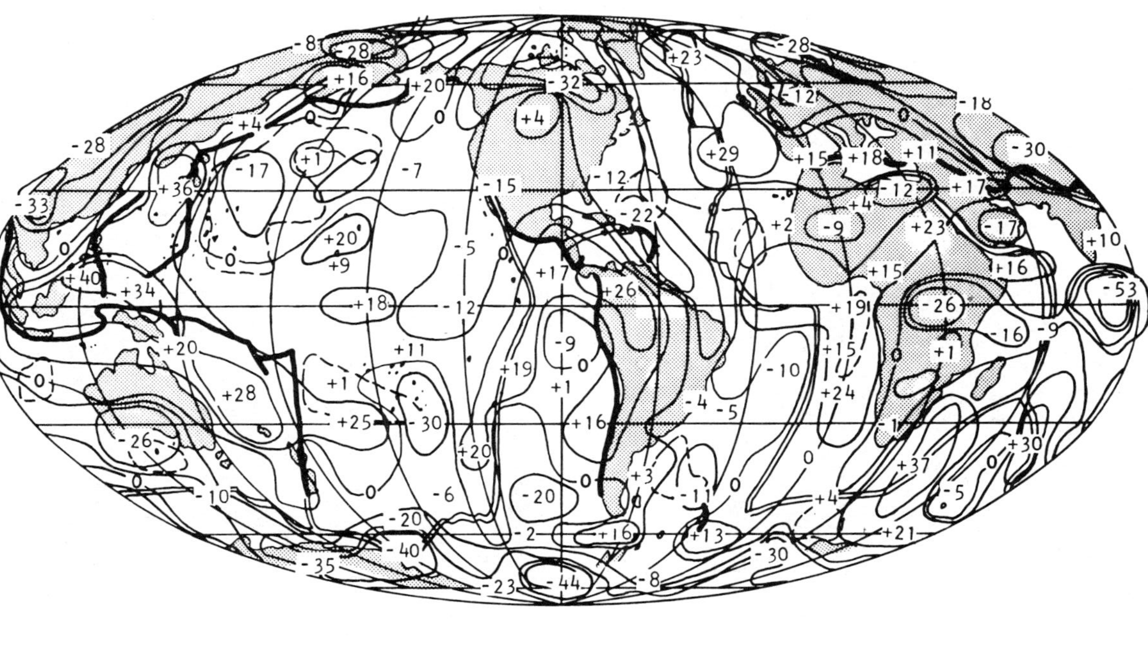

Fig. 2. Isostatic anomalies in milligals referred to an ellipsoid of flattening 1/299.8. Airy-Heiskanen compensation with nominal crustal thickness of 30 km. Calculated from Fig. 1, less the spherical harmonic coefficients for the isostatic correction of degrees 2 through 16 of Uotila (1962).

311

field have a high correlation with the corresponding harmonics of the magnetic field, provided that the latter is rotated 160° eastward. The obvious application of these facts for the purpose of interpreting upper mantle and crustal phenomena is to use a residual field. Figure 3 is the free air anomaly field calculated from spherical harmonic coefficients of degrees 6 through 16, and Fig. 4 is the corresponding isostatic anomaly field. In the four successive representations of Figs. 1 through 4, the correlation of ocean rises with positive anomalies appears more and more emphasized.

As discussed in Kaula (1969), it seems appropriate to analyze the gravity field in terms of reasonably contiguous blocks of anomaly x area, since, by the half-space application of Gauss's theorem, this quantity is directly proportional to excess mass, which in turn is a primary measure of the stresses required. Table 1 gives the 30 largest blocks in terms of free air anomalies referred to the hydrostatic figure, while Table 2 gives the 25 largest blocks in terms of isostatic anomalies referred to the fifth-degree figure. Of the 12 question marks in Table 4 of Kaula (1969), about 10 seem to be resolved. The greatest question remaining is the great negative over Antarctica; it is too large by more than a factor of 3 to be attributable to the loss of ice in recent geologic time (O'Connell, 1971).

The types given in Tables 1 and 2 are those used in Kaula (1969), with some obvious modifications.

INTERPRETATION

The principal inference from the large areas of postglacial uplift is, of course, the existence of an asthenosphere: a relatively plastic layer in the upper mantle, 80 to 400 km or more deep. Of the seven or eight major feature types, the glaciated areas are alone in being transient, with a decay time on the order of a few thousand years (O'Connell, 1971).

Since the lithosphere is not capable of supporting elastically the necessary stresses for features thousands of kilometers in extent (McKenzie, 1967), the other broad departures in the earth from equilibrium must entail flow in the asthenosphere. The asthenosphere is stiff enough, however, and the thermal conductivity of the earth is poor enough (the Prandtl number is large), that it is generally agreed that the flow is essentially *steady state* (Turcotte and Oxburgh, 1967, 1969). As indicated by magnetic reversal patterns, the present pattern of tectonic motion has persisted for about 10 m.y. (Heirtzler et al., 1968).

In a steady-state flow system, to maintain a mass excess in a particular region, there must be effectively a Lagrangian deceleration of matter entering the region and an acceleration of matter leaving it; the converse must apply to a region of mass deficiency. Mathematically, this condition requires that for the volume containing the mass excess, the surface integral

$$-\int_{surface} \rho \, \frac{\partial v}{\partial t} \cdot dn = \int_{surface} \rho v \cdot \nabla v \cdot dn > 0 \qquad (1)$$

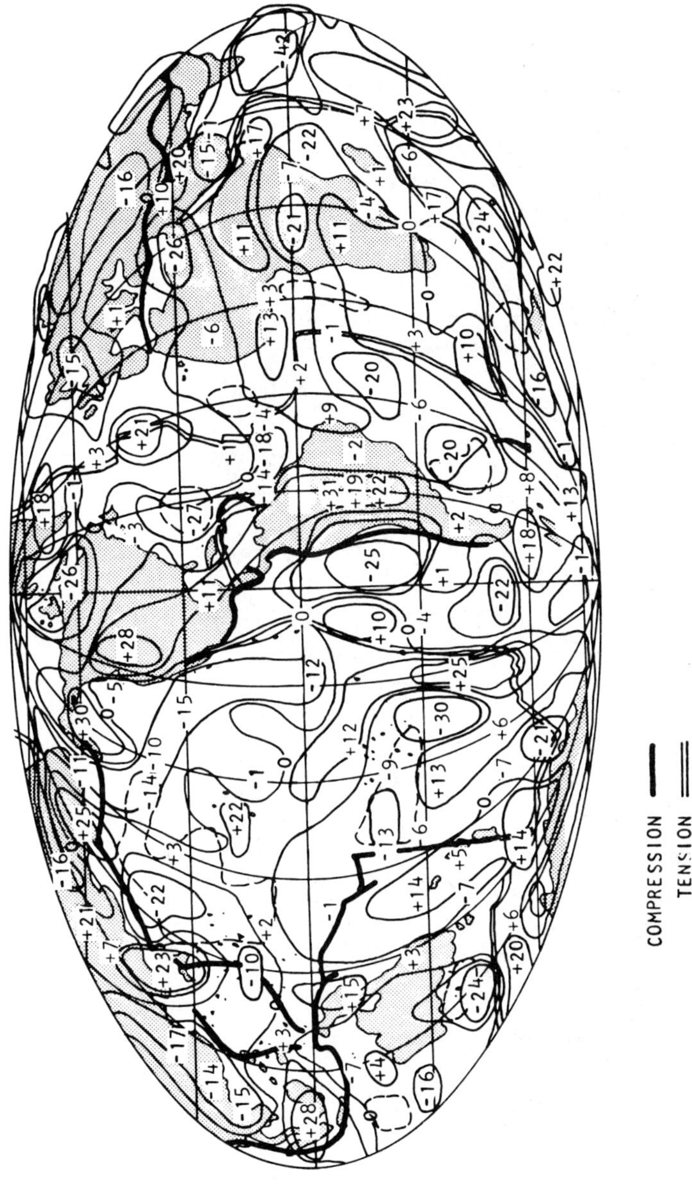

Fig. 3. Free air anomalies in milligals referred to a fifth-degree figure. Calculated from the spherical harmonic coefficients of the gravitational field of degrees 6 through 16 of Gaposhkin and Lambeck (1971).

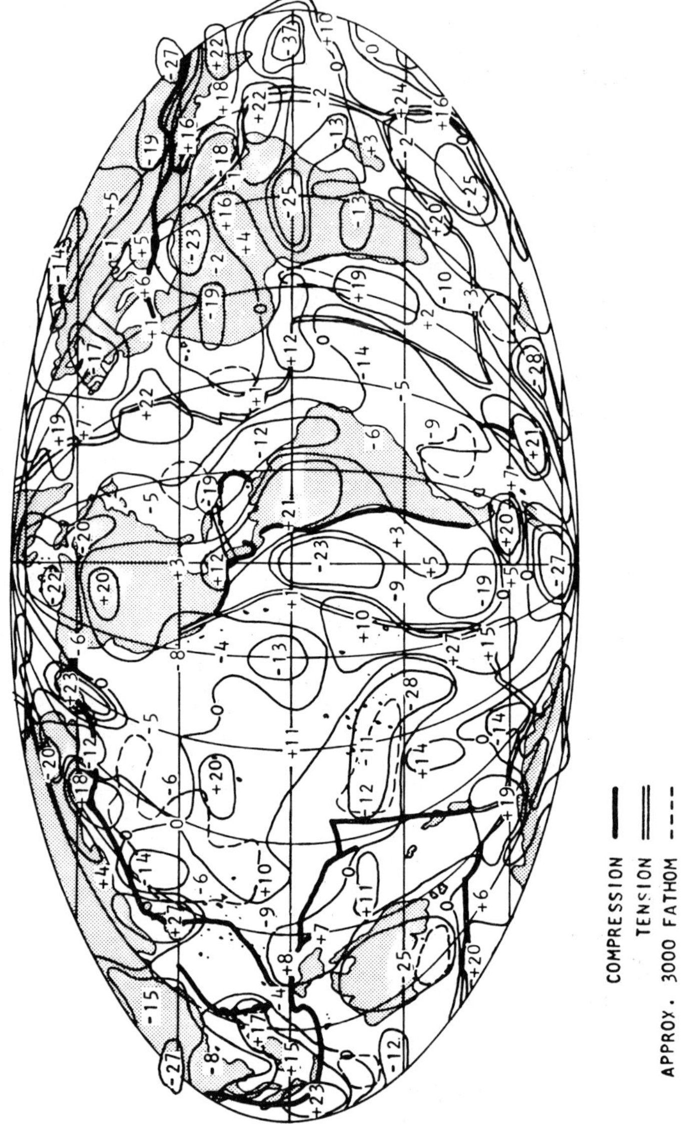

Fig. 4. Isostatic anomalies in milligals referred to a fifth-degree figure. From Fig. 3, less the spherical harmonic coefficients for the isostatic correction of degrees 6 through 16 of Uotila (1962).

314

where ρ is density, $\mathbf{v}$ is velocity, and $\mathbf{n}$ is the outward drawn unit vector normal to the surface. The uppermost boundary of this volume for purposes of gravity anomaly interpretation should be a segment of a bounding equipotential, such as the geoid. Since the compressibility of upper mantle material is slight, these "decelerations" must be accomplished by: (1) the piling up of material at the surface; (2) the replacement of less dense by more dense material at an interior interface; (3) thermal contraction; (4) transition to a denser phase; or (5) petrological fractionation in which a less dense component is left behind. The reverse of one or more of these processes is needed to accomplish an "acceleration." It is to be emphasized that mechanisms such as (3), (4), and (5) do not affect the gravity field directly by increasing the density, but rather by inducing mass transfers in accord with Eq. (1).

If the asthenosphere is a relatively thin layer, then the obvious direction to transfer matter so as to affect the external gravitational field is lateral. Vertical transfers, however, are not to be ruled out: an upward displacement of material making the density higher than the average at a shallow level, balanced by a mass deficiency at considerable depth (below about 200 km) could account for a gravity excess. From the formula for the potential arising from a spherical harmonic surface distribution of mass (Kaula, 1968, p. 67), we have for a mass excess of $\Delta\rho h$ of width L compensated at depth D:

$$\Delta g \approx 2\pi^2 G \frac{D}{L} \Delta\rho h \tag{2}$$

Then to say the data are satisfied by isostatic compensation at great depth, however, is to beg the question as to the response of the asthenosphere to the stresses that must necessarily exist at the intervening levels.

The greater effectiveness of lateral transfer also suggests that stagnation points (regions where the flow changes direction, such as ocean rises or trench and island arcs) will tend to be regions of gravity excess, while regions dominated by horizontal flow will tend to be regions of gravity deficiency.

The relationship of gravity anomalies to the flow system depends considerably on the boundary conditions. In a system of thermal convection, if the boundary is rigid, then an upcurrent is associated with a negative anomaly, because of its lower density (Runcorn, 1965). If the upper boundary is free, however, while the lower boundary remains fixed, then an upcurrent is associated with a positive anomaly because the effect of the mass pushed up at the surface outweighs the density effect (McKenzie, 1968, Pekeris, 1935).

In the case of the real earth, the question becomes to what extent the lithosphere (the layer of relative strength) and the crust (the lower density, uppermost layer of the lithosphere) act as a part of the convective flow, and to what extent they act as a restraining boundary to the flow. Manifestly, they act both roles to differing degrees in different parts of the earth. The lithosphere can even be simultaneously a rigid boundary for horizontal forces in being able to

Table 1. Areas of exceptional gravity anomaly, defined as having an area x *free air* anomaly referred to the *hydrostatic* figure more than 50 mgal x 10^6 km^2 (1.17 x 10^{21} gm) in absolute magnitude, and absolute anomaly of more than 10 mgal throughout the area

General location	Area (10^6 km^2)	Free air anomaly x area (mgal x 10^6 km^2)	Mean free air anomaly (mgal)	Type
Positive features				
1. Sumatra-Philippines-Solomons	18.9	+461.	+ 24.	Arc
2. Andes-W. Amazon Basin	9.4	206.	22.	Arc-Orogenic
3. Solomons-Tonga-Kermandec	10.4	178.	17.	Arc
4. Mid-Indian Rise-Indian Antarctic Rise	9.4	175.	19.	Rise
5. Crozet Plateau-S. Madagascar Rise	6.3	117.	19.	Rise
6. Mexico-N.W. Colombia	7.3	107.	15.	Arc
7. Carpathians-Turkey-Iran	7.2	103.	14.	Orogenic
8. Hawaii	6.1	96.	16.	Shield
9. Azores Plateau	5.8	95.	16.	Rise
10. Japan-Bonins	4.6	92.	20.	Arc
11. Atlas-Iberia-W. Mediterranean	5.5	86.	16.	Arc-Orogenic
12. West Africa-Guinea Basin	6.2	85.	14.	Orogenic
13. Ahaggar-Tibesti-Nigeria	4.7	77.	16.	Orogenic

14. Greenland-Iceland-Norwegian Sea	4.6	76.	16.	Rise
15. East Pacific Rise, N. of Easter	4.2	64.	15.	Rise
16. Walvis Rise-S.W. Africa	4.5	+ 58.	+ 13.	Rise
Negative features				
1. Antarctica	22.4	- 511.	- 23.	Glac.-Basin?
2. Siberian Platform-Turkestan-Himalayas	15.1	289.	19.	Glac.-Orogenic
3. North Canada	10.1	218.	22.	Glaciated
4. N. American-Guiana Basins	11.7	212.	18.	Basin
5. Somali Basin-Central Indian Ocean	6.2	193.	31.	Basin
6. N. Pacific Basin-N.E. Pacific Slope	12.3	175.	14.	Basin
7. W. Australian Shield-S. Australian Basin	4.6	121.	26.	Basin
8. N.W. Pacific Basin	5.3	116.	22.	Basin
9. Wharton Basin	5.6	98.	17.	Basin
10. N.W. Siberia-Aleutian Basin	5.6	87.	16.	Glac.-Basin
11. Society Islands-S.W. Pacific Basin	3.2	69.	22.	Basin
12. Congo-Kenya	5.0	60.	12.	Basin-Rift
13. Argentine Basin	3.6	54.	15.	Basin
14. Chile Rise-Pacific Antarctic Basin	2.8	- 50.	- 18.	Basin

Table 2. Areas of exceptional gravity anomaly, defined as having an area x *isostatic* anomaly referred to a *fifth-degree* figure more than 50 mgal x 10^6 km^2 (1.17 x 10^{21} gm) in absolute magnitude, and absolute anomaly of more than 10 mgal throughout the area

General location	Area (10^6 km^2)	Isostatic anomaly x area (mgal x 10^6 km^2)	Mean isostatic anomaly (mgal)	Type
Positive features				
1. Southeast Pacific Rise	6.6	+ 135.	+ 20.	Rise
2. Mid-Indian-Amsterdam-Naturaliste Ridge	6.1	100.	16.	Rise
3. Borneo-Sumatra-Cocos	5.3	98.	18.	Arc
4. Indian Peninsula-Bay of Bengal	4.7	82.	17.	Sediment?
5. S.E. Indian and MacQuarrie Rises	5.1	76.	15.	Rise
6. North Atlantic-Arctic Ocean	7.1	73.	10.	Glaciated(?) Rise
7. Azores Plateau	4.7	73.	16.	Rise
8. Indian.-Antarctic, Gaussberg Ridges	4.5	72.	16.	Rise
9. N. Andes-W. Amazon Basin	5.0	72.	14.	Arc-Orogenic
10. N.E. Georgia-S. Sandwiches-Mid-Atlantic	4.3	68.	16.	Rise-Arc

11. Japan-Bonins	4.0	66.	16.	Arc
12. Carlsberg Ridge-Gulf of Aden	3.7	65.	18.	Rise
13. South Alaska	3.6	65.	18.	Arc-Orogenic
14. Walvis Rise	4.5	+ 61.	+ 14.	Rise
Negative features				
1. Himalayas-China	8.0	- 141.	- 18.	Orogenic
2. Antarctica	8.1	140.	17.	Glaciated Basin?
3. Laccadives-Ceylon-Mid-Indian Ocean	5.0	130.	26.	Basin?
4. N. Canada-Greenland	7.5	123.	16.	Glaciated
5. Australian Shield-S. Australian Basin	7.7	113.	15.	Basin
6. N. Amer.-Guiana-Parnaiba Basins	7.3	96.	13.	Basin
7. Galapagos-Peru Basin	6.1	96.	16.	Basin
8. Society-Tuamotu-Austral Seamount	6.8	96.	14.	Basin?
9. Congo-Kenya	3.7	74.	20.	Basin-Rift
10. E. Crozet Basin-Kerguelen	2.6	53.	20.	Basin
11. N.W. Europe	3.5	- 51.	- 15.	Glaciated

act as a rigid plate in tectonic motions and a free boundary for vertical forces in not resisting convective upthrusts. The extent to which a particular portion of the lithosphere acts as a free or rigid boundary depends on its temperature, size of feature, rate of motion of material into and out of a feature, and composition, particularly its water content. The situation may be further complicated by steady surface transfers of matter: erosion and sedimentation.

How a boundary acts in the range between perfectly free and perfectly rigid depends on both (1) its elastic properties (its rigidity and thickness) and (2) its plastic properties (most simply expressed as a decay time in response to a transient loading, dependent on dimensions of the loading and stress as well as creep properties of the material). Under small stresses, the decay time of the lithosphere is very long: it is effectively acting as an elastic layer in areas of post glacial uplift. Under greater stress, however, such as in the major areas of mass excess, the effective decay time may be much shorter because of the nonlinear dependence of strain rate on stress (Weertman, 1970), as evidenced by the seismicity of these regions. Qualitatively, for both elastic and plastic behavior, we should expect that the thicker, the colder, the less hydrous the lithosphere is in a particular region, the more it will behave like a rigid boundary. Quantitatively, however, we should expect that in some cases it may be difficult even to infer the correct sign of the gravity anomaly.

The flow system for a body that has boundaries that are partly rigid, partly free, would be a difficult problem to treat rigorously. However, we might expect that usually the nature of the local boundary conditions would predominate in determining the characteristics of a particular region. We shall apply this assumption in the analysis of feature types.

Of the 11 gravity anomaly area types proposed in Kaula (1969), 6 appear to be associated with current internal activity in the earth. We shall discuss these 6 (somewhat modified) in an order suggested by their apparent relationship to the global tectonic pattern: (1) active ocean rises, (2) oceanic shield basalts still active in Quaternary, (3) basins, (4) trench and island arcs currently active, (5) current orogeny without extrusives, and (6) Cenozoic orogeny with extrusives in Quaternary.

Active Ocean Rises

The indication from the new data that these areas are generally of positive gravity anomaly is consistent with their being free boundaries over upcurrents in a convective system. Their well-known characteristics of high heat flow, shallow depth in the ocean, thin sediments, large scale volcanism, frequent moderate earthquakes, lack of a distinct Moho, and prevalence of intermediate seismic primary velocities in the range 7.2 to 7.7 km/sec are generally taken to indicate that the rises are the sites of upwelling and spreading out in a convective cycle. The intensity and uniformity of heating is apparently sufficient to prevent this mass imbalance from being large. The small temperature gradients entailed are the expected consequence of a strong temperature dependence of viscosity (Tozer, 1967; Turcotte and Oxburgh, 1969).

Oceanic Shield Basalts

With the improved data, all major oceanic positive areas appear to be associated with spreading centers except one: Hawaii. Hawaii appears to be the buildup of an appreciable mass excess by extrusive activity off the rise. This buildup is in spite of a sinking of the crust, as pointed out by Menard (1969). An approach to isostatic adjustment is also suggested by depths to the Moho somewhat greater than the oceanic average (Drake and Nafe, 1968). Apparently the lithosphere has cooled sufficiently to cause a lag in the attainment of equilibrium. This notion is corroborated by the relatively low heat flow. The existence of such a feature indicates that the asthenospheric flows that generate the required pressure do not necessarily have a simple and direct relationship to the lithospheric plate motions (McKenzie, 1969).

Hawaii is unique in that it falls midway in the 10,000-km stretch from the East Pacific Rise to the trench and island arc along the west Pacific margin: the only region in the world where the rise-continent distance exceeds 5000 km. It is also the only region in the world where the rise-basin distance exceeds 3500 km.

Basins

This commonest of the major features always occurs somewhere to the flanks of ocean rises. Landward of the basin, however, may be a trench and island arc, an orogenic belt, or a relatively quiescent continent on the same tectonic plate. This suggests that the nature of the flows associated with basins depends more on where the material came from than where it is going.

The direct source of the negative isostatic anomaly is most likely that the crust carried along in the sea floor spreading is thicker than compatible with the depth of the basin; a Moho deeper by less than a kilometer is adequate to account for the average isostatic anomaly of -14 mgal.

The underlying cause, of course, is asthenospheric withdrawal, which results in the 3-km topographic drop from the rise to the basin. Such a drop could be caused by either (1) a horizontal acceleration in the asthenosphere or (2) a downflow of a denser component of the asthenospheric material.

A horizontal acceleration in the asthenosphere at a distance from the rise equal to the typical rise-basin distance (1500 to 3500 km) seems implausible, given that the velocity at the upper boundary is maintained constant by the lithosphere. If we assume a two-dimensional flow, neglect the effect of temperature gradients on the flow, then adopting the customary stream function form (Batchelor, 1967, p. 76)

$$u = \frac{\partial \psi}{\partial y}, \qquad v = -\frac{\partial \psi}{\partial x} \qquad\qquad (3)$$

where (x, y) and (u, v) are, respectively, the horizontal and vertical position and velocity (positive downward); we obtain the biharmonic equation

$$\nabla^4 \psi = 0 \tag{4}$$

for which a solution is (Batchelor, 1967, p. 225)

$$\psi(r, \theta) = rf(\theta)$$
$$f(\theta) = A \sin\theta + B \sin\theta + C\theta \sin\theta + D\theta \cos\theta \tag{5}$$

where r is distance from the origin and θ is the angle from the x axis. Make the boundary conditions

$$u = 0, \quad v = -v_0, \quad x = 0$$
$$u = u_0, \quad v = 0, \quad y = 0 \tag{6}$$

i.e., the x axis is the asthenospheric-lithospheric boundary, u_0 is the spreading velocity, and the y axis is the vertical center of the plume of constant velocity v_0. The solution then is

$$u = u_0[1 - f(\xi)] + v_0 g(\xi)$$
$$v = v_0[c(\xi) - 1] - u_0 d(\xi) \tag{7}$$

where

$$\xi = \frac{y}{x} = \tan\theta \tag{8}$$

and

$$f(\xi) = \frac{\pi}{2} q \left(\frac{\xi}{1 + \xi^2} + \tan^{-1}\xi \right) - q \frac{\xi^2}{1 + \xi^2}$$

$$g(\xi) = q \left(\frac{\xi}{1 + \xi^2} + \tan^{-1}\xi \right) - \frac{\pi}{2} q \frac{\xi^2}{1 + \xi^2}$$

$$c(\xi) = \frac{\pi}{2} q \left(\frac{\xi}{1 + \xi^2} - \tan^{-1}\xi \right) + q \frac{\xi^2}{1 + \xi^2} + 1$$

$$d(\xi) = q\left(\frac{\xi}{1 + \xi^2} - \tan^{-1}\xi\right) + \frac{\pi}{2} q \frac{\xi^2}{1 + \xi^2} \tag{9}$$

$$q = \frac{4}{\pi^2 - 4} = 0.678\ldots \tag{10}$$

The condition imposed by Eq. (1) necessary to make the rise a mass excess and the basin a mass deficiency is that

$$\frac{\partial u}{\partial x} > 0 \tag{11}$$

or

$$u_0 \frac{\partial f}{\partial \xi} - v_0 \frac{\partial q}{\partial \xi} > 0 \tag{12}$$

Developing f and g in series of ξ, for small ξ:

$$v_0 < \frac{\pi}{2}\left(1 + \frac{4 - \pi^2}{\pi}\xi + \cdots\right)u_0 \tag{13}$$

In words, the process of lithospheric formation has to consume more than a certain fraction of the material brought up by the vertical flow v_0 if the lithosphere of velocity u_0 is to cause an acceleration in the asthenosphere by dragging the asthenosphere along with it.

The settling out of a denser component would be expected in a multicomponent laterally moving flow that was cooling. A 3-km drop requires much more than thermal contraction, however. If there is an appreciable negative gravity anomaly as well, then Eq. (2) indicates that the settling out cannot just be immediately below the basin lithosphere, but must be either (1) at several 100 km depth below the basin or (2) between the ocean rise and the basin. The process (1) could be induced by the phase transitions of olivine and pyroxene, which occur at depths of 300 to 600 km, while the process (2) might be facilitated by gabbro-to-eclogite transitions at shallower depths.

The sharpness of the crust-mantle boundary, with the 7.2 to 7.7 km/sec gap in velocities (Drake and Nafe, 1968), makes it impossible for the crust to be directly involved in causing the negative anomaly. If the crust is being carried passively along (as suggested by the lack of seismicity, volcanism, or disturbance of the sea floor), then it is hard to understand why it is thicker under the basins than on the flanks of the rises, as emphasized by Le Pichon (1969). Could it be that consolidated sediments are mistaken for basement rock? Sedimentation itself is a secondary process in explaining the gravity anomaly pattern, more a result than a cause: if the thick sediments were the driving force, then the isostatic anomalies in ocean basins would be positive, rather than negative.

Possibly includable in the category of basins caused by behavior of the lithosphere as a free boundary over flows with horizontal accelerations or settling out of denser components are two land features, Antarctica and the Congo Basin. Antarctica is an extremely large feature—large enough to require an unique explanation.

Trench and Island Arcs

The now generally accepted model of McKenzie (1969) and others of a colder, denser oceanic lithospheric slab being thrust down under a less dense but stiffer continental margin fits a simple notion of the gravity pattern: the dominant feature is the broad positive anomaly associated with the denser downthrust slab, while the secondary feature is the narrow negative belt associated with the trench caused by tensile cracking along the downward breaking line. This simple picture is based on the assumption that the applicable boundary condition of the convective flow is more "rigid" than "free": in other words, the time scale of the process is short enough that the strength of the continental lithosphere (and perhaps the oceanic lithosphere as well) significantly resists being pulled down by the downcurrent. This general idea that resistance to flow combined with densification creates positive gravity anomalies applies not only to the boundary layer, but also to deeper strata: the downthrust slab could in part be supported by stiffer matter below the asthenosphere, as suggested by Isacks and Molnar (1969) and others from seismic data.

The association of the downthrust slab with positive anomalies also suggests that the driving cause is a push from above rather than withdrawal from below. Whether this "push" is the gravitationally caused sinking of the denser oceanic lithosphere, the pressure of the spreading sea floor behind it, or the viscous drag by the sublithospheric flow, does not seem resolvable from the gravity data.

Current Orogeny Without Extrusives

The hypothesis of McKenzie (1969) that purely continent versus continent compression results in folding rather than downthrust because of the excessive bouyancy of the thicker crust is appealing as an explanation for the strongly negative gravity anomalies associated with the Asian part of the Alpide belt. The resulting pileup of lower density material results in a mass deficiency in the short run because the stiffness of the lithosphere containing low density crust enables it to push out of the way higher density asthenospheric material. In the longer run, however, the trend from "rigid" to "free" boundaries is expressed by the forcing upward of the lithosphere; geologic and geodetic indications are that the Himalayas-Turkestan complex is currently rising (Artyushkov and Mescherikov, 1969; Gansser, 1964).

The thick layers of sedimentary and metamorphic rocks constituting the upper part of the Himalayas have existed since Precambrian times. The resulting excess of radioactive material combined with low thermal conductivity will lower crustal densities. Furthermore, there may be a contribution to the negative anomaly by erosion, as corroborated by the positive features over the corresponding sedimentation basins, the Bay of Bengal and the Arabian Sea, which appear in Fig. 4.

It is possible that the foregoing suggested mechanisms are all quantitatively insufficient and that an asthenospheric withdrawal is necessary.

Cenozoic Orogeny with Extrusives

These mountain-building areas, listed as orogenic among the positive features in Table 1, are of more limited extent and closer to isostatic equilibrium. Most are associated with compressive belts of the global tectonic system, but this is not entirely so. The reason why they differ from the Himalayas-Turkestan complex in being positive may be (1) the lack of the pre-existing great thicknesses of sedimentary and metamorphic rocks or (2) the presence in the eastern Mediterranean of oceanic crust that can be "consumed" or "subducted" (McKenzie, 1970). They may also be the continental equivalents of Hawaii to some extent: the coincidence of weak features in the lithosphere with regions of excess pressure and heat in a convective system that is not directly related to surface features. Most of these areas have·positive seismic delay residuals, suggesting high temperatures to considerable depth.

DISCUSSION AND CONCLUSIONS

The gravity data now appear to be quite reconcilable with the dependence of plate tectonics on mantle convection inferred from other phenomena associated with the mid-ocean rises and the compressive belts (Isacks et al., 1968). Gravity still suffers, however, from its traditional ambiguity in being insufficient to infer the exact mechanism, such as whether the ocean basin negatives are caused by horizontal acceleration or downflow of a denser component.

The greatest feature not readily related to the global tectonic system is the Antarctic negative, much too large to be explained by glacial melting. Antarctica is five-sixths surrounded by ocean rises. Hence, either the spreading rises are migrating away from Antarctica, or Antarctica is a sink for lithospheric material. The latter seems ruled out by the complete absence of the seismicity expected with the destruction or folding of lithosphere. Given that the rises are migrating away, Antarctica must be a mass deficiency because there is not an asthenospheric flow to match the lithospheric spread: i.e., the condition of Eq. (13) applies over most of the rises around Antarctica.

Other features not well explained by the global tectonic pattern are the gravity excesses associated with extrusive flows that occur away from the ocean rises and trench and island arcs, in both oceanic and continental areas. These features appear to require higher temperatures in the asthenosphere generating excess pressures, together with weaknesses in the lithosphere allowing the extrusions. It is, however, difficult to choose whether the resulting net mass excess is a consequence of sufficient overall strength in the lithosphere to support the extruded load or of behavior as a free boundary over a horizontal deceleration or an upcurrent, the reverse of the processes that appear necessary to account for the ocean basins.

Anticipated properties of the mantle convective system that need to be better related to the gravity field are the stress- and temperature-dependence of the effective viscosity, the horizontal temperature gradients arising from variations in radiogenic heating and the contributions to driving the system by fractionations and phase transitions. All these properties are important, of course, to solution of the entire global tectonic problem.

ACKNOWLEDGMENT

I am grateful to E. M. Gaposhkin and Kurt Lambeck for providing their results in advance of publication. This work has been supported by NSF Grant GA-10963.

REFERENCES

Artyushkov, E. V., and Mescherikov, Y. A., Recent movements of the earth's crust and isostatic compensation, *in* The Earth's Crust and Upper Mantle, Am. Geophys. Union Monograph **13**, edited by P. J. Hart, pp. 379-390, 1969.

Batchelor, G. K., An Introduction to Fluid Dynamics, Cambridge Univ. Press, London, 615 pp., 1967.

Drake, C. L., and Nafe, J. E., The transition from ocean to continent from seismic refraction data, *in* The Crust and Upper Mantle of the Pacific Area, Am. Geophys. Union Monograph **15**, edited by L. Knopoff, C. L. Drake, and P. J. Hart, pp. 174-186, 1968.

Gansser, A., Geology of the Himalayas, John Wiley (Interscience), London, 289 pp., 1964.

Gaposhkin, E. M., and Lambeck, K., Earth's gravity field to the sixteenth degree and station coordinates from satellite and terrestrial data, J. Geophys. Res., **76**, 4855-4883, 1971.

Goldreich, P., and Toomre, A., Some remarks on polar wandering, J. Geophys. Res., **74**, 2555-2567, 1969.

Heirtzler, J. R., Dickson, G. O., Herron, E. M., Pitman, W. C., III, and LePichon, X., Marine magnetic anomalies, geomagnetic field reversals, and motions of the ocean floor and continents, J. Geophys. Res., **73**, 2119-2136, 1968.

Hide, R., and Malin, S. R. C., Novel correlations between global features of the earth's gravitational and magnetic fields, Nature, **225**, 605-609, 1970.

Isacks, B., and Molnar, P., Mantle earthquake mechanisms and the sinking of the lithosphere, Nature, **223**, 1121-1124, 1969.

Isacks, B., Oliver, J., and Sykes, L. R., Seismology and the new global tectonics, J. Geophys. Res., **73**, 5855-5899, 1968.

Kaula, W. M., Test and combination of satellite determinations of the gravity field with gravimetry, J. Geophys. Res., **71**, 5303-5314, 1966.

______, Geophysical implications of satellite determinations of the earth's gravitational field, Space Sci. Rev., **7**, 769-794, 1967.

______, An Introduction to Planetary Physics: the Terrestrial Planets, John Wiley & Sons, New York, 490 pp., 1968.

______, A tectonic classification of the main features of the earth's gravitational field, J. Geophys. Res., 74, 4807-4826, 1969.

Le Pichon, X., Models and structure of the oceanic crust, Tectonophysics, 7, 385-401, 1969.

McKenzie, D. P., Some remarks on heat flow and gravity anomalies, J. Geophys. Res., 72, 6261-6273, 1967.

______, The influence of the boundary conditions and rotation on convection in the earth's mantle, Geophys. J. Roy. Astron. Soc., 15, 457-500, 1968.

______, Speculations on the consequences and causes of plate motions, Geophys. J. Roy. Astron. Soc., 18, 1-32, 1969.

______, Plate tectonics of the Mediterranean region, Nature, 226, 239-243, 1970.

Menard, H. W., Growth of drifting volcanoes, J. Geophys. Res., 74, 4827-4837, 1969.

O'Connell, R. J., Pleistocene glaciation and the viscosity of the lower mantle, Geophys. J. Roy. Astron. Soc., in press, 1971.

Pekeris, C. L., Thermal convection in the interior of the earth, Monthly Notices Roy. Astron. Soc., Geophys. Supp., 3, 343-367, 1935.

Runcorn, S. K., Changes in the convective pattern in the earth's mantle and continental drift: evidence for a cold origin of the earth, Phil. Trans. Roy. Soc. London, Ser. A, 258, 228-251, 1965.

Tozer, D. C., Towards a theory of thermal convection in the earth's mantle, *in* The Earth's Mantle, edited by T. F. Gaskell, pp. 325-353, Academic Press, New York, 1967.

Turcotte, D. L., and Oxburgh, E. R., Finite amplitude convection cells and continental drift, J. Fluid Mech., 28, 29-42, 1967.

______, Convection in a mantle with variable physical properties, J. Geophys. Res., 74, 1458-1474, 1969.

Uotila, U. A., Gravity anomalies for a model earth, Ohio State Univ. Dept. Geodetic Sci. Tech. Rept., 37, 15 pp., 1962.

Weertman, J., The creep strength of the earth's mantle, Rev. Geophys. Space Phys., 8, 145-168, 1970.

25

Reprinted from *Contr. Mineralogy and Petrology* **53**:253–279 (1975)

Lead Isotope Relations in Oceanic Ridge Basalts from the Juan de Fuca-Gorda Ridge Area, N.E. Pacific Ocean

S.E. Church* and Mitsunobu Tatsumoto

Department of Geological Sciences, University of California, Santa Barbara, California 93106
and U.S. Geological Survey, Denver, Colorado 80225

Abstract. Lead isotopic analyses of a suite of basaltic rocks from the Juan de Fuca-Gorda Ridge and nearby seamounts confirm an isotopically heterogeneous mantle known since 1966. The process of mixing during partial melting of a heterogeneous mantle necessarily produces linear data arrays that can be interpreted as secondary isochrons. Moreover, the position of the entire lead isotope array, with respect to the geochron, requires that U/Pb and Th/Pb values are progressively increased over the age of the earth. Partial melting theory also dictates analogous behavior for the other incompatible trace elements. This process explains not only the LIL element character of MOR basalts, but also duplicates the spread of radiogenic lead data collected from alkali-rich oceanic basalts. This dynamic, open-system model of lead isotopic and chemical evolution of the mantle is believed to be the direct result of tectonic flow and convective overturn within the mantle and is compatible with geophysical models of a dynamic earth.

Introduction

Many petrologists have considered basaltic liquids to be the primary partial melts of the mantle. Much effort has been made to explain the variation in chemistry, petrography and tectonic setting of basaltic rocks and to fit these data into a unified theory of magma genesis. However, these types of studies are limited in that they provide information only on the conditions which persisted in the mantle immediately prior to the partial melting event. Radiogenic isotopic studies, however, provide information on the history of the mantle source region integrated over the age of the earth. The present study of the variation of lead isotopic abundances in a suite of basaltic rocks dredged from the Gorda and Juan de Fuca Ridges and nearby seamounts furnishes new data which must be incorporated into any general theory of basalt genesis.

Green and Ringwood (1967) have summarized much of the early work on basalt genesis; they proposed that MOR basalts were extensive melts of pyrolite

---

* Analytical work done while the author held a National Research Council Associateship at the Johnson Space Center, Houston, Texas.

328

at shallow levels in the mantle and that different basaltic magmas would be produced by partially melting pyrolite under different conditions of P and T. Gast (1968), using incompatible element data, proposed that alkali-rich basalts were produced by small-scale partial melting (3–6%) of the mantle in equilibrium with garnet (i.e., from depths >100 km) and that the MOR basalt (MORB) melts represented a much larger degree of partial melting at shallower levels in the mantle.

Radiogenic isotopic studies of oceanic basaltic rocks have yielded valuable information on U/Pb, Th/U, and Rb/Sr in the mantle with time as well as a general correlation of more radiogenic lead and strontium with LIL (large ion lithophile or incompatible) element enrichment (Tatsumoto, 1966a; Gast and others, 1964; $^{87}Sr/^{86}Sr$ summarized by Peterman and Hedge, 1971 and several others). The major conclusions of these studies were that the mantle was isotopically heterogeneous, and that alkali depletion of the source material for MOR basalts had occurred in the past. The determination of the time at which this depletion occurred, however, was ambiguous. Tatsumoto's (1966a) data on MOR basalts indicate an age of at least 1200 m.y. whereas the data of Gast and others (1964) suggest that local heterogeneities could exist for 3–4 b.y. These early studies however, were limited in that the sampling was from widely separated areas in the ocean basins, and extrapolation of these few data may not adequately reflect local mantle development. Consequently, we have chosen to concentrate on a small geographic area to investigate both the intraplate and interplate variations of the isotopic data before valid mantle-wide extrapolations are attempted. A detailed study of the variations of lead isotopes in a suite of oceanic basalts from the Juan de Fuca-Gorda Ridge area was planned. We will examine four major points: (1) What is the variation of lead isotopes in a suite of oceanic tholeiitic basalts collected from an area of a few square kilometers (e.g. along the Blanco Fracture Zone)? (2) Is there a geographic variation associated with the lead isotopic data which can be correlated with visible oceanic plate structure? (3) How large is the total variation of the lead isotopic data array for MOR basalts from the Juan de Fuca-Gorda Ridge area and what does it tell us about mantle history and differentiation? (4) What modifications do the lead isotopic data require for partial melting models and the origin of mantle melts?

Fig. 1, a generalized base taken from the work of McManus (1964), shows most of the sample localities used in this study. The locality coordinates, descriptions of samples, references to additional data and sample handling procedures are in the Appendix. Seamounts are identified on Fig. 1 to demonstrate the asymmetric distribution of submarine volcanoes relative to the active spreading centers at the Juan de Fuca and Gorda Ridges. Samples are readily available from most of the ridge system, but only a few seamounts in this region have been dredged. Fortunately, those that have been sampled (Cobb, Parks, Heck, Explorer, Union, Dellwood, and Hodgkins Bank) provide a good geographic grid for the project. Three samples from a sill cored at the DSDP site 177A have also been analysed. Not shown on Fig. 1 are Hodgkins Bank, located at 53°15′ N latitude, and a suite of samples dredged by the National Museum of Natural History from the area of the Blanco Fracture Zone between Parks Seamount and the south end of the Juan de Fuca Ridge.

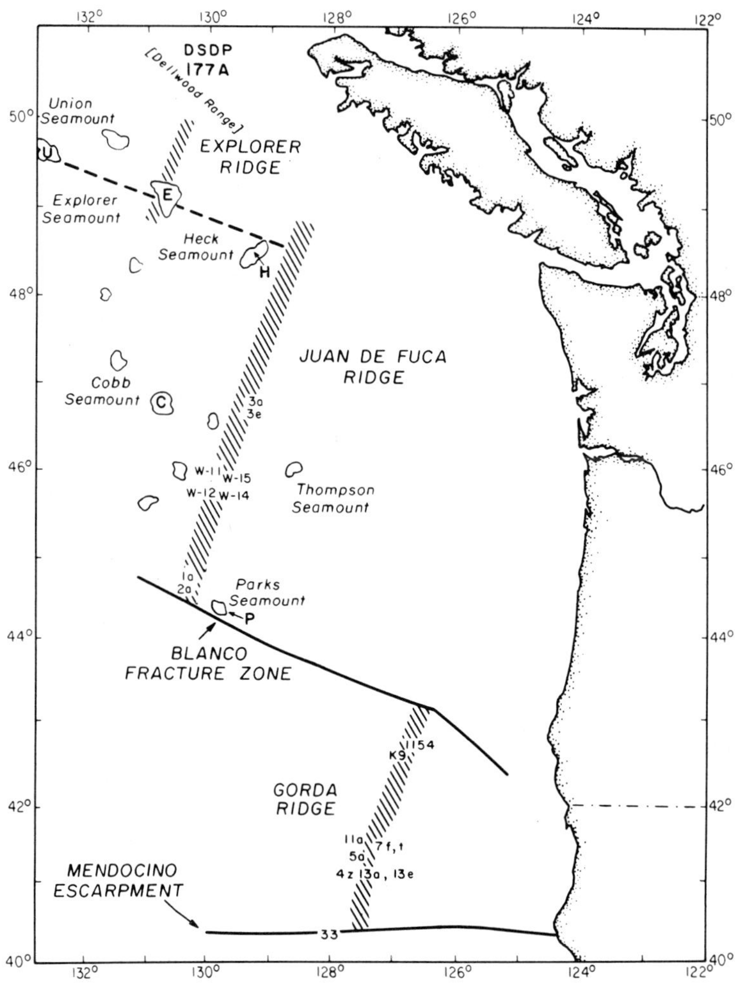

Fig. 1. Generalized map of the Juan de Fuca-Gorda Ridge area (from McManus, 1964). The locations of MOR basalt samples are shown relative to the active spreading centers (hachured line). Seamounts are designated by letter code used in Figures to follow: Cobb (*C*), Dellwood Range (*D*), DSDP site 177A, Heck (*H*), Explorer (*E*), Union (*U*), and Parks (*P*).The Heck-Explorer-Union Seamount line represents the only linear chain of seamounts sampled which is perpendicular to the ridge axis

Results

Tatsumoto (1966a) reported the initial lead isotopic data on MOR basalts. His data, corrected for thermal fractionation through the "absolute" value of the CIT reference lead (Catanzaro, 1967), are presented in Table 1. Comparisons between this set of data and the new data presented in Tables 2, 3 and 4 demonstrate both the extent of mantle heterogeneity and analytical problems. The large variations seen in these data underline the danger of assuming a specific

Table 1. Summary of previously published lead isotopic data from MOR basalts[a]

Sample	$^{206}Pb/^{204}Pb$	$^{207}Pb/^{204}Pb$	$^{208}Pb/^{204}Pb$	U	Th	Pb
AD-2	18.34	15.38	37.55	0.16	0.15	1.29
AD-3	17.70	15.38	37.00	0.10	0.29	1.15
AD-5	18.68	15.52	38.11	0.09	0.13	0.56
PD-1	18.06	15.38	37.40	0.05	0.13	0.49
PD-3	18.11	15.37	37.50	0.09	0.21	0.49
PD-4	18.37	15.42	37.80	0.07	0.11	0.53

[a] Pb isotopic data from Tatsumoto (1966a) corrected for thermal fractionation through the CIT Pb value reported by Catanzaro (1967).

Table 2. Lead isotopic and concentration data for basalts from Juan de Fuca and Gorda Ridges and Explorer Seamount

Samples	$\dfrac{^{206}Pb}{^{204}Pb}$	$\dfrac{^{207}Pb}{^{204}Pb}$	$\dfrac{^{208}Pb}{^{204}Pb}$	$\dfrac{^{238}U}{^{204}Pb}$	$\dfrac{^{232}Th}{^{204}Pb}$	$\dfrac{^{232}Th}{^{238}U}$	$\dfrac{K}{U} \times 10^4$	K^a	U	Th	Pb
Explorer Seamount											
C10-d3	18.447	15.480	37.859[b]	9.92	17.57	1.77	1.22	600	0.049	0.084	0.312
Juan de Fuca Ridge											
1a	18.518	15.492	37.893[b]	47.19	85.11	1.80	0.74	2000	0.267	0.466	0.357
3a	18.339	15.462	37.714[b]	18.05	30.23	1.67	0.74	858	0.116	0.188	0.403
3e	18.347	15.445	37.742[b]	16.06	31.57	1.97	1.09	1000	0.092	0.175	0.361
Gorda Ridge											
4z	18.345	15.476	37.805[b]	18.56	39.52	2.13	0.51	852	0.166	0.342	0.562
	18.346	15.474	37.820								
5a	18.570	15.511	37.929[b]	19.55	46.54	2.38	0.97	2600	0.188	0.434	0.608
7f	18.649	15.486	38.037[b]	25.06	61.40	2.45	0.98	5000	0.509	1.209	1.286
				27.13	64.08	2.36	0.93		0.536	1.225	1.251
7t	18.622	15.487	38.009	30.07	74.79	2.49	0.84	4300	0.509	1.225	1.071
	18.620	15.493	38.022			2.40	0.82		0.524	1.220	
11a	18.371	15.438	37.731[b]	16.76	18.94	1.13	1.10	1400	0.127	0.139	0.476
	18.354	15.430	37.732	12.17	17.16	1.41	1.30		0.108	0.148	0.557
13a	18.380	15.456	37.798[b]	30.93	42.99	1.39			0.112	0.151	0.228
	18.380	15.473	37.791	22.07	28.68	1.30			0.118	0.148	0.336
				13.51	19.02	1.41			0.105	0.143	0.490[c]
13e	18.408	15.488	37.903	9.54	25.57	2.68	2.74	1700	0.062	0.161	0.409
1154	18.288	15.454	37.633[b]	14.97	45.50	3.04	1.32	1800	0.136	0.401	0.569
	18.259	15.436	37.616	15.62	44.38	2.84	1.29		0.139	0.382	0.556
KD-9	18.410	15.481	37.859[b]	22.83	63.01	2.76	0.89	1500	0.168	0.449	0.464
				18.17	55.06	3.03	1.08		0.139	0.408	0.482

[a] Data from Kay and others (1970).
[b] Pb isotopic composition measured by the phosphoric acid-silica gel method using the double spike method (Compston and Oversby, 1969); other analyses corrected for fractionation by comparison with N.B.S. SRM 981 and 982 values (Catanzaro and others, 1968) (+0.13% mass for 1200°C run and +0.10% mass for 1280°C run).
[c] This set of concentration data determined by S.E.C.

Table 3. New lead isotopic and related chemical data for MOR basalts

Sample	$\dfrac{^{206}Pb}{^{204}Pb}$	$\dfrac{^{207}Pb}{^{204}Pb}$	$\dfrac{^{208}Pb}{^{204}Pb}$	$\dfrac{^{238}U}{^{204}Pb}$	$\dfrac{^{232}Th}{^{204}Pb}$	$\dfrac{^{232}Th}{^{238}U}$	U	Th	Pb
Juan de Fuca Ridge									
2a	18.459	15.463	37.83	13.14	29.12	2.215	0.139	0.298	0.665
234	18.475	15.468	37.90	39.56	25.65	0.648	0.405	0.254	0.644
236	18.482	15.479	37.80	30.94	27.54	0.890	0.210	0.181	0.428
	18.506	15.476	37.78						
237	18.518	15.470	37.84	31.68	26.55	0.838	0.259	0.210	0.515
240	18.417	15.456	37.74	13.80	35.06	2.54	0.137	0.337	0.625
243	18.435	15.458	38.10	10.23	23.31	2.28	0.076	0.168	0.468
	18.460	15.479	38.15						
W-11	18.786	15.548	38.26	5.99	9.58	1.60	0.029	0.045	0.309
W-12	18.574	15.495	38.02	15.97	42.63	2.67	0.147	0.381	0.582
W-14	18.574	15.483	38.01						
W-15	18.303	15.470	37.85	33.99	25.66	0.755	0.119	0.087	0.221
Gorda Ridge									
F-33	18.382	15.457	37.73	15.21	22.71	1.493	0.155	0.224	0.640
5a	18.607	15.525	38.16	10.20	26.97	2.643	0.161	0.412	1.00
7f	18.681	15.533	38.14	35.09	77.12	2.198	0.445	0.947	0.805
13e	18.416	15.497	37.92	8.31	21.82	2.625	0.061	0.155	0.463
				8.15	20.37	2.500	0.060	0.145	
1154	18.305	15.478	37.76	19.47	28.27	1.452	0.197	0.277	0.635
	18.294	15.474	37.70	19.83			0.204		0.646
KD-9	18.397	15.490	37.89	15.35	55.12	3.590	0.120	0.417	0.492
KD-11				20.72	63.91	3.085	0.152	0.454	0.462
				17.97	58.40	3.248	0.140	0.440	0.489
245	18.280	15.452	37.69	0.94	2.08	2.65	0.032	0.082	2.14
				1.09	1.95	1.78	0.036	0.062	2.07

isotopic composition of lead in MOR basalts as input into mixing models that attempt to explain the variations of lead isotopes seen in orogenic volcanic rocks.

The new lead isotopic data reported in this study are in Tables 2–4. Interlaboratory comparisons can be made by examining data from the same sample (each obtained from the supplier completely independently) and analysed in different laboratories using different techniques (see Appendix for analytical details). These comparisons can be made most directly for sample 13e. Powders were swapped for comparative measurements on this sample. The isotopic data all agree quite well, as can be seen by comparison of Tables 2 and 3. Concentration data for uranium and thorium are in quite good agreement but the lead concentration data do not agree as well as expected. The effect is to make the calculated present-day mu and kappa values higher for samples run by Church (Tables 3 and 4) than those run by Tatsumoto (Table 2); however, this is not a serious problem and does not affect the isotopic results presented here. Other samples which can be compared are KD-9, 1154, 5a, and 7f. Sample KD-9 compares very well for both the isotopic and concentration data. Sample 1154, which was also analysed by Sun (1973), compares less well, but is still within $\pm 0.1\%$ on $^{207}Pb/^{204}Pb$, which is the critical measurement. The analysis by Sun (1973)

Table 4. New lead isotopic and related chemical data for basalts from seamounts and DSDP site 177A, N.E. Pacific Ocean

Sample	$\dfrac{^{206}Pb}{^{204}Pb}$	$\dfrac{^{207}Pb}{^{204}Pb}$	$\dfrac{^{208}Pb}{^{204}Pb}$	$\dfrac{^{238}U}{^{204}Pb}$	$\dfrac{^{232}Th}{^{204}Pb}$	$\dfrac{^{232}Th}{^{238}U}$	U	Th	Pb
Cobb Seamount									
C-1	18.808	15.483	38.20	15.84	34.31	2.166	0.237	0.497	0.950
	18.810	15.475	38.14						
C-2	18.377	15.475	37.94	4.71	15.45	3.283	0.095	0.302	1.27
C-3	18.423	15.48	37.96	9.20	23.22	2.525	0.045	0.110	0.309
Dellwood Range									
D-1	18.957	15.541	38.33	32.90	84.08	2.555	0.288	0.712	0.559
D-2	19.095	15.560	38.34						
D-3	18.502	15.505	37.93						
Explorer Seamount									
E-1	18.393	15.469	37.84						
E-2	18.400	15.465	37.83	32.82	17.36	0.528	0.170	0.087	0.326
E-3	18.580	15.505	38.07						
E-4	18.400	15.478	37.82	12.10	15.12	1.250	0.077	0.093	0.4
	18.382	15.475	37.84						
E-5	18.688	15.508	38.09						
Heck Seamount									
H-1	18.290	15.494	37.67	5.05	10.18	2.016	0.021	0.041	0.260
	18.299	15.495	37.66						
H-2	18.330	15.494	37.75						
Hodgkins Bank									
B-1	18.403	15.449	37.76						
B-2	18.300	15.435	37.77	247	760	3.078	1.282	3.82	0.325
Parks Seamount									
P-227	18.297	15.447	37.70	24.69	42.47	1.720	0.248	0.413	0.630
DSDP, Leg. 18, site 177A									
7B-1	18.876	15.534	38.29						
7B-2	18.921	15.533	38.32	11.86	50.07	4.22	0.058	0.237	0.312
7B-3	18.889	15.528	38.32						
Union Seamount									
U-1	19.434	15.592	38.69	27.31	132.7	4.86	0.167	0.786	0.395

was from a third powder of this sample and lies between our analyses. Isotopic data for sample 5a agree fairly well except for the $^{208}Pb/^{204}Pb$; the lead concentration data again are higher in Table 3. The agreement for sample 7f is not as good, but could reflect weathering, as that sample was somewhat altered. Samples analysed on the NASA instrument (Church – Tables 3 and 4) tend to have slightly higher $^{208}Pb/^{204}Pb$ values ($\sim 0.12 \pm 0.05$) than the same samples analysed on the USGS instrument (Tatsumoto – Table 2). Compare also the analyses of samples 3a and 3e (Table 2) with those of samples W-11 to W-15 (Table 3)

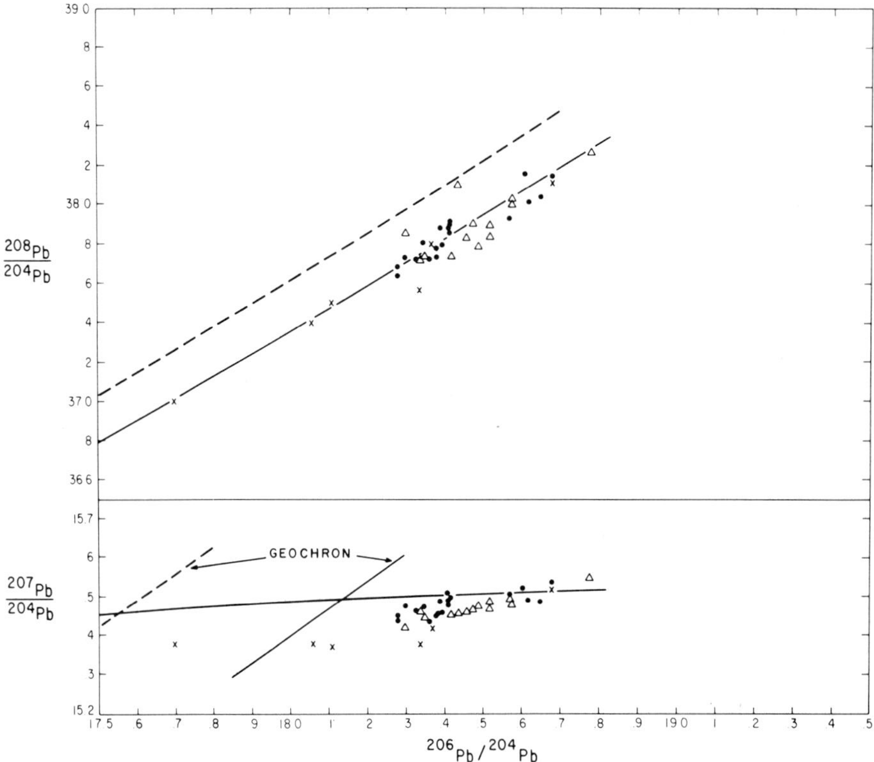

Fig. 2. Lead isotopic data from the MOR basalts from Gorda (●) and Juan de Fuca (△) Ridges (see Tables 2 and 3). Analyses of MOR basalts (X) previously published by Tatsumoto (1966a) corrected for fractionation effects (Table 1). Included for reference are two closed system growth curves using the following input parameters: For the solid curve, $T_0 = 4.55$ b.y., $\mu = 8.7$, $K = 3.7$, $^{238}U/^{235}U = 137.88$, $A_0 = 9.307$, $B_0 = 10.294$, $C_0 = 29.48$, $\lambda238 = 0.1537 \times 10^{-9}\mathrm{yr}^{-1}$, $\lambda235 = 0.9722 \times 10^{-9}\mathrm{yr}^{-1}$, $\lambda232 = 0.0488 \times 10^{-9}\mathrm{yr}^{-1}$. For the dashed curve (growth curve identical in $^{207}Pb/^{204}Pb$ versus $^{206}Pb/^{204}Pb$), $T_0 = 4.57$ b.y., $\mu = 8.0$, $K = 3.8$, $^{238}U/^{235}U = 137.88$, $A_0 = 9.307$, $B_0 = 10.294$, $C_0 = 29.48$, $\lambda238 = 0.155125 \times 10^{-9}\mathrm{yr}^{-1}$, $\lambda235 = 0.98485 \times 10^{-9}\mathrm{yr}^{-1}$, $\lambda232 = 0.049475 \times 10^{-9}\mathrm{yr}^{-1}$

and Cobb Seamount (Table 4). Possible explanations for this analytical discrepancy are discussed in the Appendix.

Figs. 2 and 3 present the results of this study in the conventional growth curve diagrams. The linear array of lead isotopic data clearly shows that the closed-system model for lead isotopic evolution in the mantle cannot approximate the chemical and physical conditions which exist in the source region and points out the isotopically heterogeneous nature of the mantle. Linear arrays of lead isotopic data of this type are usually interpreted in terms of a two-stage model. Table 5 presents two-stage model ages calculated from the slopes defined as different subsets of the data. As is immediately apparent, the lead data from MOR basalts as a whole (Table 5, cols. 4 and 11) differ significantly from data from the seamount basalts (Table 5, cols. 5 and 12). In particular, the MOR basalt data as a whole have slopes giving secondary isochron ages greater than

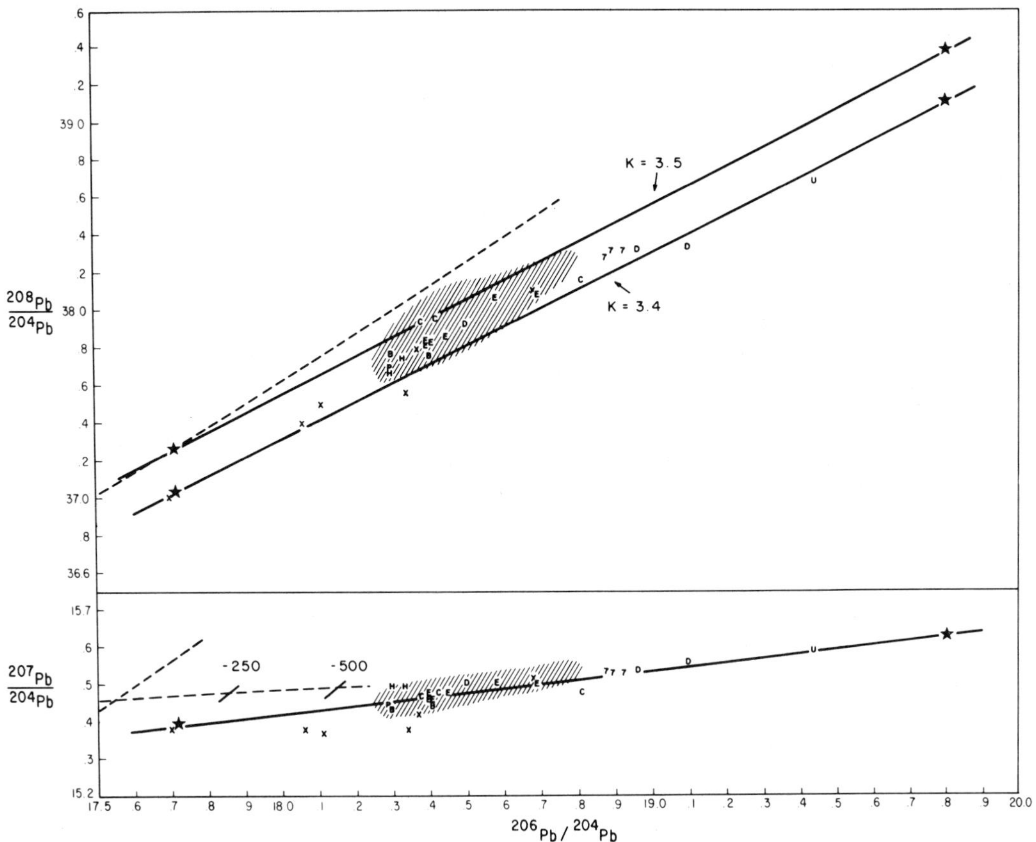

Fig. 3. Data from seamount samples analysed in this study (Table 4). The field of lead isotopic data from the MOR basalts is shown by the lined pattern; lead data from the seamount samples are designated by letter: Cobb (*C*), Dellwood Range (*D*), Explorer (*E*), Heck (*H*), Hodgkins Bank (*B*), Parks (*P*), Union (*U*), and the sill from DSDP site 177A by (7). The stars represent the average of 100 residual (nonradiogenic) and 100 interstitial (radiogenic) systems. The slope of the line matches the least squares fit to the data. Tie lines between these points are mixing lines which result during partial melting of an isotopically heterogeneous mantle. Note the somewhat lower Th/U value is required to match the data compared with the closed-system growth-curves (dashed curves)

2.0 b.y., whereas the seamount basalt data give ages of 1.75 ± 0.1 b.y. (Table 5, cols. 5 and 6). On the $^{206}Pb/^{204}Pb$ versus $^{208}Pb/^{204}Pb$ plots (Figs. 2 and 3), the MOR basalts and the seamount basalts show distinctively different patterns.

The extent of regional variation of lead isotope ratios for samples from the same ridge is shown by the Juan de Fuca MOR basalt data (Table 5, col. 8) which form a well-defined linear array with a slope giving a two-stage model age of ~ 2.5 b.y. The basalts from the Gorda Ridge (Table 5, col. 9) show approximately the same slope, but much poorer correlation. On the basis of our range of error on sample 1154, we must conclude that some of the spread in this suite may be due to analytical uncertainties; however, some of the spread in the data along the $^{207}Pb/^{204}Pb$ axis is certainly real. The uncertainty in any

Table 5. Linear regression analysis of the lead isotopic data

	^{206}Pb/^{204}Pb vs. ^{207}Pb/^{204}Pb									^{206}Pb/^{204}Pb vs. ^{208}Pb/^{204}Pb			
	1	2	3	4	5	6	7	8	9	10	11	12	13
Coefficient of correlation	0.732	0.749	0.490	0.790	0.899	0.952	0.862	0.854	0.776	0.930	0.850	0.977	0.987
Slope	0.137	0.147	0.137	0.159	0.112	0.101	0.121	0.162	0.154	0.952	0.988	0.846	0.861
Two-stage model age	2.22	2.35	2.22	2.48	1.84	1.67	2.00	2.52	2.43				

1　All data, Table 2 (12) except C10-d3.
2　Table 2, double spiked data only (10) except C10-d3.
3　Table 3, (17).
4　Tables 2 and 3 (29) except C10-d3.
5　Table 4, all data plus C10-d3 from Table 2 (21).
6　Data from seamounts at north end of Juan de Fuca Ridge only (15).
7　All data, Tables 2, 3 and 4 (50).
8　All data from Juan de Fuca samples (Tables 2 and 3) except W-11, see Appendix (12).
9　All data from Gorda samples (Tables 2 and 3) (17).
10　All double spiked data, Table 2, except C10-d3 (10).
11　All data Tables 2 and 3, except C10-d3 (29).
12　All data table 4 plus C10-d3 from Table 2 (21).
13　Data from seamounts at north end of Juan de fuca Ridge only (15).

Coefficient of correlation is the Pearson product moment correlation coefficient defined as $A/\sqrt{B*C}$; the slope is A/B where:

$$A = \sum XY - \frac{\sum X \sum Y}{N} = \sum (X - \bar{X})(Y - \bar{Y})$$

$$B = \sum X^2 - \frac{(\sum X)^2}{N} = \sum (X - \bar{X})^2$$

$$C = \sum Y^2 - \frac{(\sum Y)^2}{N} = \sum (Y - \bar{Y})^2.$$

two-stage model age computed for this array is on the order of 500 m.y. because of the limited range in ^{207}Pb/^{204}Pb values and the uncertainty associated with the measurements. Within reasonable confidence limits, there do not appear to be marked differences in ^{207}Pb/^{204}Pb in the source region for basalts from these two oceanic ridges.

Several useful comparisons between lead isotopic values obtained from seamounts and those obtained from MOR basalts from the same geographical area can be made. In particular, the sample from Parks Seamount (Table 4) is indistinguishable isotopically from similar samples dredged from along the Blanco Fracture Zone (Table 3, NMNH series 234–243) and closely resembles those dredged from the south end of the Juan de Fuca Ridge (see Fig. 1 and Tables 2 and 3). Similar comparisons can also be made at Cobb Seamount. The data (Table 3, W-series) from the central area of the Juan de Fuca Ridge compare well with the Cobb Seamount data, but data from a location directly to the east of Cobb Seamount (Table 2, analyses 3a, 3e) are not as radiogenic. Part of this discrepancy may be due to interlaboratory bias as previously discussed. Regression analyses of the data for Cobb Seamount yields a very flat slope quite unlike that seen in any other subset of data and is highly dependent upon the accuracy of the ^{207}Pb/^{204}Pb value for sample C-1. Application of an additional 0.05%/mass unit fractionation bias, which is the limit determined by repeated analyses of the N.B.S. lead standards, would greatly improve this situation, but this *a priori* assumption cannot be independently justified. The flat slope of the regression line through the Cobb Seamount data cannot be attributed to radiogenic growth of ^{206}Pb in light of the isotope dilution data in Table 4 and the recent age (1.5 and 0.5 m.y.) reported by Dymond and others (1968). Unfortunately, no MOR basalt samples from the northern end of the Juan de Fuca Ridge were available for analysis, but the good suite of seamount samples available, the very good agreement of major element chemistry of Heck Seamount with samples from the ridge in this area (Barr, 1972), and the lead isotopic character of the Heck Seamount samples suggest that this portion of the mantle may have evolved in a slightly higher ^{238}U/^{204}Pb environment than the central and southern portion of the Juan de Fuca Ridge. In light of this, examination of the variation of lead isotopes as a function of distance from the spreading center along the Heck-Explorer-Union Seamount line (Figs. 1 and 3; Table 4) is enlightening. Samples from these volcanoes appear to form a series in ^{206}Pb/^{204}Pb–^{207}Pb/^{204}Pb and ^{206}Pb/^{204}Pb–^{208}Pb/^{204}Pb space. A similar phenomenon also occurs among the samples from Dellwood Range and includes the samples from DSDP site 177A (Kulm and others, 1973). In general, there may be minor variations in ^{207}Pb/^{204}Pb space, but the larger variations are in ^{206}Pb/^{204}Pb space and are dominated by lead from the seamounts.

Discussion

Lead isotopic evolution is commonly discussed in terms of mathematical models which are thought to approximate the chemical conditions and processes in the source region of the magma. Though closed-system lead isotopic evolution cannot approximate conditions in the mantle over the age of the earth, we include such curves in Figs. 2 and 3 as a convenient frame of reference. Physically, the closed-system model assumption requires that there be no net transfer of either the parent or daughter isotopes from the time of the system's inception (assumed to be the age of the earth as inferred from meteorite studies) until the system is sampled. Many workers (e.g., Patterson, 1964; Patterson and Tatsumoto, 1964; Gast and others, 1964; Tatsumoto, 1966a) have pointed out that the closed-system model is inappropriate for modeling the growth of lead in the earth's mantle, and such a conclusion is substantiated by these new data. The observed lead isotopic array plots to the right of and below the intersection of the closed-system growth-curve—a fact which indicates a secular increase of mu (^{238}U/^{204}Pb). Initially, the mu value was less than the closed-system growth-curve value which has increased to significantly greater values in more recent times. This restricts the radiogenic buildup of ^{207}Pb/^{204}Pb early in earth history, and allows the recent radiogenic buildup of ^{206}Pb/^{204}Pb moving the array to the right of the geochron. Numerous model calculations have been made in an attempt to evaluate the effects of recent changes of the model parameters. [These include changes in the initial lead isotopic composition and the

Table 6. Constants used in the episodic multiple-stage growth model

Stage	t_n	t_{n+1}	μ	K	Pb initial values	Decay constants
1	$4.63 - 4.30$	7.2		3.7	$A_0 = 9.307$	$\lambda 238 = 0.1537 \times 10^{-9} \mathrm{yr}^{-1}$
2	$4.30 - 3.80$	7.7		3.7	$B_0 = 10.294$	$\lambda 235 = 0.9722 \times 10^{-9} \mathrm{yr}^{-1}$
3	$3.80 - 3.20$	8.3		3.7	$C_0 = 29.48$	$\lambda 232 = 0.04881 \times 10^{-9} \mathrm{yr}^{-1}$
4	$3.20 - 1.60$	$8.7 - 9.2$		3.7		$^{238}\mathrm{U}/^{235}\mathrm{U} = 137.88$
5	$1.60 - 0.0$	$9.3 - 14.0$		$3.7 - 3.4$		

age of the earth inferred from meteorite studies (Tatsumoto and others, 1973; Tilton, 1973) as well as refined measurements of the half lives of the parent isotopes involved (Jaffey and others, 1971; LeRoux and Glendenin, 1963).] The best approximation to the data array assumes episodic enrichment (increase in mu) and requires no less than five stages (see Table 6) of progressive enrichment of uranium and thorium in the system. There are numerous objections to such a model, however; the major objection is that episodic models become somewhat arbitrary when dealing with more than two stages and tend to become more of a reflection of philosophy rather than a rigorous analysis of the processes. Furthermore, an episodic system assumes instantaneous changes in the parent/ daughter isotope pairs and thus represents a static system. In light of recent geophysical data and the new global tectonics concept of plate motion (e.g., Isacks and others, 1968), a dynamic model for the growth of lead isotopes in the mantle is appropriate.

Two new concepts are important in developing a model system of lead isotope evolution which is compatible with the dynamic earth model. The first is that of continuous enrichment of the parent isotope relative to the daughter isotope (e.g., increasing the $^{238}\mathrm{U}/^{204}\mathrm{Pb}$ value) as a function of time. Such systems have been discussed in the literature (e.g., Patterson, 1964; Wasserburg, 1966; Gast, 1967; Tilton and Steiger, 1969; Sinha and Tilton, 1973) in an attempt to explain the aberration of lead model ages and radiometric ages of rocks in the earth's crust. The second concept is that of continuous mixing described by Armstrong and Hein (1973). Although we do not accept the premise that large-scale interaction of the crust and mantle throughout the earth's history has significantly affected the lead isotopes in MOR basalts and related rocks, the concept of mixing mantle subsystems as a result of convective overturn of the mantle becomes a necessary corollary for isotopic evolution in a dynamic earth. Fig. 3 presents the model array for the evolution of lead isotopes in the earth's mantle using an open-system plus mixing model (Rennick, 1973). Mixing of various subsystems will produce two very important results observed in this study. Namely, the mixing process will *necessarily* produce secondary isochrons, and it will reduce the spread of the lead isotopic array in $^{207}\mathrm{Pb}/^{204}\mathrm{Pb}$ versus $^{206}\mathrm{Pb}/^{204}\mathrm{Pb}$ space, but will increase the spread of data in $^{208}\mathrm{Pb}/^{204}\mathrm{Pb}$ versus $^{206}\mathrm{Pb}/^{204}\mathrm{Pb}$ space (Armstrong and Hein, 1973). In such a system, single-stage model ages no longer have any validity and the zero m.y. isochron will flatten to match the spread of the observed data.

In their paper on the evolution of lead and strontium isotopes, Armstrong and Hein (1973) state two deficiencies of their model. First: "regressions for 'mantle' Pb are often, but not invariably, steeper than regressions for 'crustal' Pb. This is contrary to the geologic observation that mantle-derived volcanic rocks of oceanic islands give Pb data clusters with shallower regressions than volcanic rocks in general." And second: "Our computer calculation has no detailed analog for the enrichment of ocean island volcanics in radiogenic Pb and Sr compared to abyssal volcanics." However, they suggested modifications that could produce those effects. Both objections are handled by our model of radiogenic isotopic evolution in a dynamic, open-system model *without mixing of crustal lead* into the source region for mantle-derived melts. Shallow regression lines result from mixing of a heterogeneous mantle brought about by the open-system behavior of uranium, thorium and lead. Moreover, this same process also causes chemical heterogeneity in the rubidium-strontium system and results in a range of $^{87}Sr/^{86}Sr$ values which closely matches that actually observed (i.e., from 0.7028 for the average MOR basalt, Hedge and Peterman (1970), up to 0.706 for the more alkalic basalts, Peterman and Hedge (1971)). Thus, the behavior of both radiogenic systems is adequately accounted for by the dynamic, open-system model.

Large Ion Lithophile (LIL) Element Distributions

Trace element abundances and relative variations of some of the chemically more divergent samples of MOR basalts analysed in this study are summarized in Fig. 4 and Table 7. The new REE abundances reported from samples of the basaltic sill from DSDP site 177A are the lowest yet reported from this area. The data have been selected to show the maximum variation in absolute abundance of the LIL elements and are shown in the conventional chondrite-normalized plot. The elements are grouped by valence and listed in decreasing order of ionic size. Also shown are the ionic radii of these ions in eightfold coordination (Whittaker and Muntus, 1970). Several features are noteworthy. Notice the flat REE pattern pointed out by Kay and others (1970) and the regular decrease of the alkalies and alkaline earth elements. Notice also the similar behavior of the +4 valence series. Thorium and uranium show marked depletion relative to zirconium and hafnium. All of these features can be readily explained on the basis of mantle mineralogy and crystal chemistry. The distribution coefficient data for REE in mantle mineralogy (i.e., olivine, clinopyroxene, orthopyroxene, spinel, garnet) indicate that the clinopyroxene contains most of the trace elements in the mantle minerals. Diopside, which contains both magnesium and calcium in eightfold coordination (Ca=1,20, Mg=0.97; Whittaker and Muntus, 1970), will readily accept the REE and zirconium, but does not contain a suitable lattice site for the larger ions. Thus, the source region for MOR basalts appears to have been depleted not only in alkalies and alkaline earths as postulated by Gast (1968) and Peterman and Hedge (1971), but also in thorium and uranium as shown by Tatsumoto and others (1965) and Tatsumoto (1966a). On the basis of general geochemical substitution and the data presented in Fig. 4, we would expect all ions with ionic radii 10% greater than calcium to show these effects.

Table 7. Trace element abundance data

	1a[1]	2a[1]	7f[1]	13e[1]	KD-11[1]	1154[1]	NMNH[2] 111245/95	DSDP[3] 177A 7B-2	C10d3[1]	Chondrite[4] normalization values
Cs	0.022	0.106	0.140	0.077	0.050	0.031	—	—	0.007	0.15
Rb	1.97	3.10	5.80	2.95	3.88	3.24	2.80	—	0.61	2.3
K	2000.0	2000.0	5000.0	1700.0	1400.0	1800.0	1650.0	(2490.0)	600.0	805.0
Ba	20.2	20.4	120.0	12.7	45.7	30.3	21.3	20.7	6.62	3.8
Pb	0.357	0.665	0.805	0.463	0.462	0.640	2.10	0.312	0.312	0.075
Sr	100.	75.	317.0	115.	135.	—	200.2	(200.0)	190.0	11.0
La	7.70	5.96	13.8	2.86	4.66	5.59	1.53	—	2.99	0.325
Ce	23.8	18.2	29.3	8.76	12.33	16.7	—	6.48	10.3	0.798
Nd	20.5	16.7	15.8	7.55	9.49	13.5	6.82	5.05	9.33	0.567
Sm	6.82	5.24	3.92	2.67	3.10	4.50	2.37	1.84	3.09	0.186
Eu	2.23	1.73	1.37	1.00	1.00	1.48	1.03	0.819	1.17	0.0692
Gd	8.83	6.42	4.16	3.41	3.80	—	3.7	2.68	4.07	0.255
Dy	11.1	8.65	4.63	4.72	4.98	7.52	4.25	3.15	4.54	0.305
Er	6.82	5.34	2.67	2.90	3.04	4.46	2.65	1.95	2.93	0.209
Yb	6.57	5.20	2.51	2.76	2.89	4.45	2.45	1.84	3.02	0.231
Lu	—	—	—	—	—	—	0.315	—	—	0.035
Th	0.466	0.298	0.947	0.155	0.454	0.277	0.082	0.237	0.084	0.050
U	0.267	0.139	0.445	0.061	0.152	0.197	0.032	0.058	0.049	0.013
Zr	—	154.	—	71.9	80.0	—	—	(74)	—	5.7
Hf	—	4.3	—	2.0	2.7	—	—	—	—	0.19

[1] Alkalies, alkaline earth and REE data from Kay and others (1970); U, Th, Pb, Zr and Hf from this study.

[2] All data this study; S.E.C. analyst.

[3] All data this study; REE analyses courtesy of C. Shih; K, Sr and Zr from MacLeod and Pratt (1973).

[4] U, Th, Pb values calculated from the data of this paper and that of Tatsumoto and others (1973) and Tilton (1973) assuming a terrestrial $^{238}U/^{204}Pb$ value of 8.5. For a modern common Pb having the isotopic composition of KD-9, the normalization value after correction for radiogenic growth would be 0.111 ppm. Zr and Hf normalization values from Church and others (1973); other values from Hubbard and others (1973).

[5] Abundances in glass of sample 1154: Th=0.435, U=0.185, Pb=2.08, Zr=138, Hf=3.7. Abundances in plagioclase separate from sample 1154: Th, U < 0.01, Zr < 0.1, Hf=n.d., Pb=1.40.

In his discussion of the abundance of thorium and uranium in MOR basalts, Tatsumoto (1966a) noted that the $^{232}Th/^{238}U$ values observed were generally less than 3.8, the value required by a closed-system evolution of the lead isotopes observed in the samples, and he suggested that the distribution of these elements in the magma was a function of the dual oxidation state of uranium. We have plotted the variation of thorium, uranium and lead in Fig. 5 in an attempt to correlate the recent variations in lead isotopic composition with the parent isotopes. Although the scatter is large, several points can be made. With respect to the closed-system model parameters shown in Fig. 5, most samples have mu values ($^{238}U/^{204}Pb$) which are greater than that required by the lead isotopic values observed in the samples. However, kappa ($^{232}Th/^{238}U$) is almost universally lower than the model value of 3.8. The $^{232}Th/^{204}Pb$ values scatter both above and below the model values required by their lead isotopic values. Analysis of both the chemical and isotopic data suggest an anomalous behavior of thorium

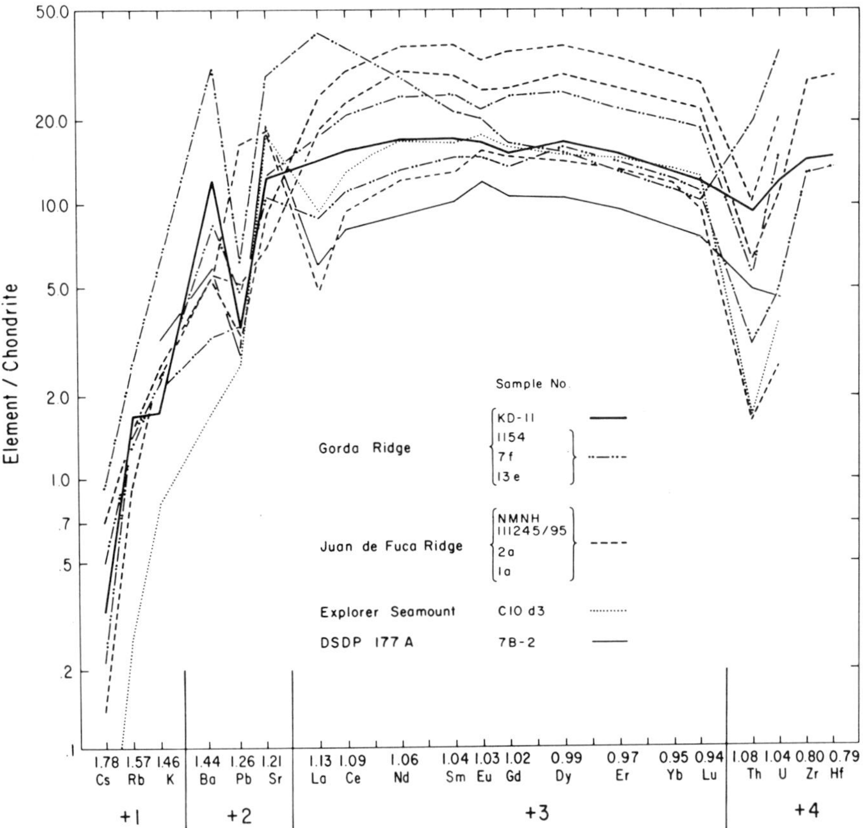

Fig. 4. Comparisons of the abundances of uranium, thorium and lead relative to the REE, alkalies, alkaline earths and zirconium. Note that these elements, like the alkalies and large alkaline earth elements show depletion relative to the REE and zirconium. Data from Table 7

or an enrichment of uranium with respect to ^{204}Pb during the final partial melting event. Because of a lack of partition coefficient data on appropriate phases, namely spinel, chromite and magnetite, we are unable to specify the phase responsible for the behavior observed. Comparisons of the behavior of these elements with other incompatible trace elements (Fig. 4) indicate a lack of a europium

Fig. 5. Comparison between abundance and isotopic data for uranium, thorium and lead. Relative to closed-system model parameters, uranium shows pronounced enrichment over that required by the lead isotopic data whereas thorium does not. However, many samples also contain adequate thorium to account for the observed lead isotopic values. Samples designated as in Figs. 2 and 3

50
40
30
20
10
238U/204Pb
u = 9.0
E
P
D
U
C
7
E
E
H
C
(a)
18 0
18.5
19 0
19.5
206Pb/204Pb

140
130
120
110
100
90
80
70
60
50
40
30
20
10
232Th/204Pb
(K·u) = 34.2
U
D
7
P
C
C
E
E
H
(b)
18 0
18.5
19 0
19.5
206Pb/204Pb

5.0
4 0
3.0
2.0
1.0
0
232Th/238U
K = 3.8
U
7
B
D
C
H
C
E
P
E
E
(c)
18 0
18.5
19.0
19.5
206Pb/204Pb

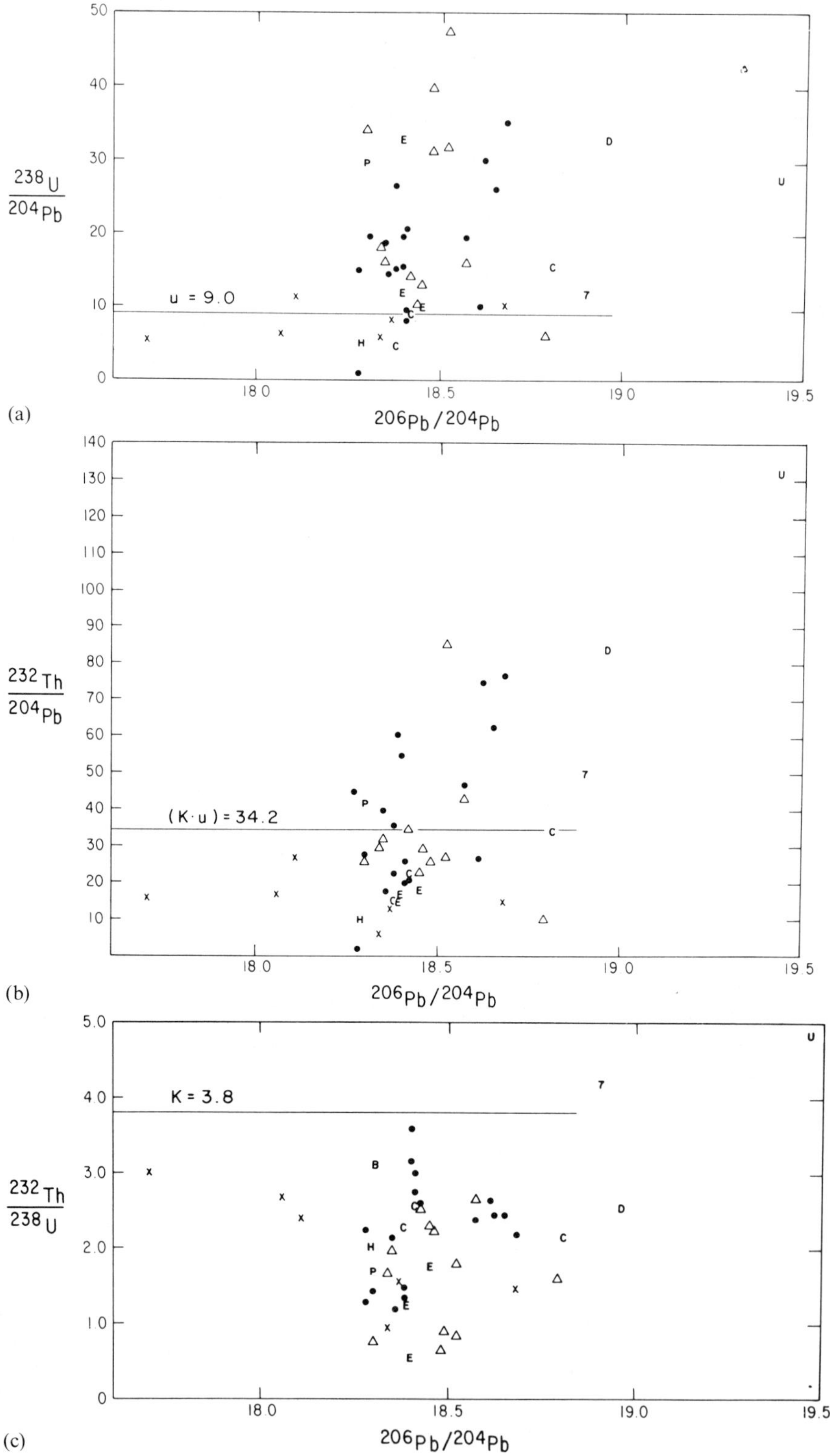

342

anomaly in the REE patterns and is evidence that a calcium-bearing phase, such as plagioclase (see footnote, Table 8) or apatite, has not caused preferential depletion of lead from the melt.

Chemical and Isopotic Evolution of the Mantle

Any model of chemical and isotopic evolution of the earth's mantle must be constrained by the following geochemical observations. (1) The mantle is isotopically heterogeneous on a local scale for both lead and strontium. (2) Tholeiitic basaltic melts, i.e. those thought to come from shallow depths in the mantle as a result of large scale partial melting (Gast, 1968; Kay and others, 1970), contain the least radiogenic lead and strontium. (3) There is a relative decrease in the abundance of trace elements of lithophilic affinity as a function of size as soon as the ionic size of the element significantly exceeds that of calcium. Thus, K, Rb, Pb, Ba, La, U, and Th relative abundances indicate depletion in the mantle source rocks. (4) Even though uranium and lead have undergone relative depletion in the mantle as a function of time, the lead isotopic data require a second mechanism which enriches uranium relative to lead on a global scale.

Our model for the chemical and isotopic evolution of the earth's mantle is a heuristic one. The solution presented here is not unique, but depends strongly on the values chosen for the partition coefficients of the elements involved, the amount of isotopic re-equilibration allowed, and the *relative* transport rates of various elements through the system. For purposes of discussion, the model presented in Fig. 3 represents the average of 100 residual (clinopyroxene) and 100 radiogenic (interstitial phase) systems where the radiogenic system is defined arbitrarily as a 1% partial melt. Partitioning of the elements is done using the equations of Shaw (1970). Uranium and thorium are allowed to diffuse into the mantle at a gradually decreasing rate of 0.15% of the original parent concentration (corrected for decay)/50 m.y. relative to a constant amount of ^{204}Pb. This results in a slowly increasing mu ($^{238}U/^{204}Pb$). We note also that the kappa ($^{232}Th/^{238}U$) required to fit the data is somewhat less than 3.8 (ca. 3.4), near that observed for KD-9. While we cannot yet specify a unique solution for the migration of these elements through the mantle, the general features of this model will not be altered as long as the enrichment of uranium, thorium and rubidium into the melt phase is more rapid than that of lead and strontium. This is substantiated by available partition coefficient data and ion substitution theory. We envision the process of tectonic flow within the mantle to be responsible for producing a "residual" and an "interstitial or radiogenic" phase in the mantle and suggest that the observed isotopic data arrays are the result of random mixing as a direct consequence of mantle convection.

The most significant question to answer concerning the application of this model to the earth's mantle is what is the scale of isotopic heterogeneity within the mantle? Does it exist on a grain sized basis as well as on a regional and structural scale? To answer this question, a brief look at the behavior of ions during partial melting is in order. In an attempt to model magma genesis, Gast

(1968) presented a partial melting formulation based on Rayleigh distribution of a dilute solute (i.e., it must obey Henry's Law) between several phases. This model was based on empirical data and assumed that volume (or lattice) diffusion was the dominant mechanism of atomic migration. No other mechanisms of trace element extraction from the host crystals were considered nor was any attempt made to evaluate the possible effect of incompatible ion concentrations along grain boundaries. Since Gast's (1968) work, our understanding of the mantle has greatly increased and the importance of lattice diffusion as an agent for mass transfer in the mantle must now be seriously questioned. Stocker and Ashby (1973) recently reviewed the data available on the rhelogical behavior of olivine. From these data, they conclude that the process of dislocation creep is the dominant mechanism of mass transfer in the upper mantle, being far more important than Nabarro-Herring creep (lattice diffusion) or Cobble creep (grain boundary diffusion). Both mechanisms will allow foreign ions within the mantle to be transfered at a much more rapid rate (5–10 $\times$) than simple lattice diffusion would allow.

How does this process bear on the problem of mantle heterogeneity? As we have pointed out above, MOR basaltic melts appear to have formed from a previously depleted reservoir. They contain low concentrations of the larger incompatible elements and they also contain less radiogenic lead and strontium than their alkalic-rich counterparts. These chemical features can be explained by our dynamic, open-system model if an interstitial, low-melting phase exists in the mantle which forms as a direct result of the tectonic stresses within the mantle. The slow, continuous extraction of the radioactive parent atoms (U, Th, Rb) relative to the daughter products (Pb and Sr) produces the radiogenic sink necessary to form the lead isotopic array. The linear array of the lead isotopic data then becomes the result of mixing during partial melting and the radiogenic character of any partial melt from the upper mantle is a random sampling, statistical problem. The position of the melt along the array depends on the past history of the source area and the degree of partial melting. Increasing the percentage of partial melting will serve only to dilute the concentrations and isotopic compositions of these elements. Finally, although a number of workers have made cases for a general correlation of alkalies and radiogenic isotopes (summarized above), there is no reason to require a perfect correlation as not all of these low-melting, interstitial phases would be equally enriched in all parent or daughter elements or alkalies.

Is there evidence that such a phase exists within the mantle? Haines and Zartman (1973) and Zartman and Tera (1973) have published data from a suite of mantle nodules which contain most of their uranium, thorium and lead in interstitial areas or along grain boundaries within the nodule. However, most petrologists favor the interpretation that these nodules are cognate xenoliths, picked up by the melt significantly above the zone of magma genesis, rather than within that zone. Whether these interstitial sites could exist at depth within the mantle cannot yet be ascertained.

Interpretation of the mantle structure and chemistry in light of this model may yield some new insight into mantle processes. Gast (1968) summarized the evidence for the spatial relationship of alkali and MOR basalts in the ridge environment based on the thermal model of Oxburgh and Turcotte (1968). He suggested that the alkali basaltic melts were produced at greater depths within the upwelling mantle. Note the lead isotopic data presented here for volcanoes which lie along the Heck-Explorer-Union Seamount line shown in Fig. 1; these data, shown in Fig. 3, show a direct correlation as a function of geographic distance from the ridge. However, how does one explain the spread in the isotopic data from a single volcano, for example Explorer or Cobb seamounts? The simplest method is by the two component mixing model we have presented here. There would be a complete mixing continuum between the alkalic basalts which are made up of only a very small percentage of partial melting of this low melting, interstitial phase and the more voluminous tholeiitic melts of the seamounts (see Fig. 12, Gast, 1968). This same process could also be involved in producing the residual mantle in the upwelling plume beneath ocean ridges by tapping off the low-melting, early-formed liquids which would be rich in incompatible and radiogenic components prior to the formation of the MOR basaltic melts at shallower levels beneath the ridge. The low velocity zone, which contains a few percent partial melt (e.g. Green, 1971) is interpreted as the decoupling zone between the oceanic crust and the convecting mantle below.

One important feature of our model has not yet been explained, namely the source of uranium and thorium for the enrichment mechanisms required by the lead isotopic data. The depth of convection within the mantle is as subjective as convection itself. However, if convection exists within the mantle (a premise we accept), most workers accept convection within the entire upper mantle (i.e. to depths of 700 km). Whether convection extends to greater depths is not important to our discussion. Two highly important geochemical phase changes take place at the 20°-discontinuity (about 350 km, Ringwood, 1972): olivine transforms to spinel, but geochemically more important, pyroxene transforms to a garnet structure. If comparisons of known element partition data on these minerals are any indication, strong exclusion of LIL elements must occur, particularly for the pyroxene to garnet transformation. This process would also result in smaller grain size (Gordon, 1971), enhancing the grain boundary transfer mechanisms discussed above. Evidence that convection does take place across this boundary comes from the lack of any compressive seismic signal as the descending slab passes through this discontinuity (Isacks and others, 1968). Cycling mantle material through this phase change would appear to be a highly efficient mechanism for extracting LIL elements from the mantle mineralogy, concentrating them in an interstitial phase. We appeal to this process for the open-system character four model, allowing a slow increase of mu as a function of time as required by the lead isotopic data.

Kay and Gast (1973) present trace element data from several alkali basalt centers and suggest that these melts are formed by a very small degree of partial melting. If our dynamic, open-system model is correct, two changes in their model are required. First, if our interstitial phase exists in the mantle and the spatial, isotopic, geochemical and crystal chemistry considerations discussed above seem to rule out any known major mantle mineral phase as a source of the radiogenic daughter and parent ions, then this interstitial phase has not been included in the model calculations carried out by Kay and Gast (1973). Second, if convection

345

across the 20°-discontinuity occurs, the fractionation of the larger REE relative to the smaller REE would alter the flat chondritic-normal distribution of the REE in the source region for alkali basalts, a basic assumption in the partial melting calculations. Tectonic stresses may act on these incipient-melt "droplets" as they migrate through the mantle causing them to coalesce, giving rise eventually to diapirs of alkalic basalt. Thus, the tectonic mechanisms may provide an explanation for the extraction of a very small percent partial melt. These would give rise to the intraplate basalts and could explain both the geophysical and geochemical characteristics ascribed to mantle plumes.

Conclusions

1. Mantle is isotopically heterogeneous and the lead isotopic data form linear arrays. The average slope of this array corresponds to a secondary isochron age of 1.75 ± 0.1 b.y.

2. Real geographic variations in $^{206}Pb/^{204}Pb$ versus $^{207}Pb/^{204}Pb$ and $^{206}Pb/^{204}Pb$ versus $^{208}Pb/^{204}Pb$ space do occur parallel to the direction of spreading (perpendicular to the strike of the ridge). Small variations in $^{207}Pb/^{204}Pb$ values at a given $^{206}Pb/^{204}Pb$ value may occur perpendicular to the direction of spreading, but the magnitude of the spread in these values is not significant in terms of the analytical uncertainly ($\pm 0.1\%$).

3. Trace-element distribution data on these rocks suggest that the low $^{232}Th/^{238}U$ values observed in MOR basalts could be simply a function of the final separation process and need not wholly reflect a source-region characteristic, although other mathematical solutions are possible. Uranium shows stronger preferential enrichment in the melt than does thorium relative to ^{204}Pb. Adequate thorium and uranium are present in most samples to support the observed lead isotopic abundances.

4. The regular decrease in the abundance of the alkalies, alkaline earths, La, Th and U relative to the REE and Zr are a direct function of ionic size. Based on these data, the source material for MOR basaltic melts should show relative depletion of all ions with ionic radii 10% greater than that of calcium as a function of increasing ionic radius. Recent rheological data on olivine raise serious question about the assumption that lattice diffusion is the dominant mechanism of ion migration in the mantle.

5. We propose a model for the isotopic evolution of lead in the earth's mantle which does not require recycling of crustal lead to account for the observed lead isotopic arrays from MOR basaltic rocks and associated seamounts. Our dynamic, open-system model assumes that both mixing and parent/daughter fractionation are a direct consequence of convective overturn of the mantle. The observed spread of both the lead and strontium isotopic data can be explained by the model. Essential features of this model are the vertical transport mechanisms of LIL element-rich, low-melting fractions and mixing of enriched crystal residua systems during magma generation. This model for mantle evolution will explain the chemical and isotopic data and the spatial relationships of alkali-rich versus MOR basaltic melts in the oceanic environment.

Acknowledgments. We thank P.W. Gast, R.W. Kay, W.G. Melson, T.E. Simkin, D.A. McManus, D.Z. Piper, the National Museum of Natural History and the Deep Sea Drilling Program, funded by the National Science Foundation, for generously supplying samples for this study. The analytical work carried out by SEC was generously supported by the National Research Council Associateship program at the Johnson Space Center. Special thanks are due the late P.W. Gast and Robin Brett for their support of this project. Numerous discussions about the project with P.W. Gast, H. Wiesmann, B.M. Bansal and L.E. Nyquist are appreciated. Unfortunately, P.W. Gast did not see the results of this study; undoubtedly this manuscript would have benefited significantly from his insight into this problem. We also thank A. Meijer and C. Shih for help with some of the supporting analytical work. Finally, we thank R.L. Armstrong, J.M. Mattinson, G.R. Tilton, C.E. Hedge and B.R. Doe for reviewing earlier versions of this manuscript. Numerous improvements have resulted. Thanks are also extended to the Dept. of Geological Sciences, University of California, Santa Barbara, California for supplying the secretarial and drafting help. Computer calculations were carried out with a grant from the Academic Senate of the University of California, Santa Barbara and the generous help and advice of Mr. W.L. Rennick.

[*Editor's Note:* The Appendix has been omitted.]

Table 8. Sample descriptions and localities

Lab sample designation	Official sample designation	Location		Depth (m.)	Source of data	General petrographic information
		Lat. (N)	Long. (W)			
Gorda Ridge						
F-33	FAN B33D	40°19′	127°59′	3925	1	Central part of sample used; some slight limonite staining visible; no phenocrysts visible in hand specimen.
4z	4z	41°24.0′	127°28.9′	2600	2	Fine-grained, fresh basalt.
5a	5a	41°24′	127°28.9′	2600	2	Sample altered along fractures; plagioclase shows FeO staining.
7f, 7t	7f, 7t	41°23.6′	127°28.2′	2500	2	Fine-grained basalt; sample 7f incipiently altered.
11a	11a	41°15.6′	127°28.2′	2500	2	Very fine-grained basalt.
13a, 13e	13a, 13e	41°15.7′	127°28.2′	2500	2	Sample 13a is glassy basalt containing phenocrystic plagioclase; sample 13e is similar, but shows considerable pervasive FeO alteration.
1154	1154	42°50′	126°55′	2200–3100	2	Glassy basalt containing ∼5% plagioclase phenocrysts.
KD-9	KD-9	42°50′	126°55′	2200–3100	2	Basaltic glass from same locality as KD-11.
KD-11	KD-11	42°50′	126°55′	2200–3100	2	Pb isotopic work reported by Sun (1973)
245	NMNH 111245/95	43°05′	127°30′	2980–3230	3	Greenstone with no identifiable phenocrysts; surface coated with thick Mn-Fe oxide rind; see Church (1973) for Pb isotopic data on these rinds.
Juan de Fuca Ridge						
1a	1a	44°20′	129°55′	4000	2	Subophitic, fine-grained basalt.
2a	2a	44°35.8′	130°18.8′	2500	2	Groundmass slightly altered.
3a, 3e	3a, 3e	44°44.2′	129°21′	2562	2	Very fine-grained glassy basalt.
234	NMNH 111234/280	44°20′	129°5″	4090–4392	3	Fresh, fine-grained basalt; some microphenocrysts of plagioclase.
236	NMNH 11236/103	44°20′	129°55′	4090–4392	3	Greenstone, not so altered as NMNH 111245/95.
237	NMNH 111237/8	44°25′	129°55′	3084	3	Sample from rock which shows some limonite staining, altered ferromagnesian minerals with blue zeolites in vesicles.

Table 8. Continued

Lab sample designation	Official sample designation	Location Lat. (N)	Long. (W)	Depth (m.)	Source of data	General petrographic information
240	NMNH 111240/18	44°40'	130°20'	2195–2200	3	Central part of glassy cobble; no phenocrysts visible.
243	NMNH 111243/4	44°20'	129°45'	3445	3	Glassy basalt with plagioclase phenocrysts.
W-11	TT029 DH011-10	46°21.8'	129°58.4'	~2600	4	Sample contains ~10% ferromagnesian phenocrysts, seemingly unaltered; appears atypical.
W-12	TT029 DH012	46°07.5'	130°00'	2750	4	Sample is a very fine-grained basalt with small plagioclase laths (~5%) visible; some incipient alteration visible.
W-14	TT029 DH014	46°05.4'	129°55'	2200	5	Sample is a very fresh basalt, somewhat vuggy and contains unaltered plagioclase phenocrysts. Some greenish coatings on inside of vugs and along fracture in sample.
W-15	TT029 DH015-1	46°16'	129°36'	2650	4	Central portion of dense, glassy basaltic pillow used; contains phenocrysts of plagioclase.
Cobb Seamount						
C-1	Cobb Pinnacle	46°46'	130°50'	35–40	5	Sample from Cobb Seamount pinnacle; see Dymond and others (1968) for petrographic data.
C-2	NMNH 113032-2	46°46'	130°50'	405	6	See Budinger (1967), Fig. 6J. Sample from dredge haul D-23, shows some incipient alteration.
C-3	NMNH 113032-1	46°46'	130°50'	405	6	See Budinger (1967), Fig. 6J. Sample from dredge haul D-23, shows some incipient alteration.
Dellwood Range						
D-1	En-70-25-7	50°17.8'	130°26.0'	~2000	7	Fresh basalt with ~3% olivine phenocrysts; shows limonite concentration in vugs, ferromagnesian minerals slightly affected.
D-2	En-70-25-9	50°36.0'	130°45.5'	1550–1770	7	Most-altered seamount sample, groundmass significantly altered, only a few scattered phenocrysts of plagioclase identifiable. Zeolites and limonite present in vesicles.
D-3	En-70-25-16	50°13.5'	130°15.0'	1600–2000	7	Freshest sample from Dellwood Range; no limonite or ferromagnesian alteration seen, few scattered phenocrysts of plagioclase present.

Explorer Seamount						
C10-d3	C10-d3	49°03.5′	130°57′	2200–3100	2	Fine-grained vesicular basalt, probably the same sample analysed by Kay and others (1970).
E-1, E-2	C10d3	49°03.5′	130°57.0′	1454	2	Samples selected from same dredge haul as C10-d3; both appeared fresh in hand specimen.
E-3	TT-063 DH035	58°59.4′	130°59.1′	1500	5	Fresh, fine-grained with fresh olivine phenocrysts present.
E-4	TT-063 DH029	49°05.0′	130°57.0′	1000–1500	5	Fresh, fine-grained basalt; shows slight alteration of ferromagnesian minerals.
E-5	TT-063 DH034	48°57.8′	131°02.0′	2300	5	Fresh, fine-grained basalt with about 3% fresh olivine phenocrysts; some limonite staining in vesicles.
Heck Seamount						
H-1	TT-053 DH010	48°16.0′	129°45.5′	2380	5	Sample is fine-grained basalt and has scarce microphenocrysts of plagioclase; ferromagnesian minerals somewhat altered.
H-2	TT-053 DH012	48°27.6′	130°09.9′	2740	5	Sample is fine-grained basalt, indistinguishable from previous sample except that the groundmass is slightly more turgid.
Hodgkins Bank						
B-1	MUK B29D	53°15′	136°46′	230	1	Finely vesicular basalt fragment containing some limonite alteration in vesicles.
B-2	MUK B30D	53°15′	136°46′	90	1	Fresh vesicular basalt with microphenocrysts of plagioclase; minor limonite staining seen in vesicles.
Parks Seamount						
P-227	NMNH 111227/5	44°15′	129°50′	2480–2655	3	Sample coated with thick (up to 1 cm) Mn rind; interior relatively unaltered having unaffected olivine phenocrysts; see Church (1973).
Union Seamount						
U-1	TT063 DH036	49°32.7′	132°44.0′	700	5	Sample is a vesicular basalt containing large phenocrysts of plagioclase; vesicles contain limonite which was easily removed during cleaning.

Table 8. Continued

Lab sample designation	Official sample designation	Location		Depth (m.)	Source of data	General petrographic information
		Lat. (N)	Long. (W)			
DSDP, Leg 18, Site 177A						
7B-1	DSDP-18-177A-24-1 (126–128 cm)	50°28.18′	130°12.30′	2006 to ocean floor	8	Sill encountered at 378–385 m depth. Sample 7B-1 contained some carbonate, but all were fairly fresh. See MacLeod and Pratt (1973) for further, more detailed petrographic data. $SiO_2 = 49.95$, $TiO_2 = 1.02$, $Al_2O_3 = 16.09$, $FeO = 7.82$, $MnO = 0.11$, $MgO = 6.88$, $CaO = 12.92$, $Na_2O = 2.10$, $K_2O = 0.02$, $H_2O^+ = 1.49$, $H_2O^- = 1.19$, $P_2O_5 = 0.10/\Sigma = 99.69$.
7B-2	DSDP-18-177A-25-2 (30–40 cm)	50°28.18′	130°12.30′		8	
7B-3	DSDP-18-177A-25-2 (108–110 cm)	50°28.18′	130°12.30′		8	

1. Samples from Scripps Institution of Oceanography, University of California, San Diego.
2. Samples furnished by R. Kay; chemical and additional petrographic data available in Kay and others (1970).
3. Samples from National Museum of Natural History, Smithsonian Institution collected and described by Melson (1969). Further work is planned on these samples. The chemistry of two samples from this collection reported in Kay and others (1970).
4. Samples supplied by D.A. McManus, Dept. of Oceanography, University of Washington, Seattle, Wash.
5. Samples supplied by D.Z. Piper and J. Kummer, Dept. of Oceanography, University of Washington, Seattle, Wash., who are working on a chemical and petrographic study of samples from the seamounts from the N.E. Pacific Ocean.
6. Samples collected and described by Budinger (1967). Samples now located at the National Museum of Natural History, Smithsonian Institution and were supplied through the courtesy of T.E. Simkin, who plans further petrographic work.
7. Samples from R.L. Chase, Institution of Oceanography, University of British Columbia, Vancouver, Canada through D.Z. Piper. Additional regional and chemical data available in Bertrand (1972) and Barr (1972).
8. Samples supplied through the assistance of the National Science Foundation. Major elements (7B-2) analysed by A. Meijer.

References

Armstrong, R.L., Hein, S.M.: Computer simulation of Pb and Sr isotope evolution of the earth's crust and upper mantle. Geochim. Cosmochim. Acta 37, 1–18 (1973)

Barnes, I.L., Murphy, T.J., Gramlich, J.W., Shields, W.R.: Lead separation by anodic deposition and isotope ratio mass spectrometry of microgram and smaller samples. Anal. Chem. 45, 1881–1884 (1973)

Barr, S.M.: Geology of the northern end of Juan de Fuca ridge and adjacent continental slope. Unpublished Ph.D. dissertation, Univ. of British Columbia, 287 p. (1972)

Bertrand, W.G.: A geological reconnaissance of the Dellwood Seamount area, northeast Pacific Ocean, and its relationship to plate tectonics. Unpublished M.Sc. thesis, Univ. of British Columbia, 151 p. (1972)

Budinger, T.F.: Cobb Seamount. Deep-Sea Res. 14, 191–201 (1967)

Catanzaro, E.J.: Absolute isotopic abundance ratios of three common lead reference samples. Earth Planet. Sci. Lett. 3, 343–346 (1967)

Catanzaro, E.J., Murphy, T.J., Shields, W.R., Garner, E.L.: Absolute isotopic abundance ratios of common, equal-atom, and radiogenic lead isotopic standards. J. Res. Natl. Bur. Std. 72A, 261–267 (1968)

Church, S.E.: Trace element abundances in ocean sediments and altered basalts — Their implications for andesite genesis (abst.). Trans. Am. Geophys. Union 53, 546 (1972)

Church, S.E.: Limits of sediment involvement in the genesis of orogenic volcanic rocks. Contrib. Mineral. Petrol. 39, 17–32 (1973)

Church, S.E., Bansal, B.M., Wiesmann, H.: The distribution of K, Ti, Zr, U and Hf in Apollo 14 and 15 materials. In: The Apollo 15 Lunar Samples, J.W. Chamberlain, C. Watkins, eds., p. 210–213. Houston: Lunar Science Inst. 1973

Church, S.E., Tilton, G.R.: Lead and strontium studies in the Cascade Mountains: Bearing on andesite genesis. Geol. Soc. Am. Bull. 84, 431–453 (1973)

Compston, W., Oversby, V.M.: Lead isotopic analysis using a double-spike. J. Geophys. Res. 74, 4338–4348 (1969)

Dymond, J.R., Watkins, N.D., Nayudu, Y.R.: Age of Cobb seamount. J. Geophys. Res. 73, 3977–3979 (1968)

Gast, P.W.: Isotopic geochemistry of volcanic rocks. In: Basalts: The Poldervaart treatise on rocks of basaltic composition 1, H.H. Hess, ed., p. 325–358. New York: John Wiley & Sons — Interscience 1967

Gast, P.W.: Trace element fractionation and the origin of tholeiitic and alkaline magma types. Geochim. Cosmochim. Acta 32, 1057–1086 (1968)

Gast, P.W., Tilton G.R., Hedge, C.E.: Isotopic composition of lead and strontium from Ascension and Gough Islands. Science 145, 1181–1185 (1964)

Gordon, R.B.: Observation of crystal plasticity under high pressure with application to the earth's mantle. J. Geophys. Res. 76, 1248–1254 (1971)

Green, D.H.: Composition of basaltic magmas as indicators of conditions of origin: application to oceanic volcanism. Phil. Trans. Roy. Soc. London, Ser. A **268**, 707–725 (1971)

Green, D.H., Ringwood, A.E.: The genesis of basaltic magmas. Contrib. Mineral. Petrol. **15**, 103–190 (1967)

Haines, E.L., Zartman, R.E.: Uranium concentration and distribution in six peridotite inclusions of probable mantle origin. Earth Planet. Sci. Lett. **20**, 45–53 (1973)

Hart, S.R.: K, Rb, Cs contents and K/Rb, K/Cs ratios of fresh and altered submarine basalts. Earth Planet. Sci. Lett. **6**, 293–303 (1969)

Hedge, C.E., Peterman, Z.E.: The strontium isotopic composition of basalts from the Gorda and Juan de Fuca Rises, northeastern Pacific Ocean. Contrib. Mineral. Petrol. **27**, 114–120 (1970)

Hubbard, N.J., Gast, P.W., Rhodes, J.M., Bansal, B.M., Weismann, H., Church, S.E.: Nonmare basalts: Part II. In: Proceedings of the Third Lunar Science Conference 2, D. Heymann, ed., p. 1161–1174. Cambridge: M.I.T. Press 1973

Isacks, B., Oliver, J., Sykes, L.R.: Seismology and the new global tectonics. J. Geophys. Res. **73**, 5855–5900 (1968)

Jaffey, A.H., Flynn, K.F., Glendenin, L.E., Bentley, W.C., Essling, A.M.: Precision measurement of the half-lives and specific activities of U^{235} and U^{238}. Phys. Rev. **C4**, 1889–1906 (1971)

Kay, R.W., Gast, P.W.: The rare earth content and origin of alkali-rich basalts. J. Geol. **81**, 653–682 (1973)

Kay, R., Hubbard, N.J., Gast, P.W.: Chemical characteristics and origin of oceanic ridge volcanic rocks. J. Geophys. Res. **75**, 1585–1614 (1970)

Kulm, L.D., Huene, R. von, Duncan, J.R., Ingle, J.C., Kling, S.A., Musich, L.F., Piper, D.J.W., Pratt, R.M., Schrader, H., Weser, O., Wise, S.W., Jr.: Site 177. In: Initial reports of the deep sea drilling project, XVIII, p. 233–285. Washington: U.S. Government Printing Office 1973

LeRoux, L.J., Glendenin, L.E.: Half-life of Th^{232}. Proc. Nat'l. Mtg. on Nucl. Energy, Pretoria, South Africa, 78 (1963)

MacLeod, N.S., Pratt, R.M.: Petrology of volcanic rocks recovered on Leg 18. In: Initial reports of the deep sea drilling project, XVIII, p. 935–945. Washington: U.S. Government Printing Office 1973

Mattinson, J.M.: Preparation of hydrochloric and nitric acids at ultralow lead levels. Anal. Chem. **44**, 1715–1716 (1972)

McManus, D.A.: Major bathymetric features near the coast of Oregon, Washington and Vancouver Island. Northwest Sci. **38**, 65–82 (1964)

Melson W.G.: Preliminary results of a geophysical study of portions of the Juan de Fuca ridge and Blanco fracture zone, 1–33. U.S. Dept. of Commerce ESSA Technical Memorandum C & GSTM 6. Springfield, Va.: Natl. Technical Information Service 1969

Oxburgh, E.R., Turcotte, D.L.: Mid-ocean ridges and geotherm distribution during mantle convection. J. Geophys. Res. **73**, 2643–2662 (1968)

Patterson, C.C.: Characteristics of lead isotope evolution on a continental scale in the earth. In: Isotopic and cosmic chemistry, 19, p. 244–268. Amsterdam: North-Holland 1964

Patterson, C.C., Tatsumoto, M.: The significance of lead isotopes in detrital feldspar with respect to chemical differentiation within the earth's mantle. Geochim. Cosmochim. Acta **28**, 1–22 (1964)

Peterman, Z.E., Hedge, C.E.: Related strontium isotopic and chemical variations in oceanic basalts. Geol. Soc. Am. Bull **82**, 493–500 (1971)

Rennick, W.L.: A model of lead isotope evolution in the earth's crust and upper mantle. Unpublished M.A. thesis, Univ. of California, Santa Barbara, 93 p. (1973)

Ringwood, A.E.: Mineralogy of the deep mantle: Current status and future developments. In: The nature of the solid earth, E.C. Robertson, ed., p. 67–92. New York: McGraw-Hill 1972

Shaw, D.M.: Trace element fractionation during anatexis. Geochim. Cosmochim. Acta **34**, 237–243 (1970)

Shields, W.R., ed.: Analytical mass spectrometry section: Summary of activities July 1969 to June 1970. U.S. Dept. of Commerce N.B.S. Technical Note 546, 120 p. Washington: U.S. Government Printing Office 1970

Sinha, A.K., Tilton, G.R.: Isotopic evolution of common lead. Geochim. Cosmochim. Acta **37**, 1823–1849 (1973)

Stocker, R.L., Ashby, F.M.: On the rheology of the upper mantle. Rev. Geophys. **11**, 391–426 (1973)

Sun, S.S.: Lead isotope studies of young volcanic rocks from oceanic, islands, mid-ocean ridges and island arcs. Unpublished Ph.D. dissertation, Columbia University, 139 p. (1973)

Tatsumoto, M.: Genetic relations of oceanic basalts as indicated by lead isotopes. Science **153**, 1094–1101 (1966a)

Tatsumoto, M.: Isotopic composition of lead in volcanic rocks from Hawaii, Iwo Jima, and Japan. J. Geophys. Res. **71**, 1721–1733 (1966b)

Tatsumoto, M., Hedge, C.E., Engel, A.E.J.: Potassium, rubidium, strontium, thorium, uranium and the ratio of strontium-87 to strontium-86 in oceanic tholeiitic basalts. Science **150**, 886–888 (1965)

Tatsumoto, M., Knight, R.J., Allegre, C.J.: Time differences in the formation of meteorites as determined from the ratio of lead-207 to lead-206. Science **180**, 1279–1283 (1973)

Tilton, G.R.: Isotopic lead ages of chondritic meteorites. Earth Planet. Sci. Lett. **19**, 321–329 (1973)

Tilton, G.R., Steiger, R.H.: Mineral ages and isotopic composition of primary lead at Manitouwadge, Ontario. J. Geophy. Res. **74**, 2118–2132 (1969)

Wasserburg, G.J.: Geochronology and isotopic data bearing on development of the continental crust. In: Advances in earth science, p 431–459. Cambridge: MIT Press 1966

Whittaker, E.J.W., Muntus, R.: Ionic radii for use in geochemistry. Geochim. Cosmochim. Acta **34**, 945–956 (1970)

Zartman, R.E., Tera, F.: Lead concentration and isotopic compositions in five peridotite inclusions of probable mantle origin. Earth Planet. Sci. Lett. **20**, 54–66 (1973)

Accepted August 28, 1975

26

Reprinted from *Canadian Jour. Earth Sci.* **9**:1318-1324 (1972)

Hairpins and Superintervals

E. IRVING AND J. K. PARK

Earth Physics Branch, Department of Energy, Mines and Resources, Ottawa, Canada

Received May 1, 1972
Accepted for publication May 18, 1972

The path of apparent polar wandering relative to North America has several sharp turning points or hairpins. These hairpins record rapid changes in the direction of motion of North America relative to the pole. The youngest of these hairpins (in the Cretaceous) corresponds approximately in time to a major "tectonic reorganization" suggested by Coney on other grounds. The older hairpins are also supposed to represent first-order tectonic breaks, and they are therefore used to divide post-Archean time into 5 superintervals. The diastrophic significance of these superintervals is briefly discussed.

Introduction

Chronologies of polarity and magnetic intervals[1] which have durations of 10^6 or 10^7 years are now being established for the past few hundred million years. In this paper a method of subdividing time into much longer units (called 'superintervals') is introduced. Chronologies of polarity and magnetic intervals are based on observations of the *polarity* of remanent magnetization, and since reversals of the geomagnetic field are a global phenomenon, reflecting changes in the core of the earth, they provide time scales applicable to the entire earth. The chronology of superintervals described below is based on observation of the *directions* of remanent magnetization. Directions of magnetization are expressed as corresponding poles, and variations in direction as

paths of apparent polar-wandering. Sharp turning points (referred to here as 'hairpins') occur in these paths, and it is the ages of these hairpins that are used to define the limits of superintervals.

The polar path described the horizontal movements of a tectonic unit (or lithospheric plate) relative to the pole, and the hairpins therefore correspond to periodic changes in the direction of these movements. If such changes of motion affect each plate independently, then the time at which hairpins occur will differ from plate to plate, and the chronologies of super-intervals will have a regional, but not a global, application. If, however, the periodic reorganizations of plate motions are a global phenomenon (and this would be the expectation from plate theory) the *time* at which changes occur (although their form may differ) will be essentially identical; that is, the chronology of superintervals will be global in its applications. There are therefore two parts to this discussion: firstly the description of such chronologies for each continental shield, and secondly the comparison of these chronologies from shield to shield. In this paper we describe, as a beginning, a *very* tentative chronology of super-intervals for the Canadian Shield. The comparison with results from other shields will be made later.

---

[1]A polarity interval is a length of time during which (aside from very short-lived reversals) the geomagnetic field had constant polarity; generally they have durations of millions of years or less. A magnetic interval typically has a duration of several tens of millions of years, and is characterized by a certain reversal frequency; the distinction made is between lengths of time during which the field reversed frequently and the polarity intervals were short, and lengths of time during which the field rarely reversed and the polarity intervals were longer. (See Fig. 4.)

Chronology of the Canadian Shield

The chronologies of Precambrian time in Canada have developed from a knowledge of the structure and radiometric ages of the Canadian Shield. About 30 years ago M. E. Wilson (1941) recognized various geological provinces. Later, Gill (1949) developed the idea of structural provinces which he thought of as regions in which the structural trends, and hence presumably the tectonic stresses operative during crustal stabilization, were approximately uniform in direction. At the same time J. T. Wilson (1949) noted that, in addition to uniformity of structure, each province had characteristic radiometric ages. Making use of developments in mapping, and especially of the age determinations of R. K. Wanless and colleagues, Stockwell (1963, 1970) refined this subdivision into structural provinces, and calculated their mean radiometric (based mostly on K–Ar) ages. These ages Stockwell considered to define the orogenies, each of which he thought of as a period of mountain building and folding, which may, or may not, be accompanied by metamorphism and granitic intrusion. That is, the radiometric ages correspond to the later phases in the classical cycle of geosynclinal theory – subsidence, deposition, folding, metamorphism, mountain building – a concept dominated by vertical motions. In view of the great success of plate theory, which seeks to explain the evolution of the lithosphere in terms of dominantly horizontal motions, it is reasonable to ask what Stockwell's chronology might mean in terms of possible long-term horizontal motion of the Canadian Shield. The movements of the Shield relative to the pole are therefore now reviewed. Readers are referred to Du Bois (1962), Black (1963), Spall (1971), and Robertson and Fahrig (1971) for earlier reviews of these movements.

Apparent Polar Wandering

Apparent polar movement relative to North America is shown in Figs. 1 to 3[2] in the usual way; the field is assumed to be a geocentric axial dipole and North America is fixed. During the later Cenozoic the pole is near the present geographic pole. In the Lower Cenozoic it moves towards Alaska, and during much of the Cretaceous it remains in the vicinity of what is nowadays the Bering Sea. The pole then doubles back upon itself during the Jurassic, reaching central Siberia in the Triassic. In the Upper Palaeozoic it is in North China, where there is apparently another hairpin bend.

The few results that are available from the early Paleozoic suggest that the pole moved into what is nowadays southeast Asia and the eastern Indian Ocean. This reconstruction of the early Paleozoic polar path given in Fig. 2 is *very* tentative. The pole appears to move across what is nowadays Australia into the central Pacific, and then executes a series of loops, first moving towards North America,

[2]The results reviewed in this paper (see footnote to Table 1 and legend to Fig. 2) are referenced by a list number. This list (Hicken *et al.* 1972, in press) will be published later this year and is in computer format. It contains the relevant statistics and bibliography and notes as to geological age etc., and it is divided into 12 sections, corresponding to the geological time units, as follows: 01 Precambrian, 02 Cambrian, 03 Ordovician, 04 Silurian, 05 Devonian, 06 Carboniferous, 07 Permian, 08 Triassic, 09 Jurassic, 10 Cretaceous, 11 Tertiary, 12 Plio-Pleistocene and Quaternary. This time number is used in Table 1 and elsewhere where necessary, but if the geological age is clear from the context (*e.g.* in legend of Fig. 2) it is omitted. Within these time sections each result is identified by a three digit number which is given, for example, in Fig. 2. Sometimes the results were obtained by averaging several determinations, often values of different authors (*e.g.* 010305 Portage Lake Lavas Fig. 2) and the details of the averaging procedures are explained in the computer listing.

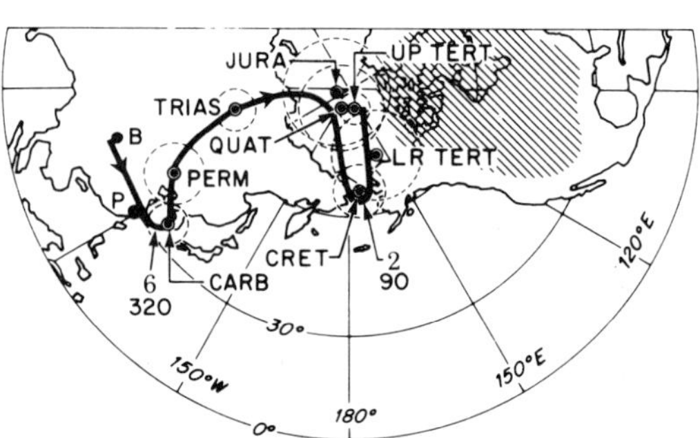

FIG. 1. Apparent polar wandering relative to North America since the late Paleozoic. Mean poles for Carboniferous and later periods (listed in Table 1) are given with their error circles ($P = 0.05$). P is the pole for the Perry Formation (Upper Devonian) and *B* that for the Bloomsburg Formation (Upper Silurian) which are the most reliable poles for the Silurian and Devonian of North America. Hairpins 2 and 6 and their approximate ages in millions of years are indicated. Polar equal-area projection.

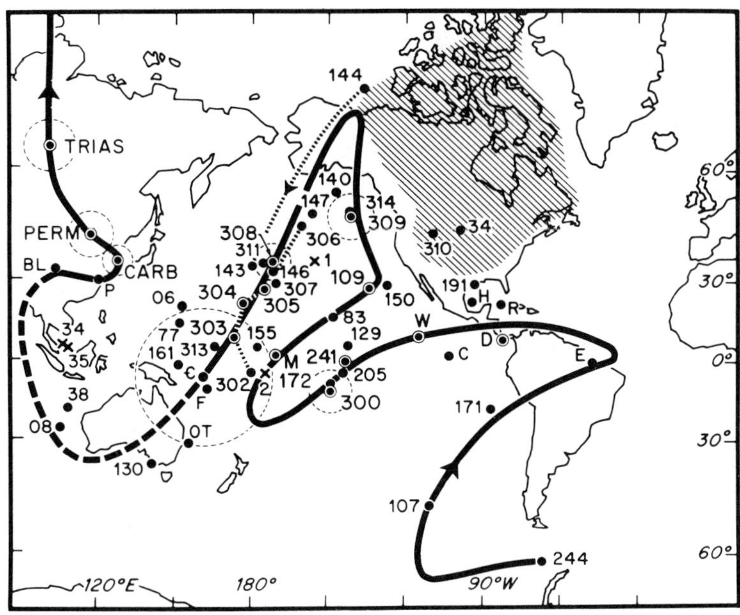

Fig. 2. Apparent polar-wandering relative to Canadian Shield Archean to Triassic. Mercator's projection is used. The exposed and probable subsurface extent of the shield is shaded. Control points that are considered to be of high reliability (they are commonly averages of several determinations) are marked by dot and exscribed circle. Crosses are results from Scotland, obtained by rotation about the pivot point (North America to Europe) of Bullard *et al.* (1965). The finely broken lines connect results whose order is known from superimposition and cross-cutting relationships. The results are identified by the directory number of Hicken *et al.* (1972, to be published) who list the numerical details and references to originals. *Devonian*: 038 Onondaga Limestone; *P* (115) Perry Formation combined. *Silurian*: *B* (027) Bloomsburg Formation; 034 and 035, Arrochar and Carabal Hill complexes 417 m.y. *Ordovician*: 006 Juniata Formation; 008 Trenton Group. *Cambrian*: mean of 019, 023, 045, 066, 078. *Precambrian*: 001 Stoer Group 935 m.y.; 002 Torridon Group 751 m.y.; 034 Baraga County dikes; 077 Eileen Sandstone; 083 Sibley Series; 107 Matachewan dikes 2500 m.y., Ontario; 109 Abitibi dikes 1230 m.y.; 129 Crocker Island Alkali Complex 1475 m.y.; 130 Allard Lake anorthosite; 140 Mamainse Point Lavas (normal) 1076 m.y., Ontario; 143 Upper Gargantua lavas, Ontario; 144 Lower Gargantua, Ontario; 146 North Shore volcanics (normal) 1115 m.y., Minnesota; 147 Osler Formation, Ontario; 150 Lower Keweenawan Lavas, Ironwood, Michigan; 155 Pike's Peak Granite 1030 m.y., Colorado; 161 Grenville dikes, Ontario; 171 Nipissing diabase 2100 m.y., Ontario; 172 Sherman Granite and associated rocks 1410 m.y.; 191 Gunflint Formation 1700–2300 m.y., Ontario; 205 Michikamau anorthosite 1400 m.y.; 241 igneous rocks of St. Francois Mountains 1300–1550 m.y.; 244 Stillwater Complex 2700 m.y., Montana; 300 Belt Series combined; 302 Jacobsville Sandstone combined; 303 Freda Sandstone and Nonesuch Shale combined 1046 m.y.; 304 Copper Harbor Lavas combined 1046–1200 m.y.; 305 Portage Lake Lavas combined 100 m.y.; 306 North Shore volcanics (reversed) combined 1115 m.y., Minnesota; 307 Beaver Bay complex combined, Minnesota; 308 Keweenawan Intrusions combined 1100 m.y.; 309 Keweenawan (reversed) combined 1100 m.y.; 310 Alona Bay Lavas combined; 311 Mamainse Point Lavas (normal) combined 1176 m.y., Ontario; 313 Franklin diabases combined 675 m.y.; 314 Mugford Volcanic Series 950 m.y. More recent data, in process of publication and not yet indexed in the directory of Hickens *et al.* (1972), are denoted by letters and referenced in the usual way: *C* Cameron Bay 1740 m.y. (Irving *et al.* 1972a); *D* Dubawnt Group 1725 m.y. (Park *et al.* 1972); *E* Et-Then Group 1250–1845 m.y. (Irving *et al.* 1972b); *F* Frontenac dikes, NW trend (Park and Irving 1972); *H* Hornby Bay Group sediments (Irving *et al.* 1972a); *M* Mackenzie diabases combined 1250 m.y. (Fahrig and Jones 1969, Irving *et al.* 1972b); *OT* Ottawa basic rocks (Irving *et al.* 1972c); *R* Port Radium Sill (Irving *et al.* 1972a); *W* Western Channel diabase 1500–1750 m.y. (Irving *et al.* 1972a). Data points *D, E, F, M, OT,* and *W* are based on detailed demagnetization studies and many sampling localities. Data points *C, H,* and *R* are based on demagnetization studies, but single sites only.

then back into the central Pacific, turning eastward across the Pacific to northern South America, and finally passing across South America into the southeast Pacific. Since the Archean the pole path seems to have undergone about two and one-half roughly sinusoidal oscillations, with periods of several hundreds of millions of years, and peak-to-peak amplitudes of the order of 90°.

In Fig. 2 poles from the Silurian (no. 34 and 35) and Precambrian (no. 1, 935 m.y. and no. 2, 751 m.y.) are plotted after rotation for closure of the Atlantic. The former fall near the pole for the Bloomsburg Formation (BL), and the latter are near the descending limb of the late Proterozoic polar path (track 2 see below). These results provide rough spotchecks of the reconstruction of the path of polar wandering relative to the Canadian Shield.

Chronology of Superintervals

The later Phanerozoic hairpins are numbered 2 and 6 (Fig. 1) because their peaks coincide with Magnetic Intervals 2 (the late

Mesozoic Normal Interval) and 6 (the late Paleozoic Reversed Interval Fig. 4). In Fig. 3 the earlier hairpins are numbered in decades 10, 20, 30, 40, and 50 to allow indexing of intermediate features that will doubtless be found. Hairpin ages have been calculated assuming a constant rate of polar movement and are given in Figs. 3 and 4. The assumption of constant polar movement may in future be proved inadequate, but the present data do not warrant more detailed treatment. The polar paths between the hairpins are referred to as 'tracks', and are numbered 1 to 5. The time corresponding to each track is called a superinterval. Their limits are given in Fig. 4.

The earlier hairpins appear to be much bolder features, and to occur less frequently than their later Phanerozoic counterparts. This may reflect the evolution of diastrophic processes, or it may be a consequence of the lower density (per unit time) of the observations for earlier time. Hairpins 10 and 50 are only poorly defined, but present evidence suggests that they correspond very approximately to the Taconic and Kenoran orogenies (Fig. 4). Hairpin 30 corresponds closely to the Elsonian orogeny of Stockwell (Fig. 4). Hairpins 20 and 40 are older than the Grenvillian and Hudsonian orogenies (as defined by the peaks in the K–Ar histograms, Fig. 4) by 150 and 250 m.y. respectively. Hairpins 6 and 2 correspond well to the Appalachian and the Columbian–Laramide orogenies of the western Cordillera, and predate the Paleozoic–Mesozoic and Mesozoic–Cenozoic boundaries by 50 and 20 m.y. respectively. The opening of the Laurasian Atlantic follows hairpin 2. It is noteworthy also that hairpins 2 and 6 are centered in magnetic intervals when few reversals occurred (Fig. 4), but the data are insufficient to determine if there is any similar correspondence for earlier time.

Hairpins 2 and 6 may be considered to subdivide track 1, and perhaps several superintervals should be inserted in the Phanerozoic. The nature of the early Paleozoic pole path needs clarifying before further subdivision is attempted.

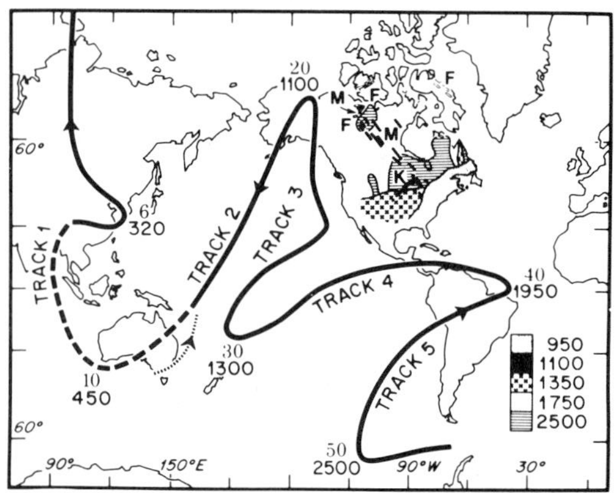

FIG. 3. Generalized polar path relative to the Canadian Shield. Hairpins 10, 20, 30, 40, and 50 are marked with their approximate ages in millions of years. The polar tracks between them are numbered. Note the speculative nature of the early Paleozoic curve. The finely broken line is the path obtained from Grenville rocks. The provinces of the Canadian Shield and its subsurface extension in the USA are shown schematically, the approximate dates of crustal stabilization are shown in the legend. *M* refers to the Mackenzie igneous rocks (1250 m.y. Fahrig and Jones 1969), *K* to the Keweenawan rocks (1100 m.y.) and *F* to the Franklin igneous rocks (675 m.y. Fahrig *et al.* 1971).

Discussion

The hairpins shown in Figs. 1 to 3 correspond to very large changes in the direction of movement of the Canadian Shield relative to

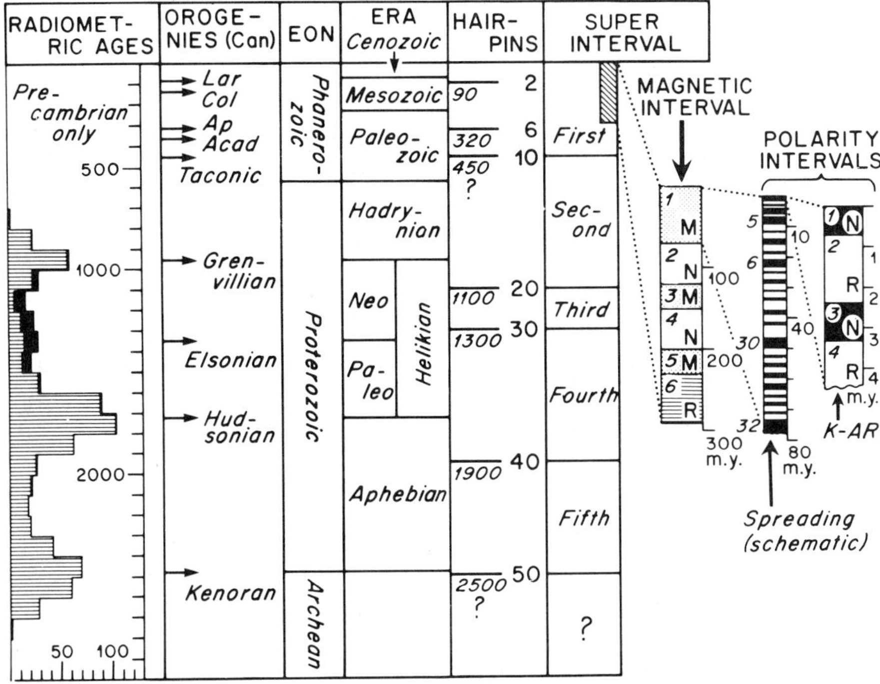

Fig. 4. Summary of paleomagnetic time-scales. On the right the scale of Polarity Intervals is given – the scale determined by K–Ar dating (recent summary by Doell *et al.* 1971) and its extension from sea-floor studies (Heirtzler *et al.* 1968). *N* denotes predominantly normal polarity, and *R* predominantly reversed polarity. The scale of Magnetic Intervals (McElhinny and Burek 1971, Irving 1972) is set out (*M* denoting mixed polarity) and further to the left the superintervals and the hairpins that define them. The geological eras and Stockwell's (1970) time scale are given in the center, and the corresponding frequency histogram of radiometric ages on the left. This histogram has been compiled from Stockwell (1970) for the exposed shield in Canada (shaded part), and from Muelberger *et al.* (1967) for the subsurface and exposed shield in the USA.

the pole. Following a time (superinterval) of motion in approximately the same direction (the speed may of course vary) the Canadian Shield has been momentarily halted (hairpin) and then the motion has been renewed in some very different direction. This seems to have occurred several times since the Archean.

Within each superinterval the approximately uniform direction of motion and the stress field that this maintains may be responsible for the approximately consistent structural trends found within each of the provinces of the shield (Stockwell 1970, Fig. IV-2), and for the linear nature of some province boundaries. It is noteworthy that tracks 2 and 3 are essentially parallel to the Grenville Front (stretches from southeastern United States to Labrador) and track 4 is parallel to the boundary between

crust stabilized at 1350 m.y. and older crust to the north (Fig. 3).

Hairpins 6 and 2 correspond to late Paleozoic and late Mesozoic orogenies and are followed by widespread (indeed worldwide) restriction of the continental seas that mark the commencement of the Mesozoic and Cenozoic. This is reminiscent of the feature, already noted, that hairpins 20 and 40 occur earlier than Grenvillian and Hudsonian orogenies. The inference is that the restriction of the epicontinental seas, and the peaks in the K–Ar histograms (Fig. 4) are a consequence of uplift caused by the reorganization of plate motions.

It is noteworthy that Coney (1971) has suggested, on the basis of entirely different evidence, that 80 m.y. ago (roughly coincident with hairpin 2) a large change in the 'absolute'

TABLE 1. North American Phanerozoic mean poles

No.	Period	N	Pole	K	A_{95}
12A	Plio–pleistocene	6	84.3 N, 154.5 E	57	8.9
12B	Plio–pleistocene	9	87.4 N, 162.3 E	17	12.7
11	Upper Tertiary	13	86.9 N, 192.9 E	256	2.6
11	Lower Tertiary	5	73.8 N, 154.3 W	45	11.5
10	Cretaceous	11	65.9 N, 174.7 W	63	5.8
9	Jurassic	4	86.8 N, 096.4 E	56	12.4
8A	Triassic	11	61.2 N, 099.8 E	102	4.5
8B	Triassic	12	63.2 N, 100.0 E	61	5.6
7	Permian	9	42.7 N, 115.9 E	68	6.3
6	Carboniferous	14	35.6 N, 126.8 E	75	4.6
5	Devonian (Perry)	4	29.4 N, 119.0 E	93	9.6
4	Silurian (Bloomsburg)	1	32.0 N, 102.0 E	—	—
2	Cambrian	5	07.4 S, 160.4 E	9	27.6

Note: The mean poles have been obtained by averaging the several individual results that are identified by the reference numbers in Hicken *et al.* (1972) through which the original descriptions may be traced: 12(A) 12047, 069, 094, 095, 136, 161; 12B 12047, 069, 094, 095, 136, 161, 098, 115, 160; 11 (Upper) 11104, 106, 107, 151, 190, 216, 217, 231, 232, 295, 320, 361, 362; 11 (Lower) 11101, 165, 212, 238, 312; 10 10017, 028, 029, 030, 035, 049, 050, 071, 072, 073, 090; 9 09034, 072, 103, 104; 8A 08040, 042, 044, 068, 089, 111, 113, 135, 165, 166, 167; 8B 08040, 042, 044, 068, 089, 111, 113, 135, 165, 166, 167, 071; 7 07049, 050, 051, 073, 094, 095, 149, 205, 206; 6 06063, 066, 087, 088, 116, 135, 136, 137, 138, 140, 170, 171, 172, 174; 2 02019, 023, 045, 066, 078.

motion of the North American plate occurred as determined by the 'hot spot' method. (Coney also suggested that there was a similar period of "tectonic reorganization" 40 m.y. ago. His maps show only a small change in direction at this later time, and there is no corresponding hairpin; it is doubtful whether the small changes are detectable paleomagnetically.) Making use of Coney's ideas regarding the interrelationship between orogenies and changes in the direction of plate motion, the following qualitative model is proposed. After a time of free-plate motion a momentary 'plate-jam' occurs, brought about either through locking of a major convergent juncture by continental lithosphere (Mackenzie 1969), or by locking of major transcurrent junctures through the development of marginal irregularities (Donaldson and Irving 1972). The forces at depth causing plate motions are still operative and the plates soon free themselves, but not without reorganization of their margins and changes in directions of motion. This freeing process is followed by the restriction of the epicontinental seas, due in part to the creation by rifting of new deep ocean, to the uplift of rifted margins, and to the rebound of deceased subduction zones – all processes that may be expected to reduce sea-level. Since most organisms live either in epicontinental seas, or upon the lowland areas that depend to some extent on such seas for their rainfall, this tectonic sequence may have set in train wholesale extinctions of life. This explains, qualitatively, the lag between the hairpins and the Phanerozoic era boundaries, and the lag between the older hairpins and the K–Ar peaks. The Precambrian–Cambrian boundary is marked by the sudden production of varied and abundant life, whereas later era boundaries correspond to wholesale extinctions, suggesting that it has other causes. The absence of a hairpin at the end of the Precambrian (Fig. 3) is therefore not unexpected. This model does not explain the correspondence between the ages of hairpins 50 and 30 and the Kenoran and Elsonian orogenies (Fig. 4). However, hairpin 50 is not well-defined, and the nature of the Elsonian orogeny is only poorly understood (the K–Ar peak, for example, is comparatively feeble, Fig. 4).

Finally, it is noted that the poles obtained from Grenville rocks do not conform to the main polar-wandering curve (Fig. 3). This feature is discussed elsewhere (Irving *et al.* 1972c; Palmer and Carmichael 1972).

Acknowledgments

Many fruitful discussions with J. A. Donaldson, J. C. McGlynn, and J. L. Roy, and criticism of the manuscript by W. A. Robertson are gratefully acknowledged.

BLACK, R. F. 1963. Palaeomagnetism of part of the Purcell system in southwestern Alberta and southeastern British Columbia. Geol. Surv. Can., Bull. 83, pp. 1–31.

BULLARD, E. C., EVERETT, J. E., and SMITH, A. G. 1965. The fit of the continents around the Atlantic Phil. Trans. Roy. Soc. **258**, pp. 41–51.

CONEY, P. J. 1971. Cordilleran tectonic transitions and motion of the North American plate. Nature **233**, pp. 462–465.

DOELL, R. R., GROMME, G. S., DALRYMPLE, G. B., and COX, A. V. 1971. Geomagnetic polarity epoch time scale. Upper Mantle Project, U.S., Program Final Report, Nat. Acad. Sci., Washington, 128 p.

DONALDSON, J. A. 1972 (In press). Grenville Front and rifting of the Canadian Shield. Nature.

DU BOIS, P. M. 1962. Palaeomagnetism and correlation of Keweenawan rocks. Geol. Surv. Can., Bull. 71, pp. 1–75.

FAHRIG, W. F., IRVING, E., and JACKSON, G. D. 1971. Paleomagnetism of the Franklin diabases. Can. J. Earth Sci. **8**, pp. 455–467.

FAHRIG, W. F. and JONES, D. L. 1969. Paleomagnetic evidence for the extent of Mackenzie igneous events. Can. J. Earth Sci. **6**, pp. 679–688.

GILL, J. E. 1949. Natural divisions of the Canadian Shield. Trans. Roy. Soc. Can. **43**, pp. 61–91.

HEIRTZLER, J. R., DICKSON, G. D., HERRON, E. M., PITMAN, W. C., and LE PICHON, X. 1968. Marine magnetic anomalies, geomagnetic field reversals, and motions of the ocean floor and continents. J. Geophys. Res. **73**, pp. 2119–2136.

HICKEN, A., IRVING, E., LAW, L. K., and HASTIE, J. 1972. Directory of paleomagnetic directions and poles: first issue. Publ. Earth Phys. Br., Ottawa. (*In press.*)

IRVING, E. 1972. Paleomagnetic stratigraphy – names or numbers. Comments on Earth Sci., Geophysics 2. (*In press*), pp. 125–130 cf.

IRVING, E., DONALDSON, J. A., and PARK, J. K. 1972*a*. Paleomagnetism of the Western Channel diabase and associated rocks, Northwest Territories. Can. J. Earth Sci. **9**, pp. 960–971.

IRVING, E., PARK, J. K., and McGLYNN, J. C. 1972*b*. Paleomagnetism of the Et-Then Group and Mackenzie diabase in the Great Slave Lake area. Can. J. Earth Sci. **9**, pp. 744–755.

IRVING, E., PARK, J. K., and ROY, J. L. 1972*c*. Palaeomagnetism and the origin of the Grenville Front. Nature **236**, pp. 344–346.

MACKENZIE, D. P. 1969. Speculations on the consequences and causes of plate motions. Geophys. J. Roy. Astron. Soc. **18**, pp. 1–32.

McELHINNY, M. W. and BUREK, P. J. 1971. Mesozoic paleomagnetic stratigraphy. Nature **232**, pp. 89–102.

MUELBERGER, W. R., DENISON, R. E., and LIDIAK, E. G. 1967. Basement rocks in continental interior of United States. Bull. Am. Assoc. Petrol. Geol. **51**, pp. 2351–2380.

PALMER, H. C. and CARMICHAEL, C. M. 1972. Grenville paleomagnetism and possible tectonic implications. (Abstr.). Trans. Am. Geophys. Un. **53**, 357 p.

PARK, J. K. and IRVING, E. 1972. Magnetism of dikes of the Frontenac axis. Can. J. Earth Sci. **9**, pp. 763–765.

PARK, J. K. IRVING, E., and DONALDSON, J. A. 1972 (In press). Paleomagnetism of the Dubawnt Group. Geol. Soc. Am. Bull.

ROBERTSON, W. A. and FAHRIG, W. F. 1971. The great Logan paleomagnetic loop – the polar wandering path from Canadian Shield rocks during the Neo-Helikian Era. Can. J. Earth Sci. **8**, pp. 1355–1364.

SPALL, H. 1971. Precambrian apparent polar wandering; evidence from North America. Earth Planet. Sci., Lett. **10**, pp. 273–280.

STOCKWELL, C. H. 1963. Second report on structural provinces, orogenies and time-classification of rocks of the Canadian Precambrian Shield. *In*: Age determinations and geological studies. Geol. Surv. Can., Pap. 62-17, pp. 123–133.

———— 1970. Geology and economic minerals of Canada. R. J. W. Douglas (*Ed.*). Geol. Surv. Can., Econ. Geol. Rep. **1**, Fifth ed., 44 p.

WILSON, J. T. 1949. Origin of continents and Precambrian history. Trans. Roy. Soc. Can. **43**, pp. 157–184.

WILSON, M. E. 1941. The Precambrian. Geol. Soc. Am., 50th Anniv. Vol., pp. 269–305.

27

Reprinted from *Phys. Earth and Planetary Interiors* **2**:393–398 (1970)

FACTORS DETERMINING THE TEMPERATURE EVOLUTION OF THERMALLY CONVECTING EARTH MODELS

D. C. TOZER

University of Toronto, Toronto, Canada

Received 3 February 1970

The thermal history of the Earth is posed as a problem in convection theory. The concept of stabilisation temperature is introduced and its significance for the cooling process is assessed. The rapidity of heat transport at higher temperatures combined with the fact that stabilisation temperatures are only half the melting temperature, cf. the classical literature of cooling from a "hot origin", makes present temperatures relatively insensitive to initial conditions.

Although the connection of tectonic processes with geothermal energy has long been suspected and discussed, we now seem to be at the threshold of a new level of understanding this energy conversion process. The purpose of this paper is to make a few brief comments on the purely thermal aspects of this problem and particularly to modify an old result that has been uncritically repeated in the textbooks.

My basic hypothesis is the assertion of the validity at all times of the equations governing thermal convection in a Newtonian viscous medium within the domain representing the Earth's interior. It is well known that the thermal evolution of such an Earth model reduces to a discussion of the evolution of a temperature distribution in accordance with conduction theory if the velocity field vanishes at all points of the interior. I have made the suggestion (TOZER, 1970) that conduction be regarded as a degenerate form of convection, the word "convection" being consistently used as the description of a model rather than an assertion of macroscopic mass transfer. This point of view seems to direct appropriate attention to the need to justify the use of conduction as a truncated form of convection theory in any particular model situation. Naturally, the use of thermal convection equations in their conventional form (LANDAU and LIFSHITZ, 1959), is not compatible with contemporaneous chemical differentiation of the material, although certain modifications can be made that will accommodate a "slow" differentiation of the material.

The primary energy for the convective process in the Earth models is taken to be a volume distribution of heat sources throughout the model interior. Over the years many different components of this heat source distribution have been suggested, e.g. radiogenic, chemical, viscous dissipation of energy associated with lunar tides and chemical differentiation, etc., but observational data has not placed close bounds on their spatial and/or temporal variations. This fact, combined with uncertainties about the energy content of the model at any particular instant and the use of conduction theory *ab initio* made the temperature distribution of the deep interior uncertain to several thousand degrees. It now seems probable that this degree of uncertainty, at least with regard to the temperature distribution of the interior in recent geological times, may well have had its main origin in the premature simplification of the heat transfer problem to one in conduction theory, rather than the uncertainties of the heat source distribution and the initial energy content for any model calculation.

The formulation of convection theory appropriate to the Earth and, with rather less certainty, to the Moon and terrestrial planets, seems to require for a number of purely geophysical reasons that the viscosity of at least large portions of their interiors be a significant function of temperature. This feature, which of course did not even appear in the conventional calculations of thermal history for these bodies, can play a dominant simplifying role in the formulation of these problems in convection theory. To gain an understanding of this point let us first consider a spherical homogeneous body without any internal heat sources whose relevant properties are constant along any adiabat, but whose viscosity decreases rapidly with increasing temperature. Let us further suppose that its surface temperature decreases from some high value so that the body as a whole is cooling. How does the temperature of the interior respond to this secular decrease of the surface temperature? Naturally the answer to this problem depends on how quickly it is done. If it is done infinitely slowly the body is always in a quasistatic (isothermal) equilibrium, and the solution is trivial. At an increasing rate of cooling a radial temperature gradient develops, and the temperature difference of the centre and surface, ΔT, increases. It is appropriate to discuss the type of solution to the resulting problem in terms of a Rayleigh number R based on ΔT and the average values of the physical parameters (see the end of this paper for a description of the symbols):

$$R = \frac{g\alpha L^3}{\kappa v}(\Delta T - \Delta T_A). \tag{1}$$

If $R < R_c \approx 10^3$, the conduction solution of the problem is appropriate, but if $R > R_c$, this particular solution is unstable, and the only acceptable, i.e. stable, solutions are those involving finite motion of the medium. As R increases from R_c, the velocity of these motions increases from zero roughly according to the relationship

$$v = \frac{\kappa}{L}(R - R_c)^{\frac{1}{2}}, \tag{2}$$

and the heat flow as

$$H = \kappa \rho c_v \Delta T \tfrac{1}{4}(R - R_c)^{\frac{1}{2}}. \tag{3}$$

Let us translate these general ideas into the simplest model of the Earth cooling from a high temperature. Let us first evaluate the condition that conduction theory be an acceptable description of the cooling process. Since the critical Rayleigh number R_c for the stability of conduction solutions is a positive number, time dependent conduction theory will certainly be an acceptable solution for cooling, regardless of the (positive) numerical values of the other quantities defining the Rayleigh number, if $\Delta T < \Delta T_A$. The quantity ΔT is approximately the (steady) rate of surface cooling multiplied by the conduction thermal time constant of the body $\tau = 0.1\, L^2/\kappa$; it may be shown that ΔT_A for

plausible Earth models $\approx 10^3$ °C. Putting $L = 6 \times 10^8$ cm and $\kappa = 10^{-2}$ cm^2/s, we have $\tau \approx 10^{11}$ y, and the condition on the cooling rate to make $\Delta T < \Delta T_A$ is that it be less than 10^{-8} °C/y. For comparison, this cooling rate corresponds to a surface heat loss from the body of $\approx 1\%$ of the present measured geothermal heat flow. At twice this rate of cooling $\Delta T - \Delta T_A \approx 10^3$ °C, and the condition for the conduction solution to be a valid stable solution to the cooling problem is:

$$R_c \approx 10^3 > \frac{g\alpha L^3}{\kappa v} \times 10^3, \tag{4a}$$

i.e.

$$v > \frac{g\alpha L^3}{\kappa}. \tag{4b}$$

N.B. ΔT is defined on the assumption of time dependent conduction theory and is merely a scaling parameter. It will not be the actual temperature difference if convective motions are occurring throughout the interior.

With $g = 10^3$ cm/s^2, $\alpha = 2 \times 10^{-5}$ °C^{-1} we have from eq. (4b) the condition $v > 4 \times 10^{26}$ cm^2/s. This is an enormous viscosity* by normal standards and much too high to be measured in the laboratory (stress relaxation times are $\approx 10^7$ y).

The next feature to introduce into our simple homogeneous model is a rate of kinematic viscosity variation with temperature (strictly speaking entropy) that is plausible for the Earth material. This relationship, like all the other constitutive parameters, need only be chosen to give accord with field observations of the characteristics of convection equation solutions. In practice, we are guided (but not compelled) by laboratory experience; naturally if our conclusions are to carry weight this guidance should be based on statistical regularities of the laboratory data rather than its details. A survey of creep behaviour indicates that in the lowest range of stress the rheology of any condensed material can be represented by a viscosity varying as follows:

$$v = v_0 T\, e^{E/kT}. \tag{5}$$

The energy E is associated with atomic diffusion and v_0 with spatial order parameters and characteristic fre-

---

* I shall not need to distinguish viscosity and kinematic viscosity in this paper since their numerical values are so similar in the c.g.s. system.

quencies of the atomic structure. As the temperature is raised above a certain characteristic temperature T_M, which is $\approx E/20k$, the long range order of the material is progressively destroyed and order is eventually only preserved over distances comparable with the atomic sizes. Although some have claimed that the long range order of Earth material for $T < T_M$ may extend over kilometres, our experience with the petrographic microscope suggests that more plausible values to choose at the beginning are in the millimetre range, with scope for revision both ways*. This particular choice of the order parameter means that the viscosity at T_M, more familiarly known as the melting or solidus temperature, has a value of (very) approximately 10^{15} cm^2/s and declines to values $< 10^5$ cm^2/s through the melting range. It should be noted that this phenomenon occurs at viscosities that are microscopic compared with those that will stabilise the conductive flow of the geothermal flux throughout the interior. The temperature at which stabilisation is lost (the stabilisation temperature) is little more than $\frac{1}{2}T_M$. Strictly speaking, the concept of stabilisation temperature is only precisely defined for homogeneous systems with viscosity independent of the local thermodynamic state and in which temperature changes due to adiabatic compression are negligible. It may be shown with experimental data that over a range of pressure appropriate to the upper mantle the viscosity of mantle material may not vary significantly along adiabats. The idea of a stabilisation entropy is then a more precise concept for this region. However, there are plausible reasons for thinking this idea also breaks down for the mantle as a whole (see below).

The preferred starting values for Earth material of T_M, and hence E, are ≈ 2000 °K at the surface rising to ≈ 5000 °K at the centre. From this it may be shown that the viscosity of material at the surface (300 °K) is $> 10^{50}$ cm^2/s. It is also to be noted that the internal energy of the Earth at the stabilisation temperature is < 400 cal/g.

We have already noted that several processes have possibly contributed to the energy content of the Earth at various times but the timing of this energy release is rather uncertain. Let us for the moment disregard the possible continued existence of heat sources in the Earth and concentrate on the magnitude of sources that observational evidence indicates made their major contribution to the energy content at least some billions of years ago. Approximately 10^4 cal/g is the specific gravitational energy associated with the formation of the Earth from diffuse material. Current doctrine asserts that more than 98% of this energy is radiated as heat energy from the surface of the growing planet, but this ignores the dissipation throughout the volume of the planet of the seismic disturbances caused by collisions. Since it may be shown that more than 50% of the mass of the Earth was due to the addition of material impacting at greater than the seismic velocities, I suspect that the contribution to the internal energy from this source could be a significant fraction, perhaps 10% of this gravitational energy. It may also be shown that chemical differentiation of the Earth material to form the core releases ≈ 500 cal/g (TOZER, 1965), and a comparable amount of energy could have been released as heat by the dissipation of lunar tides during a close approach ($\approx 3R_\oplus$) of the two bodies. The timing of core formation is uncertain, but the evidence of palaeomagnetism shows no evidence of a change in dipole moment (suitably averaged) over the last 2.6×10^9 y. The kinetics of the Moon's recession from the vicinity of the Earth is poorly known, but the great ages found for rocks collected on the lunar surface has weakened the arguments that such a close approach occurred within the last 10^9 y. All this evidence makes it plausible to suppose that at some point in time several billion years ago the Earth may well have possessed perhaps 10^3 cal/g as a result of the action of just these short-lived processes. This corresponds to an average temperature of the interior of several thousand degrees.

In keeping with our very simple model described above, let us visualise the cooling of the Earth since this time as the result of a reduction of the surface temperature by several thousand degrees over a few billion years. As a mean cooling rate of the surface this is of the order a hundred times faster than that considered above. It may be noted that if the loss of heat from the external surface were determined by a radiative balance with a universe comparable with the present one, the cooling of the material within metres of the

---

* One of the most interesting conclusions to emerge from studies of the convection problem is that the observable characteristics of the convection solutions are not very sensitive to the choice of this order parameter (TOZER, 1970).

surface would have been extremely rapid at first and some kind of surface skin of much higher viscosity than the interior would have formed in a time of the order of tens of thousands of years. The existence of sediments and crustal rocks with radiogenic ages $> 3.10^9$ y supports the idea of a quickly formed cool surface layer, but the temperature of the interior will decrease more slowly. Enormous Rayleigh numbers will be characteristic of the initial phase of cooling from such high temperatures and the associated large heat transport will keep the surface skin of cool material very thin and constantly disrupted by the convection of the interior. For example, even after solidification of the interior one can expect Rayleigh numbers $> 10^{10}$ and a radial heat transport as much as a hundred times greater than that estimated using conduction theory (TOZER, 1967); such heat transport implies a skin thickness of ≈ 100 m. At this rate of heat transport the temperature of the interior would closely approximate an isentropic temperature distribution and the mean value would be falling at $\approx 10^{-4}$ °C/y. At this rate of cooling sufficient time has existed for the stabilisation temperature to be reached, i.e. the Rayleigh number declines to $\approx 10^3$. Naturally the cooling rate steadily declines as the stabilisation temperature is approached and the behaviour approaches that predicted by conduction theory. The qualitative features of the solution after a few billion years cooling is that the surface skin has grown to a few hundred kilometres thick (due to the great decrease in radial heat flow) and that below this the stabilisation temperature is virtually "frozen" in; the amount of cooling at these depths since stabilisation is ≈ 100 °C and would increase the viscosity by perhaps a factor of ten above the stabilising value. A schematic picture of the cooling process is shown in fig. 1. It can be seen that the great decrease of viscosity with increasing energy content virtually ensures that arbitrarily large amounts of energy in excess of ≈ 300 cal/g would have had time to escape, but that the loss of the residual energy is a much slower process accomplished in a time $\approx \tau$.

The above picture of the cooling process does not give an acceptable account of the present state of the Earth for a number of reasons. This model of an Earth without energy sources has all the undesirable aspects of Kelvin's treatment of the geothermal problem in more acute form. The predicted geothermal gradient

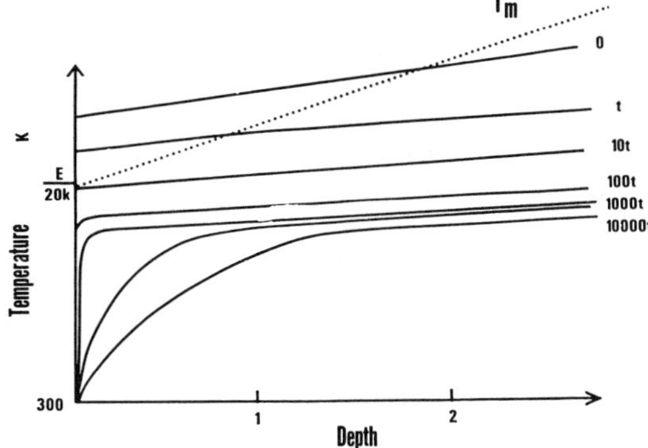

Fig. 1. Schematic temperature–depth profiles for a body containing no heat sources but having a variable viscosity (see text). The initial temperature distribution is labelled 0 and the surface temperature is fixed at 300 °K. For plausible Earth models the unit of depth is ≈ 500 km. T_M at the surface 1600 °K and $t \approx 10^5$ y.

Points to note are the rapidity of cooling in the vicinity of the melting point and the preservation of only a thin "crust", i.e. thermal boundary layer, until the temperature at depth approaches the stabilisation temperature.

at the surface is much too small and from the standpoint of modern tectonics the outer few hundred kilometres have quite the wrong viscosities ($10^{26} \rightarrow 10^{35}$ cm²/s) to explain the phenomena of isostasy, glacial rebound, sea floor spreading, etc. As far as I can see at the moment, these difficulties can all be removed by postulating the continuing existence of heat sources within the material of the Earth. It will be evident from the above model of a cooling Earth that if such heat sources are able to supply heat at a rate that exceeds ≈ 300 cal/g, in a time τ the body will be kept in a state of convection. For the case of the Earth this heat production is ≈ 0.1 erg/g y (c.f. the heat production of chondritic meteorites of 1.7 erg/g y). In a number of papers (TOZER, 1967; TOZER, 1970) I have tried to find the steady state temperature solutions of the convection equations that correspond to the observed geothermal heat flow at the external surface. It is readily shown that such a surface heat flow is only plausibly associated with temperatures that exceed the stabilisation temperature below a few tens of kilometres. In this steady state heat transport problem the great increase in heat transport that occurs as the temperature rises beyond the stabilisation temperature acts as a simplifying factor. For example, uncertainties in the exact values of the heat source distribution (only the integrated distribution is fixed by the surface heat flow and steady state condition) have relatively small effect on the temperature and hence viscosity of the region below a few tens of kilometres. My calculations indicate a viscosity associated with the steady geothermal flow of $\approx 10^{21}$ cm²/s in this region – quite small enough to interpret the phenomenon of isostacy and glacial rebound. There are also indications that this region down to depths of several hundred kilometres may be generally below 10^3 °C though much hotter in certain very limited regions (TOZER, 1970).

Actually there is evidence that one should introduce into an Earth model a further complication. It is believed that the assumption of constant viscosity along an adiabat must be modified at depths of a few hundred kilometres on account of the abnormally large increases in density that occur in this range of depth. It is readily seen that compression makes the process of atomic diffusion more difficult and on empirical grounds it is plausible that the value of E (eq. 5) should show a large increase ($\approx 60\%$) between depths of 400 and 1000 km. Also, v_0 may be expected to increase in this depth range due to progressive development of solid solution. On this account it may be shown that the viscosities for the lower mantle associated with the steady state solutions are $> 10^{28}$ cm²/s – greater than those necessary to stabilise conduction solutions.

Conclusions

The above discussion will have shown that the development of a thermal evolution theory for convecting models of the Earth has quite different features from the current theories. Although convective heat transfer is a more complicated theory, there are compensations. A tiresome feature of calculations based on conduction theory was that the thermal time constant was always much longer than the age of the Earth, and as a consequence, the values of temperature in the deep interior always depended on the thermal conditions existing at the formation of the Earth. There was and is no firm solution to that problem in sight. However, this difficulty has been largely bypassed with the convective heat transport theory. The combination of a stabilisation temperature of $< \approx 2500$ °K and the great increase in heat transport above it has the twofold effect of making present thermal conditions virtually independent of thermal conditions a few billion years ago, and

making the temperature distribution less sensitive to the choice of the details of the heat source distribution.

In studying convection in internally-heated models in which the viscosity decreases by a factor $> 10^{30}$ when the temperature changes by a factor of 10 it has become increasingly apparent that temperature is a relatively unimportant variable. By this I mean that the old conception that a knowledge of temperature was necessary to determine viscosity and hence the answers to a number of interesting geophysical questions was the wrong way of posing the problem. Nearly all the observable aspects of the convection problem are determined within close limits by a relatively crude knowledge of the viscosity–temperature relationship for Earth material. However, where observational data are related to the temperature field, the convection theory gives a more satisfactory interpretation than conduction theory (TOZER, 1970). I regard as a major success of the new theory that it predicts active convection in an outer mantle that has just the correct viscosity to explain glacial rebound kinetics and the damping of the variation of latitude. It is also very interesting that the stabilising viscosity of the lower mantle is so close to the value for the lower mantle that has been tentatively derived from the Earth's non-hydrostatic shape. If this situation is a relic of a hotter past, the present radial temperature gradient may well increase with depth in the vicinity of a depth of a few hundred kilometres. The textbook assumption in need of change is that thermal problems for the Earth can be solved with conduction theory if the temperature is less than the melting point.

List of symbols

c_p – pressure coefficient
E – activation energy
g – acceleration of gravity
H – heat flow
k – Boltzmann's constant
L – characteristic length
R – Rayleigh number
R_c – Rayleigh number above which the conduction solution is not stable to infinitesimal perturbations
T – absolute temperature
T_M – melting (solidus) temperature
ΔT – characteristic temperature difference between the centre and the surface of a cooling body

ΔT_A – maximum difference of temperature due to adiabatic compression in the body

v – characteristic velocity of convection

α – coefficient of expansion

κ – thermal diffusivity

v – kinematic viscosity

ρ – density

τ – thermal time constant according to conduction theory

References

LANDAU, L. D. and E. M. LIFSHITZ (1959), *Fluid mechanics* (Pergamon, London) 212.

TOZER, D. C. (1959), Phys. Chem. Earth 3, 414.

TOZER, D. C. (1965), Geophys. J. 9, 95.

TOZER, D. C. (1967). In: T. F. Gaskell, ed., *The Earth's mantle* (Academic Press, New York) 325.

TOZER, D. C. (1970), J. Geomagn. Geoelec., in press.

28

Reprinted from *Icarus* **26**:1-15 (1975)

The Seven Ages of a Planet

WILLIAM M. KAULA

> Our system is a stage,
> And both the Sun and planets merely players.
> They had their birth and'll have their fiery end.
> A planet in its time plays many parts,
> Its acts being seven ages. The first of these
> Is condensation: dust grains drifting to
> The nebula plane in chondrite clods. And then
> The planetesimals: breaking sometimes, but
> Most growing, though the Sun's hot breath blows gas
> Away. And then formation: sweeping up
> The bodies in its way, in fierce infalls
> To bring then full convective vigor, too hot
> For crust to form, though iron may sink and seas
> Outgas, by radioactive energy driven.
> And then comes plate tectonics: cooling leads
> To lithosphere, with many marginal breaks.
> Convective thrusts a crust create in belts
> Complex. But heating slows; the sixth age shifts
> Into the final volcanism: no
> More lithospheric spreading, only vents
> For magma, Nix Olympica or mare
> To surface, ending fractionation. Last scene
> That ends this history is quiescence: time
> Sans melt, sans plates, sans almost everything.

INTRODUCTION

Comparative planetology is in vogue with the great increase in detailed information about the surfaces of Mars and Mercury obtained by the Mariner spacecraft, following upon the Apollo exploration of the Moon and the changed view of solid Earth processes resulting from the plate tectonic revolution. However, the nature of the data has led in most comparative discussions to an emphasis on surficial processes, such as erosion and sedimentation or infalls occurring long after formation. The abstract to this paper was written as a tutorial device to emphasize a fundamental view of how and why planets evolve. By "fundamental" I mean ascribing the greatest importance to processes affecting the bulk of a planet, rather than just a surface layer. By this definition, most of the important processes tend to be rather inaccessible: they occurred either before the planets essentially completed formation, or deep within the planets since formation. The purpose of this paper is to point out and discuss certain problems of planetary evolution within a somewhat arbitrary frame, with the hope that it may guide more detailed investigations.

Given the almost universally held assumptions that (1) the planets were made from widely dispersed much smaller bodies and (2) the solar system has remained essentially isolated from outside influences since soon after formation, it follows rather plausibly that all terrestrial planets (including the Moon) in principle pass through the seven stages suggested by the abstract. These stages are:

(1) *Condensation*: solidification of grains from gas; settling to the central plane of the nebula; damping of relative motion; agglomeration into small ($\lesssim 100$ meters) bodies.

(2) *Planetesimal interactions*: collection into sizeable ($\gtrsim 100$ km) bodies; velocity dispersion increase by mutual perturbation; heating by collisions and solar effects; sweeping away of volatiles by solar outflow.

(3) *Formation*: planetary sweeping up of matter by gravitational capture; heating by infall of bodies.

(4) *Vigorous convection*: material motion driven mainly by infalls and initial inhomogeneities; gravitational separation of iron; outgassing of oceans and atmospheres; comparatively little differentiation of silicates and radioactive heat sources or crustal accumulation.

(5) *Plate tectonics*: material motion driven by radioactive heat as well as inhomogeneities; surface cooling sufficient to form a lithosphere broken along continuous lines so that it partially participates in convective circulation; considerable differentiation of silicates and accumulation of a crust; some continued moderate outgassing and recycling.

(6) *Terminal volcanism*: no plate motion, but still sufficient heating for magmatism to break through the lithosphere and make volcanic piles; moderate differentiation of silicates.

(7) *Quiescence*: thick lithosphere, no volcanism; only deep activity possible.

The first two or three stages are sometimes not thought of as part of planetary evolution. However, they must be considered if meaningful comparisons are to be made, since they determine compositional differences and initial conditions of thermal and tectonic evolution.

For any one planet, there are considerable differences in duration among the stages, while for any one stage there are considerable differences in duration among the planets. It is likely that some planets will not evolve through all stages; thus the Earth may not attain the final stage of quiescence before the Sun undertakes its terminal expansion.

Some factors accounting for the differences between planets are fairly evident. For example, the Moon's large area/mass ratio has pushed it most quickly to quiescence by radiating away its heat, while Venus' opaque atmosphere may cause it to linger still at the vigorous convective stage by raising the surface temperature [however radar imagery (Rumsey *et al.*, 1974) and gamma ray spectra (Vinogradov *et al.*, 1973) suggest an old quiescent surface]. In any case, the behavior is mainly governed by the available energy, of which only two sorts are sure to have been quantitatively sufficient in the long run: radioac-

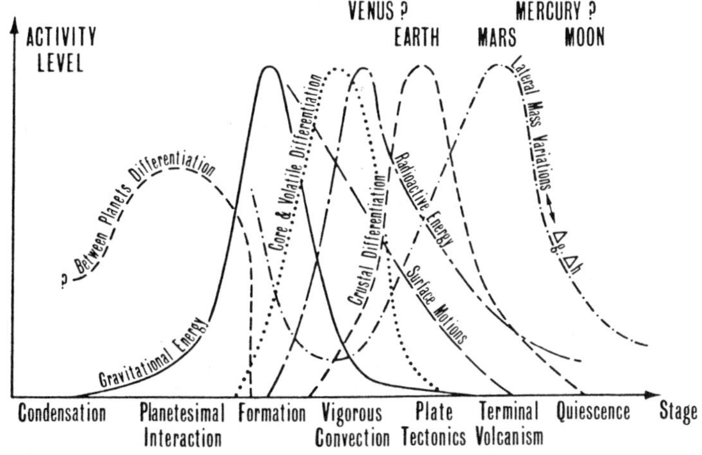

FIG. 1. Variations of energy inputs and planetary features with the stages of evolution.

369

tive and gravitational, the latter including both early infalls and subsequent mass redistribution within a planet. The manifestations of planetary activity can also be grouped in a few categories: differentiations among substances differing a few gcm^{-3} in density: e.g., volatiles from silicates, or iron from silicates; differentiations among substances differing a few $0.1g$ cm^{-3} in density: e.g., aluminum silicates from magnesium silicates; surface motions; and lateral inhomogeneties: topography, gravity anomalies, etc. Figure 1 is a rather intuitive sketch of the variations in energy inputs and the resulting variations in planetary evidences with the evolutionary stage. The current stage for each of the planets is also noted. Most planetologists would agree with most of Fig. 1. To keep this paper from becoming too bulky, I concentrate on some problems which seem most difficult, and hence about which significant differences of opinion exist. These problems can be grouped into the following categories:

(a) the relative importance of different modes of energy conversion,
(b) the influence of solar formation and evolution,
(c) the influence of the major planets and their evolution,
(d) the differentiation among the planets,
(e) the rate of planetary thermal evolution,
(f) the sequence of intraplanet differentiations, and
(g) the consequences as to planetary structure.

Modes of Energy Conversion

This section is prefatory to the next three, all of which deal with the differences in size and composition among the terrestrial planets. Explaining planetary sizes is largely a matter of understanding how relative kinetic energies of protoplanetary bodies are reduced; explaining planetary compositions is largely a matter of inferring heat sources and sinks at various stages of evolution. The principal difficulty in these explanations is essentially that the processes leading to planets are spill-overs or spin-offs from much larger scale events.

Since adequate evidence of short-lived radio isotopes is lacking (Schramm *et al.*, 1970), the significant energy sources which existed in the proto-solar system cloud were threefold:

(a) the potential energy of its extended distribution of matter,
(b) the kinetic energy of its internal motions, and
(c) the electromagnetic energy of its entrained field.

In most models, the potential energy is converted to heat which is retained when the collapsing cloud becomes sufficiently opaque (e.g., Larson, 1969); a minor portion of the kinetic energy is retained, nearly all of it as angular momentum of planetary motion; and the electromagnetic energy is either ignored or invoked to transfer momentum (e.g., Hoyle, 1960) or to heat condensed matter (Sonett *et al.*, 1970). It is also usually suggested that instabilities in the Sun led to considerable mass ejection, and hence kinetic energy ejection (e.g. Cameron, 1973).

A neglected energy source plausibly of importance to differentiation is the kinetic energy developed by perturbations of planetesimals by planet embryos. This neglect probably arises from the difficulty of modeling this velocity dispersion. The most detailed work thereon is by Safronov (1972a); models using his formulation indicate a lot of this kinetic energy could have been converted to heat through collisions in the inner solar system, if Jupiter's embryo became sufficiently large before the terrestrial planets completed formation (Kaula and Bigeleisen, 1975).

Another factor which has probably lead to underestimation of collision importance is that in a population of bodies with the usually assumed $n(m) \propto m^{-q}$ mass spectrum, the overwhelmingly most common *events* involve very small bodies and large mass ratios. However, if *equal increments of mass* are taken as the reference population, then it becomes apparent that an appreciable portion of the total mass can be significantly affected by the small fraction of events involving larger bodies and moderate mass ratios.

SOLAR INFLUENCES

The formation of the planets is, of course, a by-product of the formation of the Sun. Formation of a star as small as the Sun from even a relatively cool interstellar cloud requires the development of some sort of irregularity leading to fragmentation. Theoreticians of cloud collapse have yet to develop heuristic measures of irregularity (analogous to the hydromagnetists' "helicity"). However, some properties important to planetary formation appear in recent cloud collapse models by Larson (1969, 1972a, b, 1973). First, collapse of a spherically symmetric model (with no angular momentum) is highly nonhomologous; a back pressure develops which lengthens stellar formation time to a few times the free-fall time. Second, collapse models conserving the angular momentum of a typical cloud suggest that the normal pattern is formation of a star cluster, as is also observed. Dynamical models of systems of three or four particles of similar masses normally eject members until down to a binary in a relatively short time (Standish, 1972; Harrington, 1974). These models suggest that cloud fragments would have become well separated from their immediate neighbors before they had time to collapse into stars. Hence there would have been considerable mutual tide effects on the outer parts of the fragments, which were the proto-planetary material, with obvious implications for their angular momentum. This mechanism is consistent with Ostriker's (1972) conclusion that the solution of the solar system angular momentum problem probably lies in the initial conditions.

The slowing down of star formation because of pressures generated by shock has lead Cameron (1973) to suggest that the planets largely formed before the Sun reached appreciable size. While it is plausible that sizeable planetesimals formed before the Sun got all together, Cameron obtains a short planet formation time only by ignoring the development of velocity dispersion through mutual perturbations, equivalent to assuming nearly 100% efficiency of conversion of kinetic energy into comminution and heating upon collision. A more rational physics leads to a stretching out of planet formation time beyond solar formation time, so the more conventional model of a nebula stabilized by a central Sun appears applicable.

A solar influence required in all models is the occurrence of a solar outflow orders-of-magnitude stronger than the present solar wind, in order to sweep away excess volatiles from all parts of the planetary system except Jupiter and Saturn. This sweeping out occurred *after* Jupiter and Saturn became large enough to capture gas gravitationally, but *before* Uranus and Neptune were this large. It also places a constraint on the mass ratio of Uranus–Neptune zone planetesimals to terrestrial zone planetesimals at sweep-out time, in order that volatiles not be removed from the former while being removed from the latter. "Volatiles" in this context means ices for the outer solar system, but effectively Na_2O for the inner, the hardest-to-vaporize abundant constituent in which the Earth is appreciably depleted (Ringwood, 1966). While the eventual formation time of the outermost planets is still a matter of considerable uncertainty, it is not too severe a constraint that, given the earlier formation of Jupiter and Saturn, proto-Uranus and proto-Neptune planetesimals were sufficiently large to protect their ices from evaporation and sweep-out.

At present there is a better prospect of defining more sharply the magnitude and duration of the early solar outpourings. The principal published work thereon is by Kuhi (1964), who estimated a total mass outflux of $0.4 M_\odot$ based on flux rates derived from the spectra of six stars plus combination of T-Tauri emission-line class statistics with a model-dependent formation time of 3×10^7 years. The six stars are 1.2–4.8 bolometric magnitudes brighter than estimated for the forming Sun (Larson, 1969); two are the brightest of 27 Taurides listed by Herbig (1962), while one is the third brightest of 45 in the Orion nebula. Recent observational work suggests lower outflux rates, while theoretical work suggests shorter formation times (S. E. Strom, R. Ulrich, informal communications), so the total mass outflux may be significantly less—but still sufficient to sweep out a few times $0.001 M_\odot$ of volatiles.

Another circumstance suggesting phasing with respect to the solar sweep-out is the drastic difference in volatile retention of the Earth and the Moon, a strong indicator that, on the average, Earth material was last combined into sufficient sized bodies to prevent volatile loss significantly earlier than was lunar material. The similarity of terrestrial inert gas abundance ratios to those of the 'planetary' gases in chondrites suggests that most gas loss occurred when accretion had not gotton beyond rather small planetesimals (Wasson, 1974).

A solar effect of debatable significance is heating of planetesimals by electromagnetic induction (Sonett *et al.*, 1970, 1975). These models depend not only on the aforementioned large scale T-Tauri phase, but also on extrapolations of initial spin, spin-down rate (Kraft, 1972), magnetic field intensity, and sector structure. As emphasized by Sonett *et al.* (1975), electromagnetic and collisional heating may interact in that heating by induction has a strongly nonlinear dependence on temperature, requiring warming to about $400°C$ to raise the conductivity enough for the electromagnetic effect to dominate.

The main factors I wish to emphasize regarding solar influences are (1) the probable formation as a cloud fragment led to mutual tidal effects between fragments having significant effect on the angular momentum of proto-planetary matter; and (2) strong solar wind outflow probably removed volatiles when the proto-terrestrial planetary matter was still in the form of planetesimals.

JUPITER INFLUENCES

The absence of a planet in the asteroidal zone and the stunted growth of Mars strongly indicate that Jupiter reached significant size before the terrestrial planets were fully formed. In this context, "significant" means roughly more than 3 $M_\oplus$, at which stage the amount of matter perturbed by Jupiter to other zones would have become considerable. Monte Carlo studies obtain that, while the mass density of Jupiter-scattered material is greatest at Jupiter's orbit, the excess kinetic energy density is a maximum at the Sun: the increase in v^2 from Jupiter inward is greater than the decrease in mass [Kaula and Bigeleisen (1975). A similar effect at a later stage helps account for comparable crater densities on Mercury and the Moon]. Given sufficient mass scattered by Jupiter and given that the inner three planets were far from complete formation when scattering became significant, this effect could be adequate to explain the steep temperature gradient inferred by Lewis (1974) and Grossman (1975) from planetary densities, inferred compositions, and equilibrium condensation calculations.

Energy delivery by collision is a mechanism which results in strong correlation of heating, and hence composition, with size of both colliding planetesimals and resulting fragments. The excess of kinetic energy over that associated with momentum conservation, and hence heating, has a strong dependence on the mass ratio of the colliding bodies (Kaula and Bigeleisen, 1975). Thus the effectiveness of the mechanism depends upon the relative magnitudes of the mass spectra of the Jupiter-scattered and terrestrial-zone planetesimals. Smaller fragments from a collision suffer higher temperatures because of the positive correlation of heat and comminution energies. Furthermore, in collisions there is an inverse correlation of fragment size with radial location in the parent body (Gault and Wedekind, 1969; Chapman, 1974). Hence any preexisting differentiation in planetesimals will lead to correlations of composition with fragment size. Finally, the probability of capture by a proto-satellite swarm about an embryo planet is an inverse function of fragment size (Kaula and Bigeleisen, 1975).

The influence of Jupiter through scattering is difficult to calculate. This complexity leads to unpopularity of collisions as a heating and differentiation mechanism, and their labeling as "ad hoc", a misuse of this phrase to mean "difficult to model" rather than "improbably specific". But the very probable phasing of Jupiter's formation relative to the inner planets, and the likelihood that the excess condensates in Jupiter's zone available to scatter amounted to more than 100 $M_\oplus$ (Safronov, 1972a), make dynamical circumstances of unavoidable importance in the formation of these planets.

DIFFERENTIATION AMONG THE TERRESTRIAL PLANETS

The mechanical circumstances of formation of the terrestrial planets and the Moon were almost certainly the falling of condensates to the central plane and the initial formation of planetesimals by gravitational instability (Safronov, 1972a, Chaps. 3, 5–6; Goldreich and Ward, 1973), followed by the development of the competitive effect of velocity dispersion, but not so much as to prevent the growth of at least one planetesimal in each zone until it was sufficiently large to sweep up other bodies by gravitational capture (Safronov, 1972a, Chaps. 7–9). Superimposed upon this zone-by-zone isolated picture were the solar and Jovian influences, which swept out nearly all volatiles and prevented the formation of a planet in the asteroidal zone, and thus undoubtedly had other significant influences on terrestrial planet evolution. The major problems of inner solar system origin can be said to be the phasing and magnitude of these effects superimposed upon the "normal" evolution.

In some recent papers emphasizing the compositional aspects, effects arising from the occurrence of a planetesimal phase between dust grains and planets have been slighted. Thus in the model of Cameron (1973), the implicit assumption is made that in each zone only one planetesimal developed; the resulting low velocity dispersion allowed it to sweep up enough matter to become a planet in 1000 years or so. In discussions of the equilibrium condensation versus inhomogenous accretion processes (Lewis, 1972; Grossman and Larimer, 1974) no distinction is made between planetesimals and planets; it appears to be assumed that planets accrete the products of condensation directly. There is certainly no necessity for such accretion to prevent further thermochemical reaction of the condensates with the gas; meter-sized agglomerations suffice (Wasson, 1972).

It is impressive that the densities of bodies from Mercury to Titan (the Moon excluded) are compatible with a temperature proportionate to $1/R$, where R is distance from the Sun (Lewis, 1974). However, to infer directly that the nebula must have had a high opacity depends too much on the assumptions of similarity in time and altitude with respect to the median plane of condensation of the various bodies. Dynamical considerations cannot say anything directly about the time of condensation, but they do indicate the time of formation of various bodies, which in turn would likely be correlated with condensation time and altitude, and hence pressure. Titan and the Galilean satellites were certainly among the last 10^{-3} or so of ices to combine into bodies in the major planet zones (Harris and Kaula, 1975). Ceres and Vesta (the only asteroids for which bulk density estimates exist) might also have been late-formers, but it is more likely they are exceptional-escapers from the disrupting effects of Jupiter. The variations in reflectance spectra of asteroids (McCord and Chapman, 1975) are difficult to reconcile with any model dependent solely on distance from the sun.

Opacity probably had an influence on Jupiter and Saturn being first to form. This influence may also have caused early condensation of meteorite and asteroidal material. But the need for a steeper-than-radiative-equilibrium temperature gradient among the four terrestrial planets does not seem compelling, entirely aside from the earlier-mentioned possibility that collision energy resulting from Jupiter scattering varied inversely with distance from the Sun.

At the other extreme, models which explain differences among planets by differentiations within planets (e.g., Ringwood, 1966) perhaps overlook that efficient differentiations are known to have occurred in rather small planetesimals: the parent bodies of achondrite, stony-iron, and iron bodies. There has been no significant contradiction of the nickel-iron diffusion gradient evidence of cooling rates for iron meteorites indicative of parent bodies in the 10's of km size (Wood, 1964; Goldstein and Short, 1967; Fricker *et al.*, 1970). It may be argued that nonchondritic meteorites are a minor portion of meteoritic material, and thence that only a minor portion of all material suffered appreciable differentiation due to collision. But meteorites are surely not a systematic sampling of the asteroidal zone of the nebula. Probably almost all meteorites came from a relatively few recent collisions, as evidenced by the remarkable clusterings in composition (Wasson, 1974). Chondrites are unrepresentatively abundant because bodies not suffering earlier major collision were not ejected, and thus have persisted out of proportion in the asteroid belt, a result of kinetic energy of collision fragments being positively correlated with thermal energy density. Furthermore, among matter that has been collision-affected, irons are unrepresentatively abundant because core fragments, which contain more iron, have lower kinetic energies, on the average, than shell fragments, which contain more differentiated silicates.

The Moon is most markedly the consequence of dynamically affected differentiation, and the most marked contradiction of the nebula condensation and planetary extremes of compositional differentiation models: the contrivances to get a Moon from such models are most charitably described as awkward (Kaula and Harris, 1975). Explanations of the severe volatile and iron depletions, and the plagioclase enrichment [a more accurate term than "refractory enrichment", since, as emphasized by Grossman (1975), the Moon's surface is drastically depleted in siderophile refractories] should be sought in the circumstances that the Moon was formed from a circum-terrestrial swarm which started

when the Earth's embryo was rather small, but much of which arrived relatively late to the Earth–Moon system (Harris· and Kaula, 1975). The matter in this swarm would have been preferentially constituted of the smaller bodies coming to the Earth–Moon system, and subjected to more collisions and solar effects than planetary matter. Hence it is plausible that the Moon's bulk constitution is devolatilized matter with an excess from plagiochlase-rich planetesimal crusts and a deficiency from iron-rich planetesimal cores (Kaula, 1975a; Kaula and Bigeleisen, 1975).

It will be objected that the same processes should have lead to devolatilized satellites about the major planets, rather than the observed icy bodies. However, there are at least three great differences in circumstances. First, the tidal dissipation in the major planets is much less than in the terrestrial (Goldreich and Soter, 1966); hence the circum-planetary matter is much more systematically drawn in by the growth of the planet, and the eventual satellite matter is much more a late fraction (Harris and Kaula, 1975). Second, material being added to the swarm was more volatilerich. Third, the solar wind available to sweep out matter volatilized in collisions was weaker. Furthermore, it is plausible that the rate of disruption by infall was much lower relative to the rate of formation of satellites, as evidenced by the small eccentricities and inclinations of the orbits.

The condensation sequence and temperature gradient with respect to distance from the Sun probably determined the gross composition of the terrestrial planet zones. However, solar wind effects probably started to remove volatiles from the inner solar system before planetesimal formation was far advanced, and continued for some time. The influx of Jupiter-scattered material probably did not become significant until an appreciable Earth embryo had formed: at least 10% of the final mass. Any inhomogeneous accretion was probably on a local scale. The timing of planet growth determined its mix of Fe, FeS, silicates and volatiles, but it was crudely mixed in the planet.

Given that the Jupiter-scattered material enhanced the velocity dispersion, the time scale of formation was, if anything, longer than the 10^7 years of the isolated-zone models (Safronov, 1972a). Hence any significant heating depended on the mass influx, including some sizeable planetesimals of as much as a few-100 km size (Safronov, 1972a, b; Arrhenius *et al.*, 1974). Such large infalls also contributed significantly in a mechanical way to early planetary overturns.

The Moon was a much different matter: the shorter time scale of geocentric formation may have led to considerable near-surface heating. The size of some of the constituent moonlets may have had appreciable initial effect on the Moon (Ruskol, 1973); more likely to have been important were planetesimals which fell in at high velocity.

Rate of Planetary Thermal Evolution

The evolution of the thermal state of a planet depends, of course, on the relative magnitudes of energy gain and loss rates. The general pattern is given by the simple models of Toksöz *et al.* (1975). The basic factors are all fairly evident. On the gain side, planets start off with an energy excess, as a consequence of infalls. As discussed in the previous section, in the case of planets in heliocentric orbit the accumulation rate was probably so slow that significant heating occurred only because of the considerable size of some infalling planetesimals. In the case of the Moon, the shorter time scale of formation in geocentric orbit made considerable heating likely. An important threshold for planetary evolution is whether sufficient iron is reduced for the stimulus of infalls to trigger formation of a core. Core formation is much the most effective way to heat a body, since the energy is released internally, and not radiated away easily. Thus for the Earth, the energy released by core formation from a homogeneous body would have been about 2500 joule g^{-1} (Birch, 1965; Tozer, 1965). It is likely that the energy release was somewhat less, because the earlier infalling planetesimals constituting the Earth's center retained more carbon to reduce the iron, but in any case core formation energy release is the evident reason why the Earth apparently could not generate much stable lithosphere for 1200 m.y. (Wetherill, 1972).

Eventually radiogenic heat dominates, of course. For chondritic composition it was initially about 1.2 joule g^{-1} $(10^6 \text{yr})^{-1}$, declining to 0.2 joule g^{-1} $(10^6 \text{yr})^{-1}$ at present, sufficient to have caused melting within 1200 m.y. if the heat were not carried away. For the Earth's mantle, where the K/U ratio is much lower, the corresponding figures are 0.8 and 0.2 joule g^{-1} $(10^6 \text{yr})^{-1}$, respectively: a slower decay.

Planetary size greatly influences thermal evolution: the greater the area-to-mass ratio, the more rapid is the heat loss, and hence thermal evolution. Another obvious effect worth articulating is depth of heat sources: the deeper the sources, the slower the evolution. Depth is effectively measured with respect to the radiating level: the cloud tops, in the case of Venus.

As reiterated by Tozer (1970, 1974), heat transfer in planets which have not reached the quiescent stage is almost certainly dominated by solid state convection, which keeps temperatures somewhat below melting in a reasonably uniform material, because of the strong temperature dependence of viscosity. Also contributing to the lowering of temperatures is the presence of water. An unsettled question is whether the combination of temperature dependence of viscosity and contained heat sources makes convection in a planet oscillatory, as suggested for the earth by the geologic record. Two-dimensional computer experiments indicate that contained heat sources cause a shifting around of the flow pattern on a time scale of about 200 m.y. (McKenzie *et al.*, 1974). This result is not necessarily applicable to the three-dimensional reality, however.

The spatial character of convective flow is principally dependent on the combination of heat generation, length scale, gravity, and material properties expressed by the Rayleigh number: the higher the Rayleigh number, the smaller the cell size, all other things being equal. However, they never are equal: flow systems have a considerable hysteresis, and established planforms tend to persist when the Rayleigh number has changed, so that a different mode would be more effective in transferring heat. This principle has been applied to the Earth to justify "plumes": i.e., the persistence of a single cell system after the Rayleigh number has increased (Turcotte and Oxburgh, 1967; Morgan, 1972). However, it is more likely that the opposite applies: that the Earth and other terrestrial planets have decreased in their heat supplies, and hence that the flow patterns are shorter wavelengths than predicted by Rayleigh number considerations alone. That such circumstances can occur in the Bénard problem at moderate Rayleigh number has been demonstrated by Busse (1967). Such hangover systems would tend to be oscillatory, since they can remove more heat than is generated. It is difficult to believe, though, that they can be as regularly self-extinguishing as suggested by Tikhonov et al. (1970; see also Lubimova, 1970).

Tozer (1970), somewhat contrary to the foregoing, assets that present conditions are insensitive to intial conditions, because of the stabilizing effect of temperature dependence of viscosity. While this insensitivity is certainly true in a deterministic sense, the average character of planetary flow systems may lag significantly behind the thermal state—perhaps even to the extent that initial inhomogeneities may still be of influence. Statistical approaches to this problem, such as that by Vityazev (1972), should be developed.

When sufficient heat has been removed for a lithosphere to develop, a significant portion of the heat transfer in the outermost zone is by conduction through this lithosphere. When heat is removed to the extent that convection can no longer break the lithosphere, conduction dominates. But even in the smallest planet, the Moon, the lithosphere could not be more than 200 km or so thick now unless there had been significant upward shifting of heat sources.

A process auxiliary to both solid state convection and conduction is magmatic penetration. Magma tubes are probably secondary as heat transfer mechanisms; their contribution to thermal history is much more by upward transfer of radioactive heat sources, as incorporated in most recent models (Cassen and Reynolds, 1974; Johnston et al., 1974; Toksöz and Johnston, 1974).

This discussion of thermal evolution is incomplete because of its dependence on differentiation, to be taken up in the next section. In particular, water content of even a few 0.1% can have a major effect on effective viscosity at pressures in excess of 30 kbar (Tozer, 1974; Wood, 1975). In any case, all planets are, on the average, running down at rates determined by planet size and heat source distribution: the latter affected first by origin circumstances and later by differentiation processes.

DIFFERENTIATION WITHIN PLANETS

A simplistic notion is that the differentiation rate is proportionate to the convection rate. Dickinson and Luth (1971) took the present rate of crustal creation of the Earth (about $18 \text{km}^3 \text{yr}^{-1}$), assumed that this rate was proportionate to the rate of radiogenic heat generation, integrated back in time, and found that the mantle should have been used up in 3000 m.y. However, the optimum convective rate for some classes of differentiation may be appreciably less than the maximum rate which has occurred. It is a truism of igneous petrology that the most extreme differentiations occur in the dwindling stages of volcanism or plutonism, rather than in the earlier vigorous stages. The K_2O/Na_2O ratio is one of the principal indices of the degree of differentiation. A global analysis thereof is used by Engel et al. (1974) to infer the extent to which differentiation (sedimentary as well as igneous) is a reworking of previously differentiated material rather than the initial fractionation of more primitive matter. For Archean rocks, more than 2500 m.y. old, they find K_2O/Na_2O ratios in the range 0.4–0.6 for igneous rocks and 0.5–0.8 For sedimentary (cf. 0.1 for Cl chondrites). The Archean is followed by a transition period ending about 2000 m.y. ago, since which the ratios are 0.9–1.5 for igneous and 1.1–3.3 for sedimentary rocks, *except* for the last 250 m.y., in which the ratios first dropped sharply to Archean levels and since have risen half-way back to the Proterozoic–Paleozoic levels. This K_2O/Na_2O indicator correlates with other indicators of differentiation degree, such as relative volumes of rock types and abundances of large cations. Perhaps the Mesozoic–Cenozoic variation in the K_2O/Na_2O reflects a characteristic convective oscillation time of 250 m.y. or more: in another 500 or 1000 m.y., this wiggle in the data may be appreciably damped.

From a fluid dynamical point-of-view, one would expect that a major component differentiation involving a given density differential would occur in a vigorously convecting system only if this density differential were comparable to the differential driving the system. Thus, in early convection caused by major infalls and differentiation of an iron core, volatiles would have been degassed, but silicate differentiation would have been limited: most of the acidic and basic matter would have been swept down again with the ultrabasic main constituent. From a cosmogonic point-of-view, it is difficult to distinguish this vigorously convecting stage from the impacting process which created the planet. During this process most, but not all, volatiles would have been outgassed to form an atmosphere, and reduced iron would have separated to start core formation. The extent to which convection too vigorous to allow silicate differentiation continued depended on the amount of energy released by sinking of iron. As the initial gravitational and early radiogenic heat were removed and the convective vigor decreased, more light material was differentiated at the surface to form a crust, and the efficiency of volatile outgassing increased. Similar ideas of early but incomplete outgassing are presented by Fanale (1971) and Siever (1974).

The smaller the planet, the earlier crustal differentiation occurred, since energy was lost and convective vigor declined sooner, so that differentiates were no longer carried down again. In addition, the more primitive was the crust, since energy was lacking to sustain the recycling necessary to accomplish the more advanced differentiations. However, even in the Moon recycling of crustal matter necessarily occurred in order to produce the K_2O-enriched norites prominent in the Imbrian and Procellarian regions (Wood, 1975). The overlapping of decline in convective vigor with continued infall of planetesimals also led to higher portions of ultramafic minerals in lunar crustal rocks (Wood, 1975).

Somewhat contradictory to the foregoing notions of inefficiency of differentiation processes are the severity of some trace element enrichments and depletions measured in terrestrial and lunar rocks. However, all these abundances are peculiar in that they pertain to surface material, which is the product of magmatism at pressures negligible compared to those which exist throughout the bulk of the planets: i.e., conditions most favourable to separation of fractions of a melt, etc. Interior to planets the average differentiation of trace elements is probably appreciably less complete. An evidence thereof is the heat flow from the Earth. Typical crustal models (e.g., Gast, 1972) have enough U, Th, K to account for only one-fourth of the heat flow. The balance must come from below the crust, and hence average mantle material may have 40 times as much U, Th, K than is found in dunitic rocks which have come to the surface.

Despite the foregoing considerations, it still seems remarkable that the Moon's lithosphere is so thick, and hence that upward differentiation of large-ion lithophiles was so much more efficient than in the Earth. Perhaps convection in the Earth's mantle has always been significantly decoupled by the activation energy increase associated with the phase transition at about 650 km depth (McKenzie and Weiss, 1975).

PLANETARY STRUCTURE

We have estimates for at least four bodies of three properties which are primary—i.e., reflecting internal structure varying in the radial direction—and three more which are secondary—reflecting lateral variations in structure.

The primary quantities are mean density, moment-of-inertia, and magnetic field intensity. As mentioned earlier, the mean densities seem well explained by the variation in condensation temperatures with distance from the Sun, with the outstanding exception of the Moon. The moment-of-inertia ratio I/MR^2 varies inversely with size of the three bodies for which it is known accurately, the Earth, Mars, and the Moon, suggesting that acquisition of iron, plus volatiles and temperatures to reduce the iron, is dependent on planet size. In addition, for Mercury there is the indirect evidence of core-mantle differentiation in that the large areas of volcanism indicate that a thick silicate layer constitutes the parent material (Murray *et al.*, 1975). Of the five terrestrial bodies, only the Earth requires a predominantly iron fluid core to generate a magnetic field by dynamo action. The slow rotation of Venus may account for the absence of a marked dipole, but it is puzzling that there are not nondipole terms comparable to the Earth's if dynamos are driven by thermal convection. Possibly precessional torques (Malkus, 1968) are necessary to furnish the energy, after all. The past intensity evidenced in the lunar remanent magnetism is even more of a puzzle (Fuller, 1974). Magnetization of its forming planetesimals, lost when the interior warmed sufficiently to raise the temperature above the Curie point (Runcorn and Urey, 1973), seems most plausible.

In addition, there are seismological and other data which indicate that the Earth's lithosphere is a thin layer of 80–100 km over a relatively plastic zone, while the Moon has a lithosphere 800–1000 km thick.

The secondary properties are gravity anomalies, topographic heights, and evidences of lateral displacement. All are indirect indicators of the thermal and compositional state.

The magnitudes of gravity anomalies depend on imbalances between disturbing effects—generally dynamic—and restorative effects—generally passive. Disturbing effects are most likely generated in some sort of hot inner zone which can flow—an asthenosphere—while restorative effects most likely reside in a stiffer zone—a lithosphere. A particularly hot planet would have small gravity anomalies because its resisting lithosphere is thin, while a particularly cold planet would have small gravity anomalies because its driving convection is too weak. The gravity anomaly spectrum of the Earth is characterized by a "rule" $10^5 l^2$ for the rms magnitude of normalized potential harmonic coefficients of wavenumber l, corresponding, by Poisson's law, to a white spectrum in density irregularities, which suggests that the typical lengths of the causitive phenomena are something less than the shortest wavelength to which the rule applies. Extrapolation of the rule to the Moon under the assumption of equal stresses leads to an overestimate by two or three of lunar gravity variations, while a similar extrapolation to Mars leads to an underestimate by two or three. If lithospheric thickness is inversely proportionate to planet size, then, as suggested by Fig. 1, Mars is closest to the optimum for maximizing gravity irregularities (Kaula, 1975b).

The topography is a more obscure indicator, since in any planet which has differentiated a crust, the longer wavelength variations will tend to be isostatically compensated. This is certainly true for both the Earth and Moon, but less so for Mars. The magnitude of topographic variations, and their correlations with the gravity variations, can be used to infer crustal thickness.

So far, only the Earth among the terrestrial planets has evidence of lateral motions on its surface, suggesting that the smaller planets have already passed through the stage of convection active enough to break the lithosphere. In the case of Mars, atmospheric effects may have obliterated such evidences. Mercury and the Moon apparently passed throuth the plate tectonic stage too rapidly and too early for its traces to remain. The effective definition of plate tectonics is a circumstance where most of the boundary layer is solid, but broken along continuous lines. Passage through such a stage is unavoidable for any planet which has had a major crustal differentiation: there is no way else to get from a situation active enough to differentiate 10km or more of crust to one where the lithosphere is broken only by magma vents. Or, in other words, effective Rayleigh numbers can change with time only in a continuous manner (except by infalls implausibly large for the last 4500m.y.). The Moon has obviously had a major crustal differentiation. The principal quantitative evidences thereof on Mars are the offset of its center-of-figure from center-of-mass (Lingenfelter and Schubert, 1973), the discrepancy between its optical and dynamical oblateness, and gravity field variations which, although correlated with topography, are only half as great as would exist if the topography were rigidly supported on a homogeneous body. The principal quantitative evidence of crustal differentiation on Venus is also an offset of center-of-figure from center-of-mass, of about 1.5km (Smith et al., 1970). There are also some indications of broad topographic variations of ±3km (Campbell et al., 1972), suggesting that isostasy prevails. Gamma ray spectroscopy at the Venera 8 site indicates high K, U, Th content, comparable to terrestrial granites (Vinogradov et al., 1973), while radar imaging of 0.4% of the Venusian surface indicates craterlike structures (Rumsey et al., 1974). Both of these indicators suggest that the planet is well past the vigorously convecting stage. Mercury has some linear features indicative of overthrust in response to compression (Murray et al., 1975), but they are far from constituting a continuous net. Probably the major evidence is its areas of high albedo, and its moonlike spectral reflectance (McCord and Adams, 1972), requiring differentiation of a sialic layer.

DISCUSSION

I still like the ideas of Ringwood (1966) on terrestrial planet formation: the emphases on autoreduction due to entrainment of volatiles as the cause of core formation, on gross inhomogeneities as the cause of the mantle-core disequilibrium, and on degree of oxidation to explain differences among planets. However, a massive primordial atmosphere as he suggested raises great difficulties as to energy source for its removal (Kaula and Harris, 1975). The geologic need for such an atmosphere (Rubey, 1951) now appears to have faded (Fanale, 1971; Arrhenius et al., 1974; Siever, 1974). Hence it is desirable to infer the minimal temperatures, carbon abundances, etc., to effect core differentiation, with the hope that the volatile loss can take place at the planetesimal stage. The influence of the nebula temperature gradient through condensation on composition (Lewis, 1974; Grossman, 1975) undoubtedly was significant. However, numerous planetesimals must have existed between the initial grains and the final planets; the compositional implications of the dynamical effects thereof emphasized by Safronov (1972a) need to be taken into account. As discussed above, the evidences of earlier formation of Jupiter make it difficult to avoid the inference that planetesimal collisions were important. The phasing of this planetesimal interaction stage with respect to the "magic broom" of the Sun's T Tauri phase is also important. It is difficult to believe that this outstreaming did not start very soon after the Sun got together, and thus before the combination of planetesimals into planets had progressed very far in the inner solar system. Hence the volatile content of planetesimals should have continuously decreased as planet growth proceeded. A final problem is the physics of impacts by sizeable planetesimals; only in this way can the energy trigger early planet differentiation be delivered, if the time scale of planet formation was stretched out by velocity dispersion (Safronov, 1972a). Planetesimal collisions are also probably important in explaining various compositional inhomogeneities, such as the Moon.

Once the planets got together, their evolutions were determined almost entirely by internal conditions. A planet is a rather imperturbable object. Even a 10^{31} ergs Imbrium impact on the Moon (Kaula, 1969), while it may have fractured most of the lithosphere, did not remove more than 5×10^{-6} of the moon's mass nor heat more than 10^{-3} of it by 10 K. 10^{31} ergs was generated by radioactivity in the Moon within 3×10^{4} years. The evidences of core and crust formation do indicate very early heating from impacts much greater than Imbrium, comparable to 10^{9} yr of radiogenic heat. Hence planets are running down, on the whole, but within this general decline there have been zones which heated appreciably (particularly centers, such as the Earth's and the Moon's), or which have long retained a marginal capability for solid state convection because of the temperature dependence of viscosity emphasized by Tozer (1970, 1974). Subject areas in which further progress would help toward an understanding of the general evolutionary problem are the influence of declining sources and thermal "hysteresis" on oscillatory behaviour of convective systems, the relationship of petrological differentiation to convective vigor, and the role of relatively small H_2O and CO_2 contents.

This paper is eclectic, and doubtless reiterates some ideas suggested long ago by Shmidt (1958), Urey (1952), and others. It is motivated somewhat as an excuse to publish the poem which constitutes the abstract and somewhat as a draft for the final chapter of a textbook revision (Kaula, 1968). But I hope it also serves as a scenario to guide more serious and detailed investigations. Such scenarios are always in our minds more or less subconsciously in choosing which problems to work on and which approaches to take. It is desirable that they be made more explicit, to counterbalance the excessive influence of technical feasibility on the choices.

ACKNOWLEDGMENTS

I am grateful to D. A. Yuen (Department of Geophysics and Space Physics, UCLA) for being stubborn about why the Moon's thermal evolution differed from the Earth's, and to D. C. Kaula (Department of English, Univ. Western Ontario) for literary criticism. This work was partially supported by NASA grant NGL 05-007-002 and NSF grant GA-40749. Institute of Geophysics and Planetary Physics Publication No. 1435.

REFERENCES

ARRHENIUS, G., DE, B. R., AND ALFVÉN, H. (1974). Origin of the ocean. In *The Sea* 5 (E. Goldberg, Ed.). Wiley, New York.

BIRCH, F. (1965). Energetics of core formation. *J. Geophys. Res.* **70**, 6217–6221.

BUSSE, F. (1967). The stability of finite amplitude cellar convection and its relation to an extremum principle. *J. Fluid Mech.* **30**, 625–649.

CAMERON, A. G. W. (1973). Accumulation processes in the primitive solar nebula. *Icarus* **18**, 407–450.

CAMPBELL, D. B., DYCE, R. B., INGALLS, R. P., PETTENGILL, G. H., AND SHAPIRO, I. I. (1972). Venus: Topography revealed by radar data. *Science* **175**, 514–516.

CASSEN, P., REYNOLDS, R. T. (1974). Convection in the Moon: effect of variable viscosity. *J. Geophys. Res.* **79**, 2937–2944.

CHAPMAN, C. R. (1974). Asteroid size distribution: Implications for the origin of stony-iron and iron meteorites. *Geophys. Res. Lett.* **1**, 341–344.

DICKINSON, W. R., AND LUTH, W. C. (1971). A model for plate tectonic evolution of mantle layers. *Science* **174**, 400–404.

ENGEL, A. E. J., TISON, S. P., ENGEL, C. G., STICKNEY, D. M., AND GRAY, E. J. (1974). Crustal evolution and global tectonics: a petrogenic view. *Bull. Geol. Soc. Amer.* **85**, 843–858.

FANALE, F. P. (1971). A case for catastrophic early degassing of the Earth. *Chem. Geol.* **8**, 79–105.

FRICKER, P. E., GOLDSTEIN, J. I., AND SUMMER, A. L. (1970). Cooling rates and thermal histories of iron and stony-iron meteorites. *Geochim. Cosmochim. Acta* **34**, 475–491.

FULLER, M. (1974). Lunar magnetism. *Rev. Geophys. Space Phys.* **12**, 23–70.

GAST, P. W. (1972). The chemical composition of the Earth, the Moon, and chondritic meteorites. In *The Nature of the Solid Earth* (E. C. Robertson, Ed.), pp. 19–40. McGraw-Hill, New York.

GAULT, D. L., AND WEDEKIND, J. A. (1969). The destruction of tektites by micrometeoroid impact. *J. Geophys. Res.* **74**, 6780–6794.

GOLDREICH, P., AND SOTER, S. (1966). Q in the solar system. *Icarus* **5**, 375–389.

GOLDREICH, P., AND WARD, W. R. (1973). The formation of planetesimals. *Astrophys. J.* **183**, 1051–1061.

GOLDSTEIN, J. I., AND SHORT, J. M. (1967). The iron meteorites, their thermal history and parent bodies. *Geochim. Cosmochim. Acta* **31**, 1733–1770.

GROSSMAN, L. (1975). Chemical fractionation in the solar nebula. *Proc. Sov.-Amer. Conf. Cosmochem. Moon Planets*, Moscow, in press.

GROSSMAN, L. AND LARIMER, J. W. (1974). Early chemical history of the solar system. *Rev. Geophys. Space Phys.* **12**, 71–101.

HARRINGTON, R. S. (1974). The dynamical decay of unstable 4-body systems. *Celes. Mech.* **9**, 465–470.

HARRIS, A. W., AND KAULA, W. M. (1975). A co-accretional model of satellite formation. *Icarus* **25**, 516–524.

HERBIG, G. (1962). The properties and probmles of T Tauri stars and related objects. *Advan. Astron. Astrophys.* **1**, 47–103.

HOYLE, F. (1960). On the origin of the solar nebula. *Quart. J. Roy. Astron. Soc.* **1**, 28–55.

JOHNSTON, D. H., McGETCHIN, T. R., AND TOKSÖZ, M. N. (1974). The thermal state and internal structure of Mars. *J. Geophys. Res.* **79**, 3959–3971.

KAULA, W. M. (1968). *An Introduction to Planetary Physics: The Terrestrial Planets.* Wiley, New York.

KAULA, W. M. (1969). Interpretation of lunar mass concentrations. *Phys. Earth Plan. Int.* **2**, 123–137.

KAULA, W. M. (1975a). Mechanical processes affecting differentiation of proto-lunar material. *Proc. Sov.-Amer. Conf. Cosmochem. Moon Planets*, Moscow, in press.

KAULA, W. M. (1975b). The gravity and shape of the moon. *EOS Trans. Amer. Geophys. Union*, **56**, 309–316.

KAULA, W. M., AND BIGELEISEN, P. E. (1975). Early scattering by Jupiter and its collision effects in the terrestrial zone. *Icarus* **25**, 18–33.

KAULA, W. M., AND HARRIS, A. W. (1975). Dynamics of lunar origin and orbital evolution. *Rev. Geophys. Space Phys.* **13**, 363–371.

KRAFT, R. P. (1972). Evidence for changes in the angular velocity of the surface regions of the Sun and stars. In *Solar Wind*, pp. 276–282. NASA SP-308.

KUHI, J. V. (1964). Mass loss from T Tauri stars. *Astrophys. J.* **140**, 1409–1433.

LARSON, R. B. (1969). Numerical calculations of the dynamics of a collapsing proto-star. *Mon. Not. Roy. Astron. Soc.* **145**, 271–295.

LARSON, R. B. (1972a). The collapse of a rotating cloud. *Mon. Not. Roy. Astron. Soc.* **156**, 437–458.

LARSON, R. B. (1972b). Collapse calculations and their implications for the formation of the solar system. *Proc. Symp. Origin Solar System*, pp. 142–150. Eds. Centre Nat. Rech. Sci., Paris.

LARSON, R. B. (1973). Processes in collapsing interstellar clouds. *Ann. Rev. Astron. Astrophys.* **11**, 219–238.

LEWIS, J. S. (1972). Metal/silicate fractionation in the solar system. *Earth Planet. Sci. Lett.* **15**, 286–290.

LEWIS, J. S. (1974). The temperature gradient in the solar nebula. *Science* **186**, 440–443.

LINGENFELTER, R. E., AND SCHUBERT, G. (1973). Evidence for convection in planetary interiors from first-order topography. *The Moon* **7**, 172–180.

LUBIMOVA, E. A. (1970). Heat flow and the dynamics of the earth's interior. *Earth Physics* (Transl. *Bull. Acad. Sci. USSR*) **5**, 273–280.

MALKUS, W. V. R. (1968). Precession of the Earth as the cause of geomagnetism. *Science* **160**, 259–264.

McCORD, T. B., AND ADAMS, J. B. (1972). Mercury: interpretation of optical observations. *Icarus* **17**, 585–588.

McCORD, R. B., AND CHAPMAN, C. R. (1975). Asteroids: spectral reflectance and color characteristics. *Astrophys. J.* **195**, 553–562.

McKENZIE, D. P., ROBERTS, J. M., AND WEISS, N. O. (1974). Convection in the Earth's mantle: Toward a numerical simulation. *J. Fluid Mech.* **62**, 465–538.

McKENZIE, D. P. AND WEISS, N. (1975). Speculations on the thermal history of the earth. *Geophys. J. Roy. Astron. Soc.*, in press.

MORGAN, W. J. (1972). Plate motion and deep mantle convection. *Mem. Geol. Soc. Amer.* **132**, 7–22.

MURRAY, B. C., BELTON, M. J. S., DANIELSON, G. E., DAVIES, M. E., GAULT, D., HAPKE, B., O'LEARY, B., STROM, R. G., SUOMI, C., AND TRASK, N. (1975). Television observations of Mercury by Mariner 10. *Proc. Sov.-Amer. Conf. Cosmochem. Moon Planets.* Moscow, in press.

OSTRIKER, J. P. (1972). Hydrodynamics of the collapse: Rotation and contraction. *Proc. Symp. Origin Solar System*, pp. 154–162. Eds. Centre Nat. Rech. Sci., Paris.

PHILLIPS, R. J., AND SAUNDERS, R. S. (1975) The isostatic state of Martian topography. *J. Geophys. Res.* **80**, in press.

RINGWOOD, A. E. (1966). Chemical evolution of the terrestrial planets. *Geochim. Cosmochim. Acta* **30**, 41–104.

RUBEY, W. W. (1951). Geologic history of sea water. *Bull. Geol. Soc. Amer.* **62**, 1111–1173.

RUMSEY, H. C., MORRIS, G. A., GREEN, R. R., AND GOLDSTEIN, R. M. (1974). A radar brightness and altitude image of a portion of Venus. *Icarus* **23**, 1–7.

RUNCORN, S. K., AND UREY, H. C. (1973). A new theory of lunar magnetism. *Science* **180**, 636–638.

RUSKOL, E. L. (1973). On the model of the accumulation of the moon compatible with the data on the composition and the age of lunar rocks. *The Moon* **6**, 190–201.

SAFRONOV, V. S. (1972a). *Evolution of the Protoplanetary Cloud and Formation of the Earth and Planets.* Israel Program for Scientific Translations, Jerusalem.

SAFRONOV, V. S. (1972b). The initial state of the earth and certain features of its evolution. *Earth Phys.* (Transl. *Bull. Acad. Sci. USSR*) **7**, 35–41.

SCHRAMM, D. N., TERA, F., AND WASSERBURG, G. J. (1970). The isotopic abundance of Mg^{26} and limits on Al^{26} in the early solar system. *Earth Planet. Sci. Lett.* **10**, 44–59.

SHMIDT, O. (1958). *Four Lectures on the Theory of the Earth's Origin* (Transl. from Russian) Foreign Lang. Pub. House, Moscow.

SIEVER, R. (1974). Comparison of Earth and Mars as differentiated planets. *Icarus* **22**, 312–324.

SMITH, W. B., INGALLS, R. P., SHAPIRO, I. E. AND ASH, M. E. (1970). Surface-height variations on Venus and Mercury. *Rad. Sci.* **5**, 411–423.

SONETT, C. P., COLBURN, D. S., SCHWARTZ, K., AND KEIL, K. (1970). The melting of asteroidal parent bodies by unipolar dynamo induction from a primordial T Tauri sun. *Astrophys. Space Sci.* **7**, 446–488.

SONETT, C. P., COLBURN, D. S., AND SCHWARTZ, K. (1975). Formation of the lunar crust: an electrical source of heating. *Icarus* **24**, 231–255.

STANDISH, E. M. (1972). The dynamical evolution of triple star systems: A numerical study. *Astron. Astrophys.* **21**, 185–191.

TIKHONOV, A. N., LUBIMOVA, E. A., AND VLASOV, V. K. (1970). On the evolution of melting zones in the thermal history of the Earth. *Phys. Earth Planet. Int.* **2**, 326–331.

TOKSÖZ, M. N., AND JOHNSTON, D. K. (1974). The evolution of the Moon. *Icarus* **21**, 389–414.

TOKSÖZ, M. N., JOHNSTON, D. H., AND MINEAR, J. W. (1975). Thermal evolution and the present state of the Moon and terrestrial planets. *Lunar Science VI*, pp. 815–817.

TOZER, D. C. (1965). Thermal history of the Earth. I. The formation of the core. *Geophys. J. Roy. Astron. Soc.* **9**, 95–112.

TOZER, D. C. (1970). Factors determining the temperature evolution of thermally convecting earth models. *Phys. Earth. Planet. Int.* **2**, 393–398.

TOZER, D. C. (1974). The internal evolution of planetary-sized objects. *The Moon* **9**, 167–182.

TURCOTTE, D. L., AND OXBURGH, E. R. (1967). Finite amplitude convection cells and continental drift. *J. Fluid Mech.* **28**, 29–42.

UREY, H. C. (1952). *The Planets*, Yale Univ. Press, New Haven, Conn.

VINOGRADOV, A. P., SURKOV, Y. A., AND KIRNOZOV, F. F. (1973). The contents of uranium, thorium, and potassium rocks of Venus as measured by Venera 8. *Icarus* **20**, 253–259.

VITYAZEV, A. V. (1972). On the dynamic-statistical theory of the Earth's evolution. *Earth Physics* (Transl. *Bull. Acad. Sci. USSR*) **7**, 42–47.

WASSON, J. T. (1972). Formation of ordinary chondrites. *Rev. Geophys. Space Phys.* **10**, 711–759.

WASSON, J. T. (1974). *Meteorites.* Springer-Verlag, Heidelberg.

WETHERILL, G. W. (1972). The beginning of continental evolution. In *The Upper Mantle, Tectonophysics* (A. E. Ritsema, Ed.) **13**, 31–45.

WOOD, J. A. (1964). The cooling rates and parent planets of several iron meteorites. *Icarus* **3**, 429–459.

WOOD, J. A. (1975). A survey of lunar rock types and comparison of the crusts of Earth and Moon. *Proc. Sov.-Amer. Conf. Cosmochem. Moon Planets.* Moscow, in press.

AUTHOR CITATION INDEX

SUBJECT INDEX

About the Editor

ZVI GARFUNKEL is professor of geology at the Hebrew University of Jerusalem, Israel. He studied geology in this university, specializing in structural geology, tectonics, and regional geology of the Middle East. He received the Ph.D. (1970) for research on the Dead Sea transform. Since then he has studied various aspects of the geology of the Middle East and Southern California, concentrating on structural problems, mainly faulting, and on rifting process. He has studied extensively the structure and evolution of Israel and the surrounding areas, the Eastern Mediterranean and the formation of its passive margin, especially the Dead Sea transform, the Suez rift, and the Red Sea. He has also studied various aspects of mantle circulation, especially its relation to plate overturn and motion of subducted slabs.

Benchmark Papers in Geology Series

Editor: Rhodes W. Fairbridge, Columbia University

DOLOMITIZATION / *Donald H. Zenger and S. J. Mazzullo*
OPHIOLITIC AND RELATED MELANGES / *G. J. H. McCall*
ECONOMIC EVALUATION OF MINERAL PROPERTY /
    Sam L. VanLandingham
SUNSPOT CYCLES / *D. Justin Schove*
MINING GEOLOGY / *Willard C. Lacy*
MINERAL EXPLORATION / *Willard C. Lacy*
BASALTS / *Paul C. Ragland and John J. W. Rogers*
PHYSICAL HYDROGEOLOGY / *R. Allan Freeze and William Back*
CHEMICAL HYDROGEOLOGY / *William Back and R. Allan Freeze*
MODERN CARBONATE ENVIRONMENTS / *Ajit Bhattacharyya and*
    G. M. Friedman
FABRIC OF DUCTILE STRAIN / *Mel Stauffer*
TERRESTRIAL TRACE-FOSSILS / *William A. S. Sarjeant*
GEOLOGY OF COAL / *Charles Ross and June R. P. Ross*
NANNOFOSSIL BIOSTRATIGRAPHY / *Bilal U. Haq*
CALCAREOUS NANNOPLANKTON / *Bilal U. Haq*
RIVER NETWORKS / *Richard S. Jarvis and Michael J. Woldenberg*
STABILITY OF HEAVY MINERALS IN SEDIMENTS / *Gretchen Luepke*
STRATIGRAPHY: Foundations and Concepts / *Barbara M. Conkin and*
    James E. Conkin
VERTEBRATE PALEONTOLOGY / *Robert Schoch*
MANTLE FLOW AND PLATE THEORY / *Zvi Garfunkel*
EARTH TIDES / *J. C. Harrison*
ECONOMIC ANALYSIS OF HEAVY MINERALS IN SEDIMENTS /
    Gretchen Luepke
MODERN AND ANCIENT ALLUVIAL FAN DEPOSITS / *Tor H. Nilsen*
CONTINENTAL DRIFT / *James H. Shea*
PLATE TECTONICS / *James H. Shea*
CONTINENTAL RIFTS / *A. M. Quennell*